Association for Women in Mathematics Series

Volume 32

Series Editor

Kristin Lauter, Facebook, Seattle, WA, USA

Focusing on the groundbreaking work of women in mathematics past, present, and future, Springer's Association for Women in Mathematics Series presents the latest research and proceedings of conferences worldwide organized by the Association for Women in Mathematics (AWM). All works are peer-reviewed to meet the highest standards of scientific literature, while presenting topics at the cutting edge of pure and applied mathematics, as well as in the areas of mathematical education and history. Since its inception in 1971, The Association for Women in Mathematics has been a non-profit organization designed to help encourage women and girls to study and pursue active careers in mathematics and the mathematical sciences and to promote equal opportunity and equal treatment of women and girls in the mathematical sciences. Currently, the organization represents more than 3000 members and 200 institutions constituting a broad spectrum of the mathematical community in the United States and around the world.

Titles from this series are indexed by Scopus.

Ramla Abdellatif • Valentijn Karemaker •
Lejla Smajlovic
Editors

Women in Numbers Europe IV

Research Directions in Number Theory

Springer

Editors
Ramla Abdellatif
LAMFA
University of Picardie Jules Verne
Amiens, France

Valentijn Karemaker
Mathematical Institute
Utrecht University
Utrecht, Utrecht, The Netherlands

Lejla Smajlovic
Dept of Mathematics
Univ of Sarajevo
Sarajevo, Bosnia and Herzegovina

ISSN 2364-5733 ISSN 2364-5741 (electronic)
Association for Women in Mathematics Series
ISBN 978-3-031-52162-1 ISBN 978-3-031-52163-8 (eBook)
https://doi.org/10.1007/978-3-031-52163-8

This Springer imprint is published by the registered company Springer Nature Switzerland AG
The registered company address is: Gewerbestrasse 11, 6330 Cham, Switzerland

Foreword

Celebrating 15 Years of the Women in Numbers Network!

I am honored to write the Foreword for this volume of research articles by women in number theory from the WINE4 conference. This year we celebrate the 15th anniversary of the series of Women in Numbers (WIN) and WIN-Europe (WINE) conferences. It is a fine time to reflect on the history, motivation, and accomplishments of our WIN/WINE community and this series of research conferences.

The Founding of WIN The WIN research network for women in number theory was born in 2006, when Rachel Pries, Renate Scheidler, and I noted the lack of women speakers at number theory conferences worldwide. We applied through the usual scientific process to host the first WIN conference at Banff International Research Station (BIRS) in Canada in 2008. We decided to design a number theory conference for women, in a collaborative format, with groups and group leaders, where graduate students could apply to attend. This format is now known as the WIN model, which has spawned a whole movement of Research Collaboration Conferences for Women (RCCWs), in 25 different areas of mathematics, hosted at math institutes around the world.

Research Networks for Women The first WIN conference blossomed into an organized research community, a Research Network of women in number theory, based on the enthusiasm and hard work of the participants. Michelle Manes created an email distribution list and organized follow-up special sessions at AMS meetings; Kate Stange created and maintains a website highlighting women in number theory, womeninnumbertheory.org; we set up a Steering Committee to plan future meetings; we started a tradition of publishing a proceedings volume of the research papers from the working groups at each conference; Alina Bucur and Brooke Feigon worked together to get the first WIN-Europe conference started at CIRM in Luminy, France.

The WIN network has now run 10 conferences, 6 WIN conferences at BIRS in North America and 4 WINE conferences at institutes in Europe. The conferences involved around 400 women in number theory from around the world, and were organized by more than 25 distinct women, many of whom participated in the first WIN conference. More than 60 research articles have been published in 7 proceedings volumes and numerous journals. To keep the series of conferences going, and to manage finances, website, publications, and other matters, we decided on an inclusive, ever-growing Steering Committee for the WIN Network, consisting of all past and current organizers of WIN conferences. The Steering Committee plays an important role in the ongoing success of WIN. The committee was chaired for the first year by Rachel Pries, and then in 3-year increments by Michelle Manes, and then Kate Stange. In recent years, the European members of the Steering Committee have started to meet separately to plan the WINE conferences, under the leadership of Irene Bouw.

Spreading the WIN Model After the success of the first WIN conference, I focused my energy on spreading the WIN model to other areas of mathematics. I joined the leadership of the Association for Women in Mathematics (AWM) and worked on finding leaders to start new networks in other research areas. We encouraged NSF-funded math institutes to host RCCW conferences on the WIN model, and so a number of conferences were launched: WiSh (Women in Shape) at IPAM in 2013 and a series of three conferences at the IMA in 2014, 2015, and 2016 (Women in Math Biology (WIMB), Women in Numerical Analysis and Scientific Computing (WINASC), and Women in Computational Topology (WinCompTop)). Women in Topology (WIT), Algebraic Combinatorixx (ACxx), and Women in Noncommutative Algebra and Representation Theory (WINART) were funded at BIRS through the usual scientific competition.

The AWM leadership, Ruth Charney, Magnhild Lien, Kathryn Leonard and I, applied for grant funding to grow the effort. In 2015, we were awarded a 750,000 USD 5-year NSF grant. The goal of the AWM grant, *Career Advancement for Women Through Research-Focused Networks*, was to build and sustain Research Networks for women in many areas of mathematics to directly increase the opportunities, success, and visibility of women in mathematics. The grant helped to launch or support 25 Research Networks for Women in all areas of mathematics, sustained through conferences, websites, email listservs, and publication volumes. The networks supported more than 2,000 women researchers worldwide and organized 5–10 RCCWs per year. That same year we also launched the AWM Springer Series to publish volumes of research articles as proceedings from WIN-style conferences.

Why Publish Proceedings Volumes of Research Articles?

After the first WIN conference, I was asked by a male colleague in number theory, "But what do you do at these WIN conferences, talk about women in number theory?" I was shocked that he did not realize that we were working hard in groups at the conferences, solving important number theory research problems. But publishing volumes of research articles by the working groups provides the perfect answer for him and anyone else who doubts the mathematical quality of the conferences.

The importance of publishing these volumes of research articles cannot be overstated. The volumes provide a record of the research done by the groups and show the serious nature of the conferences, which can help to attain funding for subsequent conferences. In addition, the publication volumes play a fundamental role in the formation of the networks. To finish their papers for publication, groups formed at the workshop need to keep working together remotely after the conference. This helps to cement the collaboration and mentoring relationships. Deadlines for submissions roughly 6–9 months after the conference provide a schedule and incentive for groups to finish their results. Often graduate students and postdocs ask the group leader(s) for recommendation letters based on their contributions to the paper. Relatively quick publication of the papers is crucial at the career stage where recent PhDs are applying for jobs. At an early career stage, more experience in the publication process is helpful for professional development and a successful job search. By intervening at this "critical transition point" in the leaky pipeline, these publication groups enable more women to continue in the research mathematics profession.

Furthermore, once a group of 4–7 mathematicians has started working together on a research problem, it is important to publish at least the preliminary results, so that everyone in the group can get credit for their contributions. Those group members who have time and interest can form spin-off collaborations to continue to develop further results after the initial publication. Once the volumes are published, follow-up AWM workshops highlight the research papers in special session format. The editors have often invited some contributions from women researchers who were not at the conference, so the volumes serve to highlight work by women in the area more broadly. Many women organizers, editors, and reviewers acquire valuable experience in the publication process through their work to produce these volumes.

Some argue that it is better for women's careers to publish in journals rather than proceedings. Groups can still choose to publish results in a journal in addition to their contribution to the proceedings. In that case, the article in the proceedings may be expository. But I argue that it is still beneficial to the Research Network to publish a proceedings volume, for the other reasons listed here.

The WIN 2008 volume was published by the Fields Institute in their series, and the WIN2 2011 volume was published in the Centre de recherches mathématiques (CRM) series. Since then, all the WIN and WINE volumes have been published in the AWM Springer series. A list of the roughly 35 published volumes in the series is available on the series website.

There are now significantly more women professors of number theory at research universities than there were in 2006 when we started WIN. Informal research networks form the backbone of any mathematical community, and the WIN model gives women in many areas of mathematics the opportunity to build networks that ensure the visibility of their work in the wider research community. I know many examples where conference organizers or department chairs have used the websites created by the "Women-In-..." networks to improve diversity of speaker lists and to expand the talent pool for job searches.

The Genius of Community The WIN initiative was featured in a 2020 best-selling book [4] by Janice Kaplan, *The Genius of Women*. Kaplan celebrates the collaborative nature of this effort, designed to enable women to help other women through research mentorship. She notes that Research Networks for Women have created cohorts of women in many areas of mathematics who are empowered to work on behalf of themselves and other women.

Because building community through organizing a WIN conference requires significantly more work than organizing a normal conference in math, I would like to recognize and applaud the hard work of the WINE4 organizers Ramla Abdellatif, Valentijn Karemaker, and Ariane Mézard and Proceedings co-editor Lejla Smajlović. Organizers not only apply to the venue to host the conference, but also solicit and select group leaders and projects, make an open call for applicants to participate, assign participants to groups, and follow up after the conference by editing the proceedings volume of research articles from the groups! So, the 15th anniversary is also a great moment to thank and remember the hard work of all the WIN co-organizers of the 10 WIN conferences:

WIN Co-founders, 2008 Kristin Lauter, Rachel Pries, Renate Scheidler.

WIN Organizers

- **WIN2, 2011**: Chantal David, Matilde Lalin, Michelle Manes.
- **WINE, 2013, CIRM Luminy**: Marie-José Bertin, Alina Bucur, Brooke Feigon, Leila Schneps.
- **WIN3, 2014**: Ling Long, Rachel Pries, Katherine Stange.
- **WINE2, 2016, Lorentz Center**: Irene Bouw, Rachel Newton, Ekin Ozman.
- **WIN4, 2017**: Jennifer Balakrishnan, Chantal David, Michelle Manes, Bianca Viray.
- **WINE3, 2019, Centre Henri Lebesgue**: Sorina Ionica, Holly Krieger, Elisa Lorenzo García.
- **WIN5, 2020**: Alina Bucur, Wei Ho, Renate Scheidler.
- **WINE4, 2022, Utrecht University**: Ramla Abdellatif, Valentijn Karemaker, Ariane Mézard.
- **WIN6, 2023**: Shabnam Akhtari, Alina Bucur, Jennifer Park, Renate Scheidler.

Co-editors of the Proceedings Volumes Alina Cojocaru (WIN, WINE3), Ellen Eischen (WIN3), Jennifer Johnson-Leung (WINE2), Amanda Folsom (WIN4), Matilde Lalin (WIN4), Lejla Smajlović (WINE4), Rachel Newton, Yu-Ru Liu (WIN6).

The success of the WIN model was palpable from the very first conference: the BIRS staff said they had never experienced such energy and excitement from workshop participants, although they had been running week-long workshops 50 weeks per year for many years. I personally enjoyed my professional life and going to conferences so much more as part of the growing WIN community. I started organizing WIN reunions at other conferences in number theory and WIN reunions are now a regular occurrence. This WINE4 conference was the first of the 10 WIN or WINE conferences where I did not lead a research group. I take it as a great sign of the health and success of WIN-Europe that all WINE4 research group leaders are faculty at European universities.

To learn more about the broader Research Networks for Women effort which WIN spawned, please see historical overview articles in the AMS Notices [2], CWM Newsletter [3], European Math Society Magazine [1], and AWM 50th Anniversary Volume [5].

Happy 15th! Long live WIN/WINE!

Seattle
September 2023

Kristin Lauter
Director, FAIR Labs North America, Meta
Affiliate Professor of Mathematics, University of Washington
President, Association for Women in Mathematics (2015–2017)
Co-Founder, WIN

References

1. K. Bauer, E.W. Chambers, B. Johnson, K.E. Lauter, K. Leonard, *Research networks for women*. Eur. Math. Soc. Mag. **126**, 41–50 (2022). https://euromathsoc.org/magazine/articles/92.
2. G. Benkart, K. Lauter, S. Wiegand, *AWM at 50 and Beyond*. AMS Notices, April 2021. https://www.ams.org/journals/notices/202103/rnoti-p387.pdf.
3. G. Benkart, K. Lauter, S. Wiegand, *AWM at 50 and Beyond*. Reprinted in the CWM Newsletter, Issue 5, pp. 17–19, May 2021. https://www.mathunion.org/fileadmin/CWM/Initiatives/CWMNewsletter5.pdf.
4. J. Kaplan, *Genius of Women* (Penguin Random House, 2020). https://www.penguinrandomhouse.com/books/598507/the-genius-of-women-by-janice-kaplan/.
5. K. Lauter, *AWM and the Genius of Women: Reflections by AWM's 22nd President*. AWM 50th Anniversary Volume (Springer, 2022). https://awm-math.org/wp-content/uploads/2022/08/Lauter_Reflections50th.pdf.

Preface

This volume forms the proceedings of the workshop Women in Numbers Europe 4 (WINE4), which was held from 29 August to 2 September, 2022 in Utrecht, the Netherlands.

Organisation of the event started 3 years prior, together with co-organisers Ramla Abdellatif and Ariane Mézard and scientific advisors Irene Bouw and Ana Caraiani. We found a wonderful group of 20 talented European mathematicians willing to lead 10 exciting research projects in various areas of number theory, and were subsequently faced with the difficult task of forming project groups by choosing the workshop participants out of the many highly qualified applicants. Leading up to the event, local co-organiser Nirvana Coppola proved invaluable.

After the success of three earlier editions of Women in Numbers Europe, we were very pleased to see that also WINE4 was a highly successful and rewarding experience for all participants. Not just during the week in Utrecht, but also beforehand and afterwards, the participants carried out high-level mathematical research, much of which is represented in this proceedings volume. On the other hand, in addition to group work that took place at Utrecht University, the workshop week also consisted of social and career development activities close to the accommodation in the nearby and sunny countryside.

When we started with the organisation in 2019, nobody could have foreseen the effect of the pandemic on all aspects of our academic and personal lives. Even though the worst was over by August 2022, the virus unfortunately still prevented some participants from attending in person, and complicated the attendance of some others. We are grateful for everyone's efforts in making the event as safe and accessible as possible. Indeed, after isolation, it felt even more important to build and strengthen the women in number theory community with the collaborations that started at this workshop.

We hope the next editions of Women in Numbers (Europe) will continue to bring together number theorists from many different areas, mathematically as well as geographically, and thus support this ever-growing community.

The WINE4 workshop could not have been possible without the generous support of our many sponsors: CNRS, CMI, DIAMANT, ENS PSL, FSMP, Foundation Compositio Mathematica, GDR JC2A, Journal of Number Theory, Journal de Théorie des Nombres de Bordeaux, the Number Theory Foundation, NWO, Université de Picardie Jules Verne, and Utrecht University.

We are very happy to have received so many strong submissions to this proceedings volume, with such a wide range of topics. In particular, most project groups from WINE4 submitted a paper. Some of these are research papers: Berthé et al. describe dynamical properties of multidimensional continued fraction algorithms; Dujella et al. improve a result by Mahler on transcendence measures; Agathocleous et al. perform a cryptographic security analysis of homomorphic encryption using abelian groups; Balestrieri et al. give asymptotics for Campana points of bounded height on diagonal hypersurfaces; and Benli et al. study asymptotic densities of sets of integers with missing digits. Others are survey papers: Ernvall-Hytönen et al. survey Li-type criteria for the Generalized Riemann Hypothesis; Anscombe et al. describe the use of some local-global principles in the study of Hilbert's tenth problem; Fust et al. treat the theory of real quadratic singular moduli and families of modular forms; and Coppola et al. present an overview of techniques used in certain Diophantine problems. Yet other papers were authored by other researchers, some of whom attended earlier WIN(E) conferences and kept collaborating since then: Mase-Whitcher studied mirror constructions for K3 surfaces; Garbagnati-Salgado survey their work on elliptic fibrations of K3 surfaces; and Žunar surveys methods for studying cusp forms defined by Poincaré series.

These proceedings were edited together with Ramla Abdellatif and Lejla Smajlović. We thank our contact persons Nicholas DiBenedetto and Dahlia Fisch at Springer for their help and support with the production process.

Utrecht
September 2023

Valentijn Karemaker

Contents

On Homomorphic Encryption Using Abelian Groups: Classical Security Analysis

Eleni Agathocleous, Vishnupriya Anupindi, Annette Bachmayr, Chloe Martindale, Rahinatou Yuh Njah Nchiwo, and Mima Stanojkovski

Introduction

Homomorphic encryption is a method of encrypting plaintext that allows users to compute directly with the ciphertext. This has many interesting applications, including being able to engage in cloud computing without giving up your data to the owner of the cloud. Scientifically, the premise is easy to describe: Suppose that the plaintext and the ciphertext space both have a ring structure and that we encrypt plaintext via a map between these spaces. If this map is a ring homomorphism, then this describes a *fully homomorphic encryption* scheme. Creating such a ring homomorphism that describes secure encryption (requiring, for example, that such

E. Agathocleous
CISPA Helmholtz Center for Information Security, Saarbrücken, Germany
e-mail: eleni.agathocleous@cispa.de

V. Anupindi
RICAM Austrian Academy of Sciences, Linz, Austria
e-mail: vishnupriya.anupindi@ricam.oeaw.ac.at

A. Bachmayr
Department of Mathematics, RWTH Aachen University, Aachen, Germany
e-mail: bachmayr@mathematik.rwth-aachen.de

C. Martindale (✉)
School of Computer Science, University of Bristol, England, United Kingdom
e-mail: chloe.martindale@bristol.ac.uk

R. Y. N. Nchiwo
Department of Mathematics and Systems Analysis, Aalto University, Espoo, Finland
e-mail: rahinatou.njah@aalto.fi

M. Stanojkovski
Dipartimento di Matematica, Università di Trento, Trento, Italy
e-mail: mima.stanojkovski@unitn.it

R. Abdellatif et al. (eds.), *Women in Numbers Europe IV*, Association for Women in Mathematics Series 32, https://doi.org/10.1007/978-3-031-52163-8_1

a map should be efficiently computable and hard to invert) is, however, much harder than describing its properties. The closest the scientific community has come to constructing an example of fully homomorphic encryption is using maps based on (variants of) the *Learning With Errors* (LWE) problem from lattice-based cryptography [6, 9, 14, 22, 25]. However, although practical fully homomorphic encryption can be achieved from LWE-based constructions [7], all known such constructions are not naturally fully homomorphic: Decrypting a message that was encrypted using LWE relies on the "error" that was used in the encryption being small, and adding and especially multiplying encrypted messages together causes the error to grow. Once the error is too large, the data can no longer be decrypted, so methods such as bootstrapping need to be employed to correct this growth (see, e.g., [9]). These methods, although somewhat ad hoc, may still be the best solution for providing practical fully homomorphic encryption in the near future.

In this chapter, we explore an alternative approach for homomorphic encryption, introduced by Leonardi and Ruiz-Lopez in [19]. The interested reader may find more on homomorphic encryption in [13]. The construction of Leonardi and Ruiz-Lopez relies on something similar to the *Learning Homomorphisms with Noise* (LHN) problem introduced by Baumslag, Fazio, Nicolosi, Shpilrain, and Skeith in [2]: Roughly speaking, this is the problem of recovering a group homomorphism from the knowledge of the images of certain elements multiplied by noise. The focus of [19] is on the difficulty of constructing post-quantum secure instantiations of their primitive, but we believe the construction is interesting even in a classical setting. A big advantage of the Leonardi–Ruiz-Lopez approach over the LWE approach is that the noise, which plays the role of the errors in LWE-based homomorphic encryption, does not grow with repeated computation. As such, there is no limitation on the number of additions that can be computed on encrypted data. However, it is not clear if this construction can be extended to multiplicative homomorphic encryption. In general, extending additive homomorphic encryption to fully homomorphic encryption is a hard problem: LWE-based homomorphic encryption is the only known construction. As such, the Leonardi–Ruiz-Lopez approach is akin in some sense to the Benaloh [3] or Paillier [24] cryptosystems. In the nonabelian setting, Leonardi and Ruiz-Lopez's construction has some hopes of being post-quantum secure unlike the Benaloh or Paillier constructions, but as they explored already in their work this is nontrivial to instantiate, and our work only strengthens this claim as we show that even solvable groups may admit quantum attacks.

Our main contributions address **finite groups** and include:

1. Reducing the security of an instantiation of Leonardi and Ruiz-Lopez's public key homomorphic encryption scheme in which one of the groups is abelian (in particular, the group in which the noise lives) to the discrete logarithm problem in 2-groups (under certain plausible assumptions); this gives a polynomial-time classical attack if the 2-part of the relevant group is cyclic and a practical classical attack if it is a product of a small number of cyclic groups. See Theorem 1 and Sect. 3.4.

2. Highlighting an abelian group instantiation of Leonardi and Ruiz-Lopez's homomorphic encryption scheme where there is no known practical classical attack, namely, the product of many cyclic 2-groups. This may be of interest to the community as a new example of unbounded additively homomorphic encryption.
3. Highlighting assumptions that need to be made in order to apply any discrete logarithm-derived attack, with a view to constructing (more) examples of groups on which there is no known practical attack on Leonardi and Ruiz-Lopez's homomorphic encryption scheme.
4. A description of a quantum attack on an instantiation with solvable groups, under certain assumptions

Note that all our contributions are studying the hardness of the LHN-PKE problem, given in Definition 1, not the general LHN problem, as this is the relevant underlying "hard problem" in Leonardi–Ruiz-Lopez encryption.

The layout of this chapter is as follows: In Sect. 1, we recap the public key homomorphic encryption scheme proposed by Leonardi and Ruiz-Lopez in [19]. In Sect. 2, we discuss some simple instantiations: the abelian case and the noiseless case. In this section we also give some basic security requirements (including some recalled from [19]). In Sect. 3, we describe our reduction from the abelian group instantiation of the Leonardi–Ruiz-Lopez primitive to the (extended) discrete logarithm problem, under certain assumptions. In Sect. 4, we describe some ways of instantiating the primitive with nonabelian groups to which our attack on abelian groups would also apply. In Sect. 5, we describe our quantum attack on instantiations with solvable groups, under certain assumptions on how such groups would be represented. In Sect. 6, we outline our plans for future work.

1 Leonardi–Ruiz-Lopez Encryption

In this section, we describe the public key additive homomorphic encryption of Leonardi and Ruiz-Lopez [19, Sec. 5.2]; we will refer to this throughout this chapter as the *Leonardi–Ruiz-Lopez encryption*. Note that this chapter shows that the following encryption scheme is *not* secure (even classically) in most natural abelian instantiations, see Theorem 1, and we make no claims on the security of the nonabelian case except for some warnings, see Sects. 3.4, 4, and 5. In the course of this chapter we also discovered the necessity for some extra points in the basic Leonardi–Ruiz-Lopez encryption in order to avoid security problems even in the most general case; our additions are marked with ^{new} (both when they are completely new and when they are clarifications). These points are discussed in the later subsections. We do not reintroduce the cryptographic formalism of [19] as this is covered already by Leornardi and Ruiz-Lopez.[1]

[1] Furthermore, the introduction of security notions that, in part due to our attacks, do not hold for this scheme would distract from the narrative.

Fix three finitely generated groups G, H, and K and probability distributions ξ on G and χ on H. These data should be chosen in such a way that operations can be performed efficiently in the groups and we can sample from both distributions efficiently. A natural choice could be, for instance, to take G, H, and K finite and ξ and χ to be uniform distributions. The groups G, H and K and the distributions ξ and χ are public. These groups must also satisfy security assumptions (S1)^{new} and (S3) discussed below. In the following sections we will mostly work with finite groups, and we will always make it clear when this is the case.

For the **key generation**, Alice

- Chooses efficiently computable secret homomorphisms $\varphi\colon G \to H$ and $\psi\colon H \to K$ such that she can efficiently sample from $\ker(\psi)$ and such that the center $Z(H)$ of H is not contained in $\ker(\psi)$
- Chooses a positive integer m
- Samples elements $g_1, \ldots, g_m \in G$ via ξ and secret elements $h_1, \ldots, h_m \in \ker(\psi)$ via χ
- Chooses an element $\tau \in Z(H)\setminus\ker(\psi)$ of order 2[2] that satisfies (S4)^{new} described below

Alice computes the public key as the set

$$\{(g_1, \varphi(g_1)h_1), \ldots, (g_m, \varphi(g_m)h_m), \tau\}.$$

Note that, whereas the elements $g_1, \ldots, g_m$ are public, both φ and $h_1, \ldots, h_m$ are private (as are also ψ and $\ker(\psi)$).

For **encrypting** a one-bit message $\beta \in \{0, 1\}$, Bob chooses a natural number ℓ and then samples a word $w = w_1 \cdots w_\ell$[3] over the indices $\{1, \ldots, m\}$, and using Alice's public key, he computes

$$(g, h') = (g_{w_1} \cdots g_{w_\ell}, \varphi(g_{w_1})h_{w_1} \cdots \varphi(g_{w_\ell})h_{w_\ell}).$$

He then sends $(g, h) = (g, h'\tau^\beta)$ to Alice.

For **decrypting** (g, h), Alice computes $\nu = \psi(\varphi(g))^{-1} \cdot \psi(h) \in K$ and deduces that the message β equals 0 if ν equals 1_K (and else, β equals 1).

To see that the decryption indeed produces the correct message β, recall that $h_1, \ldots, h_m$ are contained in $\ker(\psi)$. Hence $\nu = \psi(\tau)^\beta$ and, since τ is not contained in $\ker(\psi)$, the element ν equals 1 if and only if β equals 0.

For the convenience of the reader, we give a schematic summary of the data described above:

[2] The element τ having order 2 ensures that the encryption is additive as explained in Sect. 1.1; the construction is also viable without this requirement.

[3] The sample space and method must satisfy security assumption (S2)^{new} described below.

Public	Private to Alice	Private to Bob
G, H, K, ξ, χ	$\varphi : G \to H$	
$(g_1, \varphi(g_1)h_1), \ldots, (g_m, \varphi(g_m)h_m)$	$\psi : H \to K$	$\beta \in \{0, 1\}$
$\tau \in Z(H) \setminus \ker(\psi)$	$\ker(\psi)$	word w
$(g, h) = (g, h'\tau^{\beta}) \in G \times H$	$h_1, \ldots, h_m \in \ker(\psi)$	

1.1 Homomorphic Properties

The primary selling point of Leonardi–Ruiz-Lopez encryption is that it is *unbounded additive homomorphic*. We say that an encryption function E from plaintext to ciphertext space is *additive* if the plaintext space admits an additive operator $+$ and, given encryptions $E(\beta)$ and $E(\tilde{\beta})$ of messages β and $\tilde{\beta}$, respectively, one can compute a valid encryption $E(\beta + \tilde{\beta})$ of $\beta + \tilde{\beta}$ (without the knowledge of the plaintext $\beta + \tilde{\beta}$). We say that E is *unbounded additive homomorphic* if such additions can be performed an unbounded number of times without introducing systematic decryption failures.[4]

As seen above, τ having order 2 is not necessary for successful decryption; this property is needed to make the encryption additive, as we now recall from [19].

Write the encryptions of β and $\tilde{\beta}$ sampled from $\{0, 1\}$ as

$$(g, h'\tau^{\beta}) = (g_{w_1} \cdots g_{w_\ell}, \varphi(g_{w_1})h_{w_1} \cdots \varphi(g_{w_\ell})h_{w_\ell}\tau^{\beta})$$

and

$$(\tilde{g}, \tilde{h}'\tau^{\tilde{\beta}}) = (g_{\tilde{w}_1} \cdots g_{\tilde{w}_{\tilde{\ell}}}, \varphi(g_{\tilde{w}_1})h_{\tilde{w}_1} \cdots \varphi(g_{\tilde{w}_{\tilde{\ell}}})h_{\tilde{w}_{\tilde{\ell}}}\tau^{\tilde{\beta}}),$$

respectively. Then, as τ is central in H and has order 2, we can construct a valid encryption of $\beta + \tilde{\beta}$ via the observation that

$$\begin{aligned} &\varphi(g_{w_1})h_{w_1} \cdots \varphi(g_{w_\ell})h_{w_\ell}\tau^{\beta}\varphi(g_{\tilde{w}_1})h_{\tilde{w}_1} \cdots \varphi(g_{\tilde{w}_{\tilde{\ell}}})h_{\tilde{w}_{\tilde{\ell}}}\tau^{\tilde{\beta}} \\ =&\varphi(g_{w_1})h_{w_1} \cdots \varphi(g_{w_\ell})h_{w_\ell}\varphi(g_{\tilde{w}_1})h_{\tilde{w}_1} \cdots \varphi(g_{\tilde{w}_{\tilde{\ell}}})h_{\tilde{w}_{\tilde{\ell}}}\tau^{\beta+\tilde{\beta}}; \end{aligned}$$

this encryption is given by

$$(g, h'\tau^{\beta})(\tilde{g}, \tilde{h}'\tau^{\tilde{\beta}}) = (g\tilde{g}, h'\tau^{\beta}\tilde{h}'\tau^{\tilde{\beta}}) = (g\tilde{g}, h'\tilde{h}'\tau^{\beta+\tilde{\beta}}).$$

Remark 1 Note that, for Leonardi–Ruiz-Lopez encryption to be fully homomorphic, it would also need to be *multiplicative*: That is, at the very least, given valid

[4] The number of additions is bounded in, for example, LWE-based homomorphic encryption, where the error grows too large.

encryptions of $E(\beta)$ and $E(\tilde{\beta})$, we should be able to deduce a valid encryption of $\beta\tilde{\beta}$. Since we only have one group operation on the cipher text space $G \times H$, we cannot expect to find a general strategy that provides fully homomorphic encryption for arbitrary abstract groups G and H. Our encryption function from plaintext to ciphertext space is

$$E : \{0, 1\} \to G \times H,$$

where the domain can be naturally endowed with a ring structure using addition and multiplication mod 2, but there is no natural way of adding extra data on $G \times H$ that would allow us to deduce a valid encryption of $\beta\tilde{\beta}$. We stress that in the case of LWE-based homomorphic encryption, both plaintext and ciphertext spaces come equipped with a ring structure, so the equivalent of our function E is generally taken to be a ring homomorphism.

The map E being a ring homomorphism is, however, not always strictly necessary to deduce a valid encryption of $\beta\tilde{\beta}$. If in future work we were to succeed in deducing a valid encryption of $\beta\tilde{\beta}$, we expect that this will only apply to a specific instantiation of Leonardi–Ruiz-Lopez encryption, not one for abstract groups, where the groups provide more structure that can be exploited.

1.2 Remarks on Leonardi–Ruiz-Lopez Encryption

Some remarks on the construction above:

(R1) The underlying hard problem of this encryption scheme is described as the LHN-PKE problem, so named as it is based on the Learning Homomorphisms with Noise problem (LHN) but is adapted to this Public Key Encryption (PKE) scheme.

Definition 1 Let G, H, K, ξ, and χ be as above. We define the *LHN-PKE* problem for G, H, K, ξ, and χ to be: Given G, H, K, ξ, and χ, for any

- φ sampled uniformly at random from $\mathrm{Hom}(G, H)$
- ψ sampled uniformly at random from the elements of $\mathrm{Hom}(H, K)$ whose kernel does not contain $\mathrm{Z}(H)$
- $g_1, \ldots, g_m$ sampled from G using ξ
- $h_1, \ldots, h_m$ sampled from $\ker(\psi)$ using χ
- τ sampled from the order two elements of $\mathrm{Z}(H) \setminus \ker(\psi)$ using χ
- β sampled uniformly at random from $\{0, 1\}$
- small ℓ and word $w = w_1 \cdots w_\ell$ sampled uniformly at random from $\{1, \cdots, m\}^\ell$

recover β from the following information:

- $(g_1, \varphi(g_1)h_1), \ldots, (g_m, \varphi(g_m)h_m)$

- τ
- $g = g_{w_1} \cdots g_{w_\ell}$
- $h = h'\tau^\beta = \varphi(g_{w_1})h_{w_1} \cdots \varphi(g_{w_\ell})h_{w_\ell}\tau^\beta$

(R2) In order for the encryption and decryption to work, the assumption that τ is central or of order 2 is not necessary. The reason we work under these assumptions is, as explained in Sect. 1.1, that in this case, the cryptosystem is unbounded additive homomorphic.

(R3) Once Alice has fixed the elements $g_1, \ldots, g_m$ and determined the public key, all computations inside G actually take place inside the subgroup $\langle g_1, \ldots, g_m \rangle$ that is generated by $g_1, \ldots, g_m$. So for cryptanalysis, we may and will assume that G is generated by $g_1, \ldots, g_m$, i.e., that $G = \langle g_1, \ldots, g_m \rangle$.

(R4) The work of [19] was inspired by [2], which introduces the Learning Homomorphisms with Noise problem in order to construct a symmetric primitive. However, the noise accumulates in the construction of [2] in a manner akin to the error growth in LWE constructions. Leonardi and Ruiz-Lopez also introduce a symmetric primitive in [19], but we focus on the PKE construction in this work.

(R5) The noise consists of the elements $h_1, \ldots, h_m$ that are mixed into the product h' in the second component h of the ciphertext. These elements are chosen to be in the kernel of ψ and therefore get erased during decryption. "Being contained in the kernel" of ψ can thus be thought of as an equivalent of "the error being small" in the LWE-based encryption of [25] or "the noise being small" in LHN-based encryption of [2]. The strength of Leonardi–Ruiz-Lopez encryption is that the noise does not accumulate and will not lead to systematic decryption errors, since in the decryption process, we can erase the noise neatly by applying ψ.

1.3 Cryptanalysis of the LHN-PKE Problem in the Abelian Case

In this papar we show that if G and H are finite and abelian, under some reasonable assumptions that we introduce here, the LHN-PKE problem for G and H can be reduced to the extended discrete logarithm problem (cf. Definition 2) and membership problem (cf. Definition 3) in some specific abelian 2-groups, which leads to a polynomial-time attack in case the 2-part of G is cyclic; the general asymptotic complexity is given in our main theorem below. We also attack the case with finite but nonabelian G and abelian H, again under some reasonable assumptions, in Sect. 3.4.

In each section that follows, we explicitly mention under which assumptions from the following list we are working. For a group Γ and $\gamma_1, \ldots, \gamma_m \in \Gamma$, we are interested in the following properties:

(A1) Γ is abelian.
(A2) The largest positive odd factor of the order $|\Gamma|$ of Γ is known (or easily computable).
(A3) $\gamma_1, \ldots, \gamma_m \in \Gamma$ are such that $\Gamma = \langle \gamma_1 \rangle \oplus \ldots \oplus \langle \gamma_m \rangle$.
(A4) The orders $|\gamma_1|, \ldots, |\gamma_m|$ of $\gamma_1, \ldots, \gamma_m$ are known (or easily computable).

We discuss these assumptions—why and when they are reasonable and/or needed–in Sects. 3.4.3 (for (A2) and (A4)) and 3.4.4 (for (A3)). We can now state the main theorem of this chapter:

Theorem 1 *Let $G = \langle g_1 \rangle \oplus \cdots \oplus \langle g_m \rangle$ be a finite group which satisfies (A1), (A2), and (A3). Let H be a finite group that satisfies (A1) and (A2), and for which we can efficiently compute a cyclic decomposition of its Sylow* 2*-subgroup H_2.*[5] *Let K be an arbitrary group. Let ξ and χ be probability distributions on G and H, respectively. Then there is a classical algorithm to solve* LHN-PKE *for G, H, K, ξ, and χ with asymptotic complexity*

$$\mathrm{polylog}(|G|) + \mathrm{polylog}(|H|) + O\left(\frac{\log_2(|G_2|)}{m} 2^{m/2}\right),$$

where G_2 denotes the Sylow 2*-subgroup of G.*

We will give the proof of this theorem in Sect. 3.3.

Note in particular that if $m = 1$, that is, if G is cyclic, this attack algorithm is polynomial time (in fact, as we will see, it is sufficient for G_2 to be cyclic). In Sect. 3.4, we discuss how to attack instances with nonabelian G, assuming that we can work with the abelianization $G/[G, G]$ of G. In Sect. 3.5.2, we also briefly discuss a polynomial-time quantum algorithm, complementing the existing literature for the torsion-free case [19]. In Sect. 4, we discuss instantiations in which H is not abelian (i.e., does not satisfy (A1)) but which reduce to the case in which H is abelian. In Sect. 5, we discuss how one could develop a quantum algorithm for the solvable case.

2 Simple Instantiations and Security

In this section, we describe some simple instantiations of Leonardi–Ruiz-Lopez encryption. The abelian case is a central focus of this chapter as it is much simpler to describe than the general case; the description below is for the reader who wishes only to understand the abelian case. We also describe the noiseless case in order to highlight the role that the noise plays in the encryption. Finally we discuss the

[5] Note that if we assumed (A3) for H together with a choice of generators, computing such a decomposition for H_2 would be easy, as H satisfies (A2); that is, this requirement for H_2 together with (A2) for H is weaker than (A3) for H.

requirements on the setup parameters of Leonardi–Ruiz-Lopez encryption in order to achieve security against some naive classical attacks, concluding this section with a list of properties that the groups must have in order to avoid the trivial classical attacks outlined in this section.

2.1 The Abelian Case

If H is abelian, we can rewrite $(g,h)=(g_{w_1}\cdots g_{w_\ell}, \varphi(g_{w_1})h_{w_1}\cdots\varphi(g_{w_\ell})h_{w_\ell}\cdot\tau^\beta)$ as

$$(g,h)=(g_{w_1}\cdots g_{w_\ell}, \varphi(g_{w_1})\cdots\varphi(g_{w_\ell})h_{w_1}\cdots h_{w_\ell}\cdot\tau^\beta)=(g,\varphi(g)h_{w_1}\cdots h_{w_\ell}\cdot\tau^\beta).$$

If both G and H are abelian, it makes sense to switch to the following notation: Instead of choosing indices $w_1,\ldots,w_\ell$, Bob just chooses non-negative integers $r_1,\ldots,r_m$ and encrypts β to

$$(g,h)=\left(g_1^{r_1}\cdots g_m^{r_m}, \varphi\left(g_1^{r_1}\right)\cdots\varphi\left(g_m^{r_m}\right)h_1^{r_1}\cdots h_m^{r_m}\cdot\tau^\beta\right)=\left(g,\varphi(g)h_1^{r_1}\cdots h_m^{r_m}\cdot\tau^\beta\right).$$

This system has been claimed to not be quantum secure in [19, Section 8.2], cf. also Sect. 3.5, while we discussed security in the classical sense in Sect. 1.3. Moreover, in Sect. 4 we describe instantiations of the LHN-PKE problem in which G and H are nonabelian, but the security reduces to the case in which they are.

2.2 The Noiseless Case

Let us assume that $h_1=h_2=\cdots=h_m=1$. Then the public key consists of all pairs $(g_i,\varphi(g_i))$ together with τ, and Bob would encrypt the message $\beta\in\{0,1\}$ to

$$(g,h)=(g_{w_1}\cdots g_{w_\ell}, \varphi(g_{w_1})\cdots\varphi(g_{w_\ell})\tau^\beta)=(g,\varphi(g)\tau^\beta).$$

Let Eve be an attacker who is aware of the fact that Alice decided to work in a noiseless setting. Then Eve knows all g_i's as well as their images $\varphi(g_i)$ from the public key. If she can write g as a product in $g_1,\ldots,g_m$, then she can compute $\varphi(g)$. Knowing τ from the public key, Eve can then decrypt $(g,h)=(g,\varphi(g)\tau^\beta)$. Note that even if Eve did not use the same word $w_1\cdots w_\ell$ as Bob to write g as a product in $g_1,\ldots,g_m$, she would nonetheless obtain the correct value of $\varphi(g)$.

Of course, finding such a word might still be a hard problem, even if $m=1$. For example, if G is the multiplicative group of a finite field and g_1 is a generator, finding a word in g_1 defining g is the same as solving the discrete logarithm problem, which is known to be hard for classical computers (though there are quantum algorithms to solve it, see [26, 27]). Alice, on the other hand, will probably have a closed form describing φ that does not require to write elements as products in $g_1, \ldots, g_m$ when she applies φ in the decryption process (and similarly for ψ). In the $m = 1$ example above, she would choose φ to take every element to a certain power.

For attacking the cryptosystem in the general case, a possible strategy is to construct attacks that reduce to the noiseless case. We will come back to such attacks in Sect. 2.3.

2.3 Security

In all that follows, let λ be the security parameter.[6]

2.3.1 Many Homomorphisms

First of all, note that an attacker who can guess both φ and ψ can decrypt in the same way Alice does. Denote the set of all possible choices for ψ by

$$\mathrm{Hom}(H, K)^- = \{\psi \in \mathrm{Hom}(H, K) : \mathrm{Z}(H) \not\subseteq \ker(\psi)\}.$$

To avert brute force attacks, the groups G, H, and K should be chosen in such a way that $\mathrm{Hom}(G, H)$ and $\mathrm{Hom}(H, K)^-$ are of size at least $\Theta(2^\lambda)$,[7] and φ and ψ should be sampled uniformly at random from $\mathrm{Hom}(G, H)$ and $\mathrm{Hom}(H, K)^-$, respectively. This ensures that if an attacker guesses φ, she succeeds with probability $2^{-\lambda}$ and similarly for ψ.

2.3.2 Words

As we already saw in the noiseless case, there are links between the LHN-PKE problem and the ability of an attacker to write g as an expression in the generators $g_1, \ldots, g_m$. Assume, for instance, that Eve wants to decrypt the cyphertext (g, h). She knows that g is a product in $g_1, \ldots, g_m$, and recall that $g_1, \ldots, g_m$ are public. If she knows the exact expression of g as a product in $g_1, \ldots, g_m$ that Bob used in the encryption process, then she can compute h' from the public key, erase it from $h = h'\tau^\beta$, and recover the message β. It is important to note that other than in the noiseless case, it is in general not enough to find *any* expression of g as a product

[6] Typically, we want any computations undertaken by the user to have complexity that is polynomial in λ, and an attacker who attempts to decrypt by guessing any unknowns should only succeed with probability at most $2^{-\lambda}$.

[7] We are using the Bachmann–Landau notation for complexity, see, for example, Section 1.2.11.1 of [17].

in $g_1, \ldots, g_m$, because that product will in general not produce the correct term h' yielding to a different accumulation of the noise. That is, the attacker needs to recover the *correct* word w, not just any expression of g in $g_1 \ldots, g_m$. To see this, recall that $(g, h) = (g, h'\tau^\beta)$ with

$$(g, h') = (g_{w_1} \cdots g_{w_\ell}, \varphi(g_{w_1})h_{w_1} \cdots \varphi(g_{w_\ell})h_{w_\ell}),$$

and assume that the attacker found some word $\tilde{w}_1 \ldots \tilde{w}_{\tilde{\ell}}$ such that $g = g_{\tilde{w}_1} \cdots g_{\tilde{w}_{\tilde{\ell}}}$. Using the public elements $(g_1, \varphi(g_1)h_1), \ldots, (g_m, \varphi(g_m)h_m)$, it is then possible to compute an element $\tilde{h}' = \varphi(g_{\tilde{w}_1})h_{\tilde{w}_1} \cdots \varphi(g_{\tilde{w}_{\tilde{\ell}}})h_{\tilde{w}_{\tilde{\ell}}}$, but in general $\tilde{h}' \neq h'$ and the knowledge of $\tilde{h}'$ together with h do not help to compute τ^β. In the cryptanalysis we carry out in Sect. 3.2 for finite abelian G, H, and K, we give a reduction of LHN-PKE to the extended discrete logarithm problem for finite 2-groups; our reduction circumnavigates the issue of finding the correct word.

2.3.3 An Attack on Instances with Few Normal Subgroups

The idea behind the following attack is to replace Alice's secret $\psi : H \to K$ with some new $\bar{\psi} : H \to L$ erasing the noise without erasing τ^β (for instance, L could be a quotient of H, as described below). A similar attack was also described in Section 7.2 of [19]. Assume that Eve knows a normal subgroup N of H that contains all elements $h_1, \ldots, h_m$ but does not contain τ. She can then define $\bar{\psi} : H \to H/N$ as the natural projection, and by applying $\bar{\psi}$ to all second coordinates of the elements $(g_i, \varphi(g_i)h_i)$ in the public key and to the second coordinate of the encrypted message (g, h), she can switch to the noiseless case, cf. Sect. 2.2. We deduce in particular, that there should be at least $\Theta(2^\lambda)$ normal subgroups in H, so that if an attacker guesses $\ker(\psi)$, she succeeds with probability $2^{-\lambda}$.

2.3.4 An Attack on Instances with Weak Normal Subgroups

Now suppose that an attacker can find a normal subgroup N of H that contains $\varphi(g_i)h_i$ for all $i = 1, \ldots, m$ but does not contain τ (note that these elements are all public). If efficiently computable, she can then directly apply the projection $H \to H/N$ to the second coordinate in the encrypted message (g, h) and can deduce that β equals zero if and only if she obtained the neutral element in H/N. To avoid such an attack, after the key generation process, Alice must check (assuming that she is capable of it) whether the normal closure of $\langle \varphi(g_1)h_1, \ldots, \varphi(g_m)h_m \rangle$ contains τ, and if it does not, she should choose a different key. For cryptanalysis, we may thus assume that H equals the normal closure of $\langle \varphi(g_1)h_1, \ldots, \varphi(g_m)h_m \rangle$ (as otherwise, we can just work in this smaller group). In particular, if H is abelian, we may assume that $H = \varphi(G) \ker \psi$.

2.3.5 A Summary of the Discussed Security Assumptions

We conclude this section with a list of necessary properties for security deduced from the list of naive attacks above:

(S1) $\mathrm{Hom}(G, H)$ and $\mathrm{Hom}(H, K)^-$ are of size exponential in the security parameter.
(S2) Finding the precise word w used to express g as a product in the g_i's in the encryption phase has complexity that is exponential in the security parameter.
(S3) The number of normal subgroups in H is exponential in the security parameter.
(S4) The normal closure of $\langle \varphi(g_1)h_1, \ldots, \varphi(g_m)h_m \rangle$ contains τ.

3 Cryptanalysis in the Finite Abelian Case

In this section, we prove Theorem 1 via a series of reductions of the LHN-PKE problem. Recall that Theorem 1 assumes that G and H are finite and abelian (in addition to a few other assumptions). In Sect. 3.1, we will show that this can be taken one step further by giving a reduction from G being finite and H being finite and abelian to both G and H being finite and abelian. In Sect. 3.2.1, we will give a reduction from G and H being finite and abelian to G and H being finite abelian 2-groups (under some reasonable assumptions). In Sect. 3.2.2, we will give a reduction from LHN-PKE for finite abelian G and H to the eDLP problem[8] for G and the generator membership problem[9] for H (under some reasonable assumptions).

In Sect. 3.3 we give the proof of Theorem 1, and in Sect. 3.4 we discuss the genericity and limitations of the assumptions made in Sect. 1.3, as well as the impact of the reductions made.

Leonardi and Ruiz-Lopez [19] dismissed the abelian instantiation due to an argument that there should exist a polynomial-time quantum algorithm for the LHN-PKE problem; the reduction is more complex than is suggested in [19], but their statement is true as we show in Sect. 3.5. Nevertheless, the unbounded homomorphic property of the proposed cryptosystem is sufficiently powerful that a classically secure construction would also be of great interest to the cryptographic community.

[8] See Definition 2 for a formal definition.

[9] See Definition 3 for a formal definition.

3.1 *A Reduction of* LHN-PKE *with Abelian Noise to Abelian Groups*

In this section we take G to be finite and H to be finite and abelian. We show that, when the natural projection $G \to \overline{G} = G/[G, G]$ from G to its abelianization is efficiently computable, then, for cryptanalysis, one can consider $\overline{G}$ instead of G, with the distribution $\overline{\xi}$ that is induced on $\overline{G}$ by the projection.

Lemma 1 *Assume H is abelian. Then the* LHN-PKE *problem for G, H, K, ξ, and χ is at most as hard as the* LHN-PKE *problem for $\overline{G} = G/[G, G]$, H, K, $\overline{\xi}$, and χ.*

Proof Since H is abelian, the commutator subgroup $[G, G]$ of G is contained in the kernel of φ. Define $\overline{G} = G/[G, G]$. Then $\varphi\colon G \to H$ induces a well-defined homomorphism $\overline{\varphi}\colon \overline{G} \to H$, and any ciphertext (g, h) can be interpreted as the ciphertext $(\bar{g}, h)$ in the cryptosystem given by $\overline{\varphi}\colon \overline{G} \to H$, $\psi\colon H \to K$ as before and public key given by $(\overline{g_i}, \bar{\varphi}(\overline{g_i})h_i)$ for $i = 1, \ldots, m$ together with the same τ as before. Indeed, if Bob used the word $w_1 \cdots w_\ell$ to encrypt (g, h), then the same word gives rise to the encryption $(\bar{g}, h)$ in the new cryptosystem of the same message. □

3.2 *Reductions of* LHN-PKE *in the Finite Abelian Case*

In Sect. 3.2 we discuss the LHN-PKE problem for G and H finite and abelian.

3.2.1 From Abelian Groups to 2-Groups

In this section, we take G and H to be finite groups for which (A1) and (A2) hold, i.e., we assume that both G and H are finite abelian groups and that the odd parts of $|G|$ and $|H|$ are known. We let G_2 and H_2 denote the unique Sylow 2-subgroups of G and H, respectively, and ξ_2 and χ_2 denote the distributions ξ and χ restricted to G_2 and H_2, respectively. In the new language, the elements τ and τ^β belong to H_2. In the following, we show that in the case of abelian groups, the LHN-PKE problem for the original data reduces to the LHN-PKE problem for the 2-parts G_2 and H_2 (with the induced data).

Lemma 2 *Suppose that G and H are finite groups satisfying* (A1) *and* (A2). *Then the* LHN-PKE *problem for G, H, K, ξ, and χ is at most as hard as the* LHN-PKE *problem for G_2, H_2, K, ξ_2, and χ_2.*

Proof Write the orders of G and H as

$$|G| = 2^{n_G} q_G, \ |H| = 2^{n_H} q_H,$$

where q_G and q_H are odd numbers. Let q be a common multiple of q_G and q_H that can be computed efficiently. Then for any elements $g \in G$ and $h \in H$, the orders of g^q and h^q are both powers of 2, i.e., $g^q \in G_2$ and $h^q \in H_2$. Moreover, G and H being abelian, the assignment $x \to x^q$ defines homomorphisms $G \to G_2$ and $H \to H_2$.

Equip G_2 and H_2 with the induced distributions ξ_2 and χ_2, and write $\varphi_2 \colon G_2 \to H_2$ and $\psi_2 \colon H_2 \to K$ for the restrictions of φ and ψ to G_2 and H_2, respectively. Assume that β can be recovered from $(G_2, H_2, K, \xi_2$, and $\chi_2)$. We show that β can be determined from (G, H, K, ξ, χ).

Recall that the pair $(g, h) = (g_1^{r_1} \cdots g_m^{r_m}, \varphi(g) h_1^{r_1} \cdots h_m^{r_m} \cdot \tau^\beta) = (g, h'\tau^\beta) \in G \times H$ is public. Raising both entries to their q-th power, one obtains

$$(g^q, h^q) = \left((g_1^q)^{r_1} \cdots (g_m^q)^{r_m}, \varphi(g^q) \cdot (h_1^q)^{r_1} \cdots (h_m^q)^{r_m} \cdot (\tau^q)^\beta \right)$$
$$= (g^q, (h')^q \tau^\beta) \in G_2 \times H_2.$$

From the last equation it is clear that the elements $g_1, \ldots, g_m \in G$ and $h_1, \ldots, h_m \in H$ are replaced with their q-th powers $g_1^q, \ldots, g_m^q \in G_2$ and $h_1^q, \ldots, h_m^q \in H_2$. Moreover, as $\tau = \tau^q$, it holds that $\tau \in H_2 \setminus \ker(\psi_2)$ and so the pair (τ, β) is preserved. By assumption β is determined from the data associated with the 2-parts and so the proof is complete. □

3.2.2 From LHN-PKE to eDLP and Membership

In this section, we work under Assumption (A1) of Sect. 1.3 for $\Gamma \in \{G, H\}$ and Assumptions (A3) and (A4) of Sect. 1.3 for $\Gamma = G$ and

$$\{\gamma_1, \ldots, \gamma_m\} = \{g_1, \ldots, g_m\}.$$

Under Assumptions (A1), (A3), and (A4), the following problem is well-posed.

Definition 2 The *extended discrete logarithm problem (eDLP)* for the finite abelian group $\Gamma = \langle \gamma_1 \rangle \oplus \ldots \oplus \langle \gamma_m \rangle$ is the problem of determining, for each $x \in \Gamma$, the unique vector

$$(\alpha_1, \ldots, \alpha_m) \in \mathbb{Z}/|\langle \gamma_1 \rangle|\mathbb{Z} \times \ldots \times \mathbb{Z}/|\langle \gamma_m \rangle|\mathbb{Z}$$

such that

$$x = \gamma_1^{\alpha_1} \cdots \gamma_m^{\alpha_m}.$$

Note that eDLP is just called *discrete logarithm problem* (DLP) in [30]. For a discussion of existing algorithms to solve it, we refer to Sect. 3.4.1. In the following, we additionally define a natural variation of the classical membership problem for groups.

Definition 3 Let Γ be a group. The *generator membership problem* for Γ is the problem of determining whether, given a finite subset Y of Γ and an element $x \in \Gamma$, one has $x \in \langle Y \rangle$.

Lemma 3 *Assume that G and H are both finite and abelian and that*

$$G = \langle g_1 \rangle \oplus \cdots \oplus \langle g_m \rangle$$

additionally satisfies (A3) *and* (A4). *Then, the* LHN-PKE *problem for G, H, K, ξ, and χ is at most as hard as the hardest of the* eDLP *problem for G and the generator membership problem for H.*

Proof Let us write G as $G = \langle g_1 \rangle \oplus \ldots \oplus \langle g_m \rangle$. Given the public pair

$$(g, h) = (g, h'\tau^\beta) \in G \times H$$

as in Sect. 1, let $r_1 \ldots, r_m \in \mathbb{Z}$ be the non-negative integers chosen to write

$$g = g_1^{r_1} \cdots g_m^{r_m} \quad \text{and} \quad h = \varphi(g) h_1^{r_1} \cdots h_m^{r_m} \cdot \tau^\beta.$$

For each $i \in \{1, \ldots, m\}$, set $\ell_i = \varphi(g_i) h_i$, and given that the order of $\varphi(g_i)$ divides $|\langle g_i \rangle|$, compute

$$\ell_i^{|\langle g_i \rangle|} = (\varphi(g_i) h_i)^{|\langle g_i \rangle|} = h_i^{|\langle g_i \rangle|}.$$

As a consequence, the subgroup M of H that is generated by

$$Y = \{\ell_i^{|\langle g_i \rangle|} : i = 1, \ldots, m\}$$

is contained in $\ker(\psi)$, and thus M does not contain τ.

Assume that eDLP is solvable in G. With h being as above, let $s_1, \ldots, s_m$ be such that, for any choice of i, one has $s_i \equiv r_i \mod |\langle g_i \rangle|$. Denote by t_i the integers satisfying $r_i - s_i = |\langle g_i \rangle| t_i$ and

$$x = \prod_{i=1}^{m} (\ell_i^{s_i})^{-1} \cdot h = \prod_{i=1}^{m} \ell_i^{r_i - s_i} \cdot \tau^\beta = \prod_{i=1}^{m} (\ell_i^{|\langle g_i \rangle|})^{t_i} \cdot \tau^\beta =: h_M \cdot \tau^\beta,$$

where we easily observe that $h_M \in M$. We now see that $\beta = 0$ if and only if $x \in M$, and thus, if we are able to solve the generator membership problem for H, we are also able to determine whether $\beta = 0$ or $\beta = 1$. □

3.3 Proof of Theorem 1

We are given the following public information:

1. $\{(g_1, \varphi(g_1)h_1), \ldots, (g_m, \varphi(g_m)h_m), \tau\}$
2. $(g, h) = (g, h'\tau^{\beta}) \in G \times H$

One can proceed as follows:

- By Lemma 2, as (A1) and (A2) are both satisfied for both G and H, the LHN-PKE problem for G, H, K, ξ, and χ is at most as hard as the LHN-PKE problem for G_2, H_2, K, ξ_2, and χ_2, where G_2 and H_2 are the Sylow 2-subgroups of G and H, respectively.
- As $G = \langle g_1 \rangle \oplus \cdots \oplus \langle g_m \rangle$ satisfies (A3), and by (A2) the largest odd factor q_G of $|G|$ is known or easily computable, we also have that $G_2 = \langle g_1^{q_G} \rangle \oplus \cdots \oplus \langle g_m^{q_G} \rangle$ satisfies (A3); note that some components may now be trivial, but they are necessarily still distinct as no power of g_i was contained in $\langle g_j \rangle$ for $i \neq j$. Also, as G_2 is a 2-group, checking the order of any element is polynomial time, just by repeated squaring and checking for the identity, so (A4) is also satisfied. Therefore, by Lemma 3, the LHN-PKE problem for G_2, H_2, K, ξ_2, and χ_2 is reduced to the eDLP problem in G_2 and the generator membership problem in H_2.

The complexity of the eDLP problem in a 2-group G_2 with m cyclic components is given by

$$O\left(\frac{\log\log|G_2|}{\log\log\log|G_2|}\log|G_2| + \frac{\log_2|G_2|}{m}2^{m/2}\right)$$

by [30, Cor. 1] (see Sect. 3.4.1 for more details). The membership problem for H_2 has polylogarithmic complexity by [4, Lemma 3(a)]; note that the required decomposition of H_2 discussed in [4, Section 3.1] exists by assumption.

3.4 Assumptions and Reductions

Assume in this section that G and H are finite and that H is abelian. One can then, as before under some reasonable assumptions, reduce to the abelian case of Theorem 1 as follows.

Thanks to (R3), for cryptanalysis purposes, we can replace G with

$$\tilde{G} = \langle g_1, \ldots, g_m \rangle,$$

and we do so. Moreover, in view of Lemma 1, the LHN-PKE problem on $(\tilde{G}, H, K, \xi, \chi)$ is reduced to the LHN-PKE problem on $(\overline{G}, H, K, \overline{\xi}, \chi)$, where

$\overline{G}$ denotes the abelianization of $\tilde{G}$ and $\overline{\xi}$ is as in Sect. 3.1. Assumption (A1) holds for $\overline{G}$, the elements $g_1, \ldots, g_m \in G$ are replaced with their images $\overline{g_1}, \ldots, \overline{g_m}$ in $\overline{G}$, and the homomorphism $\varphi : \tilde{G} \to H$ is replaced with the induced homomorphism $\overline{G} \to H$, which we identify with φ, for simplicity.

Assuming now that (A2) holds for H and $\overline{G}$, Lemma 2 allows to reduce the LHN-PKE problem on $(\overline{G}, H)$ to the LHN-PKE problem on $(\overline{G}_2, H_2)$, where $\overline{G}_2$ and H_2 denote the Sylow 2-subgroups of $\overline{G}$ and H. Here the elements $\overline{g_1}, \ldots, \overline{g_m}$ and $h_1, \ldots, h_m$ are replaced by

$$\overline{g_1}^q, \ldots, \overline{g_m}^q \text{ and } h_1^q, \ldots, h_m^q,$$

where q denotes the least common multiple of the odd parts of $|\overline{G}|$ and $|H|$.

Note that $\overline{G}_2 = \langle \overline{g_1}^q, \ldots, \overline{g_m}^q \rangle$, and suppose, at last, that (A3) holds for $\overline{G}_2$ together with $\overline{g_1}^q, \ldots, \overline{g_m}^q$; that is,

$$\overline{G}_2 = \langle \overline{g_1}^q \rangle \oplus \ldots \oplus \langle \overline{g_m}^q \rangle.$$

Then, assuming that we can efficiently compute in $\overline{G}_2$, (A4) also holds: The orders of each $\overline{g_i}^q$ can be found in polynomial time by repeated squaring as all of the orders are powers of 2. Then, as a consequence of Lemma 3, the LHN-PKE problem for $(\overline{G}_2, H_2)$ is reduced to the eDLP problem in $\overline{G}_2$ and the generator membership problem in H_2.

Both this more general attack and the attack on the finite abelian case presented in Theorem 1 rely in some capacity on the eDLP, the generator membership problem, and our assumptions (A1)–(A4). While the eDLP and the generator membership problem are relatively well understood in the cases we need for our attacks, their complexity depends heavily on the instantiation, and we discuss these problems in Sects. 3.4.1 and 3.4.2, respectively. Also, while assumptions (A1)–(A4) are natural assumptions to make when constructing groups that both admit efficient computation and satisfy the security requirements (S1)–(S4) of Sect. 2.3.5, there exist examples of groups where these assumptions are not satisfied or may be at odds with our security requirements. In Sects. 3.4.3 and 3.4.4 we discuss assumptions (A1)–(A4) with a view toward constructing instantiations of Leonardi–Ruiz-Lopez encryption to which our classical attack does not apply or at least is not polynomial time.

3.4.1 The eDLP in Finite Abelian p-Groups

Let p be a prime number, and let G be a finite abelian p-group given as

$$G = \langle g_1 \rangle \oplus \ldots \oplus \langle g_m \rangle,$$

where the orders of the summands are known. Let $e \in \mathbb{Z}$ be such that p^e is the exponent of G. Then, according to [30, Cor. 1], the eDLP in G can be solved using

$$O\left(\frac{\log(e+1)}{\log\log(e+2)}\log|G| + \frac{\log_p |G|}{m} p^{m/2}\right) \tag{1}$$

group operations on a classical computer. In particular, when m is polynomial in $\log\log|G|$ and p a fixed (small) prime, the eDLP has complexity polynomial in $\log|G|$. When $p = 2$, setting $n = \log_p |G|$, we deduce the following from the performance result in [30, Table 1]:

- When $m = 1, 2, 4, 8$, the counts are dominated by the first term of (1). This explains the initial cost decrease when m increases for a fixed n.
- Using Shank's algorithm, there is a possibility of improving the factor n/m by $\sqrt{n/m}$ though this is not relevant for applications: These normally require that n/m is close to 1.

To the best of our knowledge, there is no existing work on the eDLP that beats [30].

3.4.2 The Generator Membership Problem

Here we briefly discuss some known results on the generator membership problem in more generality than finite abelian 2-groups satisfying our assumptions: Let H be a finite group, and let Y be a (finite) subset of H. Below we collect some known results regarding membership testing for $\langle Y\rangle$, depending on how the group H is given.

If H is given as a subgroup of some finite $\mathrm{GL}(n, q)$ and μ denotes the largest prime dividing $|\langle Y\rangle|$ and not dividing q, then testing whether $x \in H$ belongs to $\langle Y\rangle$ can be achieved in time that is polynomial in $|Y| + n + q + \mu$; cf. [21, Theorem 3.2(2)].

If H is given as a permutation group and satisfies (A3), then membership testing in Y is NC (polylogarithmic on a polynomial number of parallel processors); cf. [1, Theorem 1.2(a)] and [23, Theorem 1].

3.4.3 On the Security Assumptions (A2) and (A4)

In Sect. 3.2.2 above we described how and under which assumptions the LHN-PKE problem for G, H, and K can be reduced to solving the eDLP in G and the membership problem in H. The necessity for setting assumptions (A2) and (A4) in particular came from the fact that there are known examples of abelian groups, some of them already being used successfully in existing cryptographic protocols, that do not have to obey them.

One type of such groups are the known RSA groups. These are given in the form $(\mathbb{Z}/N\mathbb{Z})^\times$, where $N = pq$ is hard to factor. In this case, Alice would know the factorization and hence the order of the group, but an adversary should not be able to compute it.

Another type of such groups, where even the creators of the cryptosystem might not know the group's order, is that of ideal class groups of imaginary quadratic number fields. This is an interesting category of groups for cryptography since it allows one to work in a *trustless setup*. In other words, we do not need a trusted third party to generate groups of secure order, in contrast to cryptosystems that employ RSA groups for example, where a trusted third party needs to generate a secure, i.e., hard to factor, modulus $N \in \mathbb{N}$ for the groups $(\mathbb{Z}/N\mathbb{Z})^\times$. Despite the fact that neither the structure nor the order of the ideal class group is known, the group operation is efficient and the elements of the group have a compact representation, via reduced binary quadratic forms. An excellent reference for trusted unknown-order groups is the paper by Dobson, Galbraith, and Smith [8]; they also take into account Sutherland's algorithm [29], and they propose new security parameters for cryptosystems that employ ideal class groups. In the same paper the authors also discuss other groups of unknown order that can be used, namely genus-3 Jacobians of hyperelliptic curves, initially introduced by Brent [5]. Even though these groups appear to have some computational advantages when compared to ideal class groups, they offer less security than the genus-1 and genus-2 curves, as one can see, for example, in [12].

3.4.4 Regarding Assumption (A3)

In this section, we discuss Assumption (A3). For the sake of simplicity and in view of Sect. 3.2.1, we restrict to 2-groups, but everything can be said similarly for arbitrary finite abelian groups using Lemma 2.

The case in which Assumption (A3) holds for the subgroup $\langle g_1, \ldots, g_m \rangle$ of G, i.e., $g_1, \ldots, g_m$ satisfy $\langle g_1, \ldots, g_m \rangle = \langle g_1 \rangle \oplus \ldots \oplus \langle g_m \rangle$ (equivalently $g_1, \ldots, g_m$ are *independent*), seems to be the key case for the following reasons. First of all, when sampling $g_1, \ldots, g_m$ from G uniformly, it is very likely that these are independent, at least if m is small in comparison with the number of cyclic factors in G. For example, if $G = C_2^\lambda$ is a direct product of cyclic groups of order 2, then the probability of sampling $g_1, \ldots, g_m$ that are independent is

$$\frac{(2^\lambda - 1)(2^\lambda - 2)(2^\lambda - 2^2) \cdots (2^\lambda - 2^{m-1})}{2^{m\lambda}},$$

which is very close to 1 if $m \ll \lambda$. In addition, the case in which $g_1, \ldots, g_m$ are independent can be seen as the generic case; we expect that an attacker could use similar strategies as developed in Sect. 1.3 and design an attack for the dependent case. Indeed, if $g_1, \ldots, g_m$ are dependent, the attacker will obtain more information from the public key than in the independent case.

There is also another strategy for an attack if (A3) does not hold for G and the number of cyclic factors in H is small. More precisely, assume that $|\operatorname{Hom}(H, \mathbb{F}_2)|$

is sub-exponential, or in other words, the number of maximal subgroups of H is sub-exponential. In addition, it would be necessary to list these maximal subgroups M and to be able to work in the quotients H/M efficiently. If Eve can write any element $g \in G$ as a product in $g_1, \ldots, g_m$, then she can recover the secret message β from the ciphertext (g, h) as follows. For every maximal subgroup M of H not containing τ (by assumption, there are only sub-exponentially many of these), convert h into an element $\tilde{h}$ in H/M. Decrypt $(g, \tilde{h}) \in G \times \tilde{H}$ as if the noise was erased completely in H/M (as explained in Sect. 2.2), and check on a number of self-encrypted messages if this provides a correct decryption function. Since $\ker(\psi) \neq H$, there will always be a maximal subgroup M of H that contains $h_1 \ldots, h_m$ but does not contain τ, so eventually this procedure will indeed provide a decryption function.

3.5 Comparison with the Quantum Attack

Throughout this section, assume that (A3) holds for the abelian group G together with $g_1, \ldots, g_m$. Even though we only did cryptanalysis for finite groups until now, we show in the following two subsections how one can perform quantum attacks when G is either torsion-free or finite. The torsion-free case was handled already in [19], but we include it here for completeness; the finite case is new. The mixed case can be considered as a combination of the two cases: first dealing with the free part of the group and then recovering β as explained in the finite case.

3.5.1 The Torsion-Free Case

Assume as in [19, Sec. 8.2] that $G = \langle g_1 \rangle \oplus \ldots \oplus \langle g_m \rangle$ is isomorphic to $\mathbb{Z}^m$, i.e., the orders $|\langle g_i \rangle|$ are all infinite. We briefly recall the discussion from [19, Sec. 8.2]. To this end, let $f : \mathbb{Z}^{m+1} \to G$ be defined by

$$(a_1, \ldots, a_{m+1}) \longmapsto g^{a_{m+1}} \cdot g_1^{a_1} \cdots g_m^{a_m}.$$

Then the kernel of f is equal to $\langle (r_1, \ldots, r_m, 1) \rangle$, and it can be determined, using Shor's algorithm [27], in quantum polynomial time in m, cf. [20]. Once $(r_1, \ldots, r_m, 1)$ is known, it is easy to recover β from the encrypted message (g, h) and the public information.

3.5.2 The Torsion Case

Assume in this section that G is finite and that (A4) holds in addition to (A1) and (A3). Let, moreover, $f : \mathbb{Z}^{m+1} \to G$ be defined by

$$(a_1, \ldots, a_{m+1}) \longmapsto g^{a_{m+1}} \cdot g_1^{a_1} \cdots g_m^{a_m}.$$

Then a set of generators of $\ker(f)$ can be determined in time polynomial in $\log |G|$ on a quantum computer [16, 27, 28]; see also [11]. Note that $\ker(f)$ contains

$$|\langle g_1\rangle|\mathbb{Z} \times \ldots \times |\langle g_m\rangle|\mathbb{Z} \times |\langle g\rangle|\mathbb{Z}.$$

Let now $(s_1, \ldots, s_{m+1})$ be one of the generators found. Then $g = g_1^{-s_1 s_{m+1}} \cdots g_m^{-s_m s_{m+1}}$ and so it follows that

$$g_1^{r_1+s_1 s_{m+1}} \cdots g_m^{r_m+s_m s_{m+1}} = 1.$$

This is the same as saying that, for each $1 \leq i \leq m$, one has

$$r_i \equiv -s_i s_{m+1} \bmod |\langle g_i\rangle|.$$

Modding out H by the subgroup generated by all $h_i^{|\langle g_i\rangle|}$, one can recover β and thus solve the LHN-PKE problem, cf. Lemma 3.

4 An Attack Strategy that Reduces to the Abelian Case

In this section, we assume G and H are finite. A general strategy for an attacker to solve the LHN-PKE problem could be to convert the H-part of the ciphertext as follows. Assume that Eve has access to the ciphertext

$$(g, h) = \left(g_{w_1} \cdots g_{w_\ell}, \varphi(g_{w_1})h_{w_1} \cdots \varphi(g_{w_\ell})h_{w_\ell}\tau^\beta\right) \in G \times H$$

encrypted by Bob using Alice's public key

$$\{(g_1, \varphi(g_1)h_1), \ldots, (g_m, \varphi(g_m)h_m), \tau\}.$$

Eve can then choose a subgroup $\overline{H}$ and a homomorphism $\theta\colon H \to \overline{H}$ such that $\vartheta(\tau) \neq 1$ and compute $\vartheta(h)$. Her new pair $(g, \overline{h}) = (g, \vartheta(h))$ is then of the form

$$(g_{w_1} \cdot \ldots \cdot g_{w_\ell}, \overline{\varphi}(g_{w_1})\overline{h}_{w_1} \cdot \ldots \cdot \overline{\varphi}(g_{w_\ell})\overline{h}_{w_\ell}\overline{\tau}^\beta),$$

where, for all i, we set $\overline{h_i} = \vartheta(h_i)$ and write $\overline{\varphi} = \vartheta \circ \varphi$ and $\overline{\tau} = \vartheta(\tau)$. Since $\overline{\tau} \neq 1$, this pair still contains the information on β, and if ϑ is chosen cleverly, it might be much simpler to deduce β from (g, h'). Note that in general, we cannot define a suitable counterpart $\overline{\psi}$ of ψ here, so the information that τ is not contained in the kernel of ψ cannot be directly converted into a statement on $\overline{\tau}$ and has to be considered individually (if necessary).

A special case of this strategy is to reduce, if possible, to a finite abelian group $\overline{H}$, i.e., the goal is to eventually apply the attack that we describe in Sect. 3.4 even when G and H are nonabelian.

Suppose that, given τ and H, Eve is able to find a finite abelian group $\overline{H}$ and an efficiently computable homomorphism $\vartheta : H \to \overline{H}$ such that $\vartheta(\tau) \neq 1$ and such that it can be checked that $\vartheta(\tau)$ is not contained in the subgroup generated by certain powers of $\vartheta(\varphi(g_i)h_i)$. This condition can be checked using the public key, and we will specify below which powers are sufficient.

Eve follows the following three steps based on Lemmas 1, 2, and 3.

Step 1 Let $\overline{G} = G/[G, G]$ be the abelianization of G and $\bar{\varphi}: \overline{G} \to \overline{H}$ be the homomorphism obtained from $\vartheta \circ \varphi: G \to \overline{H}$ by reducing modulo $[G, G]$. For all i, define $\bar{h}_i = \vartheta(h_i)$, set $\bar{\tau} = \vartheta(\tau)$, and, for all g in G, let $\bar{g} \in \overline{G}$ be the image of g in $\overline{G}$.

As in Lemma 1, Eve replaces the encrypted message by

$$(\bar{g}, \vartheta(h)) = (\bar{g}_{w_1} \cdots \bar{g}_{w_\ell}, \bar{\varphi}(\bar{g}_{w_1})\bar{h}_{w_1} \cdots \bar{\varphi}(\bar{g}_{w_\ell})\bar{h}_{w_\ell}\bar{\tau}^{\beta}) \in \overline{G} \times \overline{H}$$

in order to work inside abelian groups $\overline{G}$ and $\overline{H}$.

Step 2 Provided (A2) holds for the finite abelian groups $\overline{G}$ and $\overline{H}$, Eve proceeds in a similar way as in the proof of Lemma 2: First, she computes the largest positive odd factor of the group orders of $\overline{G}$ and $\overline{H}$ and finds a common multiple q of their odd parts. Then she converts the tuple $(\bar{g}, \vartheta(h))$ above into a tuple with entries inside the 2-Sylow subgroups $\overline{G}_2$ and $\overline{H}_2$ of $\overline{G}$ and $\overline{H}$ by taking both entries to their q-th powers:

$$(\bar{g}^q, \vartheta(h)^q) = (\bar{g}^q_{w_1} \cdots \bar{g}^q_{w_\ell}, \bar{\varphi}(\bar{g}^q_{w_1})\bar{h}^q_{w_1} \cdots \bar{\varphi}(\bar{g}^q_{w_\ell})\bar{h}^q_{w_\ell}\bar{\tau}^{\beta}) \in \overline{G}_2 \times \overline{H}_2.$$

Step 3 Finally, assuming (A3) and (A4) hold for $\overline{G}_2$, Eve proceeds in a similar way as in the proof of Lemma 3. For all i, define $\bar{\ell}_i = (\bar{\varphi}(\bar{g}_i)\bar{h}_i)^q = \vartheta(\varphi(g_i)h_i)^q$, and let β_i be the order of $\bar{g}_i^q$. Eve computes the elements $\bar{\ell}_i$ and the numbers β_i using the public key. Then $(\bar{\ell}_i)^{\beta_i} = (\bar{h}_i^q)^{\beta_i}$ can also be computed from the public key. Assuming the eDLP is solvable in $\overline{G}_2$ and the generator membership problem is solvable in $\overline{H}_2$ and if $\bar{\tau}$ is not contained in the subgroup M generated by $(\bar{\ell}_1)^{\beta_1}, \ldots, (\bar{\ell}_m)^{\beta_m}$, the value of β can be recovered as in the proof of Lemma 3.

Remark 2 Observe that the above attack applies to any finite group H if $\tau \notin [H, H]$ and the image of τ in $H/[H, H]$ is not contained in the subgroup generated by the images of certain powers of $\varphi(g_i)h_i$. In particular, it is advisable (but likely not sufficient) to sample τ from $[H, H]$.

5 Normal Forms and the Solvable Case

In this section we consider the case in which G and H are finite and solvable and give evidence of why, given the efficiency constraints attached to the system, the groups should not be expected to provide secure postquantum cryptosystems. This is done in analogy to Sect. 3.5, but with more iterations. Roughly speaking, since solvable groups have natural filtrations yielding abelian quotients, as we explain below, one could perform a quantum attack on solvable groups by iterating quantum attacks on abelian groups.

For the background on finite solvable groups, we refer to [15, Ch. 3].

Definition 4 For a finite group Γ, the *derived series* of Γ is the series $(\Gamma^{(i)})_{i\geq 1}$ defined inductively by

$$\Gamma^{(1)} = \Gamma \quad \text{and} \quad \Gamma^{(i+1)} = [\Gamma^{(i)}, \Gamma^{(i)}].$$

If for some index m the group $\Gamma^{(m)}$ is trivial, then Γ is said to be *solvable*.

In a finite group Γ, each quotient $\Gamma^{(i)}/\Gamma^{(i+1)}$ is abelian and, if Γ is solvable and $\Gamma^{(i)}/\Gamma^{(i+1)}$ is trivial, then $\Gamma^{(i)} = \Gamma^{(i+1)} = \Gamma^{(i+2)} = \ldots = \{1\}$.

Until the end of this section, assume that H is solvable. Then, for cryptanalysis purposes and in analogy with Lemma 1, we assume without loss of generality that G is also solvable. We let s be the *derived length* of H, i.e., s is such that $H^{(s+1)} = 1$ but $H^{(s)} \neq 1$. Since $\varphi(G^{(s+1)})$ is contained in $H^{(s+1)} = \{1\}$, in analogy to Sect. 3.1, we assume without loss of generality that $G^{(s+1)} = 1$. We remark that if $s \leq 1$, then G and H are abelian.

Remark 3 (Efficient Communication and Computation) Alice and Bob, as part of their message exchange, need to be able to communicate elements and perform operations in the groups efficiently. An often favorable approach (also proposed in [2, § 4.2]) is that of using *normal forms* of elements with respect to a *polycyclic presentation*, cf. [10, Ch. 2]. For instance, when working with G and H abelian, (A3) and (A4) holding for G are almost the same as saying that G is given by a polycyclic presentation and the expression $g = g_1^{r_1} \cdots g_m^{r_m}$ is the normal form of g with respect to this presentation. The word "almost" in the previous sentence is there to stress that the r_i's are not uniquely identified by their class modulo $|\langle g_i \rangle|$, which in turn is what happens for normal forms (see below).

We briefly explain here what it means for an element g of the solvable group G to be communicated in a *normal form* with respect to a polycyclic presentation respecting the derived filtration. To do so, for each $i \in \{1, \ldots, s\}$:

(a) Let m_i denote the minimum number of generators of $G^{(i)}/G^{(i+1)}$.
(b) Let $g_{i1}, \ldots, g_{im_i}$ be elements of $G^{(i)}$ such that

$$G^{(i)}/G^{(i+1)} = \langle g_{i1} G^{(i+1)} \rangle \oplus \cdots \oplus \langle g_{im_i} G^{(i+1)} \rangle.$$

(c) For each $j \in \{1, \ldots, m_i\}$, let o_{ij} denote the order of g_{ij} modulo $G^{(i+1)}$.

Then any element $g \in G$ can be uniquely represented by a vector

$$\delta = (\delta_{11}, \ldots, \delta_{1m_1}, \delta_{21}, \ldots, \delta_{2m_2}, \ldots, \delta_{s1}, \ldots, \delta_{sm_s})$$

of integers $0 \leq \delta_{ij} < o_{ij}$ of length $n = m_1 + \ldots + m_s$ such that

$$g = \prod_{i=1}^{s} \prod_{j=1}^{m_i} g_{ij}^{\delta_{ij}},$$

which is precisely the normal form of g with respect to the chosen generators $\{g_{ij}\}$.

Note that the data (a)–(b)–(c) mentioned above should be public for Alice and Bob to be able to share the elements with each other. If the chosen generators fit into a *polycyclic sequence*, then the group operation is performed through the *collection process* [10, § 2.2]. It should be mentioned that, though polycyclic presentations generally yield a good practical performance, it is, to the best of our knowledge, not clear whether multiplication in these presentations can always be performed in polynomial time [18]. Note that the expression of h' also depends on the vector δ:

$$h' = \prod_{i=1}^{s} \prod_{j=1}^{m_i} (\varphi(g_{ij}) h_{ij})^{\delta_{ij}}.$$

Shor's algorithm, applied on each level $G^{(i)}/G^{(i+1)}$, is polynomial in the log of the size of this quotient. Moreover, Shor's algorithm really does determine the vector δ of g because of the condition $0 \leq \delta_{ij} < o_{ij}$. In particular Eve can recover the vector δ of g and use it to compute

$$\tau^{\beta} = \left(\prod_{i=1}^{s} \prod_{j=1}^{m_i} (\varphi(g_{ij}) h_{ij})^{\delta_{ij}} \right)^{-1} \cdot h.$$

The complexity of this algorithm on each quotient $G^{(i)}/G^{(i+1)}$ is polynomial in $\log |G^{(i)}/G^{(i+1)}|$, which makes the overall complexity to be polynomial in $\log |G|$. There is therefore no separation in complexity between the algorithm being run by the adversary and the one being run by the user. This implies that for the system to be secure, we would need to have $|G| > 2^{2^{\lambda}}$, making it hard to say if it would even be possible to represent elements of G in a computer.

Remark 4 Given that the quotients of consecutive elements of the derived series are abelian, it is natural to ask whether the classical attack we designed for abelian groups could be generalized to an attack in the solvable context. It seems, however, that the nonabelian solvable case is substantially different from the abelian one. Among the limitations are:

- τ could, for instance, belong to $H^{(s)}$, while $g_1, \ldots, g_m$ will typically live in $H^{(1)} \setminus H^{(2)}$ (in the general setting we are indeed not necessarily working with normal forms).
- Without knowing φ, it is not at all clear at which depths in the derived filtration $\varphi(g_1), \ldots \varphi(g_m)$ are to be found in H (as publicly given only with noise).
- Dealing with quotients is more delicate as one always has to consider normal closures of subgroups.

6 Future Work

In future work, we plan to consider several types of finite groups G, H, and K for instantiating Leonardi–Ruiz-Lopez encryption and explore whether we can either construct attacks for the corresponding cryptosystems or prove security results.

A first candidate would be the group C_2^{λ} for all groups G, H, and K. As none of the classical attacks presented in this chapter applies in this case, Leonardi–Ruiz-Lopez encryption might prove to be classically secure for this choice. Other abelian candidates are the RSA groups and ideal class groups mentioned in Sect. 3.4.3.

As a first nonabelian example, we plan to work with certain p-groups and use strategies that allow us to circumnavigate the attacks presented in Sect. 5. In particular, it would be beneficial to work with presentations of groups that are not based on normal forms and yet allow efficient computation. The advantage of working with nonabelian groups is that it may be possible to construct a post-quantum additive homomorphic cryptosystem.

Acknowledgments We warmly thank the organizers of *Women in Numbers Europe 4* for putting together this team. We also wish to thank Chris Leonardi and Andrew Sutherland for helpful conversations around the contents of [19] and [30], respectively. In addition, we thank the three anonymous referees for their comments, which helped improve the exposition of this work.
The first author was supported by the European Union's H2020 Programme under grant agreement number ERC-669891. The second author was supported by the Austrian Science Fund, Project P34808. The fifth author was supported by the Magnus Ehrnrooth grant 336005, the Academy of Finland project grant (351271, PI Camilla Hollanti), and the MATINE project grant, Ministry of Defence Finland (2500M-0147, PI Camilla Hollanti). The last author was partially supported by the Deutsche Forschungsgemeinschaft (DFG, German Research Foundation)—Project ID 286237555—TRR 195 and by the Italian program Rita Levi Montalcini for young researchers, Edition 2020.

References

1. L. Babai, E. Luks, A. Seress, Permutation groups in NC, in *Proc. 19th ACM Symp. on Theory of Computing*, pp. 409–420 (1987)
2. G. Baumslag, N. Fazio, A.R. Nicolosi, V. Shpilrain, W.E. Skeith, III, Generalized learning problems and applications to non-commutative cryptography (extended abstract), in *Provable Security*, volume 6980 of *Lecture Notes in Comput. Sci.* (Springer, Heidelberg, 2011), pp. 324–339

3. J. Benaloh, Dense probabilistic encryption, in *Proceedings of the Workshop on Selected Areas of Cryptography*, pp. 120–128 (1994)
4. J. Bermejo-Vega, M. Van den Nest, Classical simulations of Abelian-group normalizer circuits with intermediate measurements. Quantum Inf. Comput. **14**(3–4), 181–216 (2014)
5. R.P. Brent, *Public Key Cryptography with a Group of Unknown Order*. Tech. Rep. Oxford University, 2000
6. I. Chillotti, N. Gama, M. Georgieva, M. Izabachène, TFHE: Fast fully homomorphic encryption over the torus. J. Cryptol. **33**, 34–91 (2020)
7. K. Cong, R.C. Moreno, M.B. da Gama, W. Dai, I. Iliashenko, K. Laine, M. Rosenberg, Labeled PSI from homomorphic encryption with reduced computation and communication, in *ACM-CCS 2021*, Lecture Notes in Computer Science, pp. 1135–1150 (2021)
8. S. Dobson, S. Galbraith, B. Smith, Trustless unknown-order groups. Math. Cryptol. **1**(2), 25–39 (2022)
9. L. Ducas, D. Micciancio, FHEW: Bootstrapping homomorphic encryption in less than a second, in *Advances in Cryptology - EUROCRYPT 2015*, Lecture Notes in Computer Science, pp. 617–640 (2015)
10. B. Eick, Algorithms for polycyclic groups. Habilitationsschrift, Universität Kassel, 2001
11. M. Ettinger, P. Hoyer, E. Knill, The quantum query complexity of the hidden subgroup problem is polynomial. Inform. Process. Lett. **91**(1), 43–48 (2004)
12. X. Fan, T. Wollinger, Y. Wang, Inversion-free arithmetic on genus 3 hyperelliptic curves and its implementations, in *International Conference on Information Technology: Coding and Computing (ITCC'05)* - Volume II, Las Vegas, NV, USA, 2005, pp. 642–647, Vol. 1. https://doi.org/10.1109/ITCC.2005.179
13. C. Fontaine, F. Galand, A survey of homomorphic encryption for nonspecialists. EURASIP J. Inf. Secur. **2007**, 013801 (2007)
14. C. Gentry, A. Sahai, B. Waters, Homomorphic encryption from learning with errors: Conceptually-simpler, asymptotically-faster, attribute-based, in *Advances in Cryptology - Crypto 2013*, Lecture Notes in Computer Science, pp. 75–92 (2013)
15. I.M. Isaacs, *Finite Group Theory*, volume 92 of *Graduate Studies in Mathematics* (American Mathematical Society, Providence, RI, 2008)
16. A.Y. Kitaev, Quantum computations: algorithms and error correction. Uspekhi Mat. Nauk **52**(6(318)), 53–112 (1997)
17. D.E. Knuth, *The Art of Computer Programming. Vol. 1: Fundamental Algorithms*, 3rd edn. (Addison-Wesley, Reading, MA, 1997)
18. C.R. Leedham-Green, L.H. Soicher, Symbolic collection using deep thought. LMS J. Comput. Math. **1**, 9–24 (1998)
19. C. Leonardi, L. Ruiz-Lopez, Homomorphism learning problems and its applications to public-key cryptography. Cryptology ePrint Archive, Report 2019/717, 2019. https://ia.cr/2019/717
20. S.J. Lomonaco jun, L.H. Kauffman, Quantum hidden subgroup algorithms: a mathematical perspective, in *Quantum Computation and Information. AMS Special Session*, Washington, DC, USA, January 19–21, 2000, pp. 139–202 (AMS, American Mathematical Society, Providence, RI, 2002)
21. E.M. Luks, Computing in solvable matrix groups, in *33rd Annual Symposium on Foundations of Computer Science (FOCS). Proceedings*, Pittsburgh, PA, USA, October 24–27, 1992, pp. 111–120 (IEEE Computer Society Press, Washington, DC, 1992)
22. V. Lyubashevsky, C. Peikert, O. Regev, On ideal lattices and learning with errors over rings. J. ACM (JACM) **60**(6), 1–35 (2013)
23. P. McKenzie, S.A. Cook, The parallel complexity of abelian permutation group problems. SIAM J. Comput. **16**(5), 880–909 (1987)
24. P. Paillier, Public-key cryptosystems based on composite degree residuosity classes, in *International Conference on the Theory and Application of Cryptographic Techniques (EUROCRYPT'99)*, vol. 1592, pp. 223–238 (1999)
25. O. Regev, On lattices, learning with errors, random linear codes, and cryptography. J. ACM (JACM) **56**(6), 1–40 (2009)

26. P.W. Shor, Algorithms for quantum computation: discrete logarithms and factoring, in *35th Annual Symposium on Foundations of Computer Science (Santa Fe, NM, 1994)*, pp. 124–134 (IEEE Comput. Soc. Press, Los Alamitos, CA, 1994)
27. P.W. Shor, Polynomial-time algorithms for prime factorization and discrete logarithms on a quantum computer. SIAM J. Comput. **26**(5), 1484–1509 (1997)
28. D.R. Simon, On the power of quantum computation. SIAM J. Comput. **26**(5), 1474–1483 (1997)
29. A.V. Sutherland, Order computations in generic groups, 2007. PhD Thesis, Massachusetts Institute of Technology. https://math.mit.edu/~drew/sutherland-phd.pdf
30. A.V. Sutherland, Structure computation and discrete logarithms in finite abelian p-groups. Math. Comput. **80**(273), 477–500 (2011)

A Survey of Local–Global Methods for Hilbert's Tenth Problem

Sylvy Anscombe, Valentijn Karemaker, Zeynep Kisakürek, Vlerë Mehmeti, Margherita Pagano, and Laura Paladino

1 Introduction

Hilbert's Tenth Problem (**H10**) for a ring R asks for an algorithm to decide correctly, for each $f \in \mathbb{Z}[X_1, \ldots, X_n]$, whether the diophantine equation

$$f(X_1, \ldots, X_n) = 0$$

has a solution in R. A stronger form allows polynomials with coefficients in certain recursive subrings of R.

S. Anscombe (✉)
Université Paris Cité and Sorbonne Université, CNRS, IMJ-PRG, Paris, France
e-mail: sylvy.anscombe@imj-prg.fr

V. Karemaker
Mathematisch Instituut, Universiteit Utrecht, Utrecht, The Netherlands
e-mail: V.Z.Karemaker@uu.nl

Z. Kisakürek
Mathematisches Institut der Heinrich-Heine-Universität Düsseldorf, Düsseldorf, Germany
e-mail: kisakuer@uni-duesseldorf.de

V. Mehmeti
Sorbonne Université and Université Paris Cité, CNRS, IMJ-PRG, Paris, France
e-mail: vlere.mehmeti@imj-prg.fr

M. Pagano
Mathematisch Instituut, Universiteit Leiden, Leiden, The Netherlands
e-mail: m.pagano@math.leidenuniv.nl

L. Paladino
Università della Calabria, Ponte Bucci, Italy
e-mail: laura.paladino@unical.it

R. Abdellatif et al. (eds.), *Women in Numbers Europe IV*, Association for Women in Mathematics Series 32, https://doi.org/10.1007/978-3-031-52163-8_2

Davis, Putnam, and Robinson, largely working in the 1950s and 1960s [17], approached **H10** by studying diophantine sets: A set $A \subseteq \mathbb{Z}^n$ is *diophantine* if there is a polynomial $f \in \mathbb{Z}[X_1, \dots, X_m, Y_1, \dots, Y_n]$ such that

$$(a_1, \dots, a_n) \in A \Longleftrightarrow \exists x_1 \dots x_m \; f(x_1, \dots, x_m, a_1, \dots, a_n) = 0,$$

for all $a_1, \dots, a_n \in \mathbb{Z}$. They proved that a negative solution to **H10** would follow from the existence of a diophantine subset of $\mathbb{Z}$ of 'exponential growth'. Moreover they gave a negative solution to the related 'exponential' version of **H10** for $\mathbb{Z}$ in which f is allowed to be an exponential polynomial. In 1970, Matiyasevich [45] completed the proof of the celebrated 'DPRM theorem' by finding a diophantine set of exponential growth using the Fibonacci sequence:

Theorem 1 (Davis–Putnam–Robinson–Matiyasevich) **H10** *for $\mathbb{Z}$ is unsolvable, i.e., there is no algorithm that decides correctly whether or not each multivariable polynomial has a solution in $\mathbb{Z}$.*

Since then, Hilbert's Tenth Problem has been studied in a wide range of rings, fields, and associated structures, in particular valued fields. Indeed, the study of **H10** in fields and rings brings together a variety of results and techniques from model theory, including the theories of local fields, due to Tarski [73], Ax–Kochen [2], and Ershov [28], especially uniform aspects of these theories, as well as definability in global fields, field arithmetic, and valuation theory.

In the case of function fields of positive characteristic, continuing a long line of work beginning with Pheidas [53], Eisentrager and Shlapentokh showed in [27] that **H10** – in its stronger form – is unsolvable if the field does not contain the algebraic closure of a finite field; this in particular applies to global function fields. For results over function fields of characteristic zero, see, e.g., [24, 25, 38, 39, 61]. In another direction, Rumely showed in [60] that **H10** is solvable for the ring of algebraic integers. For more comprehensive overviews of this subject, the reader is encouraged to consult [40, 66].

Perhaps the most important active area of research in this direction is related to **H10** for $\mathbb{Q}$, which is still an open problem. Poonen showed in [54] that, for certain large sets of primes S, **H10** over $\mathbb{Z}[S^{-1}]$ has a negative answer. Another major breakthrough came through the use of local–global principles to find new families of diophantine subsets of $\mathbb{Q}$. Firstly, in [57] Robinson used the Hasse–Minkowski theorem, a local–global principle on quadratic forms, to show that every valuation ring $\mathcal{O}$ of a number field K is diophantine over K and that $\mathbb{Z}$ is definable in $\mathbb{Q}$ by a first-order formula in the language of rings – see Sect. 2 for a brief introduction to the terminology of first-order logic. Subsequent steps were made in the work of Eisenträger [23] who used the Albert–Brauer–Hasse–Noether theorem, a local–global principle for central simple algebras, and of Poonen [55].

The ideas of Eisenträger and Poonen kickstarted a series of discoveries. More precisely, in [55], Poonen proved a striking theorem: $\mathbb{Z}$ is 'universal-existentially definable' in $\mathbb{Q}$; that is, it is described by a first-order formula that consists of universal quantifiers, followed by existential quantifiers, followed by a quantifier-free formula. Building on this result, Koenigsmann showed in [41] that $\mathbb{Q} \setminus \mathbb{Z}$ is diophantine in $\mathbb{Q}$; Park in [52] in turn generalized this from $\mathbb{Q}$ to general number fields. In each case, the authors make use of the Hasse–Minkowski local–global principle for quadratic forms. In a new direction, in [18], Dittmann generalized this machinery from quaternion algebras to general central simple algebras and replaced the Hasse–Minkowski theorem with the Albert–Brauer–Hasse–Noether theorem. This allowed him to show that 'irreducibility is diophantine', or more precisely: for each global field K and each $n \in \mathbb{N}$, the set $\{(a_0, \ldots, a_{n-1}) \in K^n \mid X^n + \sum_{i<n} a_i X^i$ is irreducible in $K[X]\}$ is diophantine over K. Further use has been made of the approach using central simple algebras, for example, by Daans [13, 15] and Anscombe–Dittmann–Fehm [1].

The above shows this is a growing research line of rich interaction between logic and model theory and algebra and number theory. Moreover, these results suggest natural extensions and generalizations in several directions, which we explore in the last section of this survey: On the one hand, the diophantine sets studied in these papers and the local–global behaviour they exhibit can fit into a framework of *diophantine families*, and on the other hand, one may replace the Hasse–Minkowski and Albert–Brauer–Hasse–Noether theorems with other local–global principles arising naturally in arithmetic geometry and obtain different diophantine sets.

In this chapter, we survey the stream of work following [23, 55, 57] in which local–global principles are used to great effect. After a concise introduction to the topic, we survey how the Hasse–Minkowski and Albert–Brauer–Hasse–Noether theorems have been used to prove striking new definability results in the above-mentioned works of Eisenträger, Poonen, Koenigsman, Park, Dittmann, and Daans. In particular, in Sect. 5.1 we compare the techniques and results of these papers, in which the authors have been concerned with similar questions. We hope that such an analysis will be useful for future uses of local–global principles in this study of definability in number theory, as we discuss in Sects. 5.2 and 5.3.

Roughly speaking, we say a local–global principle holds when a property is true over a global field if and only if it is true over all, or all but finitely many, completions of that global field. One of the main reasons why local–global principles can be useful in the study of arithmetic properties of global objects is that, typically, the analogous properties over complete local fields are more accessible. The above-mentioned Hasse–Minkowski and Albert–Brauer–Hasse–Noether theorems are concerned with the property of a quadratic form being isotropic, respectively, of a central simple algebra splitting (or equivalently, a Brauer group element being trivial).

In arithmetic geometry one is concerned with rational points and arithmetic invariants of algebraic varieties and may study local–global principles for the existence of rational points (see, e.g., [33, 37, 51]) and for the behaviour of arithmetic invariants such as endomorphisms and automorphisms (see, e.g., [72], [74]).

There are many other contexts where local–global questions can be considered. Indeed, local–global principles are studied also in related research areas like arithmetic dynamics (see, e.g., [3, 35]), and recently, in arithmetic geometry over a base field which is not a global field but, for instance, a function field over a number field [36], or a function field over a discretely valued field [6–8]. In the literature there are also examples of different kinds of local–global questions, not formulated in the setting of fields at all. In those cases only the basic idea of studying properties in relation to behaviour at primes is borrowed from the classical local–global setup. For example, in [71] the local behaviour of polynomial rings $R[X]$, where R is a commutative ring, is considered over its localizations at maximal ideals. In [43] some properties of the characters of a group are encoded by those of the normalizers of its p-Sylow subgroups. However, as stated above, in the present paper we will consider the classical local–global background of a global field and its completions.

This survey is intended for a broad audience, so we do not assume a detailed background in either logic or number theory. Accordingly, its structure is as follows: In Sect. 2 we briefly introduce the necessary tools and terminology from logic, and likewise in Sect. 3 we introduce the algebraic tools, in particular quaternion algebras and central simple algebras. We define the notion of a local–global principle and give the Hasse–Minkowski and Albert–Brauer–Hasse–Noether theorems as two important examples. In Sect. 4 we survey the main results and methods contained in Robinson's seminal work and the above-mentioned papers [13, 18, 23, 41, 52, 55], in which the authors adopt similar methods of proof, in particular by making use of the Hasse–Minkowski and Albert–Brauer–Hasse–Noether local–global principles. Finally, in Sect. 5, we make comparisons between the surveyed articles, in order to establish a common model-theoretic framework. We explore two further directions: On the one hand, we describe one possible model-theoretic viewpoint of diophantine sets satisfying some local–global principle, while on the other hand, we propose two other local–global principles from arithmetic geometry that could be used to study diophantine sets.

2 Prerequisites from Model Theory

We begin by introducing a few basic concepts from first-order logic.

2.1 First-Order Logic and Definable Sets

We give a brief introduction to the basic tools of mathematical logic: formulas, structures, and definable sets. Generally speaking, logic is the study of valid reasoning.

In the same spirit, the sub-discipline of mathematical logic is the study of valid symbolic reasoning within formal languages, and it includes applications to theories drawn from mathematics. Prominent examples of formal languages are the so-called first-order languages, in which a significant proportion of mathematics can be formalized. These languages are of particular importance in model theory, which studies the relationship between first-order theories and the mathematical structures that model them. In this survey we will restrict most of our discussion to the first-order language of rings.

For more information, especially for a more general perspective on model theory, the readers are encouraged to consult [34, 44].

The first-order *language of rings*, denoted $\mathfrak{L}_{\text{ring}}$, is constructed in several stages, as follows. We begin with the *symbols* of the language, which are of two types: the logical and the non-logical.

The logical symbols are

- *Connectives*: $\wedge$ for conjunction, $\vee$ for disjunction, $\neg$ for negation, $\rightarrow$ for implication, and $\leftrightarrow$ for biconditional
- *Variables*: $x_1, x_2, x_3, \ldots, y, z, \ldots$
- *Quantifiers*: $\forall$ for the universal quantifier *'for all'* and $\exists$ for the existential quantifier *'there exists'*
- *Punctuation*: $=$ for equality, and the parentheses ( )

The non-logical symbols are the *function symbols* $+$, $\cdot$, and $-$ and the *constant symbols* 0 and 1. All these symbols are combined according to the following recursive construction to form two families: the terms and the first-order formulas. The *terms* are the smallest collection of finite strings in the above symbols containing all the variables, the constant symbols 0 and 1, and such that if t_1 and t_2 are terms, then $(t_1 + t_2)$, $(t_1 \cdot t_2)$, and $(t_1 - t_2)$ are terms. The (first-order) *formulas* are the smallest collection of finite strings in the above symbols containing $t_1 = t_2$, for all terms t_1, t_2, and also containing $(\neg\varphi)$, $(\varphi \wedge \psi)$, $(\varphi \vee \psi)$, $(\varphi \rightarrow \psi)$, $(\varphi \leftrightarrow \psi)$, $(\forall x)\,\varphi$, and $(\exists x)\,\varphi$, for all formulas φ and ψ and each variable x. Informally, a first-order formula is a well-formed 'finitary' mathematical statement, where quantifiers may range over elements (as opposed to sets or other 'higher order' objects).

Examples of first-order formulas in $\mathfrak{L}_{\text{ring}}$ are

$$((1 + 1) + 1) = 0, \tag{1}$$

$$(\forall x)(x \cdot y) = (y \cdot x), \tag{2}$$

$$(\forall x)(\exists y)(y \cdot y) = x, \tag{3}$$

$$(\forall x)(((x \cdot x) + (y \cdot y)) = 1 \rightarrow (\neg(x \cdot y) = 0)). \tag{4}$$

For readability we may make various informal simplifications. For example, Equation (2) may be written as $(\forall x)x \cdot y = y \cdot x$, and (3) may be written as $(\forall x)(\exists y)y \cdot y = x$ without ambiguity. In each of the examples (2), (3), and (4) above, the variable x is in the *scope* of a quantifier: In other words, x is *bound* by that quantifier. If a variable is not bound by any quantifier, it is *free*. For example, y is free in (2) and (4). We write $\varphi(x_1, \ldots, x_n)$ for φ to signify that the free variables of φ are among $x_1, \ldots, x_n$.

An $\mathfrak{L}_{\text{ring}}$-*structure* $(R, +^R, \cdot^R, -^R, 0^R, 1^R)$ is a set R together with binary operations $+^R$, $\cdot^R$, and $-^R$ on R and constants $0^R, 1^R \in R$. Thus every ring is naturally an $\mathfrak{L}_{\text{ring}}$-structure, as we would hope, but there are plenty of $\mathfrak{L}_{\text{ring}}$-structures that are not rings. When working only with rings, further notational simplifications are possible, for example, (1) may be written as $1+1+1=0$ and (4) as $\forall x\ (x^2+y^2=1 \rightarrow xy \neq 0)$. Moreover we usually write an $\mathfrak{L}_{\text{ring}}$-structure simply as, e.g., $(R,+,\cdot,-,0,1)$, without distinguishing between the structure and the language, when there is no risk of ambiguity. We may even write simply R for $(R,+,\cdot,-,0,1)$ when the operations and constants are clear from the context.

We define a notion of 'truth': Given a ring $(R,+,\cdot,-,0,1)$ and elements $a_1,\ldots,a_n \in R$, each formula $\varphi(x_1,\ldots,x_n)$ may be interpreted by substituting each element a_i into the place of its corresponding variable x_i. Then the truth of $\varphi(a_1,\ldots,a_n)$ in R may be determined from (really, *defined by*) the usual mathematical meanings of 'and', 'or', 'implies', etc. and the universal and existential quantifiers. We write $R \models \varphi(a_1,\ldots,a_n)$ to mean that $\varphi(a_1,\ldots,a_n)$ is true in R. For example, the formula $\varphi(y)$ appearing in (2) has one free variable, namely y, and for a ring R and an element $b \in R$, $\varphi(b)$ is true in R (i.e., $R \models \varphi(b)$) if and only if b commutes multiplicatively with each element of R. For a full definition of interpretation and of the relation $\models$, see, for instance, [44, Definition 1.1.6].

A *sentence* is a formula with no free variables. The truth of a sentence φ in a structure R does not depend on any elements a_i. Thus, in a given structure, each sentence is either true or false and is not both. For example, Equations (1) and (3) are sentences. By contrast, a first-order formula $\varphi(x_1,\ldots,x_n)$ with free variables $x_1,\ldots,x_n$ defines the set of tuples $(a_1,\ldots,a_n)$ such that φ is true when each x_i is replaced by a_i. More broadly, a set $A \subseteq R^n$ is *definable* if there is an $\mathfrak{L}_{\text{ring}}$-formula $\varphi(x_1,\ldots,x_n,z_1,\ldots,z_l)$ and there are *parameters* $c_1,\ldots,c_l \in R$ such that for all $(a_1,\ldots,a_n) \in R^n$

$$(a_1,\ldots,a_n) \in A \iff R \models \varphi(a_1,\ldots,a_n,c_1,\ldots,c_l),$$

i.e., if and only if $\varphi(a_1,\ldots,a_n,c_1,\ldots,c_l)$ is true in R.

There are several special kinds of definability, distinguished by the syntactic complexity of the defining formula, as follows. A formula φ is:

- *Quantifier-free* if it contains no quantifiers
- *Diophantine* if it consists of existential quantifiers followed by a single polynomial equality with integer coefficients $f(x_1,\ldots,x_n,y_1,\ldots,y_m,z_1,\ldots,z_l)=0$
- *Existential* if it consists of existential quantifiers followed by a quantifier-free formula
- *Universal* if it consists of universal quantifiers followed by a quantifier-free formula
- *Universal–existential* if it consists of universal quantifiers followed by an existential formula

For example, the formula (1) is quantifier-free, whereas the formulas (2) and (4) are universal, and the formula (3) is universal–existential.

In the study of **H10**, we are mostly concerned with diophantine and existential formulas and the sets they define. Straight from the definition, every diophantine formula is existential. Given a ring R and a subset $A \subseteq R^n$, we say A is *existentially* definable (respectively, *diophantine*) if it is defined by an existential (respectively, diophantine) formula. Likewise, A is *universally* (respectively, *universal-existentially*) definable if it is defined by a universal (respectively, universal–existential) formula. Existentially definable sets are preserved in extensions: If $R \subseteq S$ is an extension of rings and $A \subseteq R^n$ and $B \subseteq S^n$ are both defined by the same existential formula, then $A \subseteq B$ [34, Theorem 2.4.1]. Moreover, one repeatedly uses that finite intersections and finite unions of diophantine sets are diophantine.

Remark 1 We might also consider a definition of 'definability' for sets A of a ring R that requires that the parameters $c_1, \ldots, c_l$ are elements of a given subring $C \subseteq R$. We then say that A is *definable over* C and likewise for 'existentially definable', etc. In the majority of this chapter we make no such restriction, and for clarity we often write, e.g., 'A is diophantine in R' to remind the reader both that A is diophantine and that the parameters may be any elements of R.

Remark 2 In many of the rings and fields we consider, a set is diophantine if and only if it is existentially definable. More precisely, there are several standard methods that may be used to find 'reductions' of a given existential formula to an equivalent diophantine one. These reductions work in various settings: For example, if R is a field, then

$$R \models \forall x_1 \ldots \forall x_n\ (f(x_1, \ldots, x_n) \neq 0 \leftrightarrow \exists y\ f(x_1, \ldots, x_n) \cdot y = 1),$$

so in fields we may eliminate a negation in favour of an extra existential quantifier. If R is an integral domain, then

$$\begin{aligned} R \models \forall x_1 \ldots \forall x_n\ ((f(x_1, \ldots, x_n) = 0 \vee g(x_1, \ldots, x_n) = 0) \leftrightarrow \\ f(x_1, \ldots, x_n) \cdot g(x_1, \ldots, x_n) = 0), \end{aligned}$$

for $f, g \in R[x_1, \ldots, x_n]$. Thus in integral domains disjunctions of polynomial equalities may be replaced by a single polynomial equality. On the other hand, if R is a ring that admits a total order (compatible with the addition and multiplication), then $a^2 + b^2 = 0$ if and only if $a = b = 0$, for all $a, b \in R$. Thus we have

$$\begin{aligned} R \models \forall x_1 \ldots \forall x_n\ (f(x_1, \ldots, x_n) = g(x_1, \ldots, x_n) = 0 \leftrightarrow \\ (f(x_1, \ldots, x_n))^2 + (g(x_1, \ldots, x_n))^2 = 0). \end{aligned}$$

Therefore in such a ring a conjunction of polynomial equalities may be replaced by a single polynomial equality. A similar method works whenever R is an integral domain and there is a non-constant polynomial over R with no zero in the field of fractions of R.

Examples

1. $\mathbb{R} \setminus \{0\}$ is diophantine over $\mathbb{R}$, defined by the formula $(\exists y)\ xy = 1$.
2. $\mathbb{R}_{\geq 0}$ is diophantine over $\mathbb{R}$, defined by the formula $(\exists y)\ x = y^2$.
3. $\mathbb{Q}_{\geq 0}$ is diophantine over $\mathbb{Q}$, defined by the formula $(\exists w_1)(\exists w_2)(\exists w_3)(\exists w_4)\ x = w_1^2 + w_2^2 + w_3^2 + w_4^2$, which makes use of Euler's theorem on sums of rational squares.
4. $\mathbb{N}$ is diophantine over $\mathbb{Z}$, defined by the formula $(\exists w_1)(\exists w_2)(\exists w_3)(\exists w_4)\ x = w_1^2 + w_2^2 + w_3^2 + w_4^2$.
5. $\mathbb{Z} \setminus \{0\}$ is diophantine over $\mathbb{Z}$, defined by the formula

$$(\exists w_1)(\exists w_2)(\exists w_3)(\exists w_4)\ (w_1^2+w_2^2+w_3^2+w_4^2+1-x)(w_1^2+w_2^2+w_3^2+w_4^2+1+x) = 0.$$

6. $(-1, 1)$ is diophantine over $\mathbb{R}$, defined by the formula $(\exists w)\ (1 - x^2)w^2 = 1$.

Example

In [58], Robinson showed that $\mathbb{Z}_p$ is diophantine over $\mathbb{Q}_p$: For any prime $l \neq p$, the diophantine formula $\exists y\ 1 + px^l = y^l$ defines $\mathbb{Z}_p$ in $\mathbb{Q}_p$. To see this, if $a, b \in \mathbb{Q}_p$ are such that $1 + pa^l = b^l$, then a simple calculation of p-adic valuations shows that a lies in $\mathbb{Z}_p$. Conversely, if $a \in \mathbb{Z}_p$, then $1 + pa^l$ lies in $1 + p\mathbb{Z}_p$, and by Hensel's lemma every such element is an l-th power.

In Sect. 4.1 we will recall Robinson's definition of $\mathbb{Z}$ in $\mathbb{Q}$.

2.2 Theories

Remaining in the language of rings, a *theory* is simply a set of sentences. Accordingly, for a ring R, we define the *theory of* R, denoted $\mathrm{Th}(R)$, to be the set of all $\mathfrak{L}_{\mathrm{ring}}$-sentences that hold in R. Similarly the *existential theory* of R, denoted $\mathrm{Th}_\exists(R)$, is the set of existential $\mathfrak{L}_{\mathrm{ring}}$-sentences that are true in R. A theory T *axiomatizes* the class of its models, i.e., those $\mathfrak{L}_{\mathrm{ring}}$-structures R such that $R \models \varphi$ for every $\varphi \in T$. For example, the class of fields is axiomatized by the usual axioms for fields: associativity and commutativity of additions, etc., including the axiom

$$(\forall x)(\exists y)\ (x = 0 \vee xy = 1)$$

that describes the existence of multiplicative inverses.

Given a theory T, naturally we ask ourselves if there is an algorithm (i.e., a Turing machine) that correctly determines whether or not any given sentence in the language of rings is true in all models of T. When there is such an algorithm, we say that T is *decidable*; otherwise it is *undecidable*. For example, as a consequence of Gödel's famous First Incompleteness theorem, the theory Th($\mathbb{Z}$) of the ring $\mathbb{Z}$ is undecidable. In particular, Peano's axioms are incomplete. This is in contrast with the situation for $\mathbb{R}$ and $\mathbb{C}$ whose complete theories are decidable and admit recursive axiomatizations, due to Tarski [73], and for $\mathbb{Q}_p$, due to Ax–Kochen [2], and independently by Ershov [28].

Robinson's first-order definition of $\mathbb{Z}$ in $\mathbb{Q}$ ([57], and see Theorem 4) provides us with an *interpretation* of the theory of $\mathbb{Z}$ inside the theory of $\mathbb{Q}$. More precisely, let $\psi(x)$ be a formula such that for any $a \in \mathbb{Q}$,

$$\mathbb{Q} \models \psi(a) \Longleftrightarrow a \in \mathbb{Z}.$$

Then we may translate each formula $\varphi(x)$ into a formula $\hat{\varphi}(x)$ so that

$$\mathbb{Z} \models \varphi(a_1, \ldots, a_n) \Longleftrightarrow \mathbb{Q} \models \hat{\varphi}(a_1, \ldots, a_n), \tag{5}$$

for any formula $\varphi(x)$ and any integers $a_1, \ldots, a_n$. We proceed by recursion on the recursive definition of formulas: the terms and atomic formulas remain unchanged, and the connectives also are unaltered, but we 'relativize' quantifiers, as follows. Since

$$\mathbb{Z} \models \exists x\ \varphi(x, b_1, \ldots, b_n) \Longleftrightarrow \mathbb{Q} \models \exists x\ (\psi(x) \wedge \hat{\varphi}(x, b_1, \ldots, b_n))$$

and

$$\mathbb{Z} \models \forall x\ \varphi(x, b_1, \ldots, b_n) \Longleftrightarrow \mathbb{Q} \models \forall x\ (\psi(x) \rightarrow \hat{\varphi}(x, b_1, \ldots, b_n)),$$

for all tuples $b_1, \ldots, b_n$ of integers, we define $\widehat{\exists x\ \varphi}(y_1, \ldots, y_n)$ to be $\exists x\ (\psi(x) \wedge \hat{\varphi}(x, y_1, \ldots, y_n))$ and $\widehat{\forall x\ \varphi}(y_1, \ldots, y_n)$ to be $\forall x\ (\psi(x) \rightarrow \hat{\varphi}(x, y_1, \ldots, y_n))$. This construction yields a primitive recursive function $\varphi(y_1, \ldots, y_n) \mapsto \hat{\varphi}(y_1, \ldots, y_n)$, from the set of $\mathfrak{L}_{\mathrm{ring}}$-formulas to itself satisfying (5). Thus, for every $\mathfrak{L}_{\mathrm{ring}}$-sentence φ, to check whether or not $\mathbb{Z} \models \varphi$, it suffices to check whether or not $\mathbb{Q} \models \hat{\varphi}$. By Gödel's First Incompleteness theorem, the theory of $\mathbb{Z}$ is undecidable, and therefore the theory of $\mathbb{Q}$ is undecidable

Let us return to Hilbert's Tenth Problem (**H10**) for $\mathbb{Q}$. When we translate it into the language of mathematical logic, this is the question of whether the existential theory $\mathrm{Th}_{\exists}(\mathbb{Q})$ is decidable. If $\mathbb{Z}$ is a diophantine subset of $\mathbb{Q}$, then the interpretation $\varphi(x) \mapsto \widehat{\varphi}(x)$ restricts to a map from existential sentences to existential sentences, still satisfying (5). Consequently, by Theorem 1, if $\mathbb{Z}$ is a diophantine subset of $\mathbb{Q}$, then (**H10**) is also unsolvable for $\mathbb{Q}$. This motivates the search for a diophantine definition of $\mathbb{Z}$ in $\mathbb{Q}$.

Koenigsmann's definition of $\mathbb{Z}$ in $\mathbb{Q}$ [41, Theorem 1] is by a universal formula. Using this formula, the interpretation $\varphi(x) \mapsto \hat{\varphi}(x)$ restricts to a map from existential sentences to universal–existential sentences, and so that at least the '$\forall\exists$-theory' of $\mathbb{Q}$ is undecidable, i.e., there is no algorithm that decides correctly whether or not a universal–existential $\mathfrak{L}_{\text{ring}}$-sentence is true in $\mathbb{Q}$ [41, Corollary 3].

3 Prerequisites from Algebra

In this section we first give a brief description of quaternion algebras and, more generally, central simple algebras. More details can be found, for example, in [31, Chapter 2]. Afterwards we recall some local–global principles relevant to work on Hilbert's Tenth Problem.

3.1 Quaternions and Central Simple Algebras

Throughout, we let K be a field; we will assume for simplicity that $\text{char}(K) \neq 2$. All K-algebras are assumed to be associative and to have a multiplicative identity. Moreover we consider only K-algebras of finite dimension.

Definition 1 A K-algebra A is *simple* if it has no non-trivial two-sided ideals. It is a *central simple algebra* if moreover its centre is K.

The dimension of a central simple algebra A/K is a square integer, and its positive square root is called its *degree*. A central simple algebra A/K is said to be *split* if it is isomorphic to a matrix algebra $M_n(K)$ for some n. Otherwise we call it *non-split*. For any field extension L/K, we may consider $A \otimes_K L$, which is a central simple L-algebra. For K a global field and $\mathfrak{p}$ a place of K, we will want to consider the $K_\mathfrak{p}$-algebra $A \otimes_K K_\mathfrak{p}$, where $K_\mathfrak{p}$ denotes the completion of K with respect to $\mathfrak{p}$. The set $\Delta_{A/K}$ of places $\mathfrak{p}$ of K at which $A \otimes_K K_\mathfrak{p}$ is non-split is always finite (we will also denote it simply by Δ_A, when K is clear from the context).

Definition 2 For $a, b \in K^\times$, we denote by $H_{a,b}$ the *quaternion algebra* over K with generators i and j satisfying $i^2 = a$, $j^2 = b$, and $ij = -ji$. Thus,

$$H_{a,b} = \{w + xi + yj + zij \mid w, x, y, z \in K\}.$$

Every four-dimensional central simple algebra over K is isomorphic to $H_{a,b}$ for some $a, b \in K^\times$ [4, Proposition 1]. Every quaternion algebra $H_{a,b}$ is either split, so isomorphic to the matrix algebra $M_2(K)$, or it is a division algebra, in which case it is non-split. For an extension L/K, we denote $H_{a,b}(L) := H_{a,b} \otimes_K L$. For a global field K and a place $\mathfrak{p}$ of K, we say $H_{a,b}$ is *split* at $\mathfrak{p}$ if $H_{a,b}(K_\mathfrak{p}) \simeq M_2(K_\mathfrak{p})$ and is *ramified* at $\mathfrak{p}$ otherwise.

Definition 3

1. The *reduced trace* of an element $\alpha = w + xi + yj + zij$ of $H_{a,b}$ is defined as $\operatorname{tr}(\alpha) = 2w$.
2. Its *reduced norm* is defined as $\operatorname{nr}(\alpha) = w^2 - ax^2 - by^2 + abz^2$.

Any central simple algebra A/K has reduced trace and norm maps, defined on A with values in K (cf. [31, §2.6]), which we will also denote by tr and nr, respectively. The reduced trace is K-linear, and the reduced norm is multiplicative on $A^\times$.

Let A be a degree n central simple K-algebra. For $\alpha \in A$, let $P_\alpha(X) = X^n + a_{n-1}X^{n-1} + \cdots + a_0$ be the reduced characteristic polynomial of α over K. Then $\operatorname{tr}(\alpha) = a_{n-1}$ and $\operatorname{nr}(\alpha) = (-1)^n a_0$.

Definition 4 We say that two central simple algebras A and A' over K are *Brauer equivalent* if there exist integers $m, m' > 0$ such that

$$A \otimes_K M_m(K) \simeq A' \otimes_K M_{m'}(K).$$

This defines an equivalence relation on the set of central simple algebras over K, equipped with the tensor product. The set of equivalence classes is an abelian group denoted by $\operatorname{Br}(K)$, see [9] and [70] for more details. Finally, the class of a central simple algebra A/K is trivial in $\operatorname{Br}(K)$ if and only if A is split.

3.2 Local–Global Principles

Let K be a global field and $\mathfrak{p}$ a place of K. Let M_K be the set of all places of K. A quadratic form $F(X_1, \ldots, X_n) \in K[X_1, \ldots, X_n]$ is *isotropic* over K if the equation $F = 0$ has a non-trivial solution in K.

Theorem 2 ([42, Theorem VI.3.1]) *Let $F(X_1, \ldots, X_n) \in K[X_1, \ldots, X_n]$ be a quadratic form, where K is a global field. If F is isotropic over $K_\mathfrak{p}$, for all completions $K_\mathfrak{p}$ of K, then F is isotropic over K.*

This result was proved by Minkowski in 1923 in the case when $K = \mathbb{Q}$ and by Hasse in 1924 in the case when K is a number field. Since then many mathematicians have been inspired to study similar so-called *local–global problems*, i.e., problems in which the validity of a certain property is assumed to hold in all, or all but finitely many, completions of a global field K and one asks if the same property holds in K. When the answer to a local–global problem is affirmative, one says that a local–global principle or a Hasse principle holds.

The validity of certain local–global principles implies properties of the global fields over which they are formulated. For instance, if in Theorem 2 we take $F = X^2 - a$ and $K = \mathbb{Q}$, then we get that a rational number a is a square modulo all primes $\mathfrak{p}$ if and only if it is a square in $\mathbb{Q}$. Thus we have $\mathbb{Q}^2 = \bigcap_\mathfrak{p} \mathbb{Q}_\mathfrak{p}^2$. Hence, knowing the validity of a local–global principle allows us to write certain subsets of a global field as intersections of the corresponding subsets of local fields. In this case we say that these (global) sets *satisfy a local–global principle*. Local–global problems are often formulated in the setting of algebraic varieties. In this case the validity of a local–global principle implies a description of some subsets of the K-rational points of certain varieties in terms of their $K_{\mathfrak{p}}$-rational points.

Along with the Hasse–Minkowski theorem on quadratic forms in Theorem 2, one of the most famous local–global principles is the Albert–Brauer–Hasse–Noether (ABHN) theorem for central simple algebras.

Theorem 3 (Albert, Brauer, Hasse, and Noether, 1932, cf. [47, Theorem 8.1.17]) *Let K be a global field, and let A be a central simple algebra over K. Then A splits over K if and only if A splits over $K_{\mathfrak{p}}$, for all places $\mathfrak{p}$ of K. Moreover, we have the following five-term exact sequence:*

$$0 \longrightarrow \mathrm{Br}(K) \longrightarrow \bigoplus_{\mathfrak{p}\in M_K} \mathrm{Br}(K_{\mathfrak{p}}) \longrightarrow \mathbb{Q}/\mathbb{Z} \longrightarrow 0. \tag{6}$$

Remark 3 The exact sequence (6) in the ABHN theorem describes the relation of the Brauer group of a global field K with those of all its completions $K_{\mathfrak{p}}$. More precisely:

1. The map

$$\mathrm{Br}(K) \to \bigoplus_{\mathfrak{p}\in M_K} \mathrm{Br}(K_{\mathfrak{p}})$$

is the diagonal map. Hence, every element $A \in \mathrm{Br}(K)$ has trivial image in $\mathrm{Br}(K_{\mathfrak{p}})$ for all but finitely many places $\mathfrak{p}$. The injectivity of the map is telling us that if an element in $\mathrm{Br}(K)$ has trivial image in $\mathrm{Br}(K_{\mathfrak{p}})$ for all places $\mathfrak{p}$ of K, then it has to be trivial. For every field K and every central simple algebra A/K, the class of A in $\mathrm{Br}(K)$ is trivial if and only if A is split (see Definition 4). Thus, a central simple algebra A over a global field K is split if and only if for every place $\mathfrak{p}$ of K the central simple algebra $A \otimes_K K_{\mathfrak{p}}$ is split over $K_{\mathfrak{p}}$.
2. The map

$$\bigoplus_{\mathfrak{p}\in M_K} \mathrm{Br}(K_{\mathfrak{p}}) \to \mathbb{Q}/\mathbb{Z}$$

is given by the sum of the invariant maps $\mathrm{inv}_{\mathfrak{p}} : \mathrm{Br}(K_{\mathfrak{p}}) \to \mathbb{Q}/\mathbb{Z}$. For local fields the invariant map is an isomorphism, and its construction can be found in [64, Chapter XVIII, Section 3]; for non-Archimedean real places, it is given by the isomorphism $\mathrm{Br}(\mathbb{R}) \cong \mathbb{Z}/2\mathbb{Z}$, and for non-Archimedean complex places it is the zero map. The exactness in the centre of the short exact sequence of Theorem 3 tells us that an element $(A_{\mathfrak{p}}) \in \oplus_{\mathfrak{p}\in M_K} \mathrm{Br}(K_{\mathfrak{p}})$ comes from an element $A \in \mathrm{Br}(K)$ if and only if $\sum_{\mathfrak{p}\in M_K} \mathrm{inv}_{\mathfrak{p}}(A_{\mathfrak{p}}) = 0$. Hence, Theorem 3 gives

a recipe to detect whether a (global) element in $\mathrm{Br}(K)$ with prescribed (local) behaviour at each completion of K exists.
3. Finally, if we restrict the short exact sequence to the 2-torsion elements, then the injectivity of the diagonal map from $\mathrm{Br}(K)[2]$ to $\oplus_{\mathfrak{p}}\mathrm{Br}(K_{\mathfrak{p}})[2]$ follows from Theorem 2 (Hasse–Minkowski). In fact, it is proven in [31, Proposition 1.3.2] that a quaternion algebra (a, b) splits (i.e., has trivial image in the Brauer group) if and only if the corresponding conic $ax^2 + by^2 = z^2$ has a non-trivial solution.

4 Survey

In 1949 Robinson applied the Hasse–Minkowski local–global principle for quadratic forms to show that $\mathbb{Z}$ is definable in $\mathbb{Q}$. In the decades since then, others have applied a variety of local–global principles to obtain model-theoretic definability results, and this area has been especially active in the last 15 years. Most recently the local–global principle for central simple algebras has been used in several significant papers. In this section we give a brief survey of these results.

4.1 Robinson's Results

In 1949, Julia Robinson proved in [57] that the integers are a definable subset of $\mathbb{Q}$:

Theorem 4 (Robinson, 1949) *For a rational number n to be an integer, it is necessary and sufficient that it satisfies the following formula:*

$$\forall a, b\Big(\Big\{\Big(\exists x, y, z : 2{+}bz^2{=}x^2{+}ay^2 \wedge \forall m\Big[\Big(\exists x, y, z : 2{+}abm^2{+}bz^2{=}x^2{+}ay^2\Big) \rightarrow \Big(\exists x, y, z : 2 + ab(m + 1)^2 + bz^2 = x^2 + ay^2\Big)\Big]\Big\} \rightarrow \Big(\exists x, y, z : 2 + abn^2 + bz^2 = x^2 + ay^2\Big)\Big). \tag{7}$$

To prove Theorem 4, Robinson uses Hasse–Minkowski Theorem 2 to show that if n is a rational number such that some of the quadratic forms in (7), with $m = n$, have a solution in $\mathbb{Q}$, then the denominator of n is odd and coprime to various classes of infinitely many prime numbers. She then deduces that the denominator of n is not divisible by any prime number, i.e., that n is an integer.

As described in Sect. 2.2, it follows from Theorem 4 and Gödel's First Incompleteness theorem that the $\mathfrak{L}_{\mathrm{ring}}$ theory of $\mathbb{Q}$ is undecidable. See, for example, [57, Theorems I-IV].

Theorem 4 also implies the following statement.

Corollary 1 (cf. [57, Theorem 3.2]) *Let n be a positive integer, and let $R \subseteq \mathbb{Q}^n$ be an n-ary relation between rational numbers. Let $R' \subseteq \mathbb{Z}^{2n}$ be the relation between $2n$ integers that holds for $(a_1, \ldots, a_n, b_1, \ldots, b_n)$ if and only if $b_i \neq 0$ for all $i = 1, \ldots, n$ and R holds for $\left(\frac{a_1}{b_n}, \ldots, \frac{a_n}{b_n}\right)$. Then R is definable in $\mathbb{Q}$ if and only if R' is definable in $\mathbb{Z}$.*

This corollary shows another way in which $\mathbb{Q}$ and $\mathbb{Z}$ are interpretable with respect to each other: The map $R \to R'$ is a bijection between definable subsets of $\mathbb{Q}^n$ and certain definable subsets of $\mathbb{Z}^{2n}$.

4.2 The Sets 'S'

If $\mathfrak{p}$ is a non-Archimedean place, then $R_\mathfrak{p}$ denotes the valuation ring in $K_\mathfrak{p}$; we also make use of the ring $R_{(\mathfrak{p})} := R_\mathfrak{p} \cap K$.

Over a global field K (with char $K \neq 2$) with non-Archimedean places $\mathfrak{p}$ and $\mathfrak{q}$, let $a, b, p, q \in K^*$ be such that the quaternion K-algebra $H_{a,b}$ ramifies exactly at $\mathfrak{p}$ and $\mathfrak{q}$ and such that $\mathrm{ord}_\mathfrak{p} p = \mathrm{ord}_\mathfrak{q} q = 1$ and $\mathrm{ord}_\mathfrak{p} q = \mathrm{ord}_\mathfrak{q} p = 0$. Eisenträger [23] defines the set (denoted T in *loc. cit.*)

$$\begin{aligned} S^E_{a,b,p,q}(K) :&= \left\{w \in K \mid \exists x, y, z \in K \ : \ w^2 - ax^2 - by^2 + abz^2 = pq\right\} \\ &= \{\ \mathrm{tr}(\alpha)/2 \mid \alpha \in H_{a,b}(K) : \mathrm{nr}(\alpha) = pq\ \}, \end{aligned} \tag{8}$$

i.e., the (halves of) reduced traces of elements of $H_{a,b}$ of reduced norm pq. This set plays an important role in one of her main results [23, Theorem 3.1], which states that for any non-Archimedean place $\mathfrak{p}$ of K, the set $R_{(\mathfrak{p})} := \{x \in K \mid \mathrm{ord}_\mathfrak{p}(x) \geq 0\}$ (denoted by $R_\mathfrak{p}$ in [23]) is diophantine over K.

Remark 4 Existential definability of valuation rings of global function fields was proven by Rumely [59] and extended to many non-global function fields by Shlapentokh [65], see also [66] for other diophantine definitions of valuation rings. Very recently, Daans showed in his PhD thesis [14] that certain valuation rings of the function field F of a curve are existentially definable, by using local–global principles for quadratic forms defined over F, and that these definitions can be given in a uniform way.

Following Poonen [55], for any extension $L/\mathbb{Q}$ and $a, b \in L^\times$, we define

$$\begin{aligned} S^P_{a,b}(L) :&= \left\{2w \in L \mid \exists x, y, z \in L \ : \ w^2 - ax^2 - by^2 + abz^2 = 1\right\} \\ &= \{\ \mathrm{tr}(\alpha) \mid \alpha \in H_{a,b}(L) : \mathrm{nr}(\alpha) = 1\ \}, \end{aligned} \tag{9}$$

i.e., the set of traces of elements of $H_{a,b}(L)$ of reduced norm 1. In fact, Poonen writes $S_{a,b}$ for our $S^P_{a,b}(\mathbb{Q})$. He considers the cases $L = \mathbb{Q}$ and $L = \mathbb{Q}_p$ and describes the sets $S^P_{a,b}(\mathbb{Q}_p)$ in [55, Lemma 2.1]; importantly, they depend on whether $p \in \Delta_{a,b} := \Delta_{H_{a,b}}$, i.e., on whether $H_{a,b}$ is split at p. Precisely, in [55, Lemma 2.1], he proves for all rational primes p:

- If $p \notin \Delta_{a,b}$, then $S^P_{a,b}(\mathbb{Q}_p) = \mathbb{Q}_p$.
- If $p \in \Delta_{a,b}$, then $S^P_{a,b}(\mathbb{Q}_p) \subseteq \mathbb{Z}_p$.

Remark 5 Moreover, the same lemma proves that if $p \in \Delta_{a,b}$, then $S^P_{a,b}(\mathbb{Q}_p)$ is nonempty because of the containment $\mathrm{res}^{-1}(U_p) \subseteq S^P_{a,b}(\mathbb{Q}_p)$, where $\mathrm{res} : \mathbb{Z}_p \to \mathbb{F}_p$ is the reduction-modulo-p map and $U_p = \{\mathrm{Tr}_{\mathbb{F}_{p^2}/\mathbb{F}_p}(\beta) : \mathrm{N}_{\mathbb{F}_{p^2}/\mathbb{F}_p}(\beta) = 1\} \subseteq \mathbb{F}_p$ is the set of finite field traces of elements of norm 1.

In addition, he shows (cf. [55, Lemma 2.2]) that the sets $S^P_{a,b}$ satisfy a local–global principle if either $a > 0$ or $b > 0$:

$$S^P_{a,b}(\mathbb{Q}) = \left(\bigcap_p S^P_{a,b}(\mathbb{Q}_p)\right) \cap \mathbb{Q}, \tag{10}$$

where the first intersection is taken over all prime numbers p in $\mathbb{Q}$.

Koenigsmann [41] uses the same sets $S^P_{a,b}$ as Poonen over $\mathbb{Q}$ and $\mathbb{Q}_p$. So does Park [52], for any number field K; in particular, her Lemma 2.2 is the number field analogue of [55, Lemmas 2.1 and 2.2].

Dittmann [18] extends Poonen's definition of the set $S^P_{a,b}$ from quaternion algebras to any central simple algebra A over any field K. More precisely, he defines the set (denoted $S(A/L)$ in [18])

$$S^D_A(L) := \{\mathrm{tr}(x) \mid x \in A \otimes_K L, \mathrm{nr}(x) = 1\}, \tag{11}$$

for any field extension L/K.

In [18, Lemma 2.5 and Proposition 2.6], the author describes the sets S^D_A over local fields. Namely, if A is a central simple algebra of prime degree l over a local field F with valuation ring $\mathcal{O}_F$, then:

- If A splits over F, then $S^D_A(F) = F$.
- If A is non-split over F and $l \neq 2$, then $S^D_A(F) = \mathcal{O}_F$.

A description is also given in the case $l = 2$. If K is a global field and A is of prime degree, then [18, Proposition 2.1] gives the local–global principle analogous to [55, Lemma 2.2] for the sets S^D_A. That is,

$$S^D_A(K) = \left(\bigcap_{\mathfrak{p}} S^D_A(K_{\mathfrak{p}})\right) \cap K, \tag{12}$$

where $\mathfrak{p}$ runs over all places of K.

Finally, Daans gives a similar, but slightly different, definition of S-sets in his [13]; see also [15]. For a field K and a quaternion algebra Q over K, he defines

$$S_Q^{Da}(K) := \{\mathrm{tr}(x) \mid x \in Q \backslash K, \mathrm{nr}(x) = 1\}. \tag{13}$$

See Sect. 5.1.1 for a comparison with Poonen's $S^P_{a,b}$.

4.3 The Sets 'T'

In each of the papers surveyed here, the authors define from the sets S a new family of sets T. Importantly, the new sets T should have the same 'logical complexity' as the sets S: More precisely, the sets S and T should be definable by existential formulas in the language of rings.

Following Poonen [55], for any field extension $L/\mathbb{Q}$, we define the set (denoted $T_{a,b}$ and only considered for $L = \mathbb{Q}$ in *loc. cit.*)

$$T^P_{a,b}(L) := S^P_{a,b}(L) + S^P_{a,b}(L) + \{\, 0, 1, 2, \dots, 2309 \,\}, \tag{14}$$

where $a, b \in \mathbb{Q}^\times$. In [55, Lemma 2.5] he proves that for any $a, b \in \mathbb{Q}^\times$ such that $a > 0$ or $b > 0$,

$$T^P_{a,b}(\mathbb{Q}) = \bigcap_{p \in \Delta_{a,b}} \mathbb{Z}_{(p)}, \tag{15}$$

where $\mathbb{Z}_{(p)} := \mathbb{Q} \cap \mathbb{Z}_p$, and $\Delta_{a,b} = \Delta_{H_{a,b}}$ as before.

Remark 6 For any prime $p > 11$, [55, Lemma 2.3] shows that $U_p + U_p = \mathbb{F}_p$ with U_p as in Remark 5; i.e., the finite field traces cover all of $\mathbb{F}_p$ for p large enough. To deal with the remaining primes, the set $\{0, 1, 2 \dots, 2309 = 2\cdot3\cdot5\cdot7\cdot11 - 1\}$ appears in (14), because for any $t \in \bigcap_{p \in \Delta_{a,b}} \mathbb{Z}_{(p)}$, there exists $n \in \{1, 2, \dots, 2309\}$, so that $\mathrm{red}_p(t - n) \in U_p + U_p$ for all $p \leq 11$, as well. This is used to prove the right-to-left inclusion in (15). The left-to-right inclusion is a consequence of (10).

Regarding T-sets, in the spirit of Koenigsmann, for every field extension L of $\mathbb{Q}$ we define

$$T^K_{a,b}(L) := S^P_{a,b}(L) + S^P_{a,b}(L). \tag{16}$$

In [41, Definition 4], Koenigsmann defines this set (denoted $T_{a,b}$ in *loc. cit.*) only over the field of rational numbers. Then [41, Proposition 6] strengthens [55, Lemma 2.5] by showing that for *any* $a, b \in \mathbb{Q}^\times$,

$$T^K_{a,b}(\mathbb{Q}) = \bigcap_{p \in \Delta_{a,b}} \mathbb{Z}_{(p)}, \tag{17}$$

where $\mathbb{Z}_{(\infty)} := \{x \in \mathbb{Q} \mid -4 \leq x \leq 4\}$. Whenever at least one of a and b is positive, the T-set defined in (17) coincides with the T-set defined by Poonen. For Poonen, the set $\mathbb{Z}_{(\infty)}$ does not appear, seeing as $\infty \notin \Delta_{a,b}$ when $a > 0$ or $b > 0$. See Sect. 5.1.2 for more details.

Park [52] generalizes the above to number fields K and defines the set T analogously to Koeningsmann:

$$T^{Pa}_{a,b}(K) := S^P_{a,b}(K) + S^P_{a,b}(K). \tag{18}$$

Her [52, Proposition 2.3] states that

$$T^{Pa}_{a,b}(K) = \bigcap_{\mathfrak{p} \in \Delta_{a,b}} R_{(\mathfrak{p})} \tag{19}$$

for any $a, b \in K^\times$ such that $\sigma(a) > 0$ or $\sigma(b) > 0$ for any real embedding σ of K, where $R_{(\mathfrak{p})} = \{x \in K : \mathrm{ord}_{\mathfrak{p}}(x) \geq 0\}$ is denoted by $R_{\mathfrak{p}}$ in [52] for any non-Archimedean place $\mathfrak{p}$ of K.

In [18], Dittmann gives a definition of the set T that depends on the degree of the central simple algebra A/K. Let K be a global field and A/K a central simple algebra of prime degree l. Then

$$T^D_A(K) := \begin{cases} S^D_A(K) - S^D_A(K) & \text{if } l = 2; \\ S^D_A(K) & \text{otherwise.} \end{cases} \tag{20}$$

Then [18, Proposition 2.9] is the analogue of [55, Lemma 2.4]; it proves that if A splits at all real places of K, then

$$T^D_A(K) = \bigcap_{\mathfrak{p} \in \Delta_{A/K}} R_{(\mathfrak{p})}, \tag{21}$$

where $\Delta_{A/K}$ is the finite set of places of K at which A does not split. In [1], the authors define the T-sets as in (20).

In [13], the author defines T-sets for a quaternion algebra Q over an arbitrary field K as follows:

$$T^{Da}_Q(K) := \bigcap_{\mathfrak{p} \in \Delta_Q} R_{(\mathfrak{p})}, \tag{22}$$

where Δ_Q is the set of discrete valuations on K for which $Q_{\mathfrak{p}} := Q \otimes_K K_{\mathfrak{p}}$ is non-split. Daans then shows in [13, Theorem 5.1] that for a *non-real* quaternion algebra over a global field K (i.e., a quaternion algebra that splits over all the Archimedean places of K), one obtains

$$T_Q^{Da}(K) = S_Q^{Da}(K) + S_Q^{Da}(K). \tag{23}$$

4.4 Constructing Diophantine Sets Using Local–Global Principles

We present an outline of how the Hasse–Minkowski (Theorem 2) and ABHN (Theorem 3) local–global principles appear in the papers surveyed here, beginning with [23]. We suppose throughout this section that the characteristic of K is not 2.

Remark 7 All the sets S and T appearing in the preceding sections are diophantine subsets over the relevant fields. For example, $T_{a,b}^{Pa}(K)$ is diophantine over K, for each number field K.

LGPs and Eisenträger [23] In [23], Eisenträger proves that given a global field K and a non-Archimedean place $\mathfrak{p}$, the ring $R_{(\mathfrak{p})} := \{x \in K : \mathrm{ord}_{\mathfrak{p}}(x) \geqslant 0\}$ is diophantine over K. Since $R_{(\mathfrak{p})} = R_{(\mathfrak{p})} \cap R_{(\mathfrak{q})} + R_{(\mathfrak{p})} \cap R_{(\mathfrak{l})}$ for any two additional non-Archimedean places $\mathfrak{q}$ and $\mathfrak{l}$ of K, it suffices to show that $R_{(\mathfrak{p})} \cap R_{(\mathfrak{q})}$ is diophantine over K for every two non-Archimedean places $\mathfrak{p}$ and $\mathfrak{q}$. This is done in Theorem 3.1 of *loc. cit.*, as follows:

1. By ABHN, there exists a quaternion K-algebra $H_{a,b}$ ramified *precisely* at $\mathfrak{p}$ and $\mathfrak{q}$, see 2. in Remark 3; we may assume that $a, b \in R_{(\mathfrak{p})} \cap R_{(\mathfrak{q})}$. As just before Equation (8) defining $S_{a,b,p,q}^E$, let $p, q \in K$ be such that $\mathrm{ord}_{\mathfrak{p}} p = \mathrm{ord}_{\mathfrak{q}} q = 1$ and $\mathrm{ord}_{\mathfrak{p}} q = \mathrm{ord}_{\mathfrak{q}} p = 0$. For large enough r, Eisenträger proves that $(pq)^r S_{a,b,p,q}^E(K) \subseteq R_{(\mathfrak{p})} \cap R_{(\mathfrak{q})}$.
2. Let $x \in pR_{(\mathfrak{p})} \cap qR_{(\mathfrak{q})}$, and let β be a root of the polynomial $X^2 - 2xX + pq$. The algebras $H_{a,b}(K_{\mathfrak{p}})$ and $H_{a,b}(K_{\mathfrak{q}})$ now split over the fields $K_{\mathfrak{p}}(\beta)$ and $K_{\mathfrak{q}}(\beta)$, respectively; hence $H_{a,b}(K(\beta))$ is split *everywhere* locally. By ABHN, the algebra $H_{a,b}(K(\beta))$ then splits over $K(\beta)$, see 1. in Remark 3. It follows that $K(\beta) \subseteq H_{a,b}$, so $\beta \in H_{a,b}$ and $x \in S_{a,b,p,q}^E$. Thus, $pR_{(\mathfrak{p})} \cap qR_{(\mathfrak{q})} \subseteq S_{a,b,p,q}^E$, implying that $G := p^{r+1}R_{(\mathfrak{p})} \cap q^{r+1}R_{(\mathfrak{q})} \subseteq (pq)^r S_{a,b,p,q}^E \subseteq R_{(\mathfrak{p})} \cap R_{(\mathfrak{q})}$ for some r, see also point 1. above.
3. Since G has finite index in $R_{(\mathfrak{p})} \cap R_{(\mathfrak{q})}$, by introducing coset representatives s_i, one obtains $R_{(\mathfrak{p})} \cap R_{(\mathfrak{q})} = \bigsqcup_{i=1}^t (s_i + G) = \bigcup_{i=1}^t (s_i + (pq)^r S_{a,b,p,q}^E)$. Since $S_{a,b,p,q}^E$ is diophantine over K and $R_{(\mathfrak{p})} \cap R_{(\mathfrak{q})} = s_1 + s_2 + \cdots s_t + (pq)^r S_{a,b,p,q}^E$, the set $R_{(\mathfrak{p})} \cap R_{(\mathfrak{q})}$ is diophantine over K.

   In a sense, the set $T_{a,b,p,q}^E := (pq)^r S_{a,b,p,q}^E + \{s_1, \ldots, s_t\}$ can be seen as a precursor of the T-sets of the other authors. Indeed, in Eisenträger's construction the elements $a, b, p, q \in K^\times$ are chosen such that $\Delta_{H_{a,b}} =: \Delta_{a,b} = \{\mathfrak{p}, \mathfrak{q}\}$ and

$$T^E_{a,b,p,q} = R_{(\mathfrak{p})} \cap R_{(\mathfrak{q})} = \bigcap_{\mathfrak{l} \in \Delta_{a,b}} R_{(\mathfrak{l})},$$

cf. Equations (15), (17), (19), and (21).

Remark 8 As ABHN was used only in the case of quaternions in the above, it could have been replaced by the Hasse–Minkowski theorem, as in the works of Robinson and Poonen. As mentioned in part 3., in Remark 3 it also follows from the Hasse–Minkowski Theorem that a quaternion algebra $H_{a,b}$ over a global field K is unramified if and only if it is unramified over all the completions $K_{\mathfrak{p}}$. Moreover, Robinson and Poonen work over the field of rational numbers, and for every couple of places $\mathfrak{p}$, $\mathfrak{q}$, it is possible to construct 'by hand' a quaternion algebra $H_{a,b}$ that splits exactly at $\mathfrak{p}$ and $\mathfrak{q}$.

LGPs and Poonen [55] By definition, for $s \in \mathbb{Q}$, membership of 's' in Poonen's $S^P_{a,b}$ is equivalent to the existence of an element x of the quaternion $\mathbb{Q}$-algebra $H_{a,b}$ such that $\mathrm{tr}(x) = s$ and $\mathrm{nr}(x) = 1$. Equivalently, the characteristic polynomial of x over $\mathbb{Q}$ is a quadratic polynomial $X^2 - sX + 1$. In the local–global principle (10) for $S^P_{a,b}$ where either $a > 0$ or $b > 0$, the left-to-right inclusion follows from the definition. To obtain the reverse inclusion, let $s \in \mathbb{Q} \cap \bigcap_p S^P_{a,b}(\mathbb{Q}_p)$. Then for any prime p, there exists a solution to $X^2 - sX + 1 = 0$ in $H_{a,b}(\mathbb{Q}_p)$. By the Hasse–Minkowski theorem (Theorem 2), this quadratic polynomial also has a solution in $H_{a,b}$, thus $s \in S_{a,b}(\mathbb{Q})$.

Poonen proves that $\mathbb{Z}$ is universal-existentially definable in $\mathbb{Q}$ by using the following description of $\mathbb{Z}$ in terms of the T-sets $T^P_{a,b}$ (which are in turn defined through the S-sets $S^P_{a,b}$):

$$\bigcap_{a,b \in \mathbb{Q}_{>0}} T^P_{a,b}(\mathbb{Q}) = \mathbb{Z}. \tag{24}$$

By Equation (15), see also Remark 6, it suffices to show that for any rational prime p, there exist rational numbers $a > 0$ and $b > 0$ such that the quaternion $\mathbb{Q}$-algebra $H_{a,b}$ is ramified at p. This is a direct consequence of ABHN. However, Poonen avoids using ABHN altogether, by constructing (in a simple manner) such a quaternion algebra 'by hand' in [55, Lemma 2.7]: If $p = 2$, take $a = b = 7$; otherwise, take $a = p$ and an integer b with $\mathrm{red}_p(b) \in \mathbb{F}_p^\times \backslash \mathbb{F}_p^{\times 2}$, where as before, red_p is the reduction-modulo-p map.

Remark 9 It follows from the definition of $S^E_{a,b,p,q}$ in Equation (8) that in general $S^E_{a,b,p,q}(K) \not\subseteq R_{(\mathfrak{p})}$ for $\mathfrak{p} \in \Delta_{a,b}$; this is remedied by choosing r large enough so that $p^r S^E_{a,b,p,q}(K) \subseteq R_{(\mathfrak{p})}$. By contrast, by prescribing reduced norm 1 when defining $S^P_{a,b}$ in Equation (9), it follows that $S^P_{a,b}(\mathbb{Q}_p) \subseteq \mathbb{Z}_p$ for all $p \in \Delta_{a,b}$. The removal of the factor p^r is an important step towards finding a uniform diophantine definition of $\mathbb{Z}_{(p)}$ in $\mathbb{Q}$, which is important for obtaining the description of $\mathbb{Z}$ in Equation (24), by seeing $T^P_{a,b}(\mathbb{Q})$ as an intersection of all localizations $\mathbb{Z}_{(p)}$ as in Equation (15).

LGPs and Koenigsmann [41] In Koenigsmann's [41], the situation with respect to ABHN is the same as in [55]. Given the description of the T-sets $T^K_{a,b}$ via (17), it is immediate that (24) remains true when replacing the sets $T^P_{a,b}$ with the sets $T^K_{a,b}$; cf. [41, p. 79]. See also Sect. 5.1.2.

LGPs and Park [52] The set $S^{Pa}_{a,b}$ satisfies a local–global principle in the same way as the set $S^P_{a,b}$ does. In fact, as described above, for a number field K, since $S^{Pa}_{a,b}(K)$ is defined by a quadratic form $w^2 - ax^2 - by^2 + abz^2 - 1$, the Hasse–Minkowski theorem assures an inclusion $K \cap \bigcap_{\mathfrak{p}\in\Delta_{a,b}} S^{Pa}_{a,b}(K_\mathfrak{p}) \subseteq S^{Pa}_{a,b}(K)$. The reverse inclusion follows by definition, so

$$S^{Pa}_{a,b}(K) = K \cap \bigcap_{\mathfrak{p}\in\Delta_{a,b}} S^{Pa}_{a,b}(K_\mathfrak{p}), \tag{25}$$

for all number fields K.

As above, let $R_\mathfrak{p}$ be the ring of integers of $K_\mathfrak{p}$, and let R be the ring of integers of K. As a consequence of Equation (25), we have that $T^{Pa}_{a,b}(K) \subseteq \bigcap_{\mathfrak{p}\in\Delta_{a,b}} R_\mathfrak{p}$. Let $F_\mathfrak{p}$ be the residue field of $R_\mathfrak{p}$. To prove the other inclusion, Park considers the set $U_\mathfrak{p}$ of elements $s \in F_\mathfrak{p}$ such that $x^2 - sx + 1$ is irreducible over $F_\mathfrak{p}$ and its preimage $V_\mathfrak{p}$ in $R_\mathfrak{p}$ (for $\mathfrak{p}$ lying above $p \geq 11$; for $p < 11$ the definition of $V_\mathfrak{p}$ is slightly different, cf. [52, Proposition 2.3]). She then shows that $V_\mathfrak{p} + V_\mathfrak{p} = R_\mathfrak{p}$ and uses strong approximation and the local–global principle given in Equation (25) a second time to get the other inclusion. Thus

$$\bigcap_{a,b} T^{Pa}_{a,b}(K) = \bigcap_{\mathfrak{p}\in M_{K,na}} R_{(\mathfrak{p})} = R, \tag{26}$$

where the intersection runs over $a, b \in K^\times$ such that $\sigma(a) > 0$ or $\sigma(b) > 0$ for each real Archimedean place σ and where $M_{K,na}$ is the subset of M_K containing all the non-Archimedean places.

Thus, just as Poonen and Koenigsmann do, at this point we may already deduce that the ring of integers R of a number field K is universal-existentially definable in K. The description of R given in Equation (26) is analogous to that of [55], cf. (24). She then uses class field theory to obtain a first-order universal formula defining R in K.

LGPs and Dittmann [18] In [18, Proposition 2.1], Dittmann also uses a local–global principle to prove (12). Let K be a global field and A a central simple algebra over K of prime degree l. As for Poonen, the left-to-right inclusion in (12) is seen directly from the definitions. We illustrate Dittmann's use of ABHN for proving the reverse inclusion. Let $a \in \bigcap_{\mathfrak{p}} S_A^D(K_{\mathfrak{p}}) \cap K$.

1. If A is split over L, then (12) is obtained quickly as both sides coincide with K. So assume it is not, and let $\mathfrak{p}_i$, $i = 1, 2, \ldots, n$, be the places of K where A is ramified. Then for any i, there exists $x_i \in A \otimes_K K_{\mathfrak{p}_i}$ of reduced trace a and reduced norm 1 by assumption. If $x_i \in K_{\mathfrak{p}_i}$, then $x_i^l = 1$ by the reduced norm hypothesis, so one obtains $x_i = a/l \in K$ if $l \neq \mathrm{char}(K)$, and $x_i = 1$ otherwise, from the reduced trace hypothesis. Since these quantities are both global, the reverse inclusion follows in this case.
2. Assume $x_i \notin K_{\mathfrak{p}_i}$, so that $K_{\mathfrak{p}_i}(x_i)/K_{\mathfrak{p}_i}$ is a degree l field extension splitting $A \otimes_K K_{\mathfrak{p}_i}$. Moreover, the minimal polynomial f_i of x_i has degree $(l-1)$-coefficient equal to a and constant term $(-1)^l$ by assumption. By a weak approximation argument [18, Lemma 2.4], one can assume that the field extensions of the $K_{\mathfrak{p}_i}$, obtained by adjoining a root of the respective f_i, all coincide with those obtained by adjoining a root of some common degree l polynomial $f \in K[X]$. Then, by construction, the global field extension $K[X]/(f)$ splits A, meaning it is contained in A, see 1. in Remark 3. The global element $X + (f) \in K[X]/(f)$ has reduced trace a and reduced norm 1, so we are done.

We remark that there are common points between the above proof and that of Eisenträger. One difference with [23] and [55] is that in [18] the degree of the central simple algebra is a prime number (possibly larger than 2), meaning the Hasse–Minkowski theorem is not enough, so the more general ABHN is invoked.

Dittmann then also uses ABHN to prove that for any non-Archimedean place $\mathfrak{p}$ of a global field K, the ring $R_{(\mathfrak{p})}$ is diophantine over K. This is done similarly as in Eisenträger's [23] and Poonen's [55], except that instead of constructing a suitable global central simple algebra 'by hand', ABHN is used to find two central simple algebras with prescribed local splitting behaviour, see 2. in Remark 3. More precisely, one can find central simple algebras A and A' over K such that $T_A^D(K) + T_{A'}^D(K) = R_{(\mathfrak{p})}$, thus proving the result. A predecessor of [18] is the paper [59] produced by Rumely in 1980, in which the techniques are similar with the use of norm forms instead of quadratic forms (which instead were previously used by Robinson).

LGPs and Daans [15] In [15], Daans picks up on the theme of analyzing the exact number of quantifiers used in definitions of rings of integers in global fields, following, for example, [26, 77] and the framework of [16]. Using ABHN he shows that for any finite set V of non-Archimedean places of a global field K, the ring $\bigcap_{\mathfrak{p}\in V} R_{(\mathfrak{p})}$ has an $\exists_3$-$\mathfrak{L}_{\mathrm{ring}}(K)$-definition in K, i.e., there is an existential $\mathfrak{L}_{\mathrm{ring}}$-formula $\varphi(x, y_1, \ldots, y_m)$ with three existential quantifiers and elements $b_1, \ldots, b_m \in K$ such that $K \models \varphi(a, b_1, \ldots, b_m)$ if and only if $a \in \bigcap_{\mathfrak{p}\in V} R_{(\mathfrak{p})}$. In particular, in Lemma 4.1 of *loc. cit.*, the injectivity of the first map in Equation (6) gives a description of $\bigcap_{\mathfrak{p}\in V} R_{(\mathfrak{p})}$ in terms of the splitting behaviour of a particular quaternion algebra. Then in Proposition 4.2 of *loc. cit.*, he uses Equation (6) again to prove that such a quaternion algebra always exists. In addition, in Section 5 of *loc. cit.* a global quaternion algebra with prescribed splitting is constructed using ABHN, in order to prove that for any non-empty finite set V of non-Archimedean places, the intersection $\bigcap_{\mathfrak{p}\in M_K\setminus V} R_{(\mathfrak{p})}$ is $\forall_{10}$-$\mathfrak{L}_{\text{ring}}(K)$-definable in K; likewise this means that there is a universal $\mathfrak{L}_{\text{ring}}$-formula $\varphi(x, y_1, \ldots, y_m)$ with ten universal quantifiers and elements $b_1, \ldots, b_m \in K$ such that $K \models \varphi(a, b_1, \ldots, b_m)$ if and only if $a \in \bigcap_{\mathfrak{p}\in M_K\setminus V} R_{(\mathfrak{p})}$.

5 Local–Global Methods and Definability

In the previous section, we surveyed a number of recent articles that use local–global principles to answer model-theoretic questions. In essence, the surveyed articles introduce similar notions of sets 'S', consisting of reduced traces of elements of certain simple algebras with prescribed reduced norm, and construct similar sets 'T' out of these by taking sums or differences. The sets 'T' turn out to be intersections of certain other sets; using this observation, they can be shown to be diophantine, by using local–global principles to construct global algebras with suitable properties. In this section, we make some more detailed comparisons between the methods employed in the surveyed articles in Sect. 5.1, give a more model-theoretic outlook in Sect. 5.2, and sketch two possible applications of this framework using other local–global principles in Sect. 5.3.

5.1 Comparisons Between Surveyed Articles

With a view towards both establishing a common model-theoretic framework and finding new applications of other local–global principles, we first provide some more detailed comparisons between the sets introduced in the surveyed articles.

5.1.1 Comparisons Between the Sets 'S'

Let us compare the definitions of $S^E_{a,b,p,q}$, $S^P_{a,b}$, and S^{Da}_{Q} in Equations (8), (9), and (13), respectively. We start by recalling (see Sect. 4.2) that the definition of the S-sets of Park in [52], Koenigsmann in [41], and Dittmann in [18] essentially coincides with that of Poonen in [55].

- **$S^E_{a,b,p,q}$ versus $S^P_{a,b}$**: In characteristic different from 2 the difference between taking half of the reduced trace $\text{tr}(\alpha)/2$ in Eisenträger's (8) versus taking $\text{tr}(\alpha)$ in Poonen's (9) in the definitions of the S-sets is inconsequential. In fact, the proofs of [55] would still go through if we instead define $S^P_{a,b}(L) = \{\text{tr}(\alpha)/2 \mid \alpha \in$

$H_{a,b}(L) : \mathrm{nr}(\alpha) = 1\}$ and replace $U_p = \{\mathrm{Tr}_{\mathbb{F}_{p^2}/\mathbb{F}_p}(\beta) : \mathrm{N}_{\mathbb{F}_{p^2}/\mathbb{F}_p}(\beta) = 1\} \subseteq \mathbb{F}_p$ with $U_p/2$, cf. Remark 5.

- $\boldsymbol{S^P_{a,b}}$ **versus** $\boldsymbol{S^{Da}_{H_{a,b}}}$: While Daans gives a slightly different definition of the S-sets in [13], it is comparable to that of Poonen's (9). For a quaternion algebra Q over a field K, if $x \in K$ is such that $\mathrm{nr}_Q(x) = x^2 = 1$, then $\mathrm{tr}_Q(x) = 2x \in \{\pm 2\}$, where nr_Q and tr_Q denote the reduced norm and reduced trace in Q, respectively. By [15, Proposition 3.4], Q is split over K if and only if there exists $\alpha \in Q \backslash K$ such that $\mathrm{nr}_Q(\alpha) = 1$ and $\mathrm{tr}_Q(\alpha) \in \{\pm 2\}$, meaning:

  1. If Q is split over K, then $S^{Da}_Q(K) := \{\mathrm{tr}_Q(x) \mid x \in Q \backslash K, \mathrm{nr}_Q(x) = 1\} = \{\mathrm{tr}_Q(x) \mid x \in Q, \mathrm{nr}_Q(x) = 1\}$.
  2. If Q is non-split over K, then $S^{Da}_Q(K) = \{\mathrm{tr}_Q(x) \mid x \in Q, \mathrm{nr}_Q(x) = 1\} \backslash \{\pm 2\}$.

  If K is a field extension of $\mathbb{Q}$, then there exist $a, b \in K^\times$ such that $Q = H_{a,b}$. So, if Q is split over K, then $S^{Da}_Q(K) = S^P_{a,b}(K)$, and if not, then $S^{Da}_Q(K) = S^P_{a,b}(K) \backslash \{\pm 2\}$. This difference is however minor, since the corresponding T-sets will still coincide, see Sect. 5.1.2.

5.1.2 Comparisons Between the Sets 'T'

Let us now compare the definitions of $T^P_{a,b}$, $T^K_{a,b}$, $T^{Pa}_{a,b}$, T^D_A, and T^{Da}_Q made in Equations (14), (16), (18), (20), and (22), respectively.

- $\boldsymbol{T^P_{a,b}}$ **versus** $\boldsymbol{T^K_{a,b}}$: For every $a, b \in \mathbb{Q}^\times$, it follows immediately from the definitions in Equations (14) and (16) that $T^K_{a,b}(\mathbb{Q}) \subseteq T^P_{a,b}(\mathbb{Q})$. Moreover, if $a > 0$ or $b > 0$, then the quaternion $\mathbb{Q}$-algebra $H_{a,b}$ splits over the real numbers; therefore, Equations (15) and (17) tell us that in this case actually $T^K_{a,b}(\mathbb{Q}) = T^P_{a,b}(\mathbb{Q}) = \bigcap_{p \in \Delta_{a,b}} \mathbb{Z}_{(p)}$.

  The reason for the appearance of the set $\{0, 1, 2, \ldots, 2309\}$ in the definition of $T^P_{a,b}$ in (14) is explained in Remark 6. Even though Koenigsmann removes this in his definition (16) of T-sets, Equation (17), an analogue of Poonen's (15), remains true: For any $a, b \in \mathbb{Q}^\times$,

$$T^K_{a,b}(\mathbb{Q}) = \bigcap_{p \in \Delta_{a,b}} \mathbb{Z}_{(p)}. \tag{27}$$

  The left-to-right inclusion is a consequence of the definition of $T^K_{a,b}$ and the fact that for $p \in \Delta_{a,b}$ we have $S^K_{a,b}(\mathbb{Q}_p) \subseteq \mathbb{Z}_p$ if $p \neq \infty$ and $S_{a,b}(\mathbb{R}) \subseteq \mathbb{R}$ otherwise (cf. [41, pg. 77] and [55, Lemma 2.1]). To prove the right-to-left inclusion, for $p > 11$ prime, Koenigsmann uses the same technique as Poonen, cf. Remark 6. To prove it for $p \leq 11$ or $p = \infty$, he constructs 'by hand' intermediary sets V_p such that $\emptyset \neq V_p \subseteq \mathbb{Z}_p$, where $\mathbb{Z}_\infty := [-4, 4] \subseteq \mathbb{R}$. The V_p are moreover constructed to satisfy $\mathrm{res}^{-1}(U_p) \subseteq V_p$ for all p prime, where, as in Remark 5, $\mathrm{res} : \mathbb{Z}_p \to \mathbb{F}_p$ is the reduction-modulo-p map and

$U_p = \{\mathrm{Tr}_{\mathbb{F}_{p^2}/\mathbb{F}_p}(\beta) : \mathrm{N}_{\mathbb{F}_{p^2}/\mathbb{F}_p}(\beta) = 1\} \subseteq \mathbb{F}_p$ is the set of finite field traces of elements of norm 1. Then he checks case by case that they also satisfy the property $V_p + V_p = \mathbb{Z}_p$ for all p (including ∞). Since $V_p \subseteq S^P_{a,b}(\mathbb{Q}_p)$ is an open subset, this last property ensures that $\bigcap_{p\in\Delta_{a,b}} \mathbb{Z}_{(p)} \subseteq T^K_{a,b}(\mathbb{Q})$, thus proving (27).

- $\boldsymbol{T^P_{a,b}}$ **versus** $\boldsymbol{T^{Pa}_{a,b}}$: Proposition 2.3 in [52], cf. (19), is a generalization to number fields K of Lemma 2.4 of Poonen, cf. (15). Poonen's sign condition that either $a > 0$ or $b > 0$ assures that the quaternion algebra $H_{a,b}$ splits over the real numbers, i.e., at the only Archimedean place of $\mathbb{Q}$. Park's condition, that $\sigma(a) > 0$ or $\sigma(b) > 0$ for each real Archimedean place of K, is taking care of the same phenomenon. Indeed, her condition ensures that the quaternion algebra $H_{a,b}$ splits over all the real places of K; any quaternion algebra splits at all complex places. See also Sect. 4.4 for an idea of the proof which is similar to the one described in the previous bullet point.
- $\boldsymbol{T^K_{a,b}}$ **versus** $\boldsymbol{T^D_A}$: To define the set $T^K_{a,b}$, Koeningsmann takes the sum of $S^P_{a,b}$ with itself, while to define T^D_A, Dittmann takes the difference of S^D_A with itself. Choosing our central simple algebra $A = H_{a,b}$ to be a quaternion algebra over some extension K of $\mathbb{Q}$, the sets $S^P_{a,b}(K)$ and $S^D_A(K)$ coincide. Then to compare $T^K_{a,b}(K)$ and $T^D_A(K)$, it suffices to check that $S^P_{a,b}(K) = S^D_A(K)$ is invariant under multiplication by -1. Recall that $s \in S^P_{a,b}(K)$ if and only if there exists $\alpha \in H_{a,b}$ such that $\mathrm{tr}(\alpha) = s$ and $\mathrm{nr}(\alpha) = 1$. Moreover, $\mathrm{tr}(-\alpha) = -\mathrm{tr}(\alpha)$ and $\mathrm{nr}(-\alpha) = \mathrm{nr}(\alpha) = 1$; hence $-s \in S^P_{a,b}(K)$ as well. In general, the difference of two sets has a topological interpretation: If a set V has non-empty interior in a topological ring R, then $V - V$ is a neighbourhood of zero.
- $\boldsymbol{T^{Pa}_{a,b}}$ **versus** $\boldsymbol{T^{Da}_{H_{a,b}}}$: Let Q be a non-real quaternion algebra over a number field K. Then there exist $a, b \in K^\times$ such that $Q \cong H_{a,b}$, and for any real Archimedean place σ of K, either $\sigma(a) > 0$ or $\sigma(b) > 0$. Considering Park's relation (19), by Daans' definition (22), $T^{Pa}_{a,b}(K) = T^{Da}_Q(K)$.

5.2 *From Diophantine Sets to Diophantine Families*

Although the sets $S = S^E_{a,b,p,q}, S^P_{a,b}$, etc. and $T = T^P_{a,b}, T^K_{a,b}$, etc. are often defined differently in the various papers, as described above, there are clear commonalities. Most obviously, they are all diophantine sets or rather families of diophantine sets defined with parameters describing the relevant quaternion algebras and central simple algebras. That is, for each of the S-sets S, in the papers above (and similarly for the T-sets), there is a field K, an existential $\mathfrak{L}_{\mathrm{ring}}$-formula $\varphi(x, y_1, \ldots, y_n)$, and there are elements $b_1, \ldots, b_n \in K$ such that

$$S(F) = \{a \in F \mid F \models \varphi(a, b_1, \ldots, b_n)\},$$

for each F/K.

This slightly shifts our perspective from a set that is existentially definable in a given field K, to the family of subsets of field extensions of K that are defined by the same formula, using constants from K. More precisely, we recall the following definition from [1, Section 3]: For a field K, a one-*dimensional diophantine family* over K is a map D from the class of field extensions F of K to sets given by polynomials $f_1, \dots, f_r \in K[X, Y_1, \dots, Y_m]$, for some m, that is,

$$D(F) = \{x \in F \mid \exists y_1 \dots \exists y_m \; f_1(x, y_1, \dots, y_m) = \dots = f_r(x, y_1, \dots, y_m) = 0\},$$

for every F/K. From another point of view, such diophantine families are the subsets of fields F extending K that are defined by a given existential formula

$$\varphi(x) = \exists y_1 \dots \exists y_m \; f_1(x, y_1, \dots, y_m) = \dots = f_r(x, y_1, \dots, y_m) = 0.$$

The reader will note a slight divergence from the nomenclature of Sect. 2 in that we allow $r > 1$ in the above definition: one-dimensional diophantine families are really families of existentially definable sets; however according to Remark 2 this rarely makes a difference.

In the preceding sections we discussed the various sets S and T: They both form one-dimensional diophantine families. For example, Poonen's sets $S^P_{a,b}$ form a one-dimensional diophantine family over $\mathbb{Q}$. In another case, Dittmann's sets S^D_A form a one-dimensional diophantine family over K, where K is the field over which the central simple algebra A is defined. Moreover, each of the sets S considered in Sect. 4.2 satisfies a local–global principle, namely that $S(K) = K \cap \bigcap_{\mathfrak{p} \in \Delta} S(K_\mathfrak{p})$, for a finite set Δ of places of K. Furthermore, each such S has a very simple description in each closure $K_\mathfrak{p}$: $S(K_\mathfrak{p}) = K_\mathfrak{p}$ if $\mathfrak{p} \notin \Delta$ and otherwise $S(K_\mathfrak{p}) \pm S(K_\mathfrak{p}) = R_\mathfrak{p}$, with the addition of a finite set in the case of Poonen, as in (14).

This motivates the general study of one-dimensional diophantine families satisfying similar local–global properties. Let D be a one-dimensional diophantine family over a field K, let $\Delta \subseteq M(K)$ be a finite set of (non-Archimedean) places of K, let $\tau = (e, f) \in \mathbb{N}^2$, and let $\mathcal{F}_K$ be a family of field extensions of K. Now, for any field extension F/K, not necessarily in $\mathcal{F}_K$, let $M^\tau_\Delta(F)$ be the set of places $\mathfrak{P}$ of F lying above some place $\mathfrak{p}$ of Δ of relative type $\leq \tau$, i.e., the ramification degree of the extension $\mathfrak{P}/\mathfrak{p}$ is at most e and the inertia degree divides f. Let $R^\tau_\Delta(F)$ be the holomorphy ring

$$\bigcap_{\mathfrak{P} \in M^\tau_\Delta(F)} R_{(\mathfrak{P})}$$

corresponding to $M^\tau_\Delta(F)$. For more background on this setting, see, for example, [1, Section 2]. Suppose that

1. $D(F) = F \cap \bigcap_{\mathfrak{P} \in M(F)} D(F_\mathfrak{P})$
2. $D(F_\mathfrak{P}) = F_\mathfrak{P}$ if $\mathfrak{P} \notin M^\tau_\Delta(F)$
3. $D(F_\mathfrak{P}) = R_\mathfrak{P}$ if $\mathfrak{P} \in M^\tau_\Delta(F)$

for all extensions $F \in \mathcal{F}_K$. Then $D(F) = R^{\tau}_{\Delta}(F)$ for $F \in \mathcal{F}_K$. This is the perspective of [1], with K a number field and $\mathcal{F}_K$ the finite extensions of K, and $\Delta = \{\mathfrak{p}\}$ is a singleton, to prove that $R^{\tau}_{\mathfrak{p}}(F)$ is a one-dimensional diophantine family over K, for finite extensions F/K [1, Proposition 4.8]. This approach can be immediately generalized to allow finite sets Δ of primes on K or to allow n-dimensional diophantine families for $n \geq 1$. In another direction, we may replace $M^{\tau}_{\Delta}(F)$ with the family $M^{\tau}_{\Delta}(F; a) = \{\mathfrak{P} \in M^{\tau}_{\Delta}(F) \mid a \in \mathcal{O}_{\mathfrak{P}}\}$ of open–closed subsets of $M^{\tau}_{\Delta}(F)$, given by a fixed $a \in K$. In [1, Theorem 6.5, Corollary 6.6] a sufficient condition is given for the holomorphy rings associated with $M^{\tau}_{\Delta}(F; a)$ to form a one-dimensional diophantine family over K. Finally, turning away from holomorphy rings, we may generalize condition (3), for example, by instead requiring that $D(F_{\mathfrak{P}})$ be the subset of $F_{\mathfrak{P}}$ defined by a given formula $\varphi(x)$ in the language of rings.

These considerations prompt the search for a general approach to understand the theory of diophantine families D that satisfy (1), in a suitable generality. For the sake of concreteness, let us consider the case where K has characteristic zero and $\mathcal{F}_K$ is just the singleton $\{K\}$. For our purposes, a *classical place* on K is a place $\mathfrak{p}$ lying over a p-adic place on $\mathbb{Q}$ and of relative type $\leq \tau$ for some $\tau \in \mathbb{N}^2$. Let $K^{\mathfrak{p}}$ denote a choice of p-adic closure of K with respect to $\mathfrak{p}$ – this is a substitute for the notion of completion. Let $M^{c}(K)$ be the space of classical places on K, equipped with the constructible topology given by sub-basic open sets $\{\mathfrak{p} \in M^{c}(K) \mid a \in \mathcal{O}_{\mathfrak{p}}\}$ and $\{\mathfrak{p} \in M^{c}(K) \mid a \notin \mathcal{O}_{\mathfrak{p}}\}$, for $a \in K$. Let $\mathcal{D}$ be the set of one-dimensional diophantine families D over K for which there exists a compact set $M_D \subseteq M^{c}(K)$ with M_D open–closed in $M^{c}(K)$, such that $D(K) = K \cap \bigcap_{\mathfrak{p} \in M_D} D(K^{\mathfrak{p}})$. Then, instead of the language of rings, we consider the first-order language $\mathfrak{L}$ equipped with a unary predicate symbol P_D for each $D \in \mathcal{D}$. We then form an $\mathfrak{L}$-structure $\mathcal{K} = (K, (D(K))_{D \in \mathcal{D}})$ with underlying set K in which each P_D is interpreted in the obvious way by $D(K)$. In a natural sense $\mathcal{K}$ is less complicated than the field K: Each new predicate in $\mathcal{K}$ is definable by a formula in the language of rings (in fact, by an existential formula). In model-theoretic language, we say that $\mathcal{K}$ is a *reduct* of the field K. We can be more precise: Each positive existential $\mathfrak{L}$-formula defines in $\mathcal{K}$ a set that is already existentially definable in the field K; thus the positive existential theory of $\mathcal{K}$ is a reduct of the existential $\mathfrak{L}_{\text{ring}}$-theory of K.

? Question

Under what conditions on K and relative to what information, can one axiomatize the $\mathfrak{L}$-theory of $\mathcal{K}$?

For number fields this problem seems to be extremely hard, since even to describe $\mathcal{D}$ amounts to deciding which varieties admit some form of local–global principle. On the other hand, pseudo-algebraically closed fields, pseudo-real closed fields, pseudo-p-adically closed fields, and others satisfy local–global principles for various classes of varieties: The complete $\mathfrak{L}_{\text{ring}}$-theories of such fields are already understood, and there is a large literature. For example, see [30, Chapter 12], [56], and [29]. In such cases, it seems reasonable to expect that the $\mathfrak{L}$-theory of $\mathcal{K}$ can be axiomatized in a similar way. Perhaps it is possible to find a field K – outside of the classes mentioned previously – for which we can axiomatize $\mathcal{K}$, and in so doing we might find new classes of fields of wider model-theoretic interest. Further study of these structures $\mathcal{K}$ may well shed light on the possibilities and limitations of the local–global method in the analysis of diophantine sets in global fields.

5.3 *Other Local–Global Principles*

As described in Sect. 4.4, two local–global principles were used repeatedly in the surveyed articles to obtain diophantine sets, namely, the Hasse–Minkowski and Albert-Brauer-Hasse–Noether theorems, cf. Theorems 2 and 3, respectively. In this subsection, we put forward two other local–global principles, both arising from the study of elliptic curves, which can potentially be used in a similar way to obtain other diophantine sets. For a comprehensive introduction to the arithmetic of elliptic curves, cf., e.g., [5, 69].

Elliptic curves have already been used in the context of Hilbert's Tenth Problem for global fields. For example, see [10], where the authors use a conjecture on the properties of elliptic divisibility sequences on an elliptic curve to deduce some results on the complexity of defining $\mathbb{Z}$ in $\mathbb{Q}$; see also [46, 67, 68].

5.3.1 Endomorphism Rings of Elliptic Curves

Let K be any field, and consider elliptic curves over K.

An *isogeny* between elliptic curves E_1 and E_2 is a K-morphism $E_1 \to E_2$ of algebraic groups. Any nonzero isogeny is surjective and hence has a finite kernel. The set $\mathrm{Hom}_K(E_1, E_2)$ of isogenies from E_1 to E_2 inherits a group structure from the group structure on E_2, by pointwise addition: $(\varphi + \psi)(P) := \varphi(P) + \psi(P)$. We will denote the inverse of φ in this group by $-\varphi$. There is a natural action of $\mathbb{Z}$ on $\mathrm{End}_K(E_1, E_2)$ given by $(n, \varphi) \mapsto n \cdot \varphi$, where $n \cdot \varphi$ means the sum of $|n|$ copies of $\mathrm{sgn}(n)\varphi$, making $\mathrm{End}_K(E_1, E_2)$ a $\mathbb{Z}$-module.

For an elliptic curve E/K, its *endomorphism ring* $\mathrm{End}_K(E) := \mathrm{Hom}_K(E, E)$ has an additional multiplication structure from composing isogenies. It has no zero divisors and is a torsion-free $\mathbb{Z}$-module. Moreover, it is a $\mathbb{Z}$-*order* in the *endomorphism algebra* $\mathrm{End}_K^0(E) := \mathrm{End}_K(E) \otimes_{\mathbb{Z}} \mathbb{Q}$, which means it is a lattice (i.e., a finitely generated $\mathbb{Z}$-module whose scalar extension to $\mathbb{Q}$ spans $\mathrm{End}_K^0(E)$) that is also a subring. As a consequence, one shows that the endomorphism ring $\mathrm{End}_K(E)$ is either $\mathbb{Z}$, an order in a quadratic imaginary field, or an order in a quaternion algebra. The latter moreover only occurs when $\mathrm{char}(K) > 0$ (since $\mathrm{End}_K^0(E)$ is commutative when $\mathrm{char}(K) = 0$); we call the elliptic curve *supersingular* in this case. See Sections 10.1 and 10.2 of [75] for an introduction to lattices and orders, respectively.

We may localize any $\mathbb{Z}$-lattice L in a $\mathbb{Q}$-vector space V at any prime ℓ by forming $L_\ell := L \otimes_{\mathbb{Z}} \mathbb{Z}_\ell$, which is a $\mathbb{Z}_\ell$-lattice in the $\mathbb{Q}_\ell$-vector space $V \otimes_{\mathbb{Q}} \mathbb{Q}_\ell$. For any prime ℓ, we have a natural injection of L into L_ℓ. Lattices over $\mathbb{Z}$ (or any Dedekind domain) satisfy a local–global principle [75, Theorem 9.4.9 and Lemma 9.5.3]: Given L, the map $N \mapsto (N_\ell)_{\ell-\text{prime}}$ determines a bijection between $\mathbb{Z}$-lattices N in V and the collection of tuples $(N_\ell)_\ell$ of $\mathbb{Z}_\ell$-lattices in $V \otimes_{\mathbb{Q}} \mathbb{Q}_\ell$ satisfying $L_\ell = N_\ell$ for all but finitely many primes ℓ. We note here that the choice of the lattice L is irrelevant, as for any other such L', one has $L_\ell = L'_\ell$ for all but finitely many primes ℓ, so there is no 'real' dependence on L; see also the comment and remark after [75, Theorem 9.4.9].

Similarly, we may localize a $\mathbb{Z}$-order O in a $\mathbb{Q}$-algebra R at any prime ℓ by forming $O_\ell := O \otimes_{\mathbb{Z}} \mathbb{Z}_\ell$, which is now a $\mathbb{Z}_\ell$-order in the $\mathbb{Q}_\ell$-algebra $R \otimes_{\mathbb{Q}} \mathbb{Q}_\ell$; again, for any prime ℓ, the order O naturally embeds into O_ℓ. Moreover, for $\mathbb{Z}$-lattices, being an order is a local property, i.e., checking that a given lattice is an order can be done locally at each prime. Hence, the lobal-global principle for $\mathbb{Z}$-lattices above specializes to a local–global principle for $\mathbb{Z}$-orders in R.

For $\ell \neq \mathrm{char}(K)$ a prime, the *ℓ-adic Tate module* (or *ℓ-divisible group*) of an elliptic curve E/K is $T_\ell(E) := \varprojlim_n E[\ell^n]$, where the inverse limit is taken with respect to the natural multiplication-by-ℓ maps. The Galois group $\mathrm{Gal}(\bar{K}/K)$ acts on each $E[\ell^n]$ in a way which respects the inverse limit and hence extends to the Tate module $T_\ell(E)$; cf. [69, III.7]. Due to Tate's theorem, localizations of endomorphism rings, viewed as orders in their respective endomorphism algebras, have an interpretation in terms of Tate modules. More precisely, for any two elliptic curves E_1 and E_2, there is a natural injective map $\mathrm{Hom}_K(E_1, E_2) \otimes_{\mathbb{Z}} \mathbb{Z}_\ell \hookrightarrow \mathrm{Hom}(T_\ell(E_1), T_\ell(E_2))$, where the right-hand side consists of $\mathrm{Gal}(\bar{K}/K)$-*equivariant* morphisms of $\mathbb{Z}_\ell$-modules (cf. Definition 34.1.60 of [75]). Tate's theorem shows that this map is an isomorphism when K is a finite field; Faltings showed the same when K is a number field.

When $\mathrm{char}(K) =: p > 0$ and $\ell = p$, instead of the Tate module of E one takes its *(contravariant) Dieudonné module* $M(E)$ or (categorically) equivalently its associated *p-divisible group*, cf. [76, §1.2] for both of these notions. When K is moreover finite, Tate also showed that $\mathrm{Hom}_K(E_1, E_2) \otimes_{\mathbb{Z}} \mathbb{Z}_p$ is isomorphic to $\mathrm{Hom}(M(E_2), M(E_1))$.

Putting the above together, for an elliptic curve E over a finite or number field K, we have an exact sequence

$$0 \to \mathrm{End}_K(E) \to \prod_\ell \mathrm{End}_K(E) \otimes \mathbb{Z}_\ell \simeq \prod_\ell \mathrm{End}(T_\ell(E)), \tag{28}$$

where the product runs over all primes ℓ, including possibly $\ell = \mathrm{char}(K)$, though in that case $T_\ell(E) := M(E)$ is the Dieudonné module. The image of the second map lands in the restricted product where all but finitely many entries are the identity map.

Inspired by the use of local–global principles for quaternion algebras in the surveyed articles, we aim to use the local–global behaviour of endomorphism rings of elliptic curves over finite fields and number fields mentioned above to obtain diophantine sets. To do this, it is necessary to first determine the next term in the exact sequence (28), since the above-mentioned restricted product is hard to work with. It should be noted that while endomorphism rings are orders that have reduced trace and norm maps, the exact sequence (28) is a map of *elements* of orders, so a suitable replacement for trace and norm maps is needed in order to define the analogues of the sets S and T from the articles we surveyed in Sect. 4.

5.3.2 Torsion Points on Elliptic Curves

In this subsection, we let K be a number field. As in the previous sections, we denote by M_K the set of all places of K.

Let A be a commutative connected algebraic group defined over K. In [20], the authors propose a local-to-global approach to study the divisibility of K-points of A by integers. More precisely, they ask the following:

? Question

Let $m \in \mathbb{N}$ and $P \in A(K)$. Suppose that for all but finitely many $\mathfrak{p} \in M_K$, there exists $Q_\mathfrak{p} \in A(K_\mathfrak{p})$ such that $P = mQ_\mathfrak{p}$ in $A(K_\mathfrak{p})$. Does this imply there exists $Q \in A(K)$ such that $P = mQ$ in $A(K)$?

If the answer to this question is positive, we say that A satisfies a local–global principle for m-divisibility. One can quickly reduce to the case where $m = p^n$ for p a prime number and $n \in \mathbb{N}$. There have been many recent advances in the study of this question; for more details the interested reader can consult the recent survey [19]. One of the main tools is to translate the question into a group cohomology problem, where the obstruction to the local–global principle for m-divisibility is given by a subgroup of $H^1(\mathrm{Gal}(K(A[m])/K), A[m])$, denoted by $H^1_{\mathrm{loc}}(\mathrm{Gal}(K(A[m])/K), A[m])$, considered in [20]. Similar groups were introduced before by Tate (as stated by Serre in [63]) and Sansuc [62]. The group $H^1(\mathrm{Gal}(K(A[m])/K), A[m])$ contains a subgroup which is isomorphic to the Tate–Shafarevich group $Ш(K, A[p^n])$ (see [19] for further details). The vanishing of $H^1_{\mathrm{loc}}(\mathrm{Gal}(K(A[m])/K), A[m])$ provides a sufficient condition for the local–global principle to hold [20, Proposition 2.1]. There are also known examples where it does not hold [21, 22, 48].

In the case where $A = E$ is an elliptic curve, it was shown in [20, Theorem 3.1] that the local–global principle for m-divisibility holds for $m = p$ a prime number. Later in [49], the authors proved this for all powers $m = p^n$ of every prime number

$p > \left(3^{\frac{[K:\mathbb{Q}]}{2}} + 1\right)^2$. The bound has been extended to $p \geq 5$ when $K = \mathbb{Q}$ in [50], and this is the best possible because of counterexamples produced for powers of $p = 2, 3$ in [12, 48]. In addition, in the special case when P is a torsion point of an elliptic curve E defined over a number field K, it was shown in [32] that the local–global principle for p^n-divisibility holds for all $p \geq 3$ and $n \geq 1$.

By the definition of the Tate–Shafarevich group, one can write down the following exact sequence:

$$0 \to Ш(K, E[p^n]) \to H^1(K, E[p^n]) \xrightarrow{\pi} \prod_{\mathfrak{p} \in M_K} H^1(K_{\mathfrak{p}}, E[p^n]), \tag{29}$$

which holds whenever E is a commutative algebraic group and in particular when E is an elliptic curve.

Taking the articles surveyed in Sect. 4 as motivation, we propose to use the local–global principle for p^n-divisibility, through the exact sequence (29), for the purpose of proving that certain sets, naturally occurring in arithmetic, are diophantine. In particular, we would like to work in a setting where the local–global principles hold, so that $Ш(K, E[p^n])$ vanishes. A first crucial step would then be to understand the image of the map π in (29), which would allow us to determine its cokernel and hence complete the sequence to a short exact sequence. It seems however that, so far, the cokernel of π is not well understood.

Another important point is to choose a suitable interpretation of $H^1(K, E[p^n])$ and hence suitable presentations of its elements, e.g., as p^n-coverings of E, or $E[n]$-torsors, or theta groups (cf. [11, §1]). This choice will determine the context in which the analogues of the sets S and T need to be defined, and hence in which way the model-theoretic arguments can be adapted and extended.

Acknowledgments This collaboration started at the workshop Women in Numbers Europe 4 which took place in Utrecht, the Netherlands in August 2022. S. A. was supported by GeoMod AAPG2019 (ANR-DFG). V. K. was supported by the Dutch Research Council (NWO) through grant VI.Veni.192.038. Z. K. was partially supported by the research training group 'GRK 2240: Algebro-Geometric Methods in Algebra, Arithmetic and Topology', funded by the DFG. L. P. is a member of INdAM-GNSAGA. We thank the referees for helpful comments.

References

1. S. Anscombe, P. Dittmann, A. Fehm, A p-adic analogue of Siegel's theorem on sums of squares. Math. Nachr. **293**(8), 1434–1451 (2020)
2. J. Ax, S. Kochen, Diophantine problems over local fields. II. A complete set of axioms for p-adic number theory. Am. J. Math. **87**, 631–648 (1965)
3. S. Barańczuk, On a dynamical local-global principle in Mordell-Weil type groups. Expo. Math. **35**(2), 206–211 (2017)
4. N. Bourbaki, *Eléments de mathématique. XIII. Première partie: Les structures fondamentales de l'analyse. Livre II: Algèbre, Modules et Anneax semi-simples. Chapitre VIII*. Actualités

Scientifiques et Industrielles, vol. 1261 (Hermann & Cie, Paris, 1958)
5. J.W.S. Cassels, *Lectures on Elliptic Curves*. London Mathematical Society Student Texts, vol. 24 (Cambridge University Press, Cambridge, 1991)
6. J.-L. Colliot-Thélène, D. Harbater, J. Hartmann, D. Krashen, R. Parimala, V. Suresh, Local-global principles for zero-cycles on homogeneous spaces over arithmetic function fields. Trans. Am. Math. Soc. **372**(8), 5263–5286 (2019)
7. J.-L. Colliot-Thélène, D. Harbater, J. Hartmann, D. Krashen, R. Parimala, V. Suresh, Local-global principles for tori over arithmetic curves. Algebr. Geom. **7**(5), 607–633 (2020)
8. J.-L. Colliot-Thélène, D. Harbater, J. Hartmann, D. Krashen, R. Parimala, V. Suresh, Local-global principles for constant reductive groups over semi-global fields. Michigan Math. J. **72**, 77–144 (2022)
9. J.-L. Colliot-Thélène, A.N. Skorobogatov, *The Brauer-Grothendieck Group*. Ergebnisse der Mathematik und ihrer Grenzgebiete. 3. Folge. A Series of Modern Surveys in Mathematics, vol. 71 (Springer, Cham, 2021)
10. G. Cornelissen, K. Zahidi, Elliptic divisibility sequences and undecidable problems about rational points. J. Reine Angew. Math. **613**, 1–33 (2007)
11. J.E. Cremona, T.A. Fisher, C. O'Neil, D. Simon, M. Stoll, Explicit n-descent on elliptic curves. I. Algebra J. Reine Angew. Math. **615**, 121–155 (2008)
12. B. Creutz, On the local-global principle for divisibility in the cohomology of elliptic curves. Math. Res. Lett. **23**(2), 377–387 (2016)
13. N. Daans, Universally defining finitely generated subrings of global fields. Doc. Math. **26**, 1851–1869 (2021)
14. N. Daans, Existential first-order definitions and quadratic forms. PhD Thesis in Mathematics defended at Antwerp University (2022)
15. N. Daans, *Universally defining $\mathbb{Z}$ in $\mathbb{Q}$ with 10 quantifiers*. arXiv e-prints 2301.02107 (2023)
16. N. Daans, P. Dittmann, A. Fehm, *Existential rank and essential dimension of diophantine sets*. arXiv e-prints 2102.06941 (2021)
17. M. Davis, H. Putnam, J. Robinson, The decision problem for exponential diophantine equations. Ann. Math. (2) **74**, 425–436 (1961)
18. P. Dittmann, Irreducibility of polynomials over global fields is diophantine. Compos. Math. **154**(4), 761–772 (2018)
19. R. Dvornicich, L. Paladino, Local-global questions for divisibility in commutative algebraic groups. Eur. J. Math. **8**, S599–S628 (2022)
20. R. Dvornicich, U. Zannier, Local-global divisibility of rational points in some commutative algebraic groups. Bull. Soc. Math. France **129**(3), 317–338 (2001)
21. R. Dvornicich, U. Zannier, An analogue for elliptic curves of the Grunwald-Wang example. C. R. Math. Acad. Sci. Paris **338**(1), 47–50 (2004)
22. R. Dvornicich, U. Zannier, On a local-global principle for the divisibility of a rational point by a positive integer. Bull. Lond. Math. Soc. **39**(1), 27–34 (2007)
23. K. Eisenträger, Integrality at a prime for global fields and the perfect closure of global fields of characteristic $p > 2$. J. Number Theory **114**(1), 170–181 (2005)
24. K. Eisenträger, Hilbert's tenth problem for function fields of varieties over number fields and *p-adic fields*. J. Algebra **310**(2), 775–792 (2007)
25. K. Eisenträger, *Hilbert's Tenth Problem for Function Fields of Characteristic Zero*. Model theory with applications to algebra and analysis Vol. 2. London Math. Soc. Lecture Note Ser., vol. 350 (Cambridge University Press, Cambridge, 2008), pp. 237–254
26. K. Eisenträger, T. Morrison, Universally and existentially definable subsets of global fields. Math. Res. Lett. **25**(4), 1173–1204 (2018)
27. K. Eisenträger, A. Shlapentokh, Hilbert's tenth problem over function fields of positive characteristic not containing the algebraic closure of a finite field. J. Eur. Math. Soc. **19**(7), 2103–2138 (2017)
28. J.L. Eršov, On the elementary theory of maximal normed fields. Dokl. Akad. Nauk SSSR **165**, 21–23 (1965)

29. A. Fehm, Elementary geometric local-global principles for fields. Ann. Pure Appl. Logic **164**(10), 989–1008 (2013)
30. M.D. Fried, M. Jarden, *Field Arithmetic*, 3rd edn. Ergebnisse der Mathematik und ihrer Grenzgebiete, 3. Folge, A Series of Modern Surveys in Mathematics, vol. 11 (Springer, Berlin, 2008), revised by Jarden
31. P. Gille, T. Szamuely, *Central Simple Algebras and Galois Cohomology*. Cambridge Studies in Advanced Mathematics, vol. 165 (Cambridge University Press, Cambridge, 2017)
32. F. Gillibert, G. Ranieri, On the local-global divisibility of torsion points on elliptic curves and GL_2-type varieties. J. Number Theory **174**, 202–220 (2017)
33. Y. Harpaz, A.N. Skorobogatov, Hasse principle for Kummer varieties. Algebra Number Theory **10**(4), 813–841 (2016)
34. W. Hodges, *A Shorter Model Theory* (Cambridge University Press, Cambridge, 1997)
35. L.-C. Hsia, J.H. Silverman, On a dynamical Brauer-Manin obstruction. J. Théor. Nombres Bordeaux **21**(1), 235–250 (2009)
36. K. Kato, A Hasse principle for two-dimensional global fields. J. Reine Angew. Math. **366**, 142–183 (1986). With an appendix by Jean-Louis Colliot-Thélène
37. N.M. Katz, Galois properties of torsion points on abelian varieties. Invent. Math. **62**(3), 481–502 (1981)
38. K.-H. Kim, F.W. Roush, Diophantine undecidability of $\mathbb{C}(t_1, t_2)$. J. Algebra **150**(1), 35–44 (1992)
39. K.-H. Kim, F.W. Roush, Diophantine unsolvability over p-adic function fields. J. Algebra **176**(1), 83–110 (1995)
40. J. Koenigsmann, *Undecidability in Number Theory*. Model Theory in Algebra, Analysis and Arithmetic. Lecture Notes in Math., vol. 2111 (Heidelberg, 2014), pp. 159–195
41. J. Koenigsmann, Defining $\mathbb{Z}$ in $\mathbb{Q}$. Ann. Math. (2) **183**(1), 73–93 (2016)
42. T.Y. Lam, *Introduction to Quadratic Forms over Fields*. Graduate Studies in Mathematics, vol. 67 (American Mathematical Society, Providence, RI, 2005)
43. G. Malle, *Local-Global Conjectures in the Representation Theory of Finite Groups*. Representation theory—current trends and perspectives, EMS Ser. Congr. Rep. (Eur. Math. Soc., Zürich, 2017), pp. 519–539
44. D. Marker, *Model Theory*. Graduate Texts in Mathematics, vol. 217 (Springer, New York, 2002)
45. Y.V. Matijasevič, The Diophantineness of enumerable sets. Dokl. Akad. Nauk SSSR **191**, 279–282 (1970)
46. M.R. Murty, B. Fodden, *Hilbert's Tenth Problem*. An Introduction to Logic, Number Theory, and Computability, Student Mathematical Library, vol. 88 (American Mathematical Society, Providence, RI, 2019)
47. J. Neukirch, A. Schmidt, K. Wingberg, *Cohomology of Number Fields*, 2nd edn. Grundlehren der mathematischen Wissenschaften [Fundamental Principles of Mathematical Sciences], vol. 323 (Springer, Berlin, 2008)
48. L. Paladino, On counterexamples to local-global divisibility in commutative algebraic groups. Acta Arith. **148**(1), 21–29 (2011)
49. L. Paladino, G. Ranieri, E. Viada, On local-global divisibility by p^n in elliptic curves. Bull. Lond. Math. Soc. **44**(4), 789–802 (2012)
50. L. Paladino, G. Ranieri, E. Viada, On the minimal set for counterexamples to the local-global principle. J. Algebra **415**, 290–304 (2014)
51. R. Parimala, V. Suresh, Local-global principle for classical groups over function fields of p-adic curves. Comment. Math. Helv. **97**(2), 255–304 (2022)
52. J. Park, A universal first-order formula defining the ring of integers in a number field. Math. Res. Lett. **20**(5), 961–980 (2013)
53. T. Pheidas, Hilbert's tenth problem for fields of rational functions over finite fields. Invent. Math. **103**(1), 1–8 (1991)
54. B. Poonen, Hilbert's tenth problem and Mazur's conjecture for large subrings of $\mathbb{Q}$. J. Am. Math. Soc. **16**(4), 981–990 (2003)

55. B. Poonen, Characterizing integers among rational numbers with a universal-existential formula. Am. J. Math. **131**(3), 675–682 (2009)
56. A. Prestel, *Pseudo Real Closed Fields*. Set Theory and Model Theory (Bonn, 1979). Lecture Notes in Math., vol. 872 (1981), pp. 127–156
57. J. Robinson, Definability and decision problems in arithmetic. J. Symbolic Logic **14**, 98–114 (1949)
58. J. Robinson, *The Decision Problem for Fields*. Theory of Models (Proc. 1963 Internat. Sympos. Berkeley) (North-Holland, Amsterdam, 1965), pp. 299–311
59. R.S. Rumely, Undecidability and definability for the theory of global fields. Trans. Am. Math. Soc. **262**(1), 195–217 (1980)
60. R.S. Rumely, Arithmetic over the ring of all algebraic integers. J. Reine Angew. Math. **368**, 127–133 (1986)
61. R. Miller, A. Shlapentokh, On existential definitions of c.e. subsets of rings of functions of characteristic 0. Ann. Pure Appl. Logic **173**(4), Paper No. 103076, 50 (2022)
62. J.-J. Sansuc, Groupe de Brauer et arithmétique des groupes algébriques linéaires sur un corps de nombres. J. Reine Angew. Math. **327**, 12–80 (1981)
63. J.-P. Serre, Sur les groupes de congruence des variétés abéliennes. Izv. Akad. Nauk SSSR Ser. Mat. **28**, 3–20 (1964)
64. J.-P. Serre, *Local Fields*. Graduate Texts in Mathematics, vol. 67 (Springer, New York, Berlin, 1979), translated from the French by Marvin Jay Greenberg
65. A. Shlapentokh, Hilbert's tenth problem for algebraic function fields over infinite fields of constants of positive characteristic. Pacific J. Math. **193**(2), 463–500 (2000)
66. A. Shlapentokh, *Hilbert's Tenth Problem*. Diophantine classes and extensions to global fields. New Mathematical Monographs, vol. 7 (Cambridge University Press, Cambridge, 2007)
67. A. Shlapentokh, Elliptic curves retaining their rank in finite extensions and Hilbert's tenth problem for rings of algebraic numbers. Trans. Am. Math. Soc. **360**(7), 3541–3555 (2008)
68. A. Shlapentokh, Elliptic curve points and Diophantine models of $\mathbb{Z}$ in large subrings of number fields. Int. J. Number Theory **8**(6), 1335–1365 (2012)
69. J.H. Silverman, *The Arithmetic of Elliptic Curves*, 2nd edn. Graduate Texts in Mathematics, vol. 106 (Springer, Dordrecht, 2009)
70. A.N. Skorobogatov, *Torsors and Rational Points*. Cambridge Tracts in Mathematics, vol. 144 (Cambridge University Press, Cambridge, 2001)
71. A.A. Suslin, The structure of the special linear group over rings of polynomials. Izv. Akad. Nauk SSSR Ser. Mat. **11**(2), 235–252, 477 (1977)
72. A.V. Sutherland, A local-global principle for rational isogenies of prime degree. J. Théor. Nombres Bordeaux **24**(2), 475–485 (2012)
73. A. Tarski, *A Decision Method for Elementary Algebra and Geometry* (The Rand Corporation, Santa Monica, CA, 1948)
74. I. Vogt, A local-global principle for isogenies of composite degree. Proc. Lond. Math. Soc. (3) **121**(6), 1496–1530 (2020)
75. J. Voight, *Quaternion Algebras*. Graduate Texts in Mathematics, vol. 288 (Springer, Cham, 2021)
76. W.C. Waterhouse, Abelian varieties over finite fields. Ann. Sci. École Norm. Sup. (4) **2**, 521–560 (1969)
77. G.-R. Zhang, Z.-W. Sun, $\mathbb{Q} \setminus \mathbb{Z}$ is diophantine over $\mathbb{Q}$ with 32 unknowns. Bull. Pol. Acad. Sci. Math. **70**(2), 93–106 (2022)

Campana Points on Diagonal Hypersurfaces

Francesca Balestrieri, Julia Brandes, Miriam Kaesberg, Judith Ortmann, Marta Pieropan, and Rosa Winter

1 Introduction

The study of *Campana points* on varieties has gained a lot of attention in recent years. Philosophically, sets of Campana points on a variety over a number field interpolate between the set of rational points and the set of integral points. As such, the general aim when studying Campana points is to use techniques from the study of rational points to say something about Campana points, which can in turn be used to understand integral points better. There are several definitions of Campana points

F. Balestrieri
The American University of Paris, Paris, France
e-mail: fbalestrieri@aup.edu

J. Brandes
Mathematical Sciences, University of Gothenburg and Chalmers Institute of Technology, Göteborg, Sweden
e-mail: brjulia@chalmers.se

M. Kaesberg
Mathematical Institute, University of Göttingen, Göttingen, Germany
e-mail: miriam.kaesberg@mathematik.uni-goettingen.de

J. Ortmann
Institut für Algebra, Zahlentheorie und Diskrete Mathematik, Leibniz Universität Hannover, Hannover, Germany
e-mail: ortmann@math.uni-hannover.de

M. Pieropan (✉)
Mathematical Institute, Utrecht University, Utrecht, The Netherlands
e-mail: m.pieropan@uu.nl

R. Winter
Department of Mathematics, King's College London, London, UK
e-mail: rosa.winter@kcl.ac.uk

R. Abdellatif et al. (eds.), *Women in Numbers Europe IV*, Association for Women in Mathematics Series 32, https://doi.org/10.1007/978-3-031-52163-8_3

(see Sect. 3), all having in common that they are integral points on a proper variety satisfying prescribed 'intersection conditions' with respect to a boundary divisor. When X is a hypersurface in a projective space $\mathbb{P}^n_{\mathbb{Q}}$ and the closure of X in $\mathbb{P}^n_{\mathbb{Z}}$ is a flat proper model of X over $\mathbb{Z}$, a set of Campana points on X for the boundary divisor induced by the coordinate hyperplanes of the projective space is the set of $\mathbb{Z}$-points in X for which the coordinates are *m-full* for different m (an integer a is m-full if $p|a$ implies $p^m|a$ for all primes p), see Example 1.

Many classical results and conjectures for rational points have been formulated also for Campana points; an overview of some of the following can be found in [30]. An analogue of the Mordell Conjecture for Campana points (*orbifold Mordell conjecture*) over function fields was introduced and proven in characteristic 0 in [10], and in positive characteristic in [18]. Asymptotics for the number of Campana points of bounded height were first given in [32] for linear hypersurfaces in $\mathbb{P}^n_{\mathbb{Q}}$ for $n \geq 4$, with the prescribed intersection conditions corresponding to points with squareful coordinates. The same was done for $n = 3$ in [27], while [8] deals with general m-full conditions on each coordinate for n large enough (see Sect. 2.3 for more details on these counting results). Upper and lower bounds for $n = 2$ are given in [7]. In [24], the authors gave an analogue of the Manin Conjecture for Campana points of bounded height on Fano orbifolds (*PSTVA Conjecture*) and proved it for equivariant compactifications of vector groups. The PSTVA Conjecture has also been proven for split toric varieties with torus invariant boundary [23, 26] and biequivariant compactifications of the Heisenberg group [35]. It is also compatible with the order of growth in the results in [32], [7], [8] and [27]. However, in [28] it is shown that the conjectured leading constant in [7] does not agree with the one predicted by the PSTVA Conjecture for the orbifold $(\mathbb{P}^1, \frac{1}{2}[0]+\frac{1}{2}[\infty]+\frac{1}{2}[1])$, and a counterexample to the leading constant predicted by the PSTVA Conjecture is given by counting Campana points on $\mathbb{P}^1$ with the boundary divisor $\frac{1}{2}\{37x^2+109y^2=0\}$. For the same orbifolds, [28] shows that the leading constant can be made arbitrarily small by excluding suitable thin sets from the counting problem. To the best of our knowledge, it has not yet been investigated whether the leading constants in [32], [8] and [27] agree with the PSTVA Conjecture.

The first asymptotics for the arguably harder case of *weak* Campana points are given in [29], where the author counts such points of bounded height on orbifolds associated with norm forms for Galois extensions of number fields. The same paper also contains the first asymptotics for Campana points on *singular* orbifolds. Weak Campana points also appear in [26], where the author counts diagonal quartic surfaces with a Brauer-Manin obstruction to the Hasse principle and shows that this number gives a lower bound for the number of weak Campana points in $\mathbb{P}^3$ for the boundary divisor induced by the coordinate hyperplanes. The study of the potential density of rational points (i.e., density of rational points over a finite extension of the base field) was translated to Campana points as well, with Campana first stating a conjecture in this direction for Campana points on curves over number fields in [10]. In [2], the authors extended the conjecture to a much larger class of varieties, and they showed that their conjecture recovers Lang's conjecture for rational points on varieties of general type, and the Lang-Vojta conjecture for S-integral points on varieties of logarithmic general type. The authors of [3] proved the conjecture when the boundary divisor is a general smooth divisor satisfying some extra conditions. Instances of the conjecture formulated over function fields are proven in [25] and in [15]. Finally, analogues of local-global principles for Campana points are also studied: in [22] the authors defined the Campana analogues of weak weak approximation and the Hilbert property and proved the Campana version of a result by Colliot-Thélène and Ekedahl that weak weak approximation implies the Hilbert property. A Campana analogue of the Hasse Principle is defined in [20], where the authors studied weak approximation and Brauer-Manin obstructions for Campana points.

In the setting mentioned earlier, where X is a hypersurface in projective space over $\mathbb{Q}$, counting Campana points of bounded height is equivalent to counting m-full solutions of bounded size to homogeneous equations. This counting problem lends itself very well to the *circle method*, a powerful counting method in analytic number theory. So far, the circle method has been used to count Campana points when both X and the boundary divisor are linear [7, 8, 27, 32]; we give more detail on this in Sect. 2.3. In this paper we extend the techniques in [8] to study the problem of counting Campana points on a *non-linear* diagonal hypersurface X in $\mathbb{P}^n$. Our main result is the following.

Theorem 1 *Let $k, n, m_0, \ldots, m_n$ be positive integers. Let B be a positive real number. Let $X \subset \mathbb{P}^n_{\mathbb{Q}}$ be the hypersurface given by $\sum_{i=0}^n c_i x_i^k = 0$ with $c_0, \ldots, c_n$ in $\mathbb{Z}_{\neq 0}$ and $\gcd(c_0, \ldots, c_n) = 1$. Let further D be the $\mathbb{Q}$-divisor on X given by $D = \sum_{i=0}^n (1 - \frac{1}{m_i})\{x_i = 0\}$. With the model defined in Sect. 4.1 and the height function induced by the Weil height on $\mathbb{P}^n(\mathbb{Q})$, the set of Campana $\mathbb{Z}$-points of height at most B on $X \cap \left(\mathbb{P}^n \setminus \bigcup_{i=0}^n \{x_i = 0\}\right)$ with respect to the boundary divisor D is given by*

$$N(X, D, B) = \left\{ (x_0 : \cdots : x_n) \in \mathbb{P}^n(\mathbb{Q}) \;\middle|\; \begin{array}{c} x_0, \ldots, x_n \subset \mathbb{Z}_{\neq 0}, \\ \gcd(x_0, \ldots, x_n) = 1, \\ x_i \text{ is } m_i\text{-full } \forall i \in \{0, \ldots, n\}, \\ |x_0|, \ldots, |x_n| \leq B, \\ c_0 x_0^k + \cdots + c_n x_n^k = 0 \end{array} \right\}.$$

Assume that $k \geq 2$, $2 \leq m_0 \leq \cdots \leq m_n$ and that

$$\sum_{i=0}^n \frac{1}{2 s_0(k m_i)} > 1, \tag{1}$$

where

$$s_0(m) = \min\{2^{m-1}, \tfrac{1}{2} m(m-1) + \lfloor \sqrt{2m+2} \rfloor\} \qquad \forall m \in \mathbb{N}.$$

Then there exist constants $C \geq 0$ *and* $\eta > 0$ *such that for all* $B > 0$ *we have*

$$\#N(X, D, B) = CB^{k\Gamma} + O_\eta\left(B^{k\Gamma-\eta}\right), \tag{2}$$

where $k\Gamma = \sum_{i=0}^{n} \frac{1}{m_i} - k$ *is the Fujita invariant of* (X, D) *with respect to the line bundle* $O_X(1)$.

An explicit expression for the constant C is given in (28). Further analysis of the leading constant, including determining under which conditions it is non-zero, will be carried out in a forthcoming paper.

Remark 1 For $k = 1$, a version of Theorem 1 was established in [8], and it is compatible with the PSTVA Conjecture as far as the order of growth is concerned. For $k \geq 2$, the integral model obtained by the closure of X in $\mathbb{P}^n_{\mathbb{Z}}$ that we use in Theorem 1 is not regular. However, the order of growth in Theorem 1 is still compatible with the prediction in [24, Conjecture 1.1]. In Sect. 4.2 we describe the set of Campana points on the regular model obtained by inverting all the prime numbers dividing $k \cdot c_0 \cdots c_n$. The set of Campana points with respect to this model is larger than the set in Theorem 1, but the corresponding asymptotic, which we compute in a forthcoming paper, is expected to have the same order of magnitude as (2).

Remark 2 Browning and Yamagishi [8] had the stronger condition $\sum_{i=0}^{n-1} \frac{1}{m_i(m_i+1)} \geq 1$ in their result. This seems stricter than necessary, since on the one hand it isolates the last coefficient, creating a somewhat artificial condition, and on the other hand it fails to take advantage of the strongest available bounds from [34, Section 14]. In our result, we address both of those issues in the case $k \geq 2$. In principle, the methods can be extended also to the case $k = 1$, but this would require a more careful treatment in some parts of the argument. Since the main focus of our paper is on equations of degree 2 or higher, we did not expend the extra effort that would be required to achieve a strengthening of the result of [8]. For context, in the smallest possible case that is admissible within Theorem 1, namely in the case $k = m_0 = \ldots = m_n = 2$, we have $s_0(km_i) = s_0(4) = 8$, and consequently the condition (1) translates into the bound $n \geq 16$.

Although it improves the range of applicability compared to [8], the condition in our theorem is still far from the log Fano condition $\sum_{i=0}^{n} \frac{1}{km_i} > 1$, which is the geometric condition under which it makes sense to expect an asymptotic as predicted by the PSTVA Conjecture. In fact, as far as applications of the circle method involving mean value estimates are concerned, the limit of the method is at $\sum_{i=0}^{n} \frac{1}{km_i} > 2$.

1.1 Outline of the Paper

The paper is organized as follows. In Sect. 2 we give a brief history of the circle method and explain the main ideas of the technique. We also give an overview of applications of the circle method to the problem of counting Campana points of bounded height. In Sect. 3 we define Campana points and give an overview of different definitions that have been used in the literature. We focus on diagonal hypersurfaces in Sect. 4, where we describe sets of Campana points of bounded height and set up the counting problem for Theorem 1. In Sect. 5 we develop a variant of the circle method of [8] combined with the bounds in [34, Section 14], and we finally prove Theorem 1 in Sect. 6.

1.2 Notation

Given $m \in \mathbb{N}$ and a set S of prime numbers, an integer x is *m-full outside S* if $p^m | x$ holds for every prime $p \notin S$ dividing x; we say that x is *m-full* if we can take $S = \emptyset$. To avoid confusion, we stress here that we do not consider 0 to be an element of $\mathbb{N}$.

For a number field F, and for a finite set S of places of F containing all infinite places, we denote by $\mathcal{O}_{F,S}$ the ring of S-integers, that is, all elements of F with non-negative v-adic valuation for all places $v \notin S$. We denote by ∞ the infinite place of $\mathbb{Q}$. For a prime number p, we denote by val_p the p-adic valuation.

In Sects. 4, 5, and 6, we denote tuples of integers by bold letters as follows: $\boldsymbol{c} = (c_0, \ldots, c_n) \in \mathbb{Z}^{n+1}$, and similarly we have $\boldsymbol{d}, \boldsymbol{\varepsilon}, \boldsymbol{s}, \boldsymbol{x}, \tilde{\boldsymbol{u}} \in \mathbb{Z}^{n+1}$. We further have $\boldsymbol{t} = (t_{i,r})_{0 \le i \le n, 1 \le r \le m_i - 1} \in \mathbb{N}^{\sum_i m_i - n}$ for m_i as in Theorem 1, and similarly we have $\tilde{\boldsymbol{v}} \in \mathbb{N}^{\sum_i m_i - n}$. Throughout, we assume that B is a large positive real number. For $\boldsymbol{x} \in \mathbb{Z}^{n+1}$, we write $|\boldsymbol{x}| = \max_{0 \le i \le n} |x_i|$. We denote by $\mathbb{Z}^{n+1}_{\text{prim}}$ the set given by $\{\boldsymbol{x} \in \mathbb{Z}^{n+1} : \gcd(x_0, \ldots, x_n) = 1\}$, by $\prod_p$ a product over all prime numbers, and by μ the Möbius function. For $\alpha \in \mathbb{R}$, we write $e(\alpha) = \exp(2\pi i \alpha)$, and we denote by $\|\alpha\|$ the distance between α and its closest integer.

Throughout, the implicit constants in the estimates $\ll$ and $O(\cdot)$ are allowed to depend on the fixed data k, ϵ and δ, as well as the coefficients $c_0, \ldots, c_n$ and $d_0, \ldots, d_n$, and the parameters $m_0, \ldots, m_n$ and $\tilde{m}_0, \ldots, \tilde{m}_n$. In particular, all statements involving the symbol ϵ are asserted to hold for all $\epsilon > 0$. Here, we do not track the precise 'value' of ϵ, which consequently may vary from one expression to the next.

1.3 Acknowledgements

We want to thank the organizers of the workshop Women in Numbers Europe 4 which took place in the summer of 2022 in Utrecht, the Netherlands, which is where this project originated. We thank Tim Browning for useful comments. While working on this paper Judith Ortmann was partially supported by the Caroline Herschel Programme of Leibniz University Hannover, as well as a scholarship for a research stay abroad of the Graduiertenakademie of Leibniz University Hannover since part of the work was done while visiting the University of Bath. Rosa Winter was supported by UKRI Fellowship MR/T041609/2. In the final stages of the work, Julia Brandes was supported by Project Grant 2022-03717 from Vetenskapsrådet (Swedish Science Foundation), and Marta Pieropan was supported by the grant VI.Vidi.213.019 of the Nederlandse Organisatie voor Wetenschappelijk Onderzoek. Part of the work was completed while Julia Brandes was visiting the Max Planck Institute for Mathematics in Bonn, whose generous support is also gratefully acknowledged. Last but not least, we are grateful to Sam Streeter as well as two anonymous referees for their useful comments.

2 The Circle Method

2.1 History of the Circle Method

In the 1920's Hardy and Littlewood developed the so-called Hardy–Littlewood circle method (sometimes called just 'Hardy–Littlewood method' or 'circle method'), which is an analytic method to treat additive problems in number theory. These problems deal with the representation of a large number as a sum of numbers of some specified type. The most famous additive problem is Waring's problem: Let k be a positive integer. Can every large integer N be written as a sum of a bounded number of k-th powers

$$N = x_1^k + \cdots + x_s^k, \tag{3}$$

where s is a positive integer? Let $r_{k,s}(N)$ denote the number of such representations of N. Then, Waring's problem is equivalent to showing that $r_{k,s}(N) > 0$ for some s and all sufficiently large integers N. Waring's problem was first proved by Hilbert in 1909. From 1918 to 1920, Hardy and Littlewood together with Ramanujan were the first to prove an explicit upper bound for $r_{k,s}(N)$ by using a new analytic method. Their work laid down the foundations of the circle method in its original form (see [14, 21]).

It turned out that the circle method is a very powerful method, since it can be used to solve a diverse range of number theoretic problems. For example, it is one of the most significant all purpose tools when studying rational points on higher-dimensional algebraic varieties. We refer to [6] for an overview.

In its earliest versions, the starting point of the circle method is the generating function

$$f(z) = \sum_{a \in A} z^a \quad (|z| < 1),$$

where A denotes a set of non-negative integers. Taking $f(z)$ to the sth power yields the series

$$f(z)^s = \sum_{N=0}^{\infty} R_{A,s}(N) z^N,$$

whose coefficients $R_{A,s}(N)$ encode the number of solutions of the equation

$$N = a_1 + \cdots + a_s \tag{4}$$

with $a_1, \ldots, a_s \in A$. We can isolate the Nth coefficient by means of Cauchy's integral formula, which yields

$$R_{A,s}(N) = \frac{1}{2\pi i} \int_{|z|=\rho} \frac{f(z)^s}{z^{N+1}} \mathrm{d}z \tag{5}$$

for any $\rho \in (0, 1)$. In the original (Hardy–Ramanujan) version of the circle method this integral is evaluated by dividing the circle of integration into two disjoint sets, the 'major arcs' and the 'minor arcs'. Classically, the major arcs contribute to the main term and the minor arcs to the error term.

In 1928 Vinogradov introduced a helpful simplification by transferring the problem from complex analysis to Fourier analysis.

The Fourier transform maps the set of integers $\mathbb{Z}$ onto the real unit interval $\mathbb{R}/\mathbb{Z} \simeq [0, 1)$. For any integer N we denote as before by $R_{A,s}(N)$ the number of representations of N as a sum of s elements in a finite set A. The inverse Fourier transform $\mathcal{F}^{-1} R_{A,s}$ of $R_{A,s}$ is then a function of α and is given by

$$\mathcal{F}^{-1} R_{A,s}(\alpha) = \sum_{N \in \mathbb{Z}} R_{A,s}(N) e(\alpha N) = \sum_{a_1, \ldots, a_s \in A} e(\alpha(a_1 + \cdots + a_s)) = \left(\sum_{a \in A} e(\alpha a) \right)^s.$$

It is convenient to put

$$F(\alpha) = \sum_{a \in A} e(\alpha a),$$

and the reader may note that with this definition we have $F(\alpha) = f(e(\alpha))$ where f is the generating function considered above. We can now apply the forward Fourier transform and obtain

$$\mathcal{F}\mathcal{F}^{-1} R_{A,s}(N) = \int_0^1 F(\alpha)^s e(-\alpha N) \mathrm{d}\alpha.$$

By the Fourier Inverse Theorem, we have $\mathcal{F}\mathcal{F}^{-1}R_{A,s}(N) = R_{A,s}(N)$, and thus we obtain the formula

$$R_{A,s}(N) = \int_0^1 F(\alpha)^s e(-\alpha N)\mathrm{d}\alpha. \tag{6}$$

Clearly, since we assume A to consist of non-negative integers, the equation (4) implies that without loss of generality the sum $F(\alpha)$ can be truncated at $a \leq N$, as larger values trivially cannot contribute. We remark that if the set A is allowed to include negative numbers as well, one typically includes some other truncation $a \leq B$ for some large parameter B.

2.2 The Modern Circle Method: Main Steps

We are interested in the situation when the set A is the set of k-th powers. In that case the exponential sum is given by

$$F(\alpha) = \sum_{1 \leq x \leq P} e(\alpha x^k),$$

where we put $P = \lfloor N^{1/k} \rfloor$.

The main strategy is now to try to understand the integral in (6) by studying the size of the integrand, and in particular the size of the exponential sums $F(\alpha)$ as α ranges over the unit interval. For a typical (irrational) α, the individual summands within the exponential sum $F(\alpha)$ are more or less equidistributed over the unit circle, and consequently the behaviour of the exponential sum $F(\alpha)$ is reminiscent of Brownian motion. In fact, [13, Corollary 2.2] shows that $F(\alpha) \ll P^{1/2+\epsilon}$ for all α in a set $\mathcal{L} \subset [0, 1]$ with Lebesgue measure 1. At the same time, when α is a rational number with denominator q, the summands inside $F(\alpha)$ can take at most q distinct values on the unit circle. Thus, in particular for small values of q there is significant potential for interference. The extreme case of this is when $\alpha = 0 = \frac{0}{1}$, which gives the value $F(0) = P$. Moreover, since F is continuous, this interference behaviour extends to suitably small neighbourhoods of rational numbers with small denominator.

This motivates the dissection of the interval $[0, 1]$ into major and minor arcs, where the major arcs $\mathfrak{M}$ comprise all α that are close to a rational number with small denominator, so that $F(\alpha)$ is potentially large, and the minor arcs $\mathfrak{m}$ collect the remaining α that lack a similarly strong rational approximation.

The treatment of the major arcs is by now mainly standard. By using the fact that $\alpha \in \mathfrak{M}$ has a good approximation by a rational number a/q, we can write $\alpha = a/q + \theta$ for some small θ. This leads to a factorization of the integral over the major arcs into a product of two terms: the 'singular integral' and the 'singular series', which then can be evaluated to obtain the main term. Here, the singular integral can be interpreted as the volume of the solution set of the equation (3) when viewed as a submanifold inside $\mathbb{R}^s$; this also provides the expected order of growth. Meanwhile, the singular series encodes all congruence information related to the equation (3). In fact, it can be factorized into an Euler product, where each factor describes the volume of the solution set of equation (3) when viewed as a subset of the p-adic numbers $\mathbb{Q}_p$ as p ranges over the primes. Thus, the canonical outcome of the circle method is an asymptotic formula with a main term of size $\asymp P^{s-k} \sim N^{s/k-1}$, which is modulated by a product of local factors that encode any local obstructions the problem might have.

The bottleneck of the problem is the treatment of the minor arcs. The main difficulty is that although we have good control over the size of $F(\alpha)$ in an almost-all sense, our understanding in a *pointwise* sense remains comparatively poor. Fortunately, however, for additive problems like the one in (3) the contribution to (6) that stems from the minor arcs can be estimated by

$$\int_{\mathfrak{m}} F(\alpha)^s e(-n\alpha)\mathrm{d}\alpha \leq \sup_{\alpha\in\mathfrak{m}} |F(\alpha)| \int_0^1 |F(\alpha)|^{s-1}\mathrm{d}\alpha.$$

In other words, it becomes less crucial to have strong pointwise bounds for $F(\alpha)$ if in addition we have good control over the *average* behaviour of moments of F. Since F exhibits square-root cancellation almost everywhere, this is a more tractable problem at least for small moments. For larger moments, however, this problem still presents formidable difficulties.

A major breakthrough in the field was the resolution of the main conjecture associated with Vinogradov's mean value theorem [4, 33, 34], which gives a near complete understanding of the average behaviour of moments of the related exponential sum

$$G(\alpha) = \sum_{1\leq x\leq P} e(\alpha_1 x + \alpha_2 x^2 + \cdots + \alpha_k x^k).$$

In fact, the strongest available bounds for mean values of moments of F are derived from these results (see [34, Section 14]). These are the bounds we will exploit in our arguments below.

2.3 Examples of Counting Campana Points by the Circle Method

Here, we summarize instances in which the circle method is used to count Campana points of bounded height. So far, there are only a few examples of this.

Van Valckenborgh and Browning were the first to use the circle method to count Campana points of bounded height. Much of the early investigations is centred around the set of Campana points for the boundary divisor $\frac{1}{2}\sum_{i=0}^{n}\{x_i = 0\}$, corresponding to squareful numbers. In this setting, Van Valckenborgh [32] gave an asymptotic formula for the number of integral points $(a_0 : \cdots : a_n)$ of bounded height on the hyperplane $\{\sum_{i=0}^{n} x_i = 0\}$ in $\mathbb{P}^n$ such that a_i is squareful, provided that $n \geq 4$.

The picture becomes more complicated for $n \leq 3$, as underlying geometric considerations begin to have an effect. In [7] Browning and Van Valckenborgh considered the case $n = 2$ and investigated the number of positive coprime squareful numbers x, y, z of bounded height B that satisfy the equation $x + y = z$. In particular, they give a lower bound of order $O(B^{1/2})$ for the number of corresponding Campana points, which they conjecture to be sharp. Unfortunately, establishing the corresponding upper bound turns out to be significantly harder, and here they use a method rooted in the determinant method to give an upper bound of order $O(B^{3/5+\epsilon})$. Finally, Shute [27] settled the case $n = 3$ by adapting the delta-symbol method [17]. In this context, it is noteworthy that in order to put the problem in the framework of Manin's conjecture, the author has to exclude certain accumulating special subvarieties from the count, reinforcing the principle that geometric phenomena tend to have a disproportionate effect in small dimension. We can therefore summarize that in the setting of counting Campana points with boundary divisor $\frac{1}{2}\sum_{i=0}^{n}\{x_i = 0\}$ on the hyperplane $x_0 + \cdots + x_n = 0$ only the case $n = 2$ is still open.

Meanwhile, Browning and Yamagishi generalized the work of Van Valckenborgh [32] in [8] by extending it to more general boundary divisors. Instead of squareful integers a_i they consider more generally m_i-full integers, where $m_i \geq 2$. They further generalize to hypersurfaces $c_0x_0 + \cdots + c_{n-1}x_{n-1} = x_n$ for fixed non-zero integers $c_0, \ldots, c_{n-1}$. Similarly to [32], they give an asymptotic formula for the number of Campana points of bounded height in this setting by using the circle method under the assumption that the following condition is satisfied:

$$\sum_{\substack{0\leq i\leq n\\ i\neq j}} \frac{1}{m_i(m_i+1)} \geq 1 \text{ for some } 0 \leq j \leq n.$$

It turns out that considering these Campana points can be interpreted as studying Waring's problem for mixed exponents. Browning and Yamagishi used this to further prove an asymptotic formula for Waring's problem of mixed powers.

In this paper we extend the result of [8] to diagonal hypersurfaces of degree at least 2.

3 Campana Points

In this section we introduce the notion of Campana points that we use in Theorem 1. Campana introduced the notion of 'orbifoldes géométriques' and 'orbifold rational point' in his papers [9–12]. Since then, several different definitions of Campana points appeared in the literature [1, 2], which agree with the original definition of Campana on curves. As is explained in [24], this is not the case for higher-dimensional varieties, where the different definitions can lead to significant differences in the associated counting problems. All the papers just mentioned define Campana points on a regular integral model of the variety over the ring of S-integers of the field of definition, for some finite set of places S. Very recently, Mitankin, Nakahara and Streeter gave a definition of Campana points for *non-regular* models [20]. We will use this definition in this paper, since it allows us to directly interpret the counting problem in Theorem 1 in the setting of Campana points.

Definition 1 Let F be any field. A *Campana orbifold over F* is a pair (X, D) where

- X is a smooth proper variety over F.
- D is an effective Weil $\mathbb{Q}$-divisor on X defined over F satisfying

$$D = \sum_{\alpha \in \mathcal{A}} \epsilon_\alpha D_\alpha,$$

where $\mathcal{A}$ is a finite index set, the D_α's are distinct prime divisors on X, and ϵ_α belongs to the set

$$\mathcal{M} = \left\{ 1 - \frac{1}{m} \middle| m \in \mathbb{N}, m \geq 2 \right\}.$$

Note that the D_α's are irreducible over F, but not necessarily over $\overline{F}$. We define $D_{\mathrm{red}} = \sum_{\alpha \in \mathcal{A}} D_\alpha$, and say that (X, D) is *smooth* if D_{red} is a strict normal crossing divisor on X.

Remark 3 In the existing literature, the set $\mathcal{M}$ is usually taken to include 0 and 1. For the purposes of this paper we do not need these values, hence we omit them in our definition.

Since Campana points depend on the choice of an integral model for the Campana orbifold, we need to define what a good choice for such an integral model is. From now on, we take F to be a number field.

Definition 2 Let (X, D) be a Campana orbifold over F. Let $S \subset \Omega_F$ be a finite set of places of F containing all the infinite places. A *proper integral model* of (X, D) over $O_{F,S}$ is a pair $(\mathcal{X}, \mathcal{D})$ such that $\mathcal{X}$ is a flat proper model of X over $O_{F,S}$, and $\mathcal{D} = \sum_{\alpha \in \mathcal{A}} \epsilon_\alpha \mathcal{D}_\alpha$, where $\mathcal{D}_\alpha$ is the Zariski closure of D_α in $\mathcal{X}$. If $\mathcal{X}$ is regular, we say that $(\mathcal{X}, \mathcal{D})$ is a *good integral model* of (X, D).

Finally, before we can define Campana points, we need to understand the 'intersection behaviour' of a given point $P \in X(F)$ with the divisor D.

Definition 3 Let (X, D) be a Campana orbifold over F, and let $(\mathcal{X}, \mathcal{D})$ be a proper integral model over $\mathcal{O}_{F,S}$. Take $P \in X(F)$. Since $\mathcal{X}$ is proper, we have $X(F) = \mathcal{X}(\mathcal{O}_{F,S})$ and so P extends uniquely to a point $\mathcal{P} \in \mathcal{X}(\mathcal{O}_{F,S})$. Fix $v \notin S$ and let $\mathcal{P}_v \in \mathcal{X}(\mathcal{O}_v)$ be the point corresponding to P. By definition, we can also view it as a map $\mathcal{P}_v : \mathrm{Spec}(\mathcal{O}_v) \to \mathcal{X}$. Fix $\alpha \in \mathcal{A}$. Since $\mathcal{D}_\alpha \subset \mathcal{X}$ is a closed subscheme, the fiber product $\mathcal{D}_\alpha \times_{\mathcal{X}} \mathrm{Spec}(\mathcal{O}_v) \subset \mathrm{Spec}(\mathcal{O}_v)$ is a closed subscheme corresponding to a non-zero ideal $I_{v,\mathcal{P},\alpha}$ in $\mathcal{O}_v$, using the correspondence between closed subschemes of $\mathrm{Spec}(\mathcal{O}_v)$ and ideals in $\mathcal{O}_v$. We distinguish two cases:

- If $\mathcal{P}_v \not\subseteq \mathcal{D}_\alpha$, we define the *intersection multiplicity of P and $\mathcal{D}_\alpha$ at v* to be

$$n_v(\mathcal{D}_\alpha, P) = \mathrm{length}(\mathcal{O}_v / I_{v,\mathcal{P},\alpha}).$$

  Equivalently, since $\mathcal{O}_v$ is a discrete valuation ring, say with a choice of uniformiser π_v for its maximal ideal, it follows that the non-zero ideal $I_{v,\mathcal{P},\alpha}$ can be written as

$$I_{v,\mathcal{P},\alpha} = \left(\pi_v\right)^{n_v(\mathcal{D}_\alpha, P)}.$$

- If $\mathcal{P}_v \subseteq \mathcal{D}_\alpha$, we set the *intersection multiplicity of P and $\mathcal{D}_\alpha$ at v* to be $+\infty$ (in this case, the ideal $I_{v,\mathcal{P},\alpha}$ is just the zero ideal (0) in $\mathcal{O}_v$).

  To summarize, the *intersection multiplicity of P and $\mathcal{D}_\alpha$ at v* is

$$n_v(\mathcal{D}_\alpha, P) = \begin{cases} \mathrm{length}(\mathcal{O}_v / I_{v,\mathcal{P},\alpha}) & \text{if } \mathcal{P}_v \not\subseteq \mathcal{D}_\alpha, \\ +\infty & \text{if } \mathcal{P}_v \subseteq \mathcal{D}_\alpha. \end{cases}$$

Remark 4 In practice, given a point $\mathcal{P}_v : \mathrm{Spec}(\mathcal{O}_v) \to \mathcal{X}$, and an open subset $U \subseteq \mathcal{X}$ containing the image of $\mathcal{P}_v$, such that $\mathcal{D}_\alpha|_U$ is principal and defined by a rational function $f_{\alpha,U}$ which is regular on U (i.e., $\mathcal{D}_\alpha$ is Cartier in a neighbourhood of $\mathcal{P}_v$), we have $n_v(\mathcal{D}_\alpha, P) = \mathrm{val}_v(f_{\alpha,U}(\mathcal{P}_v))$, where $f_{\alpha,U}(\mathcal{P}_v)$ is the image of $f_{\alpha,U}$ in $\mathcal{O}_v$ under the ring homomorphism that defines $\mathcal{P}_v : \mathrm{Spec}(\mathcal{O}_v) \to U$, and val_v is the v-adic valuation on $\mathcal{O}_v$.

We are now in the position to define Campana points.

Definition 4 We keep the notation as in Definition 3. For $\alpha \in \mathcal{A}$ we write ϵ_α for $1 - \frac{1}{m_\alpha}$. A point $P \in X(F)$ is a *Campana $\mathcal{O}_{F,S}$-point on* $(\mathcal{X}, \mathcal{D})$ if, for every place $v \notin S$ and every $\alpha \in \mathcal{A}$, we have either $n_v(\mathcal{D}_\alpha, P) = 0$ or $n_v(\mathcal{D}_\alpha, P) \geq m_\alpha$.

Remark 5 Different choices of the finite set of places $S \subset \Omega_F$ lead to potentially different sets of Campana points and thus to potentially different counting problems.

Example 1 Using Remark 4, we describe Campana points for $F = \mathbb{Q}$ and $X \subseteq \mathbb{P}^n_{\mathbb{Q}}$, in the setting where the divisor D_α is the restriction to X of a divisor $D'_\alpha \subseteq \mathbb{P}^n$ for all $\alpha \in \mathcal{A}$, and we have a model $\mathcal{X} \subseteq \mathbb{P}^n_{\mathbb{Z}_S}$ for some set of places S (for example, when $\mathcal{X}$ is the Zariski closure of X in $\mathbb{P}^n_{\mathbb{Z}_S}$). For all $\alpha \in \mathcal{A}$, let $f_\alpha \in \mathbb{Z}[x_0, \ldots, x_n]$ be homogeneous with coprime coefficients (i.e., with content 1) such that D_α is defined by f_α. Fix a point $P \in X(\mathbb{Q})$ and write $P = (a_0 : \cdots : a_n)$ in projective homogeneous coordinates with $a_0, \ldots, a_n \in \mathbb{Z}$ such that $\gcd(a_0, \ldots, a_n) = 1$. Let $\ell_P \in \mathbb{Z}[x_0, \ldots, x_n]$ be a linear form with $\ell_P(a_0, \ldots, a_n) = 1$. Then the image of $\mathcal{P}$ in $\mathcal{X}$ is contained in the affine patch $U = \mathbb{P}^n_{\mathbb{Z}_S} \setminus \{\ell_P = 0\}$, so for all $\alpha \in \mathcal{A}$ we can take $f_{\alpha,U} = f_\alpha/\ell_P^{\deg f_\alpha}$ and find $n_p(\mathcal{D}_\alpha, P) = \mathrm{val}_p(f_\alpha(a_0, \ldots, a_n))$ for every prime $p \notin S$. It follows that $P \in X(\mathbb{Q}) = \mathcal{X}(\mathbb{Z}_S)$ is a Campana point precisely when, for all $\alpha \in \mathcal{A}$, we have that $f_\alpha(a_0, \ldots, a_n)$ is m_α-full outside S.

4 Campana Points on Diagonal Hypersurfaces

In this section we set up the problem of counting Campana points of bounded height on diagonal hypersurfaces in $\mathbb{P}^n_{\mathbb{Q}}$ with boundary divisor induced by the coordinate hyperplanes. We do this in two settings: first by constructing a proper integral model over $\mathbb{Z}$ and then by constructing a good integral model over $\mathbb{Z}_S$ for a suitable finite set of places S.

4.1 A Proper Model

Let $F = \mathbb{Q}$ and let $k \in \mathbb{Z}_{\geq 1}$. Let $X \subset \mathbb{P}^n_{\mathbb{Q}}$ be the diagonal hypersurface given by the equation

$$\sum_{i=0}^{n} c_i x_i^k = 0,$$

where $c_0, \ldots, c_n \in \mathbb{Z}_{\neq 0}$, and $\gcd(c_0, \ldots, c_n) = 1$. We consider the effective $\mathbb{Q}$-divisor on X given by

$$D = \sum_{i=0}^{n} \epsilon_i D_i,$$

where $D_i = \{x_i = 0\} \cap X$ and ϵ_i belongs to the set $\mathcal{M}$ for all $i \in \{0, \ldots, n\}$. Then (X, D) is a Campana orbifold over $\mathbb{Q}$. Write $\epsilon_i = 1 - \frac{1}{m_i}$ for all $i \in \{0, \ldots, n\}$.

Definition 5 We denote by $(\mathcal{X}_1, \mathcal{D}_1)$ the proper integral model of (X, D) given by the same equations for X and D over $\mathbb{Z}_S$, where $S = \{\infty\}$.

Lemma 1 *Let $B > 0$ be a real number. With the model $(\mathcal{X}_1, \mathcal{D}_1)$ and the height induced by the Weil height on $\mathbb{P}^n(\mathbb{Q})$, the set of Campana $\mathbb{Z}$-points on $X \cap \left(\mathbb{P}^n \setminus \bigcup_{i=0}^n \{x_i = 0\}\right)$ of height bounded by B is given by*

$$N(X, D, B) = \left\{ (x_0 : \cdots : x_n) \in \mathbb{P}^n(\mathbb{Q}) \;\middle|\; \begin{array}{c} x_0, \ldots, x_n \in \mathbb{Z}_{\neq 0}, \\ \gcd(x_0, \ldots, x_n) = 1, \\ x_i \text{ is } m_i\text{-full } \forall i \in \{0, \ldots, n\}, \\ |x_0|, \ldots, |x_n| \leq B, \\ c_0 x_0^k + \cdots + c_n x_n^k = 0 \end{array} \right\}. \tag{7}$$

Proof Take $a_0, \ldots, a_n \in \mathbb{Z}_{\neq 0}$ with $\gcd(a_0, \ldots, a_n) = 1$, such that the point $P = (a_0 : \cdots : a_n)$ is contained in $X \cap \left(\mathbb{P}^n \setminus \bigcup_{i=0}^n \{x_i = 0\}\right)$ and of height at most B. The only non-obvious condition to check in order to prove the lemma is that P satisfies Definition 4 if and only if the third condition in $N(X, D, B)$ holds. Fix i in $\{0, \ldots, n\}$. As in Example 1, let ℓ_P be a linear form in $\mathbb{Z}[x_0, \ldots, x_n]$ with $\ell_P(a_0, \ldots, a_n) = 1$. Then the divisor $\mathcal{D}_i$ is Cartier on the open set given by $U = \mathbb{P}^n \setminus \{\ell_P = 0\}$, given by the rational function x_i/ℓ_P. We find $n_p(\mathcal{D}_i, P) = \mathrm{val}_p(a_i)$, and by definition, P is a Campana point if and only if $\mathrm{val}_p(a_i) \in \mathbb{Z}_{\geq m_i} \cup \{0\}$, which is equivalent to a_i being m_i-full. □

4.2 Bad Primes and a Regular Model

The model $(\mathcal{X}_1, \mathcal{D}_1)$ over $\mathbb{Z}$ is not a good integral model of (X, D) in general, since $\mathcal{X}_1$ is not regular as a scheme over $\mathbb{Z}$. The following lemma gives sufficient conditions on a finite set S of primes for $(\mathcal{X}_1, \mathcal{D}_1)$ to be *smooth* over $\mathbb{Z}_S$.

Lemma 2 *For S a finite set of places of $\mathbb{Q}$ containing ∞ and all primes dividing k and all c_i, the model $(\mathcal{X}_2, \mathcal{D}_2)$ of (X, D) over $\mathbb{Z}_S$ defined by the same equations as X and D over $\mathbb{Z}_S$ is smooth.*

Proof Let $\mathcal{X} \subset \mathbb{P}^n_{\mathbb{Z}}$ be the projective scheme over $\operatorname{Spec} \mathbb{Z}$ defined by the equation

$$c_0 x_0^k + \cdots + c_n x_n^k = 0.$$

Notice that the structure morphism $s : \mathcal{X} \to \operatorname{Spec} \mathbb{Z}$ is flat by [19, Proposition 4.3.9], as $\mathcal{X}$ is integral because $\gcd(c_0, \ldots, c_n) = 1$, $\operatorname{Spec} \mathbb{Z}$ is a Dedekind domain, and s is non-constant. Let S be a finite set of places of $\mathbb{Q}$ containing ∞ and all the primes p which divide $k \cdot \prod_{i=0}^n c_i$. Consider $\mathcal{X}_S = \mathcal{X} \times_{\operatorname{Spec} \mathbb{Z}} \operatorname{Spec}(\mathbb{Z}_S)$. Since fiber products preserve flatness, and since s is flat, the structure morphism $s_S : \mathcal{X}_S \to \operatorname{Spec}(\mathbb{Z}_S)$ is flat as well. Since s_S is flat, by [16, Theorem 10.2] it suffices to show that the geometric fibers of s_S are regular. The geometric generic fiber $\mathcal{X}_{\overline{\mathbb{Q}}}$ is regular by the Jacobian criterion [16, Exercise I.5.8], as $c_0, \ldots, c_n \neq 0$, and since any Noetherian scheme is regular if and only if it is regular at its closed points [19, Chapter 4,

Corollary 2.17]. For every prime number $p \notin S$, the geometric fiber $\mathcal{X}_{\overline{\mathbb{F}}_p}$ over $(p) \in \operatorname{Spec} \mathbb{Z}$ is regular by the Jacobian criterion [16, Exercise I.5.8], as $p \nmid k \cdot \prod_{i=0}^{n} c_i$, and since any Noetherian scheme is regular if and only if it is regular at its closed points [19, Chapter 4, Corollary 2.17]. □

From Lemma 2 it follows that we can always obtain a good integral model for (X, D) by setting

$$S = \left\{ p \text{ prime} : p \mid k \cdot \prod_{i=0}^{n} c_i \right\} \cup \{\infty\},$$

and defining $(\mathcal{X}_2, \mathcal{D}_2)$ over $\mathbb{Z}_S$ by the same equations for X and D_i over $\mathbb{Z}_S$. The set of Campana points on X with respect to the model $\mathcal{X}_2$ is then similar to the set in Lemma 1, but requiring the x_i-coordinate of the point to be m_i-full outside S instead of just m_i-full. This leads to a larger set of Campana points than the one in Lemma 1.

It is not obvious that in order to obtain a regular model we need to take the set S as defined above (or larger); for certain choices of X we might get regular models using smaller S as well. However, we remark that primes dividing one or more of the coefficients c_i can be potentially problematic and induce non-regular points on $\mathcal{X}_1$, as the following example shows.

Example 2 Let $k \geq 2$ be an integer. Let p be a prime not dividing k. Let $\mathcal{X}$ be defined by

$$p^k x_0^k - x_1^k + x_2^k = 0 \subset \mathbb{P}^2_{\mathbb{Z}};$$

it is clear that, e.g., the point $(1 : p : 0)$ is on $\mathcal{X}$. Notice that the Jacobian matrix over $\mathbb{F}_p$ vanishes at the point $(1 : 0 : 0)$, so $\mathcal{X}$ is not smooth in the fiber over $\mathbb{F}_p$. We now show that $\mathcal{X}$ is not regular. We work in the affine patch of $\mathbb{P}^2_{\mathbb{Z}}$ given by $\{x_0 \neq 0\}$, and we get the affine equation $p^k - y_1^k + y_2^k = 0$. Consider the maximal ideal

$$\mathfrak{m} = (p, y_1, y_2) \in \operatorname{Spec}\left(\frac{\mathbb{Z}[y_1, y_2]}{(p^k - y_1^k + y_2^k)} \right),$$

which contains, for example, the prime ideal corresponding to the point $(y_1, y_2) = (p, 0)$. Notice that $p^k - y_1^k + y_2^k$ is contained in $\mathfrak{m}$. We have that $\operatorname{Spec}\left(\frac{\mathbb{Z}[y_1, y_2]}{(p^k - y_1^k + y_2^k)}\right)$ is regular at a point corresponding to $\mathfrak{m}/(p^k - y_1^k + y_2^k)$ if and only if $p^k - y_1^k + y_2^k \notin \mathfrak{m}^2$. But, since $k \geq 2$, it is clear that $p^k - y_1^k + y_2^k \in \mathfrak{m}^2$. Hence, $\operatorname{Spec}\left(\frac{\mathbb{Z}[y_1, y_2]}{(p^k - y_1^k + y_2^k)}\right)$ is not regular at $\mathfrak{m}$.

The primes dividing the exponent k can also be potentially problematic, as the following example shows.

Example 3 Let $k \geq 2$ be an integer. Let p be a prime dividing k and write $k = p\lambda$. We consider $\mathcal{X}$ defined by

$$x_0^k - x_1^k + x_2^k = 0 \subset \mathbb{P}^2_{\mathbb{Z}},$$

with the obvious point $(x_0 : x_1 : x_2) = (1 : 1 : 0)$ on it. First of all, we notice that the Jacobian matrix over $\mathbb{F}_p$ vanishes identically at every point. We work in the affine patch of $\mathbb{P}^2_{\mathbb{Z}}$ given by $\{x_0 \neq 0\}$, and we get the affine equation $1 - y_1^k + y_2^k = 0$. Consider the maximal ideal

$$\mathfrak{m} = (p, y_1 - 1, y_2) \in \operatorname{Spec}\left(\frac{\mathbb{Z}[y_1, y_2]}{(1 - y_1^k + y_2^k)}\right),$$

which contains, for example, the prime ideal corresponding to the point (y_1, y_2) given by $(1, 0)$.

Notice that $1 - y_1^k + y_2^k$ is contained in $\mathfrak{m}$: indeed, $1 - y_1^k + y_2^k = (1 - y_1)(1 + y_1 + \cdots + y_1^{k-1}) + y_2^k$. As in the previous example, we have that $\operatorname{Spec}\left(\frac{\mathbb{Z}[y_1,y_2]}{(1-y_1^k+y_2^k)}\right)$ is regular at a point corresponding to $\mathfrak{m}/(1 - y_1^k + y_2^k)$ if and only if $1 - y_1^k + y_2^k \notin \mathfrak{m}^2$. We now show that $1 - y_1^k + y_2^k$ is in $\mathfrak{m}^2$. Since $y_2^k \in \mathfrak{m}^2$ (as $k \geq 2$), it suffices to show that $1 - y_1^k \in \mathfrak{m}^2$.

If p is odd, we notice that

$$(y_1^\lambda - 1)^p = y_1^k - 1 + p(-(y_1^\lambda)^{p-1} + y_1^\lambda) + O(p^2)$$

and thus that

$$y_1^k - 1 = (y_1^\lambda - 1)^p + p y_1^\lambda((y_1^\lambda)^{p-2} - 1) + O(p^2).$$

But $(y_1 - 1)$ is a factor of $(y_1^\lambda - 1)$, and thus $(y_1 - 1)^2$ is a factor of $(y_1^\lambda - 1)^p$. Moreover, since $\lambda(p-2) > 0$, we have that $(y_1 - 1)$ is a factor of $((y_1^\lambda)^{p-2} - 1)$. Hence, it follows that $1 - y_1^k \in \mathfrak{m}^2$.

If $p = 2$, then completing the square yields

$$(y_1^\lambda)^2 - 1 = (y_1^\lambda - 1)^2 + 2(y_1^\lambda - 1).$$

But $(y_1 - 1)$ is a factor of $(y_1^\lambda - 1)$. Hence, it follows that $1 - y_1^k \in \mathfrak{m}^2$.

In both cases, we conclude that $1 - y_1^k + y_2^k$ is contained in $\mathfrak{m}^2$ and thus that $\operatorname{Spec}\left(\frac{\mathbb{Z}[y_1,y_2]}{(1-y_1^k+y_2^k)}\right)$ is not regular at $\mathfrak{m}$.

4.3 The Associated Counting Problem

We now set up the counting problem for Theorem 1 using the proper integral model introduced in Sect. 4.1. The counting problem using the regular integral model in Sect. 4.2 will be carried out in a forthcoming paper.

As in the statement of Theorem 1 and in Sect. 4.1, we fix non-zero coprime integers $c_0, \dots, c_n$, and integers $k, m_0, \dots, m_n \geq 2$. Up to reordering the indices, we can assume that $m_0 \leq \dots \leq m_n$. We use freely the notation introduced in Sect. 1.2. We consider the counting function

$$N(B) = \#\left\{ \boldsymbol{x} \in \mathbb{Z}_{\neq 0}^{n+1} \;\middle|\; \begin{array}{c} \gcd(x_0, \dots, x_n) = 1,\ |\boldsymbol{x}| \leq B, \\ x_i \text{ is } m_i\text{-full } \forall i \in \{0, \dots, n\},\ \sum_{i=0}^{n} c_i x_i^k = 0 \end{array} \right\}$$

analogous to [8, (1.1)]. Then

$$\#N(X, D, B) = \frac{1}{2} N(B), \tag{8}$$

as every point in $N(X, D, B)$ has exactly two representatives $(x_0, \dots, x_n) \in \mathbb{Z}^{n+1}$ satisfying the conditions in (7).

In the rest of this section we rephrase the counting problem in order to apply the circle method. We follow the strategy of [8, §3].

Recall that for each $m \in \mathbb{N}$, an m-full integer $x \neq 0$ has a unique representation

$$x = \pm u^m \prod_{r=1}^{m-1} v_r^{m+r},$$

with $u, v_1, \dots, v_{m-1} \in \mathbb{N}$ such that $v_1, \dots, v_{m-1}$ are squarefree and pairwise coprime. Thus, one can rewrite $N(B)$ as the cardinality of the set of tuples $\boldsymbol{x} \in \mathbb{Z}_{\neq 0}^{n+1}$ satisfying

$$\gcd(x_0, \dots, x_n) = 1, \quad |\boldsymbol{x}| \leq B, \tag{9}$$

$$x_i = \pm u_i^{m_i} \prod_{r=1}^{m_i - 1} v_{i,r}^{m_i + r} \quad \forall i \in \{0, \dots, n\}, \tag{10}$$

$$\mu^2(v_{i,r}) = 1, \ \ \gcd(v_{i,r}, v_{i,\tilde{r}}) = 1 \ \ \forall i \in \{0, \dots, n\}, \forall r, \tilde{r} \in \{1, \dots, m_i - 1\}, r \neq \tilde{r}, \tag{11}$$

$$\sum_{i=0}^{n} c_i x_i^k = 0. \tag{12}$$

Put $\Lambda = \sum_{i=0}^{n}(m_i - 1)$. For integer vectors $\boldsymbol{d} = (d_0, \dots, d_n) \in \mathbb{Z}_{\neq 0}^{n+1}$ as well as $\boldsymbol{s} = (s_0, \dots, s_n) \in \mathbb{N}^{n+1}$ and $\boldsymbol{t} = (t_{i,r})_{0 \le i \le n, 1 \le r \le m_i - 1} \in \mathbb{N}^{\Lambda}$, let

$$N_{\boldsymbol{d}}(B, \boldsymbol{s}, \boldsymbol{t}) = \left\{ \boldsymbol{x} \in (\mathbb{N} \cap [1, B])^{n+1} \,\middle|\, \begin{array}{c} d_0 x_0^k + \cdots + d_n x_n^k = 0,\ (10),\ (11), \\ s_i | u_i,\ t_{i,r} | v_{i,r}\ \forall i \in \{0, \dots, n\}, \forall r \in \{1, \dots, m_i - 1\} \end{array} \right\},$$

and write $N_{\boldsymbol{d}}^*(B, \boldsymbol{s}, \boldsymbol{t}) = N_{\boldsymbol{d}}(B, \boldsymbol{s}, \boldsymbol{t}) \cap \mathbb{Z}_{\text{prim}}^{n+1}$. The sets $N_{\boldsymbol{d}}^*(B, \boldsymbol{s}, \boldsymbol{t})$ can be used to rewrite $N(B)$ again. It is easy to see that

$$N(B) = \begin{cases} \displaystyle\sum_{\boldsymbol{\varepsilon} \in \{\pm 1\}^{n+1}} \#N_{\boldsymbol{\varepsilon c}}^*(B, \mathbf{1}, \mathbf{1}) & \text{if } k \text{ is odd,} \\ 2^{n+1} \#N_{\boldsymbol{c}}^*(B, \mathbf{1}, \mathbf{1}) & \text{if } k \text{ is even,} \end{cases} \tag{13}$$

where we wrote $\boldsymbol{\varepsilon c} = (\varepsilon_0 c_0, \dots, \varepsilon_n c_n)$. The counting function $N_{\boldsymbol{d}}^*(B, \mathbf{1}, \mathbf{1})$ can be expressed in terms of $N_{\boldsymbol{d}}(B, \boldsymbol{s}, \boldsymbol{t})$ by means of the inclusion-exclusion principle. Indeed, it was shown in [8, Section 3] that there exists a function $\varpi : \mathbb{N}^{n+1} \times \mathbb{N}^{\Lambda} \to \mathbb{Z}$ such that

$$\#N_{\boldsymbol{d}}^*(B, \mathbf{1}, \mathbf{1}) = \sum_{\boldsymbol{s} \in \mathbb{N}^{n+1}} \sum_{\boldsymbol{t} \in \mathbb{N}^{\Lambda}} \varpi(\boldsymbol{s}, \boldsymbol{t}) \#N_{\boldsymbol{d}}(B, \boldsymbol{s}, \boldsymbol{t}). \tag{14}$$

Moreover, they show that the function ϖ has the following properties.

Lemma 3 (Lemma 3.2 in [8]) *Let* $(\boldsymbol{s}, \boldsymbol{t}) \neq (\mathbf{1}, \mathbf{1})$. *Then the following are true.*

1. $\varpi(\boldsymbol{s}, \boldsymbol{t}) = 0$ *if any one of the coordinates of* $\boldsymbol{s}$ *or* $\boldsymbol{t}$ *is divisible by* p^2.
2. $\varpi(\boldsymbol{s}, \boldsymbol{t}) = 0$ *if one of the coordinates of* $\boldsymbol{s}$ *or* $\boldsymbol{t}$ *is divisible by* p, *but there exists* $0 \le i \le n$ *with* $p \nmid s_i t_{i,1} \cdots t_{i,m_i-1}$.

Moreover, $\varpi(\boldsymbol{s}, \boldsymbol{t}) \ll (|\boldsymbol{s}||\boldsymbol{t}|)^{\epsilon}$ *for all* $\boldsymbol{s} \in \mathbb{N}^{n+1}$, $\boldsymbol{t} \in \mathbb{N}^{\Lambda}$ *and* $\epsilon > 0$, *where the implicit constant can depend on* ϵ.

For every $\boldsymbol{x} \in N_{\boldsymbol{d}}(B, \boldsymbol{s}, \boldsymbol{t})$, we write $u_i = s_i \tilde{u}_i$ and $v_{i,r} = t_{i,r} \tilde{v}_{i,r}$ for all $i \in \{0, \dots, n\}$ and $r \in \{1, \dots, m_i - 1\}$. Let

$$\gamma_i = s_i^{km_i} \prod_{r=1}^{m_i - 1} t_{i,r}^{k(m_i + r)} \tilde{v}_{i,r}^{k(m_i + r)} \qquad \forall i \in \{0, \dots, n\}. \tag{15}$$

Then

$$x_i^k = \gamma_i \tilde{u}_i^{km_i} \qquad \forall i \in \{0, \dots, n\}.$$

For a vector $\boldsymbol{\zeta} \in \mathbb{N}^{n+1}$, let

$$M_{\boldsymbol{d},\boldsymbol{\zeta}}(B^k) = \#\left\{\tilde{\boldsymbol{u}} \in \mathbb{N}^{n+1} \,\middle|\, \zeta_i \tilde{u}_i^{km_i} \leq B^k \ \forall i \in \{0,\dots,n\}, \sum_{i=0}^{n} d_i \zeta_i \tilde{u}_i^{km_i} = 0\right\}.$$

For any pair $(\boldsymbol{s},\boldsymbol{t}) \in \mathbb{N}^{n+1} \times \mathbb{N}^{\Lambda}$ and any $R \in \mathbb{R}_{>0} \cup \{\infty\}$ define the set

$$V_R(\boldsymbol{s},\boldsymbol{t}) = \left\{ \tilde{\boldsymbol{v}} \in \mathbb{N}^{\Lambda} \,\middle|\, \begin{array}{c} s_i^{km_i} \prod_{r=1}^{m_i-1} t_{i,r}^{k(m_i+r)} \tilde{v}_{i,r}^{k(m_i+r)} \leq R^k \quad \forall i \in \{0,\dots,n\} \\ \mu^2(\tilde{v}_{i,r} t_{i,r}) = 1 \quad \forall i \in \{0,\dots,n\}, r \in \{1,\dots,m_i-1\} \\ \gcd(\tilde{v}_{i,r} t_{i,r}, \tilde{v}_{i,r'} t_{i,r'}) = 1 \quad \forall i,r,r' \text{ with } r \neq r' \end{array} \right\}.$$

Thus

$$\#N_{\boldsymbol{d}}(B,\boldsymbol{s},\boldsymbol{t}) = \sum_{\tilde{\boldsymbol{v}} \in V_B(\boldsymbol{s},\boldsymbol{t})} M_{\boldsymbol{d},\boldsymbol{\gamma}}(B^k),$$

where $\boldsymbol{\gamma}$ is given by (15).

For $R \in \mathbb{R}_{>0} \cup \{\infty\}$ we define further

$$\mathcal{T}_R = \left\{ \boldsymbol{s} \in \mathbb{N}^{n+1}, \boldsymbol{t} \in \mathbb{N}^{\Lambda} \,\middle|\, \begin{array}{c} s_i^{m_i} \prod_{r=1}^{m_i-1} t_{i,r}^{m_i+r} \leq R \\ \mu^2(s_i) = \mu^2(t_{i,r}) = 1, \ \gcd(t_{i,r}, t_{i,r'}) = 1 \\ p \mid s_j t_{j,1} \cdots t_{j,m_j-1} \Rightarrow p \mid s_i t_{i,1} \cdots t_{i,m_i-1} \\ \forall i,j \in \{0,\dots,n\}, \ r,r' \in \{1,\dots,m_i-1\} \end{array} \right\}.$$

We can thus rewrite the relation (14) in the shape

$$\#N_{\boldsymbol{d}}^*(B,\mathbf{1},\mathbf{1}) = \sum_{(\boldsymbol{s},\boldsymbol{t}) \in \mathcal{T}_B} \varpi(\boldsymbol{s},\boldsymbol{t}) \sum_{\tilde{\boldsymbol{v}} \in V_B(\boldsymbol{s},\boldsymbol{t})} M_{\boldsymbol{d},\boldsymbol{\gamma}}(B^k) \tag{16}$$

due to the properties of the function ϖ in Lemma 3.

The functions $M_{\boldsymbol{d},\boldsymbol{\gamma}}(B^k)$ can be estimated via [8, Theorem 2.7] whenever $\sum_{i=0}^{n-1} \frac{1}{km_i(km_i+1)} \geq 1$. In the next section we improve the circle method result of Browning and Yamagishi to a version that applies under the weaker assumptions of Theorem 1.

5 Application of the Circle Method

Here we prove a sharper version of [8, Theorem 2.7] with parameters $N = 0$, $H = 1$, $\boldsymbol{h} = 0$. Fix integers $2 \leq \tilde{m}_0 \leq \cdots \leq \tilde{m}_n$. For $\boldsymbol{d} \in \mathbb{Z}_{\neq 0}^{n+1}$, $\boldsymbol{\zeta} \in \mathbb{N}^{n+1}$ and $\tilde{B} > 0$, put

$$\tilde{M}_{\boldsymbol{d},\boldsymbol{\zeta}}(\tilde{B}) = \#\left\{\boldsymbol{u} \in \mathbb{N}^{n+1} \,\middle|\, \zeta_i u_i^{\tilde{m}_i} \le \tilde{B}\ \forall i \in \{0,\dots,n\},\ \sum_{i=0}^{n} d_i \zeta_i u_i^{\tilde{m}_i} = 0\right\}. \tag{17}$$

Thus, if $\tilde{m}_i = km_i$ for $0 \le i \le n$ we have $M_{\boldsymbol{d},\boldsymbol{\zeta}}(B^k) = \tilde{M}_{\boldsymbol{d},\boldsymbol{\zeta}}(B^k)$.

Our next immediate goal is to establish an asymptotic formula for $\tilde{M}_{\boldsymbol{d},\boldsymbol{\gamma}}(\tilde{B})$. Here, we use the Hardy–Littlewood circle method much in the way of [8, Section 2], but we aim for a slightly sharper bound than the one presented in that paper.

For any integer $m \ge 2$, let $s_0(m)$ be a positive integer such that for all $s \ge s_0(m)$ and $\epsilon > 0$ one has

$$\int_0^1 \left| \sum_{1 \le x \le X} e(\alpha x^m) \right|^{2s} \mathrm{d}\alpha \ll X^{2s-m+\epsilon}.$$

Similarly, let $\sigma(m)$ be a positive integer with the property that for any $\beta \in [0,1)$ and any $q \in \mathbb{N}$ with $\|q\beta\| \le q^{-1}$ one has the bound

$$\left| \sum_{1 \le x \le X} e(\beta x^m) \right| \ll X^{1+\epsilon}(X^{-1} + q^{-1} + qX^{-m})^{\sigma(m)}$$

for all $\epsilon > 0$. It follows from [5, Theorem 2.1] (see also [34, Theorem 14.8] for the bound on $s_0(m)$) that the choices

$$s_0(m) = \min\{2^{m-1}, \tfrac{1}{2}m(m-1) + \lfloor\sqrt{2m+2}\rfloor\} \quad \text{and} \quad \sigma(m)^{-1} = 2s_0(m)$$

are admissible. (We remark here that even $\sigma(m)^{-1} = 2s_0(m-1)$ is admissible. Since s_0 is strictly increasing, choosing $\sigma(m)$ as above constitutes a weakening.)

Suppose for the remainder of the section that the parameters $\boldsymbol{d} \in \mathbb{Z}_{\neq 0}^{n+1}$, $\boldsymbol{\zeta} \in \mathbb{N}^{n+1}$ and $\tilde{B} > 0$ are fixed. For every $i \in \{0,\dots,n\}$, let $\tilde{B}_i = (\tilde{B}/\zeta_i)^{\frac{1}{\tilde{m}_i}}$ and

$$S_i(\alpha) = \sum_{1 \le u \le \tilde{B}_i} e(\alpha d_i \zeta_i u^{\tilde{m}_i}),$$

so that

$$\tilde{M}_{\boldsymbol{d},\boldsymbol{\zeta}}(\tilde{B}) = \int_0^1 \prod_{i=0}^{n} S_i(\alpha)\mathrm{d}\alpha.$$

For fixed $\delta > 0$, we recall the set of minor arcs from [8, §2], which is given by

$$\mathfrak{m} = [0,1] \setminus \bigcup_{\substack{0 \le a \le q \le \tilde{B}^{\delta} \\ \gcd(a,q)=1}} \{\alpha \in [0,1) : |\alpha - a/q| < \tilde{B}^{\delta-1}\}.$$

Lemma 4 *For $i \in \{0, \dots, n\}$, $0 < \delta < \frac{1}{(2n+5)\tilde{m}_n(\tilde{m}_n+1)}$ and $\epsilon > 0$, we have*

$$\sup_{\alpha \in \mathfrak{m}} |S_i(\alpha)| \ll \tilde{B}^{1/\tilde{m}_i - \delta\sigma(\tilde{m}_i)+\epsilon} \zeta_i^{-1/\tilde{m}_i + \sigma(\tilde{m}_i)}.$$

Proof The proof is identical to that of [8, Lemma 2.5] (see in particular the last display on p.1083), with the only difference in the precise meaning of the quantity $\sigma(\tilde{m}_i)$. □

Let

$$\tilde{\Theta} = \sum_{i=0}^{n} \frac{1}{2s_0(\tilde{m}_i)} - 1 \quad \text{and} \quad \tilde{\Gamma} = \sum_{i=0}^{n} \frac{1}{\tilde{m}_i} - 1. \tag{18}$$

Lemma 5 *Assume that $\tilde{\Theta} > 0$. Let $0 < \delta < \frac{1}{(2n+5)\tilde{m}_n(\tilde{m}_n+1)}$ and $\epsilon > 0$. Then*

$$\int_{\mathfrak{m}} \left| \prod_{i=0}^{n} S_i(\alpha) \right| \mathrm{d}\alpha \ll \tilde{B}^{\tilde{\Gamma} - \delta\tilde{\Theta}+\epsilon} \prod_{i=0}^{n} \zeta_i^{1/(2s_0(\tilde{m}_i))-1/\tilde{m}_i}.$$

Proof From the condition $\tilde{\Theta} > 0$ it follows that we can find $\beta_0, \dots, \beta_n \in (0,1)$ so that

$$\sum_{i=0}^{n} \frac{\beta_i}{2s_0(\tilde{m}_i)} = 1. \tag{19}$$

Take $\ell_i = \frac{2s_0(\tilde{m}_i)}{\beta_i}$ for all $1 \le i \le n$. It then follows from Hölder's inequality, the definition of $s_0(\tilde{m}_i)$ and Lemma 4 above that

$$\begin{aligned}
\int_{\mathfrak{m}} \left| \prod_{i=0}^{n} S_i(\alpha) \right| \mathrm{d}\alpha &\ll \left(\prod_{i=0}^{n} \sup_{\alpha \in \mathfrak{m}} |S_i(\alpha)|^{1-\beta_i} \right) \int_0^1 \prod_{i=0}^{n} |S_i(\alpha)|^{\beta_i} \mathrm{d}\alpha \\
&\ll \prod_{i=0}^{n} \left(\sup_{\alpha \in \mathfrak{m}} |S_i(\alpha)|^{1-\beta_i} \left(\int_0^1 |S_i(\alpha)|^{\beta_i \ell_i} \mathrm{d}\alpha \right)^{1/\ell_i} \right) \\
&\ll \tilde{B}^{\epsilon} \prod_{i=0}^{n} \left(\left(\tilde{B}^{1/\tilde{m}_i - \delta\sigma(\tilde{m}_i)} \zeta_i^{-1/\tilde{m}_i + \sigma(\tilde{m}_i)} \right)^{1-\beta_i} \left(\tilde{B}/\zeta_i \right)^{\frac{\beta_i \ell_i - \tilde{m}_i}{\ell_i \tilde{m}_i}} \right)
\end{aligned}$$

$$\ll \tilde{B}^{\omega_1+\epsilon} \prod_{i=0}^{n} \zeta_i^{\omega_{2,i}},$$

where

$$\omega_1 = \sum_{i=0}^{n} \left((1/\tilde{m}_i - \delta\sigma(\tilde{m}_i))(1-\beta_i) + \frac{\beta_i \ell_i - \tilde{m}_i}{\ell_i \tilde{m}_i} \right)$$

and

$$\omega_{2,i} = (-1/\tilde{m}_i + \sigma(\tilde{m}_i))(1-\beta_i) - \frac{\beta_i \ell_i - \tilde{m}_i}{\ell_i \tilde{m}_i} \qquad \forall i \in \{0, \dots, n\}.$$

After inserting our choice for ℓ_i, a modicum of computation shows that

$$\omega_{2,i} = -\frac{1}{\tilde{m}_i} + \sigma(\tilde{m}_i) - \beta_i \left(\sigma(\tilde{m}_i) - \frac{1}{2s_0(\tilde{m}_i)} \right) = -\frac{1}{\tilde{m}_i} + \frac{1}{2s_0(\tilde{m}_i)},$$

where in the last step we took advantage of our choice $(2s_0(\tilde{m}_i))^{-1} = \sigma(\tilde{m}_i)$. Similarly, we have

$$\omega_1 = \sum_{i=0}^{n} \left(\frac{1}{\tilde{m}_i} - \frac{\beta_i}{2s_0(\tilde{m}_i)} - \delta\sigma(\tilde{m}_i)(1-\beta_i) \right) = \tilde{\Gamma} - \delta\tilde{\Theta},$$

where the last equality follows from (19). □

Combining Lemma 5 with [8, Lemma 2.2] yields the desired asymptotic formula. Define

$$\begin{aligned} \mathfrak{S}_{\boldsymbol{d},\zeta} &= \sum_{q=1}^{\infty} \frac{1}{q^{n+1}} \sum_{\substack{a \pmod q \\ (a,q)=1}} \prod_{i=0}^{n} \sum_{r=1}^{q} e(ad_i \zeta_i r^{m_i}/q) \\ \mathfrak{J}_{\boldsymbol{d}} &= \int_{-\infty}^{\infty} \prod_{i=0}^{n} \left(\int_0^1 e(\lambda d_i \xi^{m_i}) d\xi \right) d\lambda, \end{aligned} \tag{20}$$

noting that these coincide with the definitions of $\mathfrak{S}_{\boldsymbol{d};\zeta}(\boldsymbol{h}, H; N)$ and $\mathfrak{J}_{\boldsymbol{d}}$ of [8, Section 2.1] with the choices $\boldsymbol{h} = \boldsymbol{0}$, $H = 1$ and $N = 0$. Thus, we obtain the following.

Theorem 2 *Let* $2 \le \tilde{m}_0 \le \cdots \le \tilde{m}_n$ *be positive integers such that*

$$\sum_{i=0}^{n} \frac{1}{2s_0(\tilde{m}_i)} > 1 \qquad \textit{and} \qquad \sum_{i=0}^{n} \frac{1}{\tilde{m}_i} > 3.$$

Let $\boldsymbol{d} \in \mathbb{Z}_{\neq 0}^{n+1}$, $\boldsymbol{\zeta} \in \mathbb{N}^{n+1}$, $\tilde{B} > 0$. *Let* $0 < \delta < \frac{1}{(2n+5)\tilde{m}_n(\tilde{m}_n+1)}$ *and* $\epsilon > 0$. *With the notation introduced in* (17) *and* (18), *we have*

$$\tilde{M}_{\boldsymbol{d},\boldsymbol{\zeta}}(\tilde{B}) = \frac{\mathfrak{S}_{\boldsymbol{d},\boldsymbol{\zeta}}\mathfrak{J}_{\boldsymbol{d}}}{\prod_{i=0}^n \zeta_i^{\frac{1}{\tilde{m}_i}}}\tilde{B}^{\tilde{\Gamma}} + O\left(E_1(\boldsymbol{\zeta}) + \frac{\tilde{B}^{\tilde{\Gamma}-\delta}E_2(\boldsymbol{\zeta})}{\prod_{i=0}^n \zeta_i^{\frac{1}{\tilde{m}_i}}} + \tilde{B}^{\tilde{\Gamma}-\delta\tilde{\Theta}+\epsilon}\prod_{i=0}^n \zeta_i^{-\frac{1}{\tilde{m}_i}+\frac{1}{2s_0(\tilde{m}_i)}}\right),$$

and the error terms are given by

$$E_1(\boldsymbol{\zeta}) = \frac{\prod_{i=0}^n \tilde{B}_i}{\tilde{B}}\left(\frac{1}{\tilde{B}_0} + \cdots + \frac{1}{\tilde{B}_n}\right)\tilde{B}^{(2n+5)\delta}$$

and

$$E_2(\boldsymbol{\zeta}) = \sum_{q=1}^{\infty} q^{1-\Gamma+\epsilon}\prod_{i=0}^n \gcd(\zeta_i, q)^{\frac{1}{\tilde{m}_i}}.$$

6 Proof of the Main Theorem

In this section we conclude the proof of Theorem 1. We start with some numerical conditions that follow from the assumptions of Theorem 1.

Lemma 6 *The assumptions* $k, m_0, \ldots, m_n \geq 2$ *and* $\sum_{i=0}^n \frac{1}{2s_0(km_i)} > 1$ *imply*

$$\frac{1}{2s_0(km_i)} \leq \frac{1}{km_i} - \frac{1}{k(m_i+1)} \quad \forall i \in \{0, \ldots, n\} \quad \text{and} \quad \sum_{i=0}^n \frac{1}{km_i} > 3.$$

Proof The first inequality in the statement can be rearranged to

$$km_i(m_i+1) \leq 2s_0(km_i).$$

Now, suppose first that $km_i \geq 6$, so that $s_0(km_i) = \frac{1}{2}km_i(km_i-1) + \lfloor\sqrt{2km_i+2}\rfloor$. In that situation, the above bound becomes

$$km_i(m_i+1) \leq km_i(km_i-1) + 2\lfloor\sqrt{2km_i+2}\rfloor,$$

which can be rearranged to

$$km_i((k-1)m_i-2) + 2\lfloor\sqrt{2km_i+2}\rfloor \geq 0,$$

which is clearly satisfied for $k \geq 2$ and $m_i \geq 2$. This settles all cases in which $k \geq 3$ or $m_i \geq 3$. The remaining case $k = m_i = 2$ can be checked by hand.

Finally, the second statement follows upon observing that for $i \in \{0, \ldots, n\}$, $2s_0(km_i) \geq 3km_i$, as $km_i \geq 4$. □

Now we continue the proof of Theorem 1 from where we left off at the end of Sect. 4.3.

Proposition 1 *Let $\boldsymbol{d} \in \mathbb{Z}_{\neq 0}^{n+1}$. Fix integers $m_0, \ldots, m_n \geq 2$ and $k \geq 2$, and put $\Gamma = \sum_{i=0}^{n} \frac{1}{km_i} - 1$. There exists a real number $\eta > 0$ such that*

$$\# N_{\boldsymbol{d}}^{*}(B, \mathbf{1}, \mathbf{1}) = C_{\boldsymbol{d}} B^{k\Gamma} + O(B^{k\Gamma - \eta}), \tag{21}$$

where

$$C_{\boldsymbol{d}} = \mathfrak{J}_{\boldsymbol{d}} \sum_{(\boldsymbol{s},\boldsymbol{t}) \in \mathcal{T}_{\infty}} \varpi(\boldsymbol{s}, \boldsymbol{t}) \sum_{\tilde{\boldsymbol{v}} \in V_{\infty}(\boldsymbol{s},\boldsymbol{t})} \mathfrak{S}_{\boldsymbol{d},\boldsymbol{\gamma}} \prod_{i=0}^{n} \gamma_i^{-\frac{1}{km_i}} \tag{22}$$

with $\mathfrak{S}_{\boldsymbol{d},\boldsymbol{\gamma}}$ and $\mathfrak{J}_{\boldsymbol{d}}$ as in (20), *and $\boldsymbol{\gamma}$ defined via* (15).

Proof Our strategy is to apply Theorem 2, with $\tilde{B} = B^k$ and $\tilde{m}_i = km_i$ for $0 \leq i \leq n$, to estimate the sets $M_{\boldsymbol{d},\boldsymbol{\gamma}}(B^k)$ in (16). Note that our choice of the parameters ensures that $M_{\boldsymbol{d},\boldsymbol{\gamma}}(B^k) = \tilde{M}_{\boldsymbol{d},\boldsymbol{\gamma}}(\tilde{B})$, and that the quantity $\tilde{\Gamma}$ coincides with the Γ defined in the statement of the proposition. Thus, for every $\epsilon > 0$ and every $0 < \delta < \frac{1}{(2n+5)km_n(km_n+1)}$ we obtain

$$\begin{aligned} &\# N_{\boldsymbol{d}}^{*}(B, \mathbf{1}, \mathbf{1}) \\ &= \sum_{(\boldsymbol{s},\boldsymbol{t}) \in \mathcal{T}_B} \varpi(\boldsymbol{s}, \boldsymbol{t}) \sum_{\tilde{\boldsymbol{v}} \in V_B(\boldsymbol{s},\boldsymbol{t})} \left[\frac{\mathfrak{S}_{\boldsymbol{d},\boldsymbol{\gamma}} \mathfrak{J}_{\boldsymbol{d}}}{\prod_{i=0}^{n} \gamma_i^{\frac{1}{km_i}}} B^{k\Gamma} \right. \\ &\quad \left. + O\left(E_1(\boldsymbol{\gamma}) + \frac{B^{k\Gamma - k\delta} E_2(\boldsymbol{\gamma})}{\prod_{i=0}^{n} \gamma_i^{\frac{1}{km_i}}} \right) + O\left(B^{k\Gamma - k\delta\Theta + \epsilon} \prod_{i=0}^{n} \gamma_i^{-\frac{1}{km_i} + \frac{1}{2s_0(km_i)}} \right) \right], \end{aligned} \tag{23}$$

where $\Theta = \sum_{i=0}^{n} \frac{1}{2s_0(km_i)} - 1$. We observe that by [8, p. 1093], the leading constant is

$$\sum_{(\boldsymbol{s},\boldsymbol{t}) \in \mathcal{T}_B} \varpi(\boldsymbol{s}, \boldsymbol{t}) \sum_{\tilde{\boldsymbol{v}} \in V_B(\boldsymbol{s},\boldsymbol{t})} \mathfrak{S}_{\boldsymbol{d},\boldsymbol{\gamma}} \mathfrak{J}_{\boldsymbol{d}} \prod_{i=0}^{n} \gamma_i^{-\frac{1}{km_i}} = C_{\boldsymbol{d}} + O(B^{-\eta}), \tag{24}$$

where $C_{\boldsymbol{d}}$ is the expression given in (22) and η is some suitable positive number.

It thus remains to bound the error terms. Put

$$F_1(B) = B^{-k\Gamma} \sum_{(s,t)\in\mathcal{T}_B} \varpi(s,t) \sum_{\tilde{v}\in V_B(s,t)} E_1(\boldsymbol{\gamma}),$$

$$F_2(B) = \sum_{(s,t)\in\mathcal{T}_B} \varpi(s,t) \sum_{\tilde{v}\in V_B(s,t)} \frac{B^{-k\delta} E_2(\boldsymbol{\gamma})}{\gamma_i^{\frac{1}{km_i}}}$$

and

$$F_3(B) = B^{-k\delta\Theta+\epsilon} \sum_{(s,t)\in\mathcal{T}_B} \varpi(s,t) \sum_{\tilde{v}\in V_B(s,t)} \prod_{i=0}^{n} \gamma_i^{-\frac{1}{km_i}+\frac{1}{2s_0(km_i)}},$$

then the desired result will follow if we can show that $F_j(B) \ll B^{-\eta}$ for some $\eta > 0$.

We now begin with the estimation of $F_1(B)$. Upon inserting the definition of $E_1(B)$ from the statement of Theorem 2, one can show by a simple computation (see also [8, (3.6)]), using the last statement of Lemma 3, that

$$F_1(B) = B^{k(2n+5)\delta+\epsilon} \sum_{l=0}^{n} B^{-\frac{1}{m_l}} \sum_{(s,t)\in\mathcal{T}_B} \sum_{\tilde{v}\in V_B(s,t)} \prod_{\substack{i=1\\ i\neq l}}^{n} \gamma_i^{-\frac{1}{km_i}}.$$

For $0 \le i \le n$, we introduce the notation

$$w_i = \tilde{v}_{i,1}^{m_i+1} \cdots \tilde{v}_{i,m_i-1}^{2m_i-1} \quad \text{and} \quad \tau_i = s_i^{m_i} \prod_{r=1}^{m_i-1} t_{i,r}^{m_i+r},$$

so that $\gamma_i = w_i^k \tau_i^k$. In that notation we can write

$$F_1(B) = B^{k(2n+5)\delta+\epsilon} \sum_{l=0}^{n} B^{-\frac{1}{m_l}} \sum_{(s,t)\in\mathcal{T}_B} \sum_{\tilde{v}\in V_B(s,t)} \prod_{\substack{i=1\\ i\neq l}}^{n} (w_i\tau_i)^{-\frac{1}{m_i}}$$

$$\ll B^{k(2n+5)\delta+\epsilon} \sum_{l=0}^{n} B^{-\frac{1}{m_l}} \sum_{(s,t)\in\mathcal{T}_B} \left(\sum_{\tilde{v}_{l,1},\dots,\tilde{v}_{l,m_l-1}:w_l\le\frac{B}{\tau_l}} 1 \right)$$

$$\cdot \prod_{\substack{i=1\\ i\neq l}}^{n} \tau_i^{-\frac{1}{m_i}} \sum_{\tilde{v}_{i,1},\dots,\tilde{v}_{i,m_i-1}:w_i\le\frac{B}{\tau_i}} \left(\frac{1}{w_i}\right)^{\frac{1}{m_i}}.$$

Using the estimates

$$\sum_{v_1^{m+1}\cdots v_{m-1}^{2m-1}\le\frac{B}{\tau}} 1 \ll \sum_{v_2,\ldots,v_m=1}^{\infty}\left(\frac{B/\tau}{v_2^{m+2}\cdots v_{m-1}^{2m-1}}\right)^{\frac{1}{m+1}} \ll \left(\frac{B}{\tau}\right)^{\frac{1}{m+1}}$$

and

$$\sum_{v_1^{m+1}\cdots v_{m-1}^{2m-1}\le\frac{B}{\tau}} \left(\frac{1}{v_1^{m+1}\cdots v_{m-1}^{2m-1}}\right)^{\frac{1}{m}} \ll 1$$

from [8, p. 1090] within the above bound, it follows that

$$\begin{aligned} F_1(B) &\ll B^{k(2n+5)\delta+\epsilon}\sum_{l=0}^{n} B^{-\frac{1}{m_l}} \sum_{(s,t)\in\mathcal{T}_B}\left(\frac{B}{\tau_l}\right)^{\frac{1}{m_l+1}} \prod_{\substack{i=1\\ i\neq l}}^{n} \tau_i^{-\frac{1}{m_i}} \\ &\ll B^{-\frac{1}{m_n(m_n+1)}+k(2n+5)\delta+\epsilon} \sum_{(s,t)\in\mathcal{T}_B}\prod_{i=0}^{n}\tau_i^{-\frac{1}{(m_i+1)}}. \end{aligned}$$

The sum can be bounded by a slight modification of the argument in [8, (3.8)]. Write $\sigma_i = s_i \prod_{r=1}^{m_i-1} t_{i,r}$, then clearly $\sigma_i^m \le \tau_i$, and thus

$$\begin{aligned} \sum_{(s,t)\in\mathcal{T}_B}\prod_{i=0}^{n}\tau_i^{-\frac{1}{(m_i+1)}} &\le \sum_{(s,t)\in\mathcal{T}_B}\prod_{i=0}^{n}\sigma_i^{-\frac{m_i}{m_i+1}} \\ &= \prod_p\left(1+\sum_{\substack{(s,t)\in\mathcal{T}_B\\(s,t)\neq(\mathbf{1},\mathbf{1})}} p^{-\sum_{i=0}^{n}\mathrm{val}_p(\sigma_i)\frac{m_i}{m_i+1}}\right) \\ &\le \prod_p\left(1+\prod_{i=0}^{n}p^{-\frac{m_i}{m_i+1}}\sum_{\substack{(s,t)\in\mathcal{T}_B\\ \mathrm{val}_p(\sigma_i)\ge 1}} 1\right) \\ &\le \prod_p\left(1+\prod_{i=0}^{n}p^{-\frac{m_i}{m_i+1}}(2m_i-1)\right) \ll 1, \end{aligned}$$

as $\sum_{i=0}^{n}\frac{m_i}{m_i+1} > 1$ under the assumption $m_0,\ldots,m_n \ge 2$. Thus altogether we obtain the bound

$$F_1(B) \ll B^{-\frac{1}{m_n(m_n+1)}+k(2n+5)\delta+\epsilon}, \tag{25}$$

which is sufficient for our purposes, provided that δ was taken small enough.

We now turn to $F_2(B)$. As before, we rewrite the quantity under consideration by inserting the definition of $E_2(B)$ (see also [8, (3.6)]), whereupon we can use the upper bound

$$F_2(B) = B^{-k\delta+\epsilon} \sum_{(s,t)\in\mathcal{T}_B} \sum_{\tilde{v}\in V_B(s,t)} \sum_{q=1}^{\infty} q^{1-\Gamma+\varepsilon} \prod_{i=0}^{n} \frac{\gcd(\gamma_i, q)^{\frac{1}{km_i}}}{\gamma_i^{\frac{1}{km_i}}}$$

$$\leq B^{-k\delta+\epsilon} \sum_{q=1}^{\infty} q^{1-\Gamma+\varepsilon} f_1(q) f_2(q),$$

with

$$f_1(q)= \sum_{(s,t)\in\mathcal{T}_\infty} \prod_{i=0}^{n} \left(\frac{\gcd(\tau_i^k, q)}{\tau_i^k}\right)^{\frac{1}{km_i}} \text{ and } f_2(q)= \sum_{\tilde{v}\in V_B(\mathbf{1},\mathbf{1})} \prod_{i=0}^{n} \left(\frac{\gcd(w_i^k, q)}{w_i^k}\right)^{\frac{1}{km_i}}.$$

Since $\tilde{v}_{i,1}, \ldots, \tilde{v}_{i,m_i-1}$ are pairwise coprime for $\tilde{\boldsymbol{v}} \in V_B(\mathbf{1}, \mathbf{1})$, we can estimate the second expression via

$$f_2(q) \leq \prod_{i=0}^{n} \prod_{r=1}^{m_i-1} \sum_{\tilde{v}_{i,r}\leq B^{\frac{1}{m_i+r}}} \mu^2(\tilde{v}_{i,r}) \frac{\gcd(\tilde{v}_{i,r}^{km_i+kr}, q)^{\frac{1}{km_i}}}{\tilde{v}_{i,r}^{(m_i+r)/m_i}} \ll q^\epsilon$$

for all $\epsilon > 0$, where the last bound follows from [8, (3.9)].

For $f_1(q)$ we proceed as in [8, p. 1091]. Let

$$\mathcal{T} = \left\{\overline{\tau} = (\overline{\tau}_0, \ldots, \overline{\tau}_n) \in \mathbb{N}^{n+1} : \begin{array}{l} \mathrm{val}_p(\overline{\tau}_i) \in \{0, m_i, m_i+1, \ldots, 3m_i-1\} \\ p|\overline{\tau}_l \implies p|\overline{\tau}_j \quad \forall p, \forall i, j \in \{0, \ldots, n\} \end{array}\right\}.$$

For every $\overline{\tau} \in \mathcal{T}$ there is a unique pair $(s, t) \in \mathcal{T}_\infty$ such that $\overline{\tau}_i = s_i^{m_i} \prod_{r=1}^{m_i-1} t_{i,r}^{m_i+r}$ for all $i \in \{0, \ldots, n\}$. Then

$$f_1(q) \leq \sum_{\overline{\tau}\in\mathcal{T}} \prod_{i=0}^{n} \left(\frac{\gcd(\overline{\tau}_i^k, q)}{\overline{\tau}_i^k}\right)^{\frac{1}{km_i}}$$

$$\leq \prod_p \left(1 + \prod_{i=0}^{n} \sum_{m_i\leq\alpha_i\leq 3m_i-1} p^{(\min\{k\alpha_i, \mathrm{val}_p(q)\}-k\alpha_i)/km_i}\right) = O(1),$$

as the local contribution to the product is $1+O(p^{-(n+1)})$ if $p \nmid q$, and $O(1)$ if $p \mid q$. Thus

$$F_2(B) \ll B^{-k\delta+\epsilon} \sum_{q=1}^{\infty} q^{1-\Gamma+\epsilon} \ll B^{-k\delta+\epsilon} \tag{26}$$

for all $\epsilon > 0$, as $\sum_{i=0}^{n} \frac{1}{km_i} > 3$ by Lemma 6.

Lastly, we turn our attention to $F_3(B)$. By Lemma 6 we have

$$\sum_{(s,t)\in\mathcal{T}_B} \sum_{\tilde{v}\in V_B(s,t)} \prod_{i=0}^{n} \gamma_i^{-\frac{1}{km_i}+\frac{1}{2s_0(km_i)}} \leq \sum_{(s,t)\in\mathcal{T}_B} \left(\sum_{\tilde{v}\in V_B(s,t)} \prod_{i=0}^{n} w_i^{-\frac{1}{m_i+1}} \right) \prod_{i=0}^{n} \tau_i^{-\frac{1}{m_i+1}}$$

$$\ll \left(\prod_{i=0}^{n} \sum_{\tilde{v}_{i,1}^{m_i+1}\cdots\tilde{v}_{i,m_i-1}^{2m_i-1}\leq B} \left(\tilde{v}_{i,1}^{m_i+1}\cdots\tilde{v}_{i,m_i-1}^{2m_i-1}\right)^{-\frac{1}{m_i+1}} \right) \sum_{(s,t)\in\mathcal{T}_B} \prod_{i=0}^{n} \tau_i^{-\frac{1}{m_i+1}} \ll B^{\epsilon}$$

for all $\epsilon > 0$, where the last estimate follows from [8, (3.8)] and the bound $\ll \log B$ for the sum over $v_{i,r}$ in [8, p. 1092]. Thus

$$F_3(B) \ll B^{-k\delta\Theta+\epsilon} \tag{27}$$

for all $\epsilon > 0$.

The desired conclusion follows now upon combining (23) with (24) as well as the bounds (25), (26) and (27). □

Theorem 1 is now immediate upon combining Lemma 1, (8), (13) and Proposition 1. In particular, the leading constant in (2) is

$$C = \begin{cases} \frac{1}{2}\sum_{\boldsymbol{\varepsilon}\in\{\pm1\}^{n+1}} C_{\boldsymbol{\varepsilon c}} & \text{if } k \text{ is odd,} \\ 2^n C_{\boldsymbol{c}} & \text{if } k \text{ is even.} \end{cases} \tag{28}$$

Remark 6 Note that the condition that $k \geq 2$ enters in two places. On the one hand, we need $\Gamma > 3$ to control the main term in Theorem 2, and on the other hand it plays a role in the estimation of $F_3(B)$, via the application of Lemma 6. By implementing a suitable pruning argument in the treatment of the minor arcs in the proof of Theorem 2, it seems likely that the last error term in that theorem can be improved, giving rise to a corresponding quantity F_3 with better convergence properties. As for the main term, the condition $\Gamma > 3$ can presumably be relaxed by means of the techniques described in [31, Chapter 4]. Consequently, the expectation is that extending Proposition 1, and thus Theorem 1, for $k = 1$ should be a question of determination rather than any potential structural obstacles.

References

1. D. Abramovich, Birational geometry for number theorists, in *Arithmetic Geometry*, volume 8 of *Clay Math. Proc.* (Amer. Math. Soc., Providence, RI, 2009), pp. 335–373
2. D. Abramovich, A. Várilly-Alvarado, Campana points, Vojta's conjecture, and level structures on semistable abelian varieties. J. Théor. Nombres Bordeaux **30**(2), 525–532 (2018)
3. D. Brotbek, Y. Deng, Kobayashi hyperbolicity of the complements of general hypersurfaces of high degree. Geom. Funct. Anal. **29**(3), 690–750 (2019)
4. J. Bourgain, C. Demeter, L. Guth, Proof of the main conjecture in Vinogradov's mean value theorem for degrees higher than three. Ann. Math. (2) **184**(2), 633–682 (2016)
5. J. Brandes, S.T. Parsell, The Hasse principle for diagonal forms restricted to lower-degree hypersurfaces. Algebra Number Theory **15**(9), 2289–2314 (2021)
6. T. Browning, *Quantitative Arithmetic of Projective Varieties*, 1st edn. (Birkhäuser Basel, 2009)
7. T.D. Browning, K. Van Valckenborgh, Sums of three squareful numbers. Exp. Math. **21**(2), 204–211 (2012)
8. T. Browning, S. Yamagishi, Arithmetic of higher-dimensional orbifolds and a mixed Waring problem. Math. Z. **299**(1-2), 1071–1101 (2021)
9. F. Campana, Orbifolds, special varieties and classification theory. Ann. Inst. Fourier (Grenoble) **54**(3), 499–630 (2004)
10. F. Campana, Fibres multiples sur les surfaces: aspects geométriques, hyperboliques et arithmétiques. Manuscripta Math. **117**(4), 429–461 (2005)
11. F. Campana, Orbifoldes géométriques spéciales et classification biméromorphe des variétés kählériennes compactes. J. Inst. Math. Jussieu **10**(4), 809–934 (2011)
12. F. Campana, Special manifolds, arithmetic and hyperbolic aspects: a short survey, in *Rational points, rational curves, and entire holomorphic curves on projective varieties*, volume 654 of *Contemp. Math.* (Amer. Math. Soc., Providence, RI, 2015), pp. 23–52
13. C. Chen, I.E. Shparlinski, New bounds of Weyl sums. Int. Math. Res. Not. IMRN **11**, 8451–8491 (2021)
14. H. Davenport, T.D. Browning, *Analytic Methods for Diophantine Equations and Diophantine Inequalities*. Cambridge Mathematical Library (Cambridge University Press, Cambridge, 2005)
15. N. Garcia-Fritz, On the conjectures of Vojta and Campana over function fields with explicit exceptional sets (2022)
16. R. Hartshorne, *Algebraic Geometry*. Graduate Texts in Mathematics, vol. 52 (Springer, New York, Heidelberg, 1977)
17. D.R. Heath-Brown, A new form of the circle method, and its application to quadratic forms. J. Reine Angew. Math. **481**, 149–206 (1996)
18. S. Kebekus, J.V. Pereira, A. Smeets, Failure of the Brauer–Manin principle for a simply connected fourfold over a global function field, via orbifold Mordell. Duke Math. J. **171**(17), 3515–3591 (2022)
19. Q. Liu, *Algebraic Geometry and Arithmetic Curves*, volume 6 of *Oxford Graduate Texts in Mathematics* (Oxford University Press, Oxford, 2002). Translated from the French by Reinie Erné, Oxford Science Publications
20. V. Mitankin, M. Nakahara, S. Streeter, Semi-integral Brauer-Manin obstruction and quadric orbifold pairs (2022)
21. M.B. Nathanson, *Additive Number Theory the Classical Bases*, 1st edn. (Springer, New York, NY, 1996)
22. M. Nakahara, S. Streeter, Weak approximation and the Hilbert property for Campana points (2020)
23. M. Pieropan, D. Schindler, Hyperbola method on toric varieties (2020)
24. M. Pieropan, A. Smeets, S. Tanimoto, A. Várilly-Alvarado, Campana points of bounded height on vector group compactifications. Proc. Lond. Math. Soc. (3) **123**(1), 57–101 (2021)

25. E. Rousseau, A. Turchet, J. Tzu-Yueh Wang, Nonspecial varieties and generalised Lang-Vojta conjectures. Forum Math. Sigma **9**, Paper No. e11, 29 (2021)
26. T. Santens, Diagonal quartic surfaces with a Brauer-Manin obstruction. Compos. Math. **159**(4), 659–710 (2023)
27. A. Shute, Sums of four squareful numbers (2021)
28. A. Shute, On the leading constant in the Manin-type conjecture for Campana points. Acta Arith. **204**(4), 317–346 (2022)
29. S. Streeter, Campana points and powerful values of norm forms. Math. Z. **301**(1), 627–664 (2022)
30. S. Tanimoto, Campana points, Height zeta functions, and log Manin's conjecture, in *Problems and Prospects in Analytic Number Theory*, volume 2196 (RIMS Kokyuroku, 2021), pp. 26–37
31. R.C. Vaughan, *The Hardy-Littlewood Method*, 2nd edn. Cambridge Tracts in Mathematics (Cambridge University Press, Cambridge, 1997)
32. K. Van Valckenborgh, Squareful numbers in hyperplanes. Algebra Number Theory **6**(5), 1019–1041 (2012)
33. T.D. Wooley, The cubic case of the main conjecture in Vinogradov's mean value theorem. Adv. Math. **294**, 532–561 (2016)
34. T.D. Wooley, Nested efficient congruencing and relatives of Vinogradov's mean value theorem. Proc. Lond. Math. Soc. (3) **118**(4), 942–1016 (2019)
35. H. Xiao, Campana points on biequivariant compactifications of the Heisenberg group. Eur. J. Math. **8**(1), 205–246 (2022)

Sums of Proper Divisors with Missing Digits

Kübra Benli, Giulia Cesana, Cécile Dartyge, Charlotte Dombrowsky, and Lola Thompson

1 Introduction

Let $s(n)$ denote the *sum-of-proper-divisors function*, i.e., $s(n) = \sum_{d|n, d<n} d$. The function $s(n)$ has been studied since the time of the ancient Greeks. Indeed, the *perfect* numbers are those integers n for which $s(n) = n$. The Greeks also spoke of integers as being *deficient* if $s(n) < n$ and *abundant* if $s(n) > n$.

K. Benli
Department of Mathematics and Computer Science, University of Lethbridge, Lethbridge, AB, Canada
e-mail: kubra.benli@uleth.ca

G. Cesana
Department of Mathematics and Computer Science, Division of Mathematics, University of Cologne, Cologne, Germany
e-mail: gcesana@math.uni-koeln.de

C. Dartyge
Institut Élie Cartan, Université de Lorraine, Vandœuvre-lès-Nancy Cedex, France
e-mail: cecile.dartyge@univ-lorraine.fr

C. Dombrowsky
Science Mathematisch Instituut, Mathematisch Instituut, Universiteit Leiden, Leiden, Netherlands
e-mail: c.k.l.dombrowsky@math.leidenuniv.nl

L. Thompson (✉)
Mathematics Institute, Universiteit Utrecht, Utrecht, Netherlands
e-mail: l.thompson@uu.nl

R. Abdellatif et al. (eds.), *Women in Numbers Europe IV*, Association for Women in Mathematics Series 32, https://doi.org/10.1007/978-3-031-52163-8_4

It is natural to wonder how the function s behaves when applied to various inputs. One surprising feature is that s can map sets of asymptotic density[1] zero to sets with positive asymptotic density. For example, consider the set $\mathcal{A}$ as the set of numbers pq, where p and q are distinct primes.

Then $\#(\mathcal{A} \cap [1, x]) \ll x(\log\log x)/\log x$ implies that $\mathcal{A}$ has asymptotic density zero. Moreover, Montgomery and Vaughan [20] showed that the number of even integers less than x which are not the sum of two primes is at most $x^{1-\delta}$ for some $\delta > 0$, and recently, Pintz [22] proved that $\delta = 0.28$ is an admissible explicit value. Since $s(pq) = p + q + 1$ for distinct primes p and q, it follows that $s(\mathcal{A})$ contains almost all odd numbers, which allows one to conclude that $s(\mathcal{A})$ has asymptotic density 1/2.

In addition to studying the images of various sets $\mathcal{A}$ under the function s, one can also consider what happens when taking $s^{-1}(\mathcal{A})$. The preimages can also be a bit surprising. For example, there are sets of integers with positive density whose preimages under s have asymptotic density zero. In fact, Erdős [10] demonstrated that there are sets $\mathcal{A}$ of positive asymptotic density for which $s^{-1}(\mathcal{A})$ is empty.

In 1992, Erdős, Granville, Pomerance, and Spiro made the following conjecture that we will later refer to as the EGPS conjecture.

Conjecture 1.1 ([11, Conjecture 4]) Let $\mathcal{A}$ be a set of integers with asymptotic density zero. Then $s^{-1}(\mathcal{A})$ also has asymptotic density zero.

The full EGPS conjecture is still open today. However, it has been confirmed for certain specific sets $\mathcal{A}$. The following theorems give some examples of these sets. First, Pollack took $\mathcal{A}$ to be the set of primes.

Theorem 1.2 ([24, Theorem 1.11]) *If $\mathcal{A}$ is the set of primes, then $\#(s^{-1}(\mathcal{A}) \cap [1, x]) = O\left(\frac{x}{\log x}\right)$, for all $x \geq 2$.*

In the opposite direction, for sets of integers with "many" prime factors, Troupe proved the following theorem.

Theorem 1.3 ([30, Theorem 1.3]) *Let $\omega(m)$ denote the number of distinct prime factors of an integer m. For any fixed $\epsilon > 0$, if*

$$\mathcal{A} = \{m : |\omega(m) - \log\log m| > \epsilon \log\log m\},$$

then $s^{-1}(\mathcal{A})$ has asymptotic density zero.

In other words, not only are numbers with a lot more than the "normal" number of prime factors rare, their preimages under s are rare as well. This theorem implies

---

[1] If $\mathcal{S}$ is a subset of the natural numbers, then the *asymptotic density* of $\mathcal{S}$ is given by

$$\lim_{x\to\infty} \frac{1}{x}\#\{n \leq x : n \in \mathcal{S}\},$$

provided that this limit exists.

that $\log\log n$ is the normal order of $\omega(s(n))$. Very recently, Pollack and Troupe [27] have improved this, showing that $\omega(s(n))$ satisfies an Erdős–Kac-type theorem. This signifies that $\omega(s(n))$ asymptotically has a normal distribution with mean and variance $\log\log n$.

There are several other sets $\mathcal{A}$ whose preimages under s have been studied in recent years. For example, Pollack considered the case where $\mathcal{A}$ is a set of palindromes.

Theorem 1.4 ([23, Theorem 1]) *If $\mathcal{A}$ is the set of palindromes in any given base, then $s^{-1}(\mathcal{A})$ has asymptotic density zero.*

In another work, Troupe took $\mathcal{A}$ to be the set of integers representable as a sum of two squares.

Theorem 1.5 ([31, Theorem 1.2]) *Let $\mathcal{A}$ be the set of integers $n \leq x$ that can be written as a sum of two squares. Then $\#(s^{-1}(\mathcal{A}) \cap [1, x]) \asymp \frac{x}{\sqrt{\log x}}$.*

Note that Troupe obtains both upper and lower bounds; most of the aforementioned papers only obtain upper bounds. In general, it is difficult to obtain nontrivial lower bounds. Another recent result, due to Pollack and Singha Roy, shows that k-th power-free values of n and $s(n)$ are equally common.

Theorem 1.6 ([26, Theorem 1.3]) *Fix $k \geq 4$. On a set of integers with asymptotic density 1,*

$$n \textit{ is } k\textit{-free} \iff s(n) \textit{ is } k\textit{-free}.$$

The squarefree and cubefree cases remain conjectural. In a slightly different direction, there is a recent result by Lebowitz-Lockard, Pollack, and Singha Roy, which shows that the values of $s(n)$ (for composite n) are equidistributed among the residue classes modulo p for small primes p.

Theorem 1.7 ([17, Theorem 1.3]) *Fix $A > 0$. As $x \to \infty$, the number of composite $n \leq x$ with $s(n) \equiv a \pmod p$ is $(1 + o(1))x/p$, for every residue class $a \pmod p$ with $p \leq (\log x)^A$.*

In the present chapter, we consider the preimages under s of a new set $\mathcal{A}$, namely, the set of integers with missing digits. We will elaborate more on these integers in Sect. 2. In particular, we will give some historical background and define the notation carefully. For now, we briefly discuss our main results. We show that the EGPS conjecture holds for sets of integers with missing digits. Moreover, we prove a quantitative version of this result.

Theorem 1.8 *Fix $g \geq 3$, $\gamma \in (0, 1)$, and a nonempty set $\mathcal{D} \subsetneq \{0, 1, \ldots, g - 1\}$. For all sufficiently large x, the number of $n \leq x$ for which $s(n)$ has all of its digits in base g restricted to digits in $\mathcal{D}$ is $O(x \exp(-(\log\log x)^\gamma))$.*

Note that $s(p) = 1$ for all primes p. Then whenever the set $\mathcal{D}$ contains 1, it follows that the size of the preimage set of $\mathcal{D}$ has $\pi(x)$, the prime counting function, as a lower bound. Since $\pi(x) \sim \frac{x}{\log x}$ as $x \to \infty$, we see that Theorem 1.8 is essentially optimal in the sense that the constant $\gamma \in (0, 1)$ could not be replaced by a constant strictly greater than 1.

Our main tool in the proof of Theorem 1.8 is an upper bound for the number of positive integers $n \le x$ such that $g^k \nmid \sigma(n)$ when g^k is a large integer. Note that it was proved in [33, Hauptsatz 2] that the set of such integers has asymptotic density zero for fixed modulus g^k. By [28, Theorem 2], such a result can be made uniform in the modulus, see [24, Lemma 2.1], stated as Lemma 3.3 below. By sacrificing the uniformity, we obtain the following stronger upper bound.

Lemma 1.9 *Let $g \ge 3$ be a given integer. Let $\gamma, \delta \in (0, 1)$ and $A > 0$ also be given. Then for integers $k \in \left[\log_3 x, A(\log_2 x)^\gamma\right]$, we have*

$$\sum_{\substack{n \le x \\ g^k \nmid \sigma(n)}} 1 \ll x \exp\left(-\left(\log_2 x\right)^\delta\right),$$

where the constant implied by the $\ll$ notation depends on the choices of g, A, γ, and δ, and $\log_j x$ denotes the jth iterate of the logarithm function.

As this lemma does not rely on any facts about integers with missing digits, it may also be useful in other contexts.

The proofs of all of the aforementioned theorems make crucial use of the special forms that the numbers in these sets possess. The methods do not generalize to arbitrary sets with asymptotic density zero. In a different direction, the EGPS conjecture has also been shown to hold for all sets of $n \le x$ with cardinality up to about $x^{1/2}$. More precisely we have the following theorem by Pollack, Pomerance, and the last author, where the result is uniform in the choice of the set as long as its size is small.

Theorem 1.10 ([25, Theorem 1.2]) *Let $\epsilon = \epsilon(x)$ be a fixed function tending to 0 as $x \to \infty$. If $\mathcal{A} \subset \mathbb{N}$ with $\#(\mathcal{A} \cap [1, x]) \ll x^{1/2+\epsilon(x)}$, then $s^{-1}(\mathcal{A})$ has asymptotic density zero.*

As a corollary, one can obtain (for example) that if $\mathcal{A}$ is the set of squares up to x, then $s^{-1}(\mathcal{A})$ has asymptotic density zero. Similarly, one can use Theorem 1.10 to prove that the preimage of a set of integers with "many" missing digits has density 0. For example, if we remove at least half of the possible digits, then the size of the set of integers with missing digits will be $O\left(\sqrt{x}\right)$, and thus this is handled by Theorem 1.10. One can also deduce Theorem 1.4 as a corollary of Theorem 1.10, since the number of palindromes less than x is $O\left(\sqrt{x}\right)$.

Notation and Conventions Throughout this chapter, we will write $\#S$ to denote the number of elements in a set S. We will use $\sigma(n)$ to denote the *sum-of-divisors function*, defined by $\sigma(n) := \sum_{d|n} d$; $\varphi(n)$ to denote the *Euler φ-function* for a positive integer n, defined by $\varphi(n) := \#\{1 \le j \le n : \gcd(n, j) = 1\}$, id to denote the *identity function*, and $\omega(n)$ to denote the *number of distinct prime factors* of an integer n. We let $\mu(n)$ be the *Möbius function* defined as

$$\mu(n) := \begin{cases} (-1)^r, & \text{if } n = p_1 \cdots p_r \text{ with distinct primes } p_i, \\ 0, & \text{if there exists a prime } p \text{ such that } p^2 \mid n. \end{cases}$$

For arithmetic functions G and H, the *convolution* $G * H$ is defined by

$$G * H(n) := \sum_{ab=n} G(a)H(b),$$

for any positive integer n.

For two real functions F and G where G is a nonnegative valued function, we say that $F(x) = O(G(x))$ if there exist a positive constant C and a real number x_0 such that $|F(x)| \leq C|G(x)|$ for all $x \geq x_0$ and $F(x) = o(G(x))$ as $x \to a$ if

$$\lim_{x\to a} \frac{F(x)}{G(x)} = 0.$$

We will also at times use Vinogradov's notation $\ll$ as an alternative to Landau's Big O notation. Namely, $F \ll G$ denotes that $F(x) = O(G(x))$. Similarly, $\gg$ is used to denote the parallel notion with the inequalities reversed in the Big O definition. We write $F \asymp G$ when there are positive constants C_1 and C_2 such that $C_1|F| < |G| < C_2|F|$. Furthermore we let $\log_j(x)$ denote the jth iterate of the natural logarithm of x, e.g., $\log_3 x = \log\log\log x$.

2 Integers with Missing Digits

Let $g \in \mathbb{N}$, $g \geq 3$. We consider the base g expansion of a positive integer n,

$$n = \sum_{j\geq 0} \varepsilon_j(n) g^j,$$

with coefficients $\varepsilon_j(n) \in \{0, \ldots, g-1\}$. Note that this sum is finite. For a proper subset $\mathcal{D} \subsetneq \{0, \ldots, g-1\}$ such that $0 \in \mathcal{D}$ and an arithmetic function f taking positive integer values, we put

$$\mathcal{W}_{f,\mathcal{D}} := \left\{ n : f(n) = \sum_{j\geq 0} \varepsilon_j(f(n)) g^j, \varepsilon_j(f(n)) \in \mathcal{D} \right\} \tag{2.1}$$

as the set of integers n where the digits of $f(n)$ are restricted in the set $\mathcal{D}$.

In addition, for any $x \geq 1$ and any set of integers A, let $A(x)$ denote the set $A \cap [1, x]$. So for any $x \geq 2$, we put

$$\mathcal{W}_{f,\mathcal{D}}(x) := \left(f^{-1}(\mathcal{W}_{\mathcal{D}})\right)(x) = \left\{n \le x : f(n) = \sum_{j\ge 0} \varepsilon_j(f(n))g^j, \varepsilon_j(f(n)) \in \mathcal{D}\right\}$$

as a finite subset of $\mathcal{W}_{f,\mathcal{D}}$.

In the particular case $f = \mathrm{id}$, we simply write $\mathcal{W}_{\mathcal{D}}$ (resp., $\mathcal{W}_{\mathcal{D}}(x)$) in place of $\mathcal{W}_{\mathrm{id},\mathcal{D}}$ (resp., $\mathcal{W}_{\mathrm{id},\mathcal{D}}(x)$). The elements of $\mathcal{W}_{\mathcal{D}}$ are frequently referred to as *integers with missing digits* (or *integers with restricted digits*). In this survey we will also use the terminology proposed by Mauduit, who referred to them as *ellipsephic*[2] *integers*.

Since the set $\mathcal{D}$ is a proper subset of $\{0, \dots, g-1\}$ with $0 \in \mathcal{D}$, we have

$$\#\mathcal{W}_{\mathcal{D}}\left(g^N - 1\right) = |\mathcal{D}|^N, \tag{2.2}$$

and the elements of $\mathcal{W}_{\mathcal{D}}$ form a sparse set, i.e.,

$$\lim_{N\to\infty} \frac{\#\mathcal{W}_{\mathcal{D}}\left(g^N\right)}{g^N} = 0.$$

In other words, the set $\mathcal{W}_{\mathcal{D}}$ has asymptotic density zero.

When $0 \notin \mathcal{D}$, we can adapt the definition (2.1) by setting

$$\mathcal{W}_{f,\mathcal{D}} := \left\{n : f(n) = \sum_{j=0}^{N} \varepsilon_j(f(n))g^j, \varepsilon_j(f(n)) \in \mathcal{D},\ N \in \mathbb{N}\right\},$$

and in this case we have $\#\mathcal{W}_{\mathcal{D}}(g^N - 1) = \sum_{\ell=0}^{N-1} |\mathcal{D}|^{\ell+1} = |\mathcal{D}|(|\mathcal{D}|^N - 1)/(|\mathcal{D}| - 1)$.

In this section, as we survey the literature, we concentrate on the case with $0 \in \mathcal{D}$ in order to avoid some complications in the statements of some results. However, the interested reader can find some results related to the sets $\mathcal{W}_{\mathcal{D}}$ with $0 \notin \mathcal{D}$ in [1]. In our main theorem, we do not require $0 \in \mathcal{D}$.

If we set $g = 10$ and $\mathcal{D} = \{3, 6, 9\}$, then any number in $\mathcal{W}_{f,D}$ is divisible by 3. However, if we exclude similar trivial obstructions, we expect that the sequence of ellipsephic integers behaves like the sequence of the natural numbers. A first question could be whether these integers are well distributed in arithmetic progressions. Erdős, Mauduit, and Sárközy [12] give an affirmative answer. Their result is valid under the following two natural hypotheses for the set $\mathcal{D} = \{d_1, d_2, \dots, d_t\}$, namely

$$d_1 = 0 \in \mathcal{D} \text{ and } \gcd(d_2, \dots, d_t) = 1. \tag{2.3}$$

[2] The origin of this nomenclature comes from the fusion of the two Greek words "ellipsis" = missing and "psiphic" = digit.

For $a, q \in \mathbb{Z}$ such that $\gcd(q, g(g-1)) = 1$, we introduce the set of the ellipsephic integers congruent to a modulo q by

$$\mathcal{W}_{\mathcal{D}}(x, a, q) := \{n \in \mathcal{W}_{\mathcal{D}}(x) : n \equiv a \ (\mathrm{mod}\, q)\}.$$

Erdős, Mauduit, and Sárközy proved that if $\mathcal{D}$ satisfies (2.3), then there exist constants $c_1 := c_1(g, t) > 0$ and $c_2 := c_2(g, t) > 0$ such that

$$\left| \#\mathcal{W}_{\mathcal{D}}(x, a, q) - \frac{\#\mathcal{W}_{\mathcal{D}}(x)}{q} \right| = O\left(\frac{\#\mathcal{W}_{\mathcal{D}}(x)}{q} \exp\left(-c_1 \frac{\log x}{\log q} \right) \right), \tag{2.4}$$

for all $a \in \mathbb{Z}$ and $q \le \exp(c_2\sqrt{\log x})$ such that $\gcd(q, g(g-1)) = 1$ (see [12, Theorem 1]). This result was improved by Konyagin [15] in 2001 and by Col [6] in 2009.

The papers [12] and [13] provide several interesting applications of these equidistribution results and give lists of open problems that inspired various research projects.

Another interesting aspect is the normal order of some arithmetic functions along ellipsephic integers. Banks and Shparlinski [3] obtained various results in this direction. In particular they studied the average values on $\mathcal{W}_{\mathcal{D}}$ of the Euler φ-function and the sum-of-divisors function.

Equation (2.4) can be seen as an analogue of the Siegel–Walfisz Theorem, stated below as Lemma 3.1, for primes in arithmetic progressions. Such theorems are not applicable when the modulus q is a power of x. However, in many applications, it is sufficient to use an equidistribution result that is averaged over the moduli q. In doing so, one is able to extend the range of q. One such application can be seen in work by the third author and Mauduit [7, 8] and independently by Konyagin [15]. They proved that there exists an $\alpha := \alpha(g, \mathcal{D})$ such that (2.4) holds for almost all $q < x^{\alpha}$ satisfying $\gcd(q, g(g-1)) = 1$. Such results combined with sieve methods imply the existence of ellipsephic integers with few prime factors. For example, in [7] it is proved that there exist infinitely many $n \in \mathcal{W}_{\{0,1\}}$ with at most $k_g = (1 + o(1))8g/\pi$ prime factors as $g \to \infty$.

The problem of the existence of infinitely many primes with missing digits has been solved recently by Maynard [18, 19]. In particular, he proved the following spectacular result, given by [18, Theorem 1.1]. Let $a_0 \in \{0, \ldots, 9\}$. The number of primes $p \le x$ with no digit a_0 in their base 10 expansions is

$$\asymp \frac{x^{\frac{\log 9}{\log 10}}}{\log x}.$$

He also gives a condition to determine whether there are finitely or infinitely many n such that $P(n) \in \mathcal{W}_{\mathcal{D}}$, for any given non-constant polynomial $P \in \mathbb{Z}[X]$, large enough base g, and $\mathcal{D} = \{0, \ldots, g-1\} \setminus \{a_0\}$ (see [19, Theorem 1.2]). The papers [18] and [19] also provide deep results when the number of excluded digits is ≥ 2.

We end this short discussion on ellipsephic integers by giving an incomplete list of recent references on this subject containing many other interesting results, namely [1, 4, 5, 6, 16].

In the present chapter, we are particularly interested in the preimages of the sets of ellipsephic integers under $s(n)$, namely $\mathcal{W}_{s,\mathcal{D}}(x) = \left(s^{-1}\left(\mathcal{W}_{\mathcal{D}}\right)\right)(x)$. With this setup, our main result can be reformulated in the following manner: Let $g \geq 3$ be an integer and $\mathcal{D} \subsetneq \{0, \ldots, g-1\}$ be a nonempty proper subset. Let $0 < \gamma < 1$ be given. Then for all sufficiently large x, we have

$$\#\mathcal{W}_{s,\mathcal{D}}(x) = \#\left(s^{-1}\left(\mathcal{W}_{\mathcal{D}}\right)\right)(x) \ll x\exp(-(\log_2 x)^{\gamma}).$$

3 Preliminary Lemmata for Theorem 1.8

We start with a well-known deep result on primes in arithmetic progressions.

Lemma 3.1 (Siegel–Walfisz Theorem, [29, Theorem II.8.17, page 376]) *For any constant* $A > 0$ *and uniformly for* $x \geq 3$, $1 \leq q \leq (\log x)^A$, *and* $\gcd(a, q) = 1$, *we have*

$$\sum_{\substack{p \leq x \\ p \equiv a \bmod q}} \log p = \frac{x}{\varphi(q)} + O\left(x\exp\left(-c\sqrt{\log x}\right)\right),$$

where $c = c(A)$ *is a strictly positive constant.*

Remarks We will apply this lemma for some $q = g^{\ell}$ with $\ell \approx \log_3 x$. Norton [21] and Pomerance [28] independently proved that there exists a constant $C > 0$ such that for all $x \geq 3$ and for all integers q, a with $\gcd(a, q) = 1, q > 0$ we have[3]

$$\left|\sum_{\substack{p \leq x \\ p \equiv a \bmod q}} \frac{1}{p} - \frac{\log_2 x}{\varphi(q)}\right| \leq C.$$

We could apply these results instead of the Siegel–Walfisz Theorem in the proof of Lemma 1.9. Another remark is when q has the special form $q = g^c$ for some fixed g (as is the case in our application); Elliott [9] and then Baker and Zhao [2] proved that it is possible to have asymptotic estimates even when the size of g^{ℓ} is a power of x. The latter two also showed that if $q = g^{\ell}$ with fixed g, then it is possible to obtain a result similar to Lemma 3.1 in the range $g^{\ell} \leq x^{5/12-\varepsilon}$ with $\varepsilon > 0$ arbitrarily small.

[3] The interested reader may find a more precise formulation in [21, Section 6] and in [28, Remark 1].

Recall that a *multiplicative function* f is a function that satisfies $f(uv) = f(u)f(v)$ whenever $\gcd(u, v) = 1$. In particular, if f is a multiplicative function which is not identically zero, then we have $f(1) = 1$. Next, we quote the following technical result for the average order of a multiplicative function.

Lemma 3.2 ([29, Corollary III.3.6, page 457]) *Let* λ_1 *and* λ_2 *be constants such that* $\lambda_1 > 0$ *and* $0 \le \lambda_2 < 2$. *For any multiplicative function* f *such that*

$$0 \le f(p^\nu) \le \lambda_1 \lambda_2^{\nu-1}$$

for all primes p *and for* $\nu = 1, 2, 3, \ldots$, *we uniformly have*

$$\sum_{n \le x} f(n) \ll x \prod_{p \le x} \left(1 - \frac{1}{p}\right) \sum_{\nu \ge 0} \frac{f(p^\nu)}{p^\nu} \tag{3.1}$$

for all $x \ge 1$. *The implicit constant in* (3.1) *is less than*

$$4(1 + 9\lambda_1 + \lambda_1\lambda_2/(2 - \lambda_2)^2).$$

Lemma 3.3 ([24, Lemma 2.1]) *Let* $x \ge 3$. *Let* q *be a positive integer. Then*

$$\sum_{\substack{n \le x \\ q \nmid \sigma(n)}} 1 \ll \frac{x}{(\log x)^{1/\varphi(q)}},$$

uniformly in q.

Note that by sacrificing the uniformity, we will be able to obtain a better bound for the number of positive integers $n \le x$ such that $g^k \nmid \sigma(n)$ when g^k is a large integer.

To prepare for the proof of such a result, we first prove the following observation.

Lemma 3.4 *Let* m *be a positive integer. Then every positive integer* n *can be written uniquely as* $n = ab$ *with* $\gcd(a, b) = 1$ *and*

$$\mu^2(a) = 1,\ p \mid a \text{ implies } p \equiv -1 \pmod m \text{ and}$$
$$p \mid b \text{ implies } p^2 \mid b \text{ or } p \not\equiv -1 \pmod m.$$

Proof Let $n > 1$ as the result is vacuously true when $n = 1$. Assume that n has the following prime factorization:

$$n = p_1^{e_1} p_2^{e_2} \cdots p_j^{e_j},$$

with $p_i \ne p_j$ if $i \ne j$. Changing the order if needed, without loss of generality assume that $p_1, \ldots, p_J \equiv -1 \pmod m$ with $e_1 = \cdots = e_J = 1$ and for $k =$

$J+1,\ldots,j$ either $p_k \not\equiv -1 \pmod m$ or $e_k > 1$. Then, we choose

$$a = p_1 \ldots p_J$$

and

$$b = \frac{n}{a} = p_{J+1}^{e_{J+1}} \cdots p_j^{e_j}.$$

This finishes the proof. □

Remark If we remove the condition that $\gcd(a,b) = 1$ in Lemma 3.4, then the decomposition $n = ab$ is not unique. For example, if n has a prime divisor q with $q \equiv -1 \bmod m$ and $q^3 \mid n$, then we can also choose

$$a = q \prod_{\substack{p \equiv -1 \bmod m \\ p \| n}} p \quad \text{and} \quad b = \frac{n}{a}.$$

We are now ready to prove our key lemma.

Proof of Lemma 1.9 Let $\ell \le k$ be chosen later. By Lemma 3.4, an integer n can be written in a unique way as $n = ab$ with $\gcd(a,b) = 1$, $\mu^2(a) = 1$, and $p \mid a$ implies $p \equiv -1 \pmod{g^\ell}$ and b such that

$$p \mid b \Rightarrow p^2 \mid b \text{ or } p \not\equiv -1 \pmod{g^\ell}.$$

Suppose that $n = ab$, written in the above form, is a positive integer such that $g^k \nmid \sigma(n)$. Then we claim that n has at most $m := \lfloor k/\ell \rfloor$ prime factors p such that $p \equiv -1 \pmod{g^\ell}$ and $p^2 \nmid n$. Indeed, if n has more than k/ℓ prime factors $p \equiv -1 \pmod{g^\ell}$ with $p^2 \nmid n$, say $p_1, p_2, \ldots p_{m+1}$, then $\sigma(n) = (p_1+1)(p_2+1)\ldots(p_{m+1}+1)K$, where K is a positive integer. Since each $p_i \equiv -1 \pmod{g^\ell}$, this implies $\sigma(n) = c_1 g^\ell c_2 g^\ell \ldots c_{m+1} g^\ell K = c_1 c_2 \ldots c_{m+1} K g^{(m+1)\ell}$ with positive integers c_i. Since $(m+1)\ell \ge k$, this would imply that $g^k \mid \sigma(n)$, which contradicts our assumption. Thus for such n, we have $\omega(a) \le m$. Combining what we noted above, we obtain

$$\sum_{\substack{n \le x \\ g^k \nmid \sigma(n)}} 1 \le \sum_{\substack{ab \le x \\ p \mid a \Rightarrow p \equiv -1 \pmod{g^\ell} \\ \omega(a) \le m \\ p \mid b \Rightarrow p^2 \mid b \text{ or } p \not\equiv -1 \pmod{g^\ell}}} \mu^2(a).$$

To find an upper bound for the sum on the right-hand side, we first deal with the condition $\omega(a) \le m$. To do that, we use Rankin's method, replacing 1 with a nonnegative quantity that is at least 1 when $\omega(a) \le m$. Namely, let $t \in (0,1)$ be a parameter so that $t^{\omega(a)-m} > 0$ for any positive integer a and $t^{\omega(a)-m} \ge 1$ if $\omega(a) \le m$. Then we get

$$\sum_{\substack{n\le x\\ g^k\nmid\sigma(n)}} 1 \le \sum_{\substack{ab\le x\\ p|a\Rightarrow p\equiv -1 \pmod{g^\ell}\\ p|b\Rightarrow p^2|b \text{ or } p\not\equiv -1 \pmod{g^\ell}}} \mu^2(a)t^{\omega(a)-m}$$

$$\le t^{-m} \sum_{\substack{ab\le x\\ p|a\Rightarrow p\equiv -1 \pmod{g^\ell}\\ p|b\Rightarrow p^2|b \text{ or } p\not\equiv -1 \pmod{g^\ell}}} \mu^2(a)t^{\omega(a)}.$$

We can rewrite the sum on the right in terms of multiplicative functions

$$\sum_{\substack{ab\le x\\ p|a\Rightarrow p\equiv -1 \pmod{g^\ell}\\ p|b\Rightarrow p^2|b \text{ or } p\not\equiv -1 \pmod{g^\ell}}} \mu^2(a)t^{\omega(a)} = \sum_{n\le x}\sum_{ab=n} G(a)H(b) = \sum_{n\le x} G * H(n),$$

where G and H are multiplicative functions defined by

$$G(p) := \begin{cases} t, & \text{if } p \equiv -1 \pmod{g^\ell},\\ 0, & \text{if } p \not\equiv -1 \pmod{g^\ell},\end{cases}$$

and $G(p^\nu) := 0$ for all $\nu \ge 2$, and

$$H(p) := \begin{cases} 0, & \text{if } p \equiv -1 \pmod{g^\ell},\\ 1, & \text{if } p \not\equiv -1 \pmod{g^\ell},\end{cases}$$

and $H(p^\nu) := 1$ for all $\nu \ge 2$.

Let $f = G*H$. The function f is also multiplicative with $f(p) = G(p)+H(p)$, which gives

$$f(p) = \begin{cases} t, & \text{if } p \equiv -1 \pmod{g^\ell},\\ 1, & \text{otherwise.}\end{cases}$$

For $\nu \ge 2$, we have

$$f\left(p^\nu\right) = \sum_{ab=p^\nu} G(a)H(b) = G(p)H(p^{\nu-1}) + H\left(p^\nu\right)$$

$$= \begin{cases} 1, & \text{if } \nu = 2,\\ t+1, & \text{if } \nu \ge 3 \text{ and } p \equiv -1 \pmod{g^\ell},\\ 1, & \text{if } \nu \ge 3 \text{ and } p \not\equiv -1 \pmod{g^\ell}.\end{cases}$$

Now, we can apply Lemma 3.2 with $\lambda_1 = 2$ and $\lambda_2 = 1$ to f and obtain

$$\sum_{\substack{n\le x\\ g^k\nmid\sigma(n)}} 1 \ll t^{-m}x \prod_{\substack{p\le x\\ p\equiv -1 \ (\mathrm{mod}\, g^\ell)}} \left(1-\frac{1}{p}\right)\left(1+\frac{t}{p}+\frac{1}{p^2}+(t+1)\sum_{\nu\ge 3}\frac{1}{p^\nu}\right) \prod_{\substack{p\le x\\ p\not\equiv -1 \ (\mathrm{mod}\, g^\ell)}} \left(1-\frac{1}{p}\right)\sum_{\nu\ge 0}\frac{1}{p^\nu}.$$

Next, we note that for any prime p,

$$\left(1-\frac{1}{p}\right)\sum_{\nu\ge 3}\frac{1}{p^\nu} = \left(1-\frac{1}{p}\right)\frac{1}{p^3}\sum_{\nu\ge 0}\frac{1}{p^\nu} = \left(1-\frac{1}{p}\right)\frac{1}{p^3}\frac{1}{(1-1/p)} = \frac{1}{p^3}$$

and

$$\left(1-\frac{1}{p}\right)\sum_{\nu\ge 0}\frac{1}{p^\nu} = \left(1-\frac{1}{p}\right)\frac{1}{(1-1/p)} = 1$$

which yield

$$\sum_{\substack{n\le x\\ g^k\nmid\sigma(n)}} 1 \ll t^{-m}x \prod_{\substack{p\le x\\ p\equiv -1 \ (\mathrm{mod}\, g^\ell)}} \left(1+\frac{t-1}{p}+\frac{1-t}{p^2}+\frac{t}{p^3}\right).$$

Furthermore, we see that

$$\begin{aligned}\prod_{\substack{p\le x\\ p\equiv -1 \ (\mathrm{mod}\, g^\ell)}} \left(1+\frac{t-1}{p}+\frac{1-t}{p^2}+\frac{t}{p^3}\right) &= \prod_{\substack{p\le x\\ p\equiv -1 \ (\mathrm{mod}\, g^\ell)}} \left(1+\frac{t-1}{p}\right) \\ &\quad \left(1+\frac{1}{1+\frac{t-1}{p}}\frac{1-t}{p^2}+\frac{1}{1+\frac{t-1}{p}}\frac{t}{p^3}\right) \\ &= \prod_{\substack{p\le x\\ p\equiv -1 \ (\mathrm{mod}\, g^\ell)}} \left(1+\frac{t-1}{p}\right) \prod_{\substack{p\le x\\ p\equiv -1 \ (\mathrm{mod}\, g^\ell)}} \left(1+\frac{1-t}{p(p+t-1)}+\frac{t}{p^2(p+t-1)}\right),\end{aligned}$$

where the second product is bounded by a constant. So we obtain

$$\sum_{\substack{n\le x\\ g^k\nmid\sigma(n)}} 1 \ll t^{-m}x \prod_{\substack{p\le x\\ p\equiv -1 \ (\mathrm{mod}\, g^\ell)}} \left(1+\frac{t-1}{p}\right).$$

In order to bound the product above, we use the Taylor–Young formula for $\log(1-X)$ for $|X|\le 1/2$, and the Siegel–Walfisz Theorem, stated in Lemma 3.1. So, if g^ℓ is less than a power of $\log x$, we have uniformly for $t\in(0,1)$,

$$
\begin{aligned}
\log\left(\prod_{\substack{p\le x\\ p\equiv -1 \pmod{g^{\ell}}}}\left(1+\frac{t-1}{p}\right)\right) &= \sum_{\substack{p\le x\\ p\equiv -1 \pmod{g^{\ell}}}} \log\left(1+\frac{t-1}{p}\right)\\
&= \sum_{\substack{p\le x\\ p\equiv -1 \pmod{g^{\ell}}}} \left(\frac{t-1}{p}+O\left(\frac{1}{p^2}\right)\right)\\
&= -(1-t)\sum_{\substack{p\le x\\ p\equiv -1 \pmod{g^{\ell}}}} \frac{1}{p}+O\left(\sum_{p\le x}\frac{1}{p^2}\right)\\
&= -\frac{(1-t)\log_2 x}{\varphi(g^{\ell})}+O(1),
\end{aligned}
$$

where the last equality above is deduced from the Siegel–Walfisz Theorem after partial summation or more directly by the results of Norton [21] and Pomerance [28] mentioned just after Lemma 3.1. Thus, we obtain

$$
\sum_{\substack{n\le x\\ g^k \nmid \sigma(n)}} 1 \ll \frac{x}{t^m(\log x)^{(1-t)/\varphi(g^{\ell})}}.
$$

It remains to choose ℓ and t. Recall that $k\in\left[\log_3 x, A(\log_2 x)^{\gamma}\right]$. With our choices, we would like to have $t^m(\log x)^{(1-t)/\varphi(g^{\ell})}\to\infty$.

Let $\alpha,\alpha'\in\mathbb{R}$ such that $0<\alpha'<\alpha<1$ and $\gamma,\delta<\alpha-\alpha'$. We let

$$
\ell := \left\lfloor (1-\alpha)\frac{\log_3 x}{\log g}\right\rfloor,
$$

$$
t := 1-\frac{1}{(\log_2 x)^{\alpha'}}.
$$

Note that we then have $\ell\le k$, $g^{\ell}\le(\log_2 x)^{1-\alpha}$ and

$$
1\le m\le\frac{k}{\ell}\le A\frac{(\log_2 x)^{\gamma}}{\left\lfloor (1-\alpha)\frac{\log_3 x}{\log g}\right\rfloor}\le 2A\frac{(\log_2 x)^{\gamma}}{(1-\alpha)\frac{\log_3 x}{\log g}}\le(\log_2 x)^{\gamma}
$$

for x large enough.

With these choices, we get

$$
t^m(\log x)^{(1-t)/\varphi(g^{\ell})}\ge t^m(\log x)^{(1-t)/g^{\ell}}
$$

$$\gg \left(1 - \frac{1}{(\log_2 x)^{\alpha'}}\right)^{(\log_2 x)^{\gamma}} (\log x)^{\frac{1}{(\log_2 x)^{\alpha'+1-\alpha}}}$$
$$= \exp\left((\log_2 x)^{\gamma} \log\left(1 - \frac{1}{(\log_2 x)^{\alpha'}}\right) + \frac{\log_2 x}{(\log_2 x)^{\alpha'+1-\alpha}}\right)$$
$$= \exp\left((\log_2 x)^{\gamma} \log\left(1 - \frac{1}{(\log_2 x)^{\alpha'}}\right) + \frac{1}{(\log_2 x)^{\alpha'-\alpha}}\right).$$

Since $\gamma < \alpha - \alpha'$, the first term in the exponential above is at most one half of the second in the absolute value. Hence, for all $x \geq x_0$ with $x_0 = x_0(\alpha, \alpha', \gamma)$, we have

$$\sum_{\substack{n \leq x \\ g^k \nmid \sigma(n)}} 1 \ll x \exp\left(-\frac{1}{2}(\log_2 x)^{\alpha-\alpha'}\right) \ll x \exp\left(-(\log_2 x)^{\delta}\right),$$

where we have used the condition $\delta < \alpha - \alpha'$. □

4 Proof of Theorem 1.8

Recall our basic setup: We have $g \in \mathbb{N}$, $g \geq 3$, $\mathcal{D} \subsetneq \{0, 1, \ldots, g-1\}$ nonempty, x sufficiently large, and $0 < \gamma < 1$. For a positive integer $k \in \left[\frac{(\log_2 x)^{\gamma}}{\log(g/|\mathcal{D}|)}, 2\frac{(\log_2 x)^{\gamma}}{\log(g/|\mathcal{D}|)}\right]$, we write

$$\#\mathcal{W}_{s,\mathcal{D}}(x) = \sum_{\substack{n \in \mathcal{W}_{s,\mathcal{D}}(x) \\ \sigma(n) \equiv 0 \pmod{g^k}}} 1 + \sum_{\substack{n \in \mathcal{W}_{s,\mathcal{D}}(x) \\ \sigma(n) \not\equiv 0 \pmod{g^k}}} 1 =: S_1 + S_2.$$

We will now work on the upper bounds for S_1 and S_2 separately. To start with, we use a rather weak bound for S_2, where we drop the condition on the digit restriction on $s(n)$. Then we apply Lemma 1.9 with $\gamma = \delta$ to obtain

$$S_2 \leq \sum_{\substack{n \leq x \\ g^k \nmid \sigma(n)}} 1 = O\left(x \exp\left(-\left(\log_2 x\right)^{\gamma}\right)\right).$$

Next, we focus on finding an upper bound for S_1. Following a similar setting as in [23], for $s(n) = \sum_{j=0}^{N} \varepsilon_j(s(n)) g^j$, for some $N \geq 1$, we put

$$B := \sum_{j=0}^{k-1} \varepsilon_j(s(n))g^j$$

as the number formed by the k-rightmost digits of $s(n)$ such that $s(n) \equiv B \pmod{g^k}$. Note that this implies $B \leq g^k - 1$. Let $n \in \mathcal{W}_{s,\mathcal{D}}(x)$ with $\sigma(n) \equiv 0 \pmod{g^k}$. Then, since $g^k \mid \sigma(n)$, we have

$$n = \sigma(n) - s(n) \equiv -s(n) \equiv -B \pmod{g^k}.$$

So, we can relax the condition on S_1 with a congruence condition as follows in the case $0 \in \mathcal{D}$; namely,

$$S_1 = \sum_{\substack{n \in \mathcal{W}_{s,\mathcal{D}}(x) \\ \sigma(n) \equiv 0 \pmod{g^k}}} 1 \leq \sum_{\substack{n \leq x \\ n \equiv -B \pmod{g^k} \\ B \in \mathcal{W}_{\mathcal{D}}(g^k - 1)}} = \sum_{B \in \mathcal{W}_{\mathcal{D}}(g^k-1)} \sum_{\substack{n \leq x \\ n \equiv -B \pmod{g^k}}} 1 \leq |\mathcal{D}|^k \left\lfloor \frac{x}{g^k} \right\rfloor + |\mathcal{D}|^k,$$

where in the last inequality we used, for M a positive integer, $b \in \mathbb{Z}$ and $X \geq M$, that

$$\#\{1 \leq a \leq X : a \equiv b \pmod{M}\} \leq \lfloor X/M \rfloor + 1$$

along with (2.2). When $0 \notin \mathcal{D}$, we proceed in the same way, but we use the fact that $\#W_{\mathcal{D}}(g^k - 1) = (|\mathcal{D}|^{k+1} - |\mathcal{D}|)/(\mathcal{D}| - 1)$ instead of $|\mathcal{D}|^k$.

Inserting our choice of k yields

$$S_1 \ll x \exp(-k \log(g/|\mathcal{D}|)) \ll x \exp\left(-(\log_2 x)^{\gamma}\right).$$

Thus we obtain the following upper bound:

$$\#\mathcal{W}_{s,\mathcal{D}}(x) = \#s^{-1}(\mathcal{W}_{\mathcal{D}})(x) \ll x \exp(-(\log_2 x)^{\gamma}),$$

as desired. □

Remark In our proof of Theorem 1.8, we showed that the digits only need to be missing from the last $(\log_2 x)^{\gamma}$ digits of n. Since almost all $n \leq x$ have about $\log x / \log g$ digits in base g, this means that we do not need to impose any restrictions on the vast majority of the digits of n.

5 Some Remarks on a Lower Bound on $\#\mathcal{W}_{s,\mathcal{D}}(x)$

As indicated earlier, as soon as we have $1 \in \mathcal{D}$, then the elements in $\mathcal{W}_{s,\mathcal{D}}$ are at least as frequent as the primes, since $s(p) = 1$ for all primes p. In the case when

$\mathcal{D} = \{0, \ldots, g-1\} \setminus \{a_0\}$ for some $a_0 \in \{1, \ldots, g-1\}$ and g large enough, we can prove that $s(n)$ takes on infinitely many different values in $\mathcal{W}_\mathcal{D}$ by adapting the ideas used in [19]. As already remarked in the introduction, if p and q are two distinct primes, then $s(pq) = p+q+1$. A variant of the Strong Goldbach conjecture, which has been proven, then tells us that the image of s contains almost all odd numbers. It thus remains for us to prove that a positive proportion of ellipsephic integers can be expressed as 1 plus a sum of two primes. We follow the arguments of [19, Sections 6–8] which imply that

$$\sum_{n \in \mathcal{W}_\mathcal{D}(g^N - 1)} \sum_{p+q+1=n} \log p \log q = (c(\mathcal{D}) + o(1))(g|\mathcal{D}|)^N, \tag{5.1}$$

$$\text{with } c(\mathcal{D}) := \frac{g}{\varphi(g)^2} \# \Big\{ (b_1, b_2) \in \{0, \ldots, g-1\}^2 : \gcd(b_1 b_2, g) = 1 \text{ and } \exists\, d \in \mathcal{D} \text{ such that } b_1 + b_2 + 1 \equiv d \pmod g \Big\}. \tag{5.2}$$

When g is large enough, $c(\mathcal{D}) > 0$. Indeed, the set on the right-hand side of (5.2) is nonempty. It contains at least $(b_1, b_2) = (1, g-1)$ if $a_0 \neq 1$, and $(b_1, b_2) = (1, 1)$ in the case $a_0 = 1$.

The proof of (5.1) consists of reproducing Sections 6–8 in [19], with one additional prime variable which we handle trivially. The only small difference is in the computation of the main term coming from the major arcs at the end of the proof of [19, Lemma 7.2].

Let $r(n)$ denote the weight

$$r(n) = \sum_{p_1 + p_2 + 1 = n} \log p_1 \log p_2.$$

We can apply an upper bound sieve to detect the primes p_1 such that $n - 1 - p_1$ is also prime. By [14, Theorem 3.11] and the remark that appears afterwards, for any odd integer $n \in [\sqrt{x}, x]$ we have that

$$r(n) \ll (\log x)^2 \#\{p \le n,\ n - 1 - p = p'\} \ll x \prod_{\substack{p>2 \\ p \mid n-1}} \frac{p-1}{p-2} \le c_1 x \log\log x, \tag{5.3}$$

for some absolute $c_1 > 0$. By (5.1) and (5.3), we deduce that

$$\#\{n \in \mathcal{W}_{\mathcal{D}}(x) : n - 1 \text{ is a sum of two primes}\} \geq \frac{c(\mathcal{D}) - o(1)}{c_1 \log\log x} \#\mathcal{W}_{\mathcal{D}}(x) . \tag{5.4}$$

This implies that $s(n)$ takes infinitely many different values in $\mathcal{W}_{\mathcal{D}}$. In this short section, we wanted to highlight a result which is a direct consequence of the proofs presented in [19]. However, detecting ellipsephic primes is a considerably more difficult problem than detecting ellipsephic integers n such that $n-1$ is a sum of two primes. It is thus certainly possible to have a more precise lower bound than (5.4) for more general sets $\mathcal{W}_{\mathcal{D}}$ and with small base g. Also, we can probably remove the $\log\log x$ in the denominator in (5.4) by combining the arguments in [32, Section 3.2] with [19].

Acknowledgments We would like to thank the Women In Numbers Europe 4 conference organizers for providing us with the opportunity to work together. We would also like to thank the referees for their helpful comments, which greatly improved this work. The first and third authors were supported by ANR grant ANR-20-CE91-0006 ArithRand. The first author was also supported by a PIMS postdoctoral fellowship at the University of Lethbridge. The last author was supported by the Max Planck Institute for Mathematics during the early stages of this project.

References

1. K. Aloui, Sur les entiers ellipséphiques: somme des chiffres et répartition dans les classes de congruence. (French), Period. Math. Hungar. **70**(2), 171–208 (2015)
2. R.C. Baker, L. Zhao, Gaps of smallest possible order between primes in an arithmetic progression. Int. Math. Res. Not. IMRN **23**, 7341–7368 (2016)
3. W.D. Banks, I.E. Shparlinski, Arithmetic properties of numbers with restricted digits. Acta Arith. **112**(4), 313–332 (2004)
4. K.D. Biggs, Efficient congruencing in ellipsephic sets: the quadratic case. Acta Arith. **200**(4), 331–348 (2021)
5. K.D. Biggs, Efficient congruencing in ellipsephic sets: the general case. Int. J. Number Theory **19**(1), 169–197 (2023)
6. S. Col, Diviseurs des nombres ellipséphiques. (French), Period. Math. Hungar. **58**(1), 1–23 (2009)
7. C. Dartyge, C. Mauduit, Nombres presque premiers dont l'écriture en base r ne comporte pas certains chiffres. (French) J. Number Theory **81**(2), 270–291 (2000)
8. C. Dartyge, C. Mauduit, Ensembles de densité nulle contenant des entiers possédant auplus deux facteurs premiers. (French), J. Number Theory **91**(2), 230–255 (2001)
9. P.D.T.A. Elliott, Primes in progressions to moduli with a large power factor. Ramanujan J. **13**, 241–251 (2007)
10. P. Erdős, Über die Zahlen der Form $\sigma(n) - n$ und $n - \varphi(n)$. (German) Elem. Math. **28**, 83–86 (1973)
11. P. Erdős, A. Granville, C. Pomerance, C. Spiro, *On the normal behavior of the iterates of some arithmetic functions*. Analytic number theory (Allerton Park, IL, 1989), pp. 165–204. Progr. Math. vol. 85 (Birkhäuser Boston, Boston, MA, 1990)
12. P. Erdős, C. Mauduit, A. Sárközy, On arithmetic properties of integers with missing digits. I. Distribution in residue classes. J. Number Theory **70**(2), 99–120 (1998)

13. P. Erdős, C. Mauduit, A. Sárközy, On arithmetic properties of integers with missing digits. II. Prime factors. Paul Erdős memorial collection. Discrete Math. **200**(1–3), 149–164 (1999)
14. H. Halberstam, H.-E. Richert, *Sieve Methods*. London Mathematical Society Monographs, vol. 4 (Academic Press [Harcourt Brace Jovanovich, Publishers], London, New York, 1974), xiv, 364 pp.
15. S. Konyagin, Arithmetic properties of integers with missing digits: distribution in residue classes. Period. Math. Hungar. **42**(1-2), 145–162 (2001)
16. S. Konyagin, C. Mauduit, A. Sárközy, On the number of prime factors of integers characterized by digit properties. Period. Math. Hungar. **40**(1), 37–52 (2000)
17. N. Lebowitz-Lockard, P. Pollack, A. Singha Roy, Distribution mod p of Euler's totient function and the sum of proper divisors. Michigan Math. J. **74**(1), 143–166 (2024)
18. J. Maynard, Primes with restricted digits. Invent. Math. **217**(1), 127–218 (2019)
19. J. Maynard, Primes and polynomials with restricted digits. Int. Math. Res. Not. **14**, 1–23 (2022)
20. H.L. Montgomery, R.C. Vaughan, The exceptional set in Goldbach's problem. Acta Arith. **27**, 353–370 (1975)
21. K.K. Norton, On the number of restricted prime factors of an integer. I. Ill. J. Math. **20**(4), 681–705 (1976)
22. J. Pintz, A new explicit formula in the additive theory of primes with applications, II. The exceptional set for the Goldbach problems. arXiv: 1804.09084v2
23. P. Pollack, Palindromic sums of proper divisors. Integers **15A**, Paper No. A13, 12 pp. (2015)
24. P. Pollack, Some arithmetic properties of the sum of proper divisors and the sum of prime divisors. Ill. J. Math. **58**(1), 125–147 (2014)
25. P. Pollack, C. Pomerance, L. Thompson, Divisor-sum fibers. Mathematika **64**(2), 330–342 (2018)
26. P. Pollack, A. Singha Roy, Powerfree sums of proper divisors. Colloq. Math. **168**(2), 287–295 (2022)
27. P. Pollack, L. Troupe, Sums of proper divisors follow the Erdős-Kac law. Proc. Am. Math. Soc. **151**, 977–988 (2023)
28. C. Pomerance, On the distribution of amicable numbers. J. Reine Angew. Math. **293**, 217–222 (1977)
29. G. Tenenbaum, *Introduction to Analytic and Probabilistic Number Theory*, 3rd edn. Translated from the 2008 French edition by Patrick D.F. Ion. Graduate Studies in Mathematics, vol. 163 (American Mathematical Society, Providence, RI, 2015), xxiv, 629 pp.
30. L. Troupe, On the number of prime factors of values of the sum-of-proper-divisors function. J. Number Theory **150**, 120–135 (2015)
31. L. Troupe, Divisor sums representable as a sum of two squares. Proc. Am. Math. Soc. **148**(10), 4189–4202 (2020)
32. R.C. Vaughan, *The Hardy-Littlewood Method*, 2nd edn. Cambridge Tracts in Mathematics, vol. 125 (Cambridge University Press, Cambridge, 1997), xiv, 232 pp.
33. G.N. Watson, Über Ramanujansche Kongruenzeigenschaften der Zerfällungszahlen. (I). (German) Math. Z. **39**(1), 712–731 (1935)

Rational Approximations, Multidimensional Continued Fractions, and Lattice Reduction

V. Berthé, K. Dajani, C. Kalle, E. Krawczyk, H. Kuru, and A. Thevis

1 Introduction

Continued fraction type expansions aim (among other properties) at providing increasingly good rational Diophantine approximations of real numbers. More precisely, a multidimensional continued fraction is expected to produce simultaneous better and better rational approximations with the same denominator $\mathbf{p}^{(n)}/q^{(n)} = (p_1^{(n)}/q^{(n)}, \ldots p_d^{(n)}/q^{(n)})_{n\in\mathbb{N}}$ for d-tuples $\boldsymbol{\alpha} = (\alpha_1, \ldots, \alpha_d)$ of real numbers, with the fractions $p_i^{(n)}/q^{(n)}$ converging to α_i for each $1 \leq i \leq d$.

The usual regular continued fractions are known to provide extremely good (and even the best) rational approximations for positive real numbers [41, 89]. The

V. Berthé
Université Paris Cité, CNRS, Paris, France
e-mail: berthe@irif.fr

K. Dajani
Department of Mathematics, Utrecht University, Utrecht, The Netherlands
e-mail: k.dajani1@uu.nl

C. Kalle
Mathematisch Instituut, Leiden University, Leiden, The Netherlands
e-mail: kallecccj@math.leidenuniv.nl

E. Krawczyk
Faculty of Mathematics and Computer Science, Institute of Mathematics, Jagiellonian University, Kraków, Poland

H. Kuru
Faculty of Engineering and Natural Sciences, Sabancı University, Istanbul, Turkey
e-mail: hamidekuru@sabanciuniv.edu

A. Thevis (✉)
Institut für Mathematik, Goethe-Universität Frankfurt, Frankfurt am Main, Germany
e-mail: thevis@math.uni-frankfurt.de

R. Abdellatif et al. (eds.), *Women in Numbers Europe IV*, Association for Women in Mathematics Series 32, https://doi.org/10.1007/978-3-031-52163-8_5

situation is more complicated in higher dimension. Indeed, there is no canonical extension of regular continued fractions to higher dimensions (see Sect. 4.1), and the zoology of existing algorithms is particularly rich (see Sect. 4.5 as an illustration). The main advantage of most classical (unimodular) continued fractions is that they can be expressed as dynamical systems whose ergodic study has already been well understood (such as described in Sect. 4.4). Ergodic theory allows a precise description of the long-range statistical properties of the expansions that are produced (e.g. their mean behavior). Indeed, ergodic theory extends basic laws of large numbers in probability by dropping the assumption of intertemporal independence. Thus, it relates spatial averages $\int_X f\,d\mu$ to time averages $\frac{1}{n}\sum_{0\leq i<n} f \circ T^i$ along trajectories; in other words, the system has the same behavior when averaged over time as averaged over the whole space.

However, the main disadvantages of these algorithms lie, firstly, in the fact that the behavior of the continued fraction expansion of a given d-tuple $\boldsymbol{\alpha}$ can be difficult to grasp (it might not behave in a generic way), and secondly, in the quality of the rational approximations that are produced. Indeed, the convergence of multidimensional continued fractions is governed by their (first and second) Lyapunov exponents (see [105]), which describe the asymptotic behavior of the singular values of large products of matrices, under the ergodic hypothesis. More precisely, their approximation exponent can be expressed as $1 - \frac{\lambda_2}{\lambda_1}$ according to [105] (λ_1 and λ_2 being the two largest Lyapunov exponents of the associated dynamical system). It has to be compared with Dirichlet's exponent $1 + 1/d$ (see Theorem 1 in Sect. 3.1). However, there is numerical evidence [38] that the second Lyapunov exponent is not even negative in higher dimensions for most classical algorithms such as the Jacobi–Perron [20, 124, 134], Brun [34–36], or Selmer [142] algorithms, which prevents strong convergence of these algorithms. In a nutshell, strong (resp. weak) convergence refers to the convergence of quantities of the type $|||q^{(n)}\boldsymbol{\alpha}|||$, where $|||\cdot|||$ stands for the distance to the nearest integer (resp., $||\alpha - \mathbf{p}^{(n)}/q^{(n)}||$). In other words, these algorithms converge weakly usually, but they fail to have strong convergence (see Sect. 3.3 for the definitions of weak and strong convergence).

In terms of the quality of rational approximations that are produced, there is a second strategy which relies on lattice reduction, where rational approximations are obtained by exhibiting short vectors in a lattice attached to some given d-tuple $\boldsymbol{\alpha}$. Lattice reduction algorithms aim to find reduced bases of Euclidean lattices, formed by short and almost orthogonal vectors. The most celebrated one is the LLL algorithm, designed by Lenstra, Lenstra and Lovász in 1982 [108]. It relies heavily on the use of Gram–Schmidt orthogonalization. Its overall algorithmic structure is simple and yet, its general probabilistic behavior is far from being understood; this includes the gap between its practical performances and its proved worst-case estimates. Hence, although finding rational approximations works quite well in practice, the average behavior of such a strategy is not well understood. In particular, the lack of a description of reduction algorithms as dynamical systems prevents the use of tools from ergodic theory.

In the expository part of the paper, we focus on the two main classes of algorithms that produce rational approximations as discussed above. The first type of algorithms can be expressed via dynamical systems defined on a compact set (usually of the form $[0, 1]^d$). Such an algorithm associates with some given vector an infinite sequence of matrices, and one can consider the quality of convergence of this product of matrices. The most classical examples of such algorithms are the Jacobi–Perron [20, 68, 124, 134], the Brun [34–36], or the Selmer algorithms (the last of which is conjugate on the absorbing simplex to Mönkemeyer's algorithm [111, 123]). They are described, e.g., in [31, 105, 138]. The second type of algorithm is based on lattice reduction algorithms, such as the LLL algorithm (see Sect. 5). We focus on these two families since they share as a common feature the fact that they rely on a choice of a basis of the integer lattice $\mathbb{Z}^d$.

We then illustrate in Sect. 7 the dynamical approach with the ergodic study of a version of the Jacobi–Perron algorithm based on the use of the nearest integer part for the partial quotients. One motivation for studying the nearest integer Jacobi–Perron algorithm is to confirm the idea that working with the nearest integer part improves the quality of continued fraction algorithms, such as indicated numerically by the experimental results from Steiner [145] (see Sect. 7.2). The partial quotients produced by the usual Jacobi–Perron algorithm satisfy a simple Markovian rule. In the case of the nearest integer Jacobi–Perron algorithm the description of the admissible sequences of digits is much more involved. Hence, a simple modification—such as changing the choice of the integer part—leads to much more delicate conditions for the description of the algorithm. As a first step toward a theoretical confirmation of the above-mentioned estimates, we prove the existence of a Markov partition for the nearest integer Jacobi–Perron algorithm and suggest a possible procedure for proving the existence of a finite ergodic invariant measure absolutely continuous with respect to Lebesgue measure.

Let us sketch the contents of this paper. Section 2 recalls basic notions concerning classical continued fractions. We present their main properties that we will use as a guideline for possible generalizations to the higher-dimensional case. We then focus in Sect. 3 on the two main strategies that can be used for producing rational approximations in an effective way. Section 4 deals with the classical dynamical unimodular continued fraction algorithms. Algorithms based on the lattice reduction algorithms and homogeneous dynamics are considered in Sect. 5. We briefly discuss applications and possible ways to improve algorithms in Sect. 6. Lastly, in Sect. 7 we focus on the nearest integer Jacobi–Perron algorithm. We describe its associated Markov partition and provide a strategy for proving the existence of an absolutely continuous invariant measure.

2 Continued Fractions

In this section we briefly recall the main properties of the usual regular continued fractions. They will serve as a guideline for the discussion on the higher-dimensional case. For general references on continued fractions, see, e.g., [24, 51, 74, 75, 89]. For any positive real number $\alpha \in [0, 1]$, its continued fraction expansion is

$$\alpha = \cfrac{1}{a_1 + \cfrac{1}{a_2 + \cfrac{1}{a_3 + \ddots}}},$$

where the digits a_n are positive integers, called *partial quotients*. The rational numbers p_n/q_n, where p_n, q_n are coprime positive integers defined as

$$\frac{p_n}{q_n} = \cfrac{1}{a_1 + \cfrac{1}{a_2 + \ddots + \cfrac{1}{a_n}}},$$

are called *convergents*. The sequence of rational numbers p_n/q_n approximates α up to an error of order $1/q_n^2$: one has

$$|\alpha - p_n/q_n| \le \frac{1}{q_n^2} \text{ for all } n.$$

Dynamically, continued fraction expansions can be obtained by applying the Gauss map $T_G : [0, 1] \to [0, 1]$ defined by

$$T_G(0) = 0 \quad \text{and} \quad T_G(\alpha) = \{1/\alpha\} \text{ if } \alpha \neq 0,$$

where $\{\cdot\}$ is the fractional part of a real number. If for an $\alpha \in (0, 1]$ we write $T_G(\alpha) = \{1/\alpha\} = \frac{1}{\alpha} - \lfloor \frac{1}{\alpha} \rfloor = \frac{1}{\alpha} - a_1$, then $\alpha = \frac{1}{a_1 + T_G(\alpha)}$. Now, by setting

$$a_n = \left\lfloor \frac{1}{T_G^{n-1}(\alpha)} \right\rfloor$$

for $n \ge 1$, one gets the digits in the continued fraction expansion of α.

For $\alpha \notin (0, 1]$, one sets $a_0 = \lfloor \alpha \rfloor$ and one considers the digits generated by T_G for $\alpha - a_0$. Alternatively we can use the following form for $\gamma_0 := 1/\alpha$ in $\mathbb{R}^+$, obtained by setting, for $n \ge 0$,

$$\begin{cases} a_n &= \lfloor \gamma_n \rfloor \\ \gamma_{n+1} &= \frac{1}{\gamma_n - a_n}. \end{cases}$$

One then has $\gamma_n = \frac{1}{T_G^{n-1}(\alpha)}$ for all $n \geq 1$.

Note that the Gauss map is closely related to Euclid's algorithm: starting with two (coprime) positive integers $\ell^{(0)}$ and $\ell^{(1)}$, Euclid's algorithm works by subtracting as often as possible the smallest of both numbers from the largest one (that is, one performs the Euclidean division of the largest one by the smallest); this yields $\ell^{(0)} = \ell^{(1)} \lfloor \frac{\ell^{(0)}}{\ell^{(1)}} \rfloor + \ell^{(2)}$, $\ell^{(1)} = \ell^{(2)} \lfloor \frac{\ell^{(1)}}{\ell^{(2)}} \rfloor + \ell^{(3)}$, etc., until we reach $\ell^{(m+1)} = 1 = \gcd(\ell^{(0)}, \ell^{(1)})$. By setting, for $n \in \mathbb{N}$, $\alpha^{(n)} = \frac{\ell^{(n)}}{\ell^{(n+1)}}$ and $a_n = \lfloor \alpha^{(n)} \rfloor$, one gets $\alpha^{(n-1)} = a_{n-1} + \frac{1}{\alpha^{(n)}}$, and

$$\alpha^{(0)} = \frac{\ell^{(0)}}{\ell^{(1)}} = a_0 + \cfrac{1}{a_1 + \cfrac{1}{a_2 + \cfrac{1}{a_3 + \ddots + \cfrac{1}{a_{m-1} + \frac{1}{a_m}}}}}.$$

Let us now revisit the action of the Gauss map in matricial terms. Let $\alpha \in [0, 1]$. For all n, we have

$$\begin{aligned} \begin{bmatrix} \alpha \\ 1 \end{bmatrix} &= \alpha T_G(\alpha) \cdots T_G^{n-1}(\alpha) \begin{bmatrix} 0 & 1 \\ 1 & a_1 \end{bmatrix} \cdots \begin{bmatrix} 0 & 1 \\ 1 & a_n \end{bmatrix} \begin{bmatrix} T_G^n(\alpha) \\ 1 \end{bmatrix} \\ &= \alpha T_G(\alpha) \cdots T_G^{n-1}(\alpha) \begin{bmatrix} p_{n-1} & p_n \\ q_{n-1} & q_n \end{bmatrix} \begin{bmatrix} T_G^n(\alpha) \\ 1 \end{bmatrix}, \end{aligned}$$

by using the classical relations between convergents and partial quotients, namely $q_{-1} = 0$, $p_{-1} = 1$, $q_0 = 1$, $p_0 = 0$, and, for all n,

$$q_{n+1} = a_{n+1} q_n + q_{n-1}, \quad p_{n+1} = a_{n+1} p_n + p_{n-1}.$$

The matrix $\begin{bmatrix} p_{n-1} & p_n \\ q_{n-1} & q_n \end{bmatrix}$ is a square matrix with integer entries that has determinant of absolute value 1, which is to say that it is *unimodular*. We denote the set of unimodular matrices by $\mathrm{GL}(d, \mathbb{Z})$. (We also use the following standard notation: $\mathrm{GL}(d, \mathbb{R})$ stands for the set of $d \times d$ invertible matrices with real entries, $SL(d, \mathbb{N})$ stands for the set of $d \times d$ matrices of determinant 1 with non-negative integer coefficients.) Note that the entries of this matrix are even positive. To understand the convergence of such a sequence of matrices one can use generalizations of the Perron–Frobenius theorem, such as Theorem 2 below.

Dynamically, the Gauss map T_G goes with the map

$$A_G : [0,1] \to \mathrm{GL}(2,\mathbb{N}),\ \alpha \mapsto \begin{bmatrix} 0 & 1 \\ 1 & \lfloor 1/\alpha \rfloor \end{bmatrix}. \tag{1}$$

Such a map is called a *cocycle* in the terminology of (random) dynamical systems (see, e.g., [10, 11, 151]). Let us set $A^{(n)} := A(T_G^n)(\alpha)$ for all positive n. The line directed by the vector $(\alpha, 1)$ in $\mathbb{R}^2$ belongs to the sequence of nested cones $A^{(1)} \cdots A^{(n)}\mathbb{R}_+^2$, i.e.,

$$(\alpha, 1) \in \bigcap_n A^{(1)} \cdots A^{(n)}\mathbb{R}_+^2 = \bigcap_n A(T_G)(\alpha) \cdots A(T_G^n)(\alpha)\mathbb{R}_+^2.$$

One has even more: this sequence of nested cones converges to the line directed by $(\alpha, 1)$. The *convergence* is said to be *weak* if the angles between the (column) vectors of the product matrices $A^{(1)} \cdots A^{(n)}$ tend to 0 (as $n \to \infty$), and *strong* if the distances between the vectors tend to 0. Continued fractions can thus be seen as producing dynamically infinite convergent sequences of unimodular matrices. We revisit the notions of convergence of infinite products of matrices in Sect. 3.3 in more detail.

3 On Simultaneous Approximation

This section is devoted to simultaneous rational approximations. In Sect. 3.1 we first recall Dirichlet's theorem. We then describe in Sect. 3.2 the two main strategies for producing such approximations that we consider in this survey. We focus on the quality of approximations in Sect. 3.3.

3.1 Dirichlet's Bound

Let us recall Dirichlet's theorem; it can be obtained as a direct application of the pigeonhole principle (see, e.g., [74]) or of Minkowski's first theorem.

Theorem 1 (Dirichlet's Theorem) *For any $(\alpha_1, \cdots, \alpha_d) \in \mathbb{R}^d$ and any Q, there exists a positive integer q with $q \le Q^d$ and integers p_i such that*

$$\max_{1 \le i \le d} |q\alpha_i - p_i| < \frac{1}{Q}.$$

Theorem 1 immediately implies that for any $(\alpha_1, \cdots, \alpha_d) \in \mathbb{R}^d$ the system of inequalities

$$\left|\frac{p_i}{q} - \alpha_i\right| < \frac{1}{q^{1+\frac{1}{d}}}, \quad \text{for } i = 1, 2, \ldots, d,$$

admits infinitely many integer solutions $(p_1, \ldots, p_d, q)$.

The exponent $1 + 1/d$ is optimal as shown in [125], see also [41, 136]. The Dirichlet's theorem provides the existence of "good" approximations. One can thus make an exhaustive search, though this is not an efficient algorithmic method. See [121, Chapter 6] for a discussion on effective methods. See also [104], where the computational complexity concerning simultaneous Diophantine approximations is investigated. When the dimension d is fixed, [104] gives algorithms which, for a given N, find a good rational approximation with denominator $1 \leq q \leq N$ with respect to a specified accuracy, or which find all best approximations (in the sense of Definition 1 below) located in $[1, \cdots, N]$ in polynomial time (using methods based on the LLL algorithm). Note that the following problem is proved to be NP-hard: for a given vector $\boldsymbol{\alpha} \in \mathbb{Q}^d$, positive integer N and accuracy s_1/s_2, is there an integer Q with $1 \leq Q \leq N$ such that $|||Q\boldsymbol{\alpha}||| \leq s_1/s_2$? (The distance to the nearest integer is expressed here with respect to the supremum norm.) In a similar flavor, see also [69] concerning the problem of finding integer relations, and [25].

Continued fractions are known to provide good (and even the best) rational approximations of a given real number α in $[0, 1]$ (see, e.g., [41, 89]). One would desire to have similar algorithms yielding good rational approximations with the same denominator of d-tuples of positive real numbers. That is, for a given $\boldsymbol{\alpha} = (\alpha_1, \cdots, \alpha_d) \in [0, 1]^d$, one looks for sequences of positive integers $(q_n)_n$ and positive integer d-tuples $\mathbf{p}^{(n)} = (p_1^{(n)}, \cdots, p_d^{(n)})_n$ such that

$$\lim p_i^{(n)}/q_n = \alpha_i, \quad i = 1, \cdots, d$$

with a good quality of rational approximation of $\boldsymbol{\alpha}$. Geometrically, this corresponds to looking for approximations of a line in $\mathbb{R}^{d+1}$ by points in $\mathbb{Z}^{d+1}$. Dual problems consist of looking for small values of linear forms and small linear relations.

More precisely, given a norm $||\cdot||$ on $\mathbb{R}^d$, let $|||\cdot|||$ stand for the distance to the nearest integer. The usual norms that are considered are the supremum and the Euclidean norm. The quality of the approximation is measured by $\frac{1}{q^{(n)}}|||q^{(n)}\boldsymbol{\alpha}|||$, to be compared with Dirichlet's bound, i.e., $|||q^{(n)}\boldsymbol{\alpha}|||$ has to be compared with $(q^{(n)})^{-1/d}$. One can thus consider the approximation exponent

$$\limsup_n -\frac{\log \|\boldsymbol{\alpha} - \frac{\mathbf{p}_i^{(n)}}{q_i^{(n)}}\|}{\log q^{(n)}}$$

and compare it to the Dirichlet's bound $1 + 1/d$.

3.2 How to produce Rational Approximations

We focus here on two main approaches for producing rational approximations. The first one is based on the generation of infinite convergent sequences of matrices obtained dynamically by iteration of a map acting on a compact space, as illustrated with the Gauss map T_G for usual continued fractions in Sect. 2. This will be discussed further in Sect. 4. The second one is based on the existence of small vectors picked in well chosen lattices; we will discuss it in Sect. 5.

The first strategy associates with some element $\boldsymbol{\alpha} = (\alpha_1, \cdots, \alpha_d) \in \mathbb{R}^d$ a sequence of square matrices $(A^{(n)})_{n\in\mathbb{N}}$ of size $d+1$ with integer entries. It can be produced for instance via a dynamical system (X, T) with a map A as follows (see also (1)):

$$T\colon X \to X,\ A\colon X \to GL(d+1, \mathbb{Z}),\ \text{and } A^{(n)} = A(T^n(x)). \tag{2}$$

If the matrices belong to $GL(d+1, \mathbb{Z})$, then the corresponding algorithm is called *unimodular*. Matrices $A^{(n)}$ play the role of partial quotients and the product matrices $A^{(1)} \cdots A^{(n)}$ produce convergents. Convergents aim at providing Diophantine approximations (via their column vectors) of the direction $(\boldsymbol{\alpha}, 1)$. We write

$$A^{(1)} \cdots A^{(n)} = \begin{bmatrix} p_{1,1}^{(n)} & \cdots & p_{1,d+1}^{(n)} \\ \vdots & \ddots & \vdots \\ p_{d,1}^{(n)} & \cdots & p_{d,d+1}^{(n)} \\ q_1^{(n)} & \cdots & q_{d+1}^{(n)} \end{bmatrix}. \tag{3}$$

The last element of each column of $A^{(1)} \cdots A^{(n)}$ is a denominator for the associated simultaneous rational approximation. The first d rows of the convergent matrices are meant to provide the numerators of the simultaneous approximations, i.e., one considers the following vector

$$\left(\frac{p_{j,1}^{(n)}}{q_j^{(n)}}, \cdots, \frac{p_{j,d}^{(n)}}{q_j^{(n)}}\right),$$

whose entries are the *nth convergents* of $\boldsymbol{\alpha}$.

The convergence of these matrices means that they contract in the direction of the vector $(\boldsymbol{\alpha}, 1)$. We discuss more precisely their convergence in Sect. 3.3 by considering products $A^{(1)} \cdots A^{(n)}$ as n goes to infinity.

We now describe the second approach based on the existence of small vectors in well chosen lattices, as described in the seminal paper [108]. This approach yields a very fruitful compromise between the quality of approximation (a good approximation is deduced from a small vector) and the efficiency (this small vector is obtained in polynomial time). Let $\boldsymbol{\alpha} = (\alpha_1, \cdots, \alpha_d) \in \mathbb{R}^d$ be a vector to approximate. One works here again in a $d+1$-dimensional space, by introducing a one-parameter family of lattices $(\Lambda_t)_{t>0}$ with positive parameter t tending to 0. More precisely, let Λ_t be the lattice generated by the columns of the matrix

$$M_t := \begin{bmatrix} 1 & 0 & \cdots & 0 & -\alpha_1 \\ 0 & 1 & \cdots & 0 & -\alpha_2 \\ \vdots & \vdots & \ddots & \vdots & \vdots \\ 0 & 0 & \cdots & 1 & -\alpha_d \\ 0 & 0 & \cdots & 0 & t \end{bmatrix}.$$

Note that $\det(M_t) = t$, hence, the lattice Λ_t changes at each step of the algorithm. Let us stress the fact that this strategy differs from the previous one where, in the unimodular case, one works with bases of the fixed lattice $\mathbb{Z}^{d+1}$. We will take t small, the parameter Q of Dirichlet's theorem being connected to t as follows: $Q = t^{-\frac{1}{d+1}}$.

One of the main features of the LLL algorithm is that it produces in polynomial time a non-zero vector $\mathbf{b} = (b_1, \cdots, b_{d+1})$ of the lattice Λ_t such that

$$||\mathbf{b}||_2 \leq 2^{d/4} \det(M_t)^{1/(d+1)} = 2^{d/4} t^{1/(d+1)}. \tag{4}$$

Note that the geometry of numbers, and more precisely Minkowski's first theorem, guarantees the existence of a "small" non-zero vector $\mathbf{x} \in \Lambda_t$, i.e., such that

$$||\mathbf{x}||_2 \leq \sqrt{(d+1)(d+5)/4}\,(\mathrm{vol}(\Lambda_t))^{1/(d+1)} = \sqrt{(d+1)(d+5)/4}\, t^{1/(d+1)}. \tag{5}$$

Let $(\mathbf{e}_i)_{i=1,\ldots,d+1}$ stand for the canonical basis of $\mathbb{Z}^{d+1}$. There exist integers $p_1, \ldots, p_d, q$ such that

$$\begin{aligned} \mathbf{b} &= p_1\mathbf{e}_1 + p_2\mathbf{e}_2 + \cdots + p_d\mathbf{e}_d + q(-\alpha_1\mathbf{e}_1 - \cdots - \alpha_d\mathbf{e}_d + t\mathbf{e}_{d+1}) \\ &= (p_1 - q\alpha_1)\mathbf{e}_1 + \cdots + (p_d - q\alpha_d)\mathbf{e}_d + qt\mathbf{e}_{d+1}. \end{aligned}$$

One deduces from (4) that for all $1 \leq i \leq d$

$$|p_i - \alpha_i q| \leq 2^{d/4} t^{1/(d+1)}$$

and

$$qt \leq 2^{d/4} t^{1/(d+1)}, \text{ i.e., } t^{\frac{1}{d+1}} \leq \frac{2^{1/4}}{q^{1/d}}.$$

For all i, we deduce that

$$|p_i - \alpha_i q| \leq \frac{2^{(d+1)/4}}{q^{1/d}},$$

with

$$|q| \leq 2^{d/4} t^{-d/(d+1)} = 2^{d/4} Q^d.$$

The quality of approximation is the quality that is expected (according to Dirichlet's theorem) up to a multiplicative factor $2^{(d+1)/4}$ which depends exponentially on the dimension. We could have used the inequality (5) which would have given a different multiplicative factor, but the same quality ($q^{1/d}$). Nevertheless, the interest of a lattice reduction algorithm such as the LLL algorithm is that the small vector that is used is found in polynomial time.

For a given $t > 0$, the smallest vector of the lattice Λ_t (or a small vector in the sense of (5)) produces a good approximation. One can ask whether it is possible to devise a continued fraction algorithm from this. Note that one has to recompute everything from the beginning when one changes t. This will be discussed further in Sect. 5. Algorithms defined dynamically with maps such in (2) are on the contrary called memory-less (see Sect. 4).

3.3 Convergence and Lyapunov Exponents

One important feature of the first strategy (based on the generation of infinite products of matrices) from Sect. 3.2 is that we are still able to measure the quality of approximation that is produced in view of Dirichlet's theorem. To do this, we need to measure the quality of convergence of infinite products of matrices; this can be done using Lyapunov exponents. One can either measure the convergence of a given product of matrices or consider the generic behavior of products of matrices generated by a dynamical system. Lyapunov exponents allow, among other things, the description of the growth of the logarithm of the angles between column vectors of products of matrices $M_0 \cdots M_n$.

We state now the definitions concerning convergence. The norm $\|\cdot\|$ refers to the Euclidean norm. Let $M = (M_n)_{n\in\mathbb{N}}$ be a sequence of square matrices of size d and let $\boldsymbol{\ell} \in \mathbb{R}^d$ be a non-zero vector. Let $(\mathbf{e}_1, \cdots, \mathbf{e}_d)$ stand for the canonical basis of $\mathbb{R}^d$. We say that M is *weakly convergent* to $\boldsymbol{\ell}$ if

$$\lim_{n\to+\infty} \left\| \frac{M_0 \cdots M_{n-1}\mathbf{e}_i}{\|M_0 \cdots M_{n-1}\mathbf{e}_i\|} - \frac{\boldsymbol{\ell}}{\|\boldsymbol{\ell}\|} \right\| = 0 \quad \text{for all } i \in \{1, \ldots, d\}.$$

We say that M is *strongly convergent* to $\boldsymbol{\ell}$ if

$$\lim_{n\to+\infty} \mathrm{d}(M_0 \cdots M_{n-1}\mathbf{e}_i, \mathbb{R}\boldsymbol{\ell}) = 0 \quad \text{for all } i \in \{1, \ldots, d\},$$

where the distance d refers to the usual associated distance of a point to a line, i.e., the distance between a point $\mathbf{x}$ and the set $\mathbb{R}\boldsymbol{\ell}$ is the infimum of $\{\|\mathbf{x} - r\boldsymbol{\ell}\| \mid r \in \mathbb{R}\}$.

Lastly, we say that M is *exponentially convergent* to $\boldsymbol{\ell}$ if there exist $C, \gamma > 0$ such that

$$\mathrm{d}(M_0 \cdots M_{n-1}\mathbf{e}_i, \mathbb{R}\boldsymbol{\ell}) < Ce^{-\gamma n} \quad \text{for all } n \in \mathbb{N} \text{ and for all } i \in \{1, \ldots, d\}.$$

Exponential convergence is a first step in the direction of good rational approximations. The constant γ then has to be compared with Dirichlet's bound.

Theorem 2 below gives a sufficient condition for the sequence of cones $M_0 \cdots M_n \mathbb{R}_+^d$ to nest down to a single line as n tends to infinity for square matrices M_i with non-negative entries. It can be seen as a generalization of the classical Perron–Frobenius theorem. This statement is particularly useful in the present context since multidimensional continued fraction algorithms often generate non-negative matrices. Hence, weak convergence is usually not an issue for multidimensional continued fraction algorithms.

Theorem 2 ([64, pp. 91–95]) *Let $(M_n)_n$ be a sequence of non-negative integer matrices of size d. Assume that there exist a strictly positive matrix B and indices $j_1 < k_1 \leq j_2 < k_2 \leq \cdots$ such that $B = M_{j_1} \cdots M_{k_1-1} = M_{j_2} \cdots M_{k_2-1} = \cdots$. Then,*

$$\bigcap_{n \in \mathbb{N}} M_0 \cdots M_{n-1}\mathbb{R}_+^d = \mathbb{R}_+\boldsymbol{\ell} \quad \textit{for some positive vector } \boldsymbol{\ell} \in \mathbb{R}_+^d.$$

Let us now focus on strong convergence by first recalling the definition of the Lyapunov exponents of a sequence of matrices. For a matrix M in $\mathrm{GL}(d, \mathbb{R})$, the singular values $\delta_1, \ldots, \delta_d$ are the eigenvalues of the matrix $\left({}^tMM\right)^{1/2}$. Let us order these (positive and real) values as $\delta_1 \geq \delta_2 \geq \cdots \geq \delta_d$. Given a sequence $\mathbf{M} = (M_n)_{n \in \mathbb{N}}$ of matrices in $\mathrm{GL}(\mathrm{d}, \mathbb{R})$, the *$i$-th Lyapunov* exponent θ_i is then defined as the limit

$$\theta_i := \lim_{n \to \infty} \frac{1}{n} \log(\delta_i(n)),$$

if this limit exists, with $\delta_i(n)$ being the i-th singular value of $M_0 \cdots M_{n-1}$. The Lyapunov exponents can also be defined recursively using exterior powers (see for example [11, Proposition 3.2.7]) by

$$\theta_1 + \cdots + \theta_k = \lim_{n \to \infty} \frac{1}{n} \log\|\wedge^k (M_0 \cdots M_{n-1})\|, \quad k = 1, \ldots, d,$$

provided that the limit exists. One has $\theta_1 \geq \theta_2 \geq \cdots \geq \theta_d$. A sufficient condition for strong convergence can then be stated as follows (see [11, Proposition 3.4.2 (ii)] and [2]): Let $\mathbf{M} = (M_n)_{n \in \mathbb{N}}$ be a sequence of non-negative matrices in $\mathrm{GL}(d, \mathbb{N})$ for which the Lyapunov exponents exist; if $\mathbf{M}$ satisfies the growth condition $\limsup_{n \to \infty} \frac{1}{n} \log\|M_n\| \leq 0$ together with the condition $\theta_1 > 0 > \theta_2$, then strong convergence holds.

Lyapunov exponents can be also defined for random products of matrices, where the randomness can be provided by putting some distribution on the set of matrices or by iterating measurable dynamical systems that produce matrices with a cocyle map such as in (1). The first results in this direction were stated for sequences of independent random matrices with a given distribution function, with the Furstenberg–Kesten theorem (see [61, 65]); these results have then been refined via Kingman's subadditive ergodic theorem and lastly via Oseledets' multiplicative ergodic theorem (see for instance [10, 11, 151]) proving that the limits (involved in the definition of Lyapunov exponents) exist almost surely and take almost everywhere the same value.

Let us give a flavor of such results. We will come back to this also in Sect. 4.4. We first recall a few elements from ergodic theory. The (left) *shift* S acts on a sequence $(M_n)_{n\in\mathbb{N}}$ as $S((M_n)_n) = (M_{n+1})_n$ (i.e., the first term of the sequence $(M_n)_n$ is deleted). Let $\mathcal{M}$ be a finite set of matrices in $\mathrm{GL}(d,\mathbb{Z})$. Let $D \subset \mathcal{M}^{\mathbb{N}}$ be a closed shift-invariant subset of $\mathrm{GL}(d,\mathbb{Z})^{\mathbb{N}}$. A probability measure ν on D is called shift-invariant if $\nu(S^{-1}B) = \nu(B)$ for every measurable set $B \subset D$. A shift-invariant probability measure on D is *ergodic* if any shift-invariant measurable set has either measure 0 or 1. We have seen (see (2)) that such a set of matrices can be obtained by considering a measurable map T acting on some compact metric space and a measurable map $A : X \to \mathrm{GL}(d,\mathbb{Z})$. The sequences of matrices are then of the form $A(T^n(x))$ and one studies the existence and the almost everywhere behavior of limits of the form

$$\lim_{n\to\infty} \frac{1}{n} \log \|A(x) \cdots A(T^{n-1}(x))\|.$$

Theorem 3 ([11]) *Let $D \subset \mathcal{M}^{\mathbb{N}}$ be a closed shift-invariant subset of* $\mathrm{GL}(d,\mathbb{Z})^{\mathbb{N}}$ *together with a shift-invariant measure ν. Assume that (D, S, ν) is ergodic. Let $A : D \to \mathrm{GL}(d,\mathbb{Z})$, $\mathbf{M} = (M_n)_n \mapsto M_0$. Assume that A is log-integrable, i.e.,*

$$\int_D \log \max\{\|A(\mathbf{M})\|_\infty, \|A(\mathbf{M})^{-1}\|_\infty\} d\nu(\mathbf{M}) < \infty. \tag{6}$$

Then the quantities θ_i which are recursively defined by

$$\theta_1 + \cdots + \theta_k = \lim_{n\to\infty} \frac{1}{n} \log\|\wedge^k (M_0 \cdots M_{n-1})\|, \quad k = 1, \ldots, d, \tag{7}$$

exist and do not depend on $\mathbf{M}$ for almost all $\mathbf{M} \in D$. Here, $\wedge^k$ stands for the k-fold wedge product.

4 Higher-Dimensional Dynamical Continued Fractions

We now focus on multidimensional continued fraction algorithms. Their non-canonicity is discussed in Sect. 4.1. Section 4.2 recalls what is usually expected from a continued fraction algorithm. In Sect. 4.3 we focus on algorithms obtained by iteration of a dynamical system, which yield the so-called memory-less algorithms. The interest of their dynamical description is highlighted in Sect. 4.4. Lastly, we give examples of such algorithms in Sect. 4.5.

4.1 Non-Canonicity in Higher Dimension

The aim of this section is to present several facts sustaining the claim that there is no canonical multidimensional continued fractions algorithm.

Firstly, (usual) continued fractions rely on Euclid's algorithm: starting with two numbers, one subtracts the smallest from the largest (see Sect. 2). If we start with at least three numbers, it is not clear how to decide which operation has to be performed, hence the diversity of existing generalizations. For instance, Brun's algorithm can be described as subtracting the second largest entry from the largest one. See Sect. 4.5 for more details.

Moreover, the specific algebraic structure of $SL(2,\mathbb{N})$ plays an important role for one-dimensional continued fractions algorithms. Indeed, the matrices produced in the case of usual continued fractions are unimodular matrices with non-negative integer coefficients (see Sect. 2). The algebraic structure of $SL(2,\mathbb{N})$ is particularly simple: $SL(2,\mathbb{N})$ is a free and finitely generated monoid; it admits

$$\begin{bmatrix} 1 & 0 \\ 1 & 1 \end{bmatrix} \quad \text{and} \quad \begin{bmatrix} 1 & 1 \\ 0 & 1 \end{bmatrix} \tag{8}$$

as generators; any matrix in $SL(2,\mathbb{N})$ thus admits a unique decomposition in terms of the matrices given in (8). This decomposition is a matricial translation of Euclid's algorithm, and the continued fraction expansion of α can be recovered from the unique decomposition of matrices

$$(-1)^n \begin{bmatrix} p_{n+1} & q_{n+1} \\ p_n & q_n \end{bmatrix}, \quad n \geq 0$$

in the free monoid $SL(2,\mathbb{N})$. This explains why most one-dimensional continued fraction algorithms are closely related.

The situation is completely different for $SL(3,\mathbb{N})$ which is not finitely generated. Consider, e.g. the family of matrices

$$M_n := \begin{bmatrix} 1 & 0 & n \\ 1 & n-1 & 0 \\ 1 & 1 & n-1 \end{bmatrix}.$$

According to [131, Chap. 12] these matrices are undecomposable for $n \geq 3$: they are not equal to an even permutation matrix, and, for any pair of matrices $A, B \in SL(3, \mathbb{N})$ such that $M_n = AB$, A or B is an even permutation matrix.

Another approach for generalizing continued fractions could rely on properties of best approximation.

Definition 1 A rational number p/q is said to be a *best approximation* of a real number α if every p'/q' with $1 \leq q' \leq q$, $p/q \neq p'/q'$ satisfies

$$|q\alpha - p| < |q'\alpha - p'|.$$

Convergents in the continued fraction expansion of α and best approximations are known to coincide [41, 89]. Nevertheless, this notion is not so satisfying for defining continued fractions in higher dimensions as stressed in [101, 102]. Firstly, best approximations depend on the choice of a norm [101], and secondly, the unimodularity property is lost. More precisely, one has the following.

Definition 2 Let $\boldsymbol{\alpha} \in [0, 1]^d$. Let $||\cdot||$ be a given norm in $\mathbb{R}^d$ and let $|||\cdot|||$ denote the distance to the nearest integer. The *sequence of best approximations* of $\boldsymbol{\alpha}$ with respect to the norm $||\cdot||$ is defined as the increasing sequence of non-negative integers $(q^{(n)})_{n\in\mathbb{N}}$ such that $|||q^{(n)}\boldsymbol{\alpha}||| < |||q\boldsymbol{\alpha}|||$ for any q with $1 \leq q < q^{(n)}$.

The existence of an infinite sequence of best approximations can be derived in a classic way from Dirichlet's theorem or from Minkowski's first theorem. Best approximations fail to be unimodular [102]. More precisely, consider the square matrix M_n of size $d+1$ whose rows are given by successive best approximations vectors $\mathbf{v}_n = (p_1^{(n)}, \cdots, p_d^{(n)}, q^{(n)})$ providing $|||q^{(n)}\boldsymbol{\alpha}|||$. Let D_n stand for the determinant of this matrix. It is proved in [102] that for any norm in dimension $d \geq 2$, there exists $\boldsymbol{\alpha} \in \mathbb{R}^d$, with $\dim_{\mathbb{Q}}[1, \alpha_1, \cdots, \alpha_d] = d+1$, such that for any positive integer N, there exists an n for which $D_n = D_{n+1} = \cdots = D_{n+N} = 0$. Arbitrarily large determinants can even occur in dimension $d = 2$ with the supremum norm. For more on best approximations and multidimensional continued fractions, see the survey [46]; see also [115] and Sect. 5.

4.2 What is expected?

We briefly recall here the main properties expected from a continued fraction expansion. For a general discussion on the quest for suitable higher-dimensional continued fractions, see [31]. See also [67] for a discussion on their limitations.

The original motivation for multidimensional continued fractions (and in particular the Jacobi–Perron algorithm) arises from a question posed by Hermite to Jacobi in 1848: how to generalize continued fractions to higher dimensions in order to characterize algebraic irrationals as having a periodic expansion. This is still a very challenging problem in number theory. For more details, see, e.g., [20, 88, 138]. Using such a characterization, one could hope to determine fundamental units (e.g. of a cubic number field) and to solve Diophantine equations (as the ones described in Sect. 6). The formalism of multidimensional continued fractions based on the Klein polyhedra and sails developed in [9, 12, 95, 98] is well-suited for the detection of periodic expansions in terms of algebraic number fields by providing generalized Lagrange's theorem. For more on the subject and its history, see, e.g., [99, 100], the references in [86], and the book [87], which also includes a review of various generalizations of continued fractions. More generally, still from an arithmetic viewpoint, a multidimensional continued fraction algorithm is also expected to detect linear relations between $1, \alpha_1, \cdots, \alpha_d$.

We already discussed the fact that a continued fraction algorithm is expected to yield simultaneous better and better rational approximations with the same denominator for d-tuples $\boldsymbol{\alpha} = (\alpha_1, \cdots, \alpha_d)$ in $[0, 1]^d$, in an effective way and with a good approximation quality. More precisely, it has to produce a sequence of positive integers $(q^{(n)})_n$ such that the distance to the nearest integer $|||q^{(n)}\boldsymbol{\alpha}|||$ converges exponentially fast to 0 with respect to $q^{(n)}$, and ideally in $(q^{(n)})^{-\frac{1}{d}}$ (as predicted by Dirichlet's theorem). We have seen in Sect. 4.1 that the sequence of best approximations (see Definition 2) depends heavily on the chosen norm and that the associated transformations are no longer unimodular [102, 115]. However, it is possible to use the action of the diagonal flow on the space of unimodular lattices to better understand their behavior [42, 92, 106]. See also Sect. 5.

From a dynamical viewpoint, continued fraction algorithms are also expected to have reasonable ergodic properties such as the ones described in Sect. 4.4. For instance, we would like to have control over the (almost sure) behaviors concerning the growth of the convergents, the distribution of the partial quotients, or the speed of convergence via Lyapunov exponents. Famous examples of algorithms expressed dynamically as piecewise linear fractional transformations are the Jacobi–Perron, Brun or Selmer algorithms. Their description is the object of Sects. 4.3 and 4.5 below.

4.3 Dynamical Continued Fraction Algorithms

We recall here the main concepts related to dynamical continued fraction algorithms, expanding the brief description from Sect. 3.2. This dynamical formalism covers the most classical multidimensional continued fraction algorithms discussed in the classical references [31, 138, 146]; see also [105] for results concerning their Lyapunov exponents.

We consider here algorithms that produce the sequence of matrices $(A^{(n)})_{n\in\mathbb{N}}$ in a dynamical way. We take mostly a measure-theoretical viewpoint: the algorithms will be defined almost everywhere with respect to the Lebesgue measure on $[0, 1]^d$. We focus on the unimodular case, since this allows a geometric interpretation in terms of bases of the integer lattice $\mathbb{Z}^{d+1}$.

Let $X \subset [0, 1]^d$. (Usually X is simply $[0, 1]^d$ but some algorithms can be also defined on sets of the form $\{x = (x_1, \cdots, x_d) \in [0, 1]^d \mid 0 \le x_1 \le \cdots \le x_d \le 1\}$.) A d-dimensional unimodular *continued fraction map* over X is given by measurable maps

$$T\colon X \to X, \quad A\colon X \to GL(d+1, \mathbb{Z}), \quad \theta\colon X \to \mathbb{R}$$

such that for a.e. $\boldsymbol{\alpha} \in X$:

$$\begin{bmatrix}\boldsymbol{\alpha}\\ 1\end{bmatrix} = \theta(\boldsymbol{\alpha})A(\boldsymbol{\alpha})\begin{bmatrix}T(\boldsymbol{\alpha})\\ 1\end{bmatrix}. \tag{9}$$

The maps A and T play the main role in the algorithm; the role played by the map θ is minor: it serves as a renormalization.

The associated *continued fraction algorithm* consists of iteratively applying the map T to a vector $\boldsymbol{\alpha} \in X$. This yields a sequence of matrices $(A(T^n(\boldsymbol{\alpha})))_{n\ge 1}$, called the *continued fraction expansion* of $\boldsymbol{\alpha}$.

One has

$$\begin{bmatrix}\boldsymbol{\alpha}\\ 1\end{bmatrix} = \theta(\boldsymbol{\alpha})\theta(T(\boldsymbol{\alpha}))\ldots\theta(T^{n-1}(\boldsymbol{\alpha}))A(\boldsymbol{\alpha})A(T(\boldsymbol{\alpha}))\ldots A(T^{n-1}(\boldsymbol{\alpha}))\begin{bmatrix}T^n(\boldsymbol{\alpha})\\ 1\end{bmatrix}.$$

Such an algorithm is said to be "without memory." Indeed, the $(n+1)$-th step of the algorithm depends only on the map T and on the value $T^n(\boldsymbol{\alpha})$. This is in contrast with the algorithms based on lattice reduction that we will discuss in Sect. 5.

In most classical examples of such algorithms the continued fraction map T is piecewise continuous, or even a piecewise homography (see Sect. 4.5). Further, many of them are *positive*, that is, all the matrices appearing as the images of the map A are non-negative.

We illustrate this formalism with the regular continued fraction case. Here θ is the identity. Indeed considering

$$\begin{aligned}\begin{bmatrix}\alpha\\ 1\end{bmatrix} &= \alpha T_G(\alpha)\cdots T_G^{n-1}(\alpha)\begin{bmatrix}0 & 1\\ 1 & a_1\end{bmatrix}\cdots\begin{bmatrix}0 & 1\\ 1 & a_n\end{bmatrix}\begin{bmatrix}T_G^n(\alpha)\\ 1\end{bmatrix}\\ &= \alpha T_G(\alpha)\cdots T_G^{n-1}(\alpha)\begin{bmatrix}p_{n-1} & p_n\\ q_{n-1} & q_n\end{bmatrix}\begin{bmatrix}T_G^n(\alpha)\\ 1\end{bmatrix},\end{aligned}$$

one gets

$$\alpha\, T_G(\alpha) \cdots T_G^{n-1}(\alpha) = |q_{n-1}\alpha - p_{n-1}|,$$

where p_n/q_n stands for the n-th convergent of α, and T_G stands for the Gauss map. We see with this example that the map θ, which allows the renormalization with respect to the last coordinate (set to 1), is of an arithmetic nature, in the sense that the multiplicative cocycle associated with θ, i.e., $\theta(T_G^{n-1}(\alpha)) \cdots \theta(T_G(\alpha))\theta(\alpha)$, yields the arithmetic quantity $|q_{n-1}\alpha - p_{n-1}|$.

We now revisit this formalism more geometrically in terms of lattice bases following [31]. One approximates the vectorial line in $\mathbb{R}^{d+1}$ directed by the non-zero vector $(\boldsymbol{\alpha}, 1)$ by a sequence of integer lattice bases $(\mathbf{b}^{(n)})_{n\in\mathbb{N}}$ of $\mathbb{Z}^{d+1}$, namely the $d+1$ columns of the matrices $A(\boldsymbol{\alpha}) \cdots A(T^{n-1}(\boldsymbol{\alpha}))$. The lattice bases generate cones that are expected to converge toward the line directed by $(\boldsymbol{\alpha}, 1)$. Moreover, if the algorithm is positive, then these cones are nested and $(\boldsymbol{\alpha}, 1)$ belongs to the positive cone generated by the vectors $\mathbf{b}_i^{(n)}$, $i = 1, \ldots, d+1$, i.e.,

$$(\boldsymbol{\alpha}, 1) \in \left\{ \sum_{1\le i\le d+1} \lambda_i \mathbf{b}_i^{(n)} \mid \lambda_i \geq 0 \text{ for all } i = 1, \cdots, d+1 \right\}.$$

In fact, (9) implies that the coefficients $(\lambda_i)_{1\le i\le d+1}$ are proportional to the vector $(T^n(\boldsymbol{\alpha}), 1)$.

The algorithm thus produces a sequence of bases of the lattice $\mathbb{Z}^{d+1}$ that all determine a homogeneous cone in $\mathbb{R}^{d+1}$ that contains the ray $\{\lambda(\boldsymbol{\alpha}, 1) \mid \lambda \geq 0\}$. This is the viewpoint developed for instance in [31]; in fact, the algorithms there are designed in this way.

We recall that the continued fraction map T acts on parameters $\boldsymbol{\alpha}$ living in an ambient d-dimensional space. In view of what precedes, it is also natural to start directly with a general line $\boldsymbol{\ell} = (\ell_1, \cdots, \ell_{d+1})$ in $\mathbb{R}^{d+1}$ and to have an algorithm with such a line as an input. But then, there is no canonical way to "projectivize" the algorithm, i.e., to go from the line $\boldsymbol{\ell}$ to a d-dimensional vector $\boldsymbol{\alpha}$ working in a compact set X on which a dynamical system T acts. One can set, e.g., $\alpha_i = \ell_i/\ell_{d+1}$ for $i = 1, \cdots, d$ (e.g., if the line belongs to the positive cone of $\mathbb{R}^{d+1}$ and if ℓ_{d+1} is the largest entry). Usual ways to go from some $\boldsymbol{\ell} \in \mathbb{R}_+^{d+1}$ to some $\boldsymbol{\alpha} \in [0, 1]^d$ consists in setting $\ell_{d+1} = 1$, and working with entries $\ell_i \in [0, 1]^d$ for $1 \leq i \leq d$, or else, in working with the simplex $\sum_{i=1}^{d+1} \ell_i = 1$, with $\ell_i \geq 0$ for all i. See for instance [138] for more details. One can also choose to work directly on the projective space $\mathbb{P}(\mathbb{R}^d)$ by associating with each element $[y_1 : y_2 : \cdots : y_{d-1} : y_d]$ the representative defined by $\max y_i = 1$ and by working with projectivizations of matrices in $GL(d, \mathbb{Z})$. This possibility of having different choices for a same piecewise $d+1$-dimensional linear map explains the abundance of existing algorithms (as illustrated in Sect. 4.5).

Let us discuss now the possible steps of the algorithms, i.e., the operations that can be performed on the bases together with the choices allowed for partial quotient matrices $A^{(n)}$. An algorithm is said to be *additive* if all the matrices belong to a finite set, i.e., the map A from (2) takes finitely many values. The terminologies additive vs. multiplicative or division vs. subtractive algorithm are also commonly used; see, e.g., [31] where an algorithm is said to be subtractive if $c^{(n)} = 1$ in (10), and additive if $c^{(n)}$ is chosen as the maximal possible number allowing the line to stay within the positive cone generated by the convergent vectors $\mathbf{b}_i^{(n)}$ for $i = 1, \ldots, d$.

As an illustration, the additive version of the Gauss map is given by the Farey map

$$x \mapsto \begin{cases} \dfrac{x}{1-x} & \text{for } 0 \leq x \leq 1/2, \\ \\ \dfrac{1-x}{x} & \text{for } 1/2 \leq x \leq 1. \end{cases}$$

Dynamically, this creates non-trivial changes; indeed, the Gauss map is known to have a finite ergodic invariant measure (see (11) below), which is not the case for the Farey map. More generally, additive and multiplicative versions for a same type of rule can lead to very distinct behaviors. See for instance [31], where it is shown that the multiplicative form of Selmer's algorithm is not an acceleration of its additive version. The underlying cause of this is the fact that the group of matrices generated by positive transvections and permutations is not commutative.

One convenient way to get an additive algorithm is to restrict the range of the map A to the set of elementary and permutation matrices with entries in $\{0, 1\}$. A matrix is called *elementary* if it has 1's on the diagonal, one entry equal to 1 elsewhere, and all other entries equal to 0. In geometric terms (still following the geometric formalism from [31]) the allowed operations on the bases at each step n are of elementary types (they correspond to integer transvections): for every n, there exist $i \neq j$ (with i, j depending on n) and $c^{(n)} \in \mathbb{N}$ such that

$$\mathbf{b}_i^{(n+1)} = \mathbf{b}_i^{(n)} + c^{(n)}\mathbf{b}_j^{(n)}, \ \mathbf{b}_k^{(n+1)} = \mathbf{b}_k^{(n)} \text{ for } k \neq i. \tag{10}$$

This restriction is not a severe one and most of the algorithms discussed in the present survey enter this framework, by allowing also permutation rules between the vectors. Algorithms for which the choice of the coefficients i, j and $c^{(n)}$ depend only on the cofactors of $\boldsymbol{\ell}$ with respect to $\mathbf{b}^{(n)}$, i.e., the integers $a_i^{(n)}$ such that

$$\boldsymbol{\ell} = a_1^{(n)}\mathbf{b}_1^{(n)} + \cdots + a_{d+1}^{(n)}\mathbf{b}_{d+1}^{(n)},$$

are called *vectorial* in [31]. They are memory-less algorithms. In particular, with the notation of (2), if, for some $\boldsymbol{\alpha}$ the matrix $A(\boldsymbol{\alpha})$ is equal to an elementary matrix, then the entries of the vector $(T(\boldsymbol{\alpha}), 1)$ are obtained (modulo the renormalization produced by the map θ) by subtracting an entry from another one; it is obtained by performing subtractions.

4.4 The Effectiveness of Ergodic Theory

Having an underlying dynamical system offers a wide range of mathematical tools, including ergodic theory and thermodynamic formalism via transfer operators. Ergodic theorems describe the limiting behavior of ergodic sums of the form $\frac{1}{n}\sum_{k=0}^{n-1} f \circ T^k$. In probabilistic terms, the random variables $f \circ T^n$ satisfy the strong law of large numbers under the ergodic hypothesis on T. Ergodic sums $\frac{1}{n}\sum_{k=0}^{n-1} f \circ T^k$ allow the expression of a wide set of algorithmic and arithmetic parameters, and ergodic theorems allow the understanding of their mean behaviors. Besides ergodic theory, transfer operators provide a description of the evolution of probability density functions, by transporting the action of the map T from a dynamical system to the densities. If initial conditions (x_i) for trajectories are distributed according to a probability density function, then the new collection of points $(T(x_i))$ is distributed according to a new probability density function, obtained by applying a transfer operator.

Ergodic theory has been quite an effective tool in understanding the typical behavior of various expansions of numbers and their quality of approximation. To name a few, we mention β-expansions, Lüroth series and multidimensional continued fractions.

We first illustrate it with the Gauss map. For general ergodic aspects of the Gauss map, see, e.g., [24, 51, 57]. The statistics of occurrences of partial quotients in continued fraction expansions are deduced from the ergodic theorem applied to the Gauss map, and having an ergodic absolutely continuous invariant measure (a so-called *a.c.i.m.*) then provides metric results that hold almost everywhere with respect to the Lebesgue measure.

To explain this in more detail, let us recall basic ergodic properties of the Gauss map. See also Sect. 3.3 where first definitions have been given for shift spaces. We endow the dynamical system $([0, 1], T_G)$ with a structure of a measure-theoretic dynamical system. A *measure-theoretic dynamical system* is defined as a system $(X, T, \mu, \mathcal{B})$, where μ is a probability measure defined on the σ-algebra $\mathcal{B}$ of subsets of X, and $T : X \to X$ is a measurable map which preserves the measure μ, i.e. $\mu(T^{-1}(B)) = \mu(B)$ for all $B \in \mathcal{B}$. In this case one also says that the measure μ is *T-invariant*. Here, we endow $([0, 1], T_G)$ with the Gauss measure μ_G which is a Borel probability measure absolutely continuous with respect to the Lebesgue measure defined via the following density function

$$\mu_G = \frac{1}{\log 2} \int \frac{1}{1+\alpha} d\alpha. \tag{11}$$

One checks that this measure is T_G-invariant. The Gauss map is *ergodic* with respect to the Gauss measure, that is, every Borel subset B of $[0, 1]$ such that $T_G^{-1}(B) = B$ has either zero or full measure. This implies that almost all orbits are dense in $[0, 1]$.

Ergodicity yields furthermore the following striking convergence result. Indeed, measure-theoretic ergodic dynamical systems satisfy *Birkhoff's ergodic theorem*, also called *individual ergodic theorem*, which relates spatial means to temporal means.

Theorem 4 (Birkhoff's Ergodic Theorem) *Let $(X, T, \mu, \mathcal{B})$ be an ergodic measure-theoretic dynamical system. Let $f \in L^1(X, \mathbb{R})$. Then the sequence $(\frac{1}{n}\sum_{k=0}^{n-1} f \circ T^k)_{n\geq 0}$ converges a.e. to $\int_X f\, d\mu$:*

$$\forall f \in L^1(X, \mathbb{R})\,, \quad \frac{1}{n}\sum_{k=0}^{n-1} f \circ T^k \xrightarrow[n\to\infty]{\mu-a.e.} \int_X f\, d\mu\,.$$

For example in the case of continued fractions, using this theorem one is able to describe the distribution of the digits.

Theorem 5 *For a.e. $\alpha \in [0, 1]$ the digit j occurs in the continued fraction expansion of α with density $\frac{1}{\log 2}(2\log(1+j) - \log j - \log(2+j))$.*

One can even show that the digits have the mixing property; one can as well give a short and elegant proof of Lévy's theorem which states that

$$\lim_{n\to\infty} \frac{1}{n} \log q_n(\alpha) = \frac{\pi^2}{12\log 2} \tag{12}$$

for Lebesgue almost every point α. With the help of a dynamical construction called the *natural extension* (which is basically a way to make the dynamics invertible), one can describe the asymptotic behavior of the approximation coefficients

$$\Theta_n(\alpha) = q_n^2|\alpha - \frac{p_n}{q_n}|.$$

As an illustration, we mention the ergodic proof of Bosma, Jager and Wiedijk [26] of the famous Doeblin-Lenstra conjecture which says that for almost all α the limit

$$\lim_{n\to\infty} \frac{1}{n}\#\{1 \leq j \leq n \mid \Theta_j(\alpha) \leq z\}\,, \quad \text{where } 0 \leq z \leq 1,$$

exists and equals the distribution function $F(z)$ given by

$$F(z) = \begin{cases} \dfrac{z}{\log 2} & 0 \leq z \leq \frac{1}{2}, \\[2ex] \dfrac{1}{\log 2}(1 - z + \log 2z) & \frac{1}{2} \leq z \leq 1. \end{cases} \tag{13}$$

In [77], Jager used ergodic tools to describe the simultaneous distribution of two consecutive Θ's by determining for almost all α the exact value of

$$\lim_{n\to\infty} \frac{1}{n} \#\{1 \le j \le n : \Theta_{j-1}(\alpha) \le z_1, \Theta_j(\alpha) \le z_2\}, \text{ where } 0 \le z_1, z_2 \le 1.$$

As a third example we mention that it can be proven with the help of ergodic theory that for each real irrational number α and each integer $n \ge 1$ one has

$$\min(\Theta_{n-1}, \Theta_n, \Theta_{n+1}) < \frac{1}{\sqrt{a_{n+1}^2 + 4}} \tag{14}$$

and

$$\max(\Theta_{n-1}, \Theta_n, \Theta_{n+1}) > \frac{1}{\sqrt{a_{n+1}^2 + 4}}. \tag{15}$$

For a generalization of these results in the setting of best approximations, see [42, 43], and also Sect. 5. Inequality (14) is a generalization of a result by Borel [30], which states that

$$\min(\Theta_{n-1}, \Theta_n, \Theta_{n+1}) < \frac{1}{\sqrt{5}}.$$

A great number of people independently found (14), see for example [28, 122, 143]. Inequality (15) is due to Tong [149]. In fact, ergodic theoretic methods yield easy proofs of generalizations of a great number of classical results by Fujiwara, Segre and others like LeVeque, Szüzs, and Segre; see [51, 78] for more results and details.

In a different direction, ergodic theoretic methods can also be used to prove (generalizations of) Lochs' theorem which compares the amount of information given by the digits of different types of number expansions. The original statement of Lochs [109] compared the regular continued fraction digits to decimal digits. More precisely, for an $\alpha \in (0, 1)$, let $(a_k)_{k\ge 1}$ denote its regular continued fraction digits and $(d_k)_{k\ge 1}$ its decimal digits and, for each $n \ge 1$, let $m_n(\alpha)$ denote the largest number of digits a_k that can be determined from knowing $d_1, \dots, d_n$. Then Lochs proved the following statement.

Theorem 6 *For a.e.* $\alpha \in (0, 1)$, $\lim_{n\to\infty} \frac{m_n(\alpha)}{n} = \frac{6 \log 2 \log 10}{\pi^2}$.

Lochs' theorem was placed in a dynamical setting in [18] and developed further in [48]. It became apparent that the number $\frac{6 \log 2 \log 10}{\pi^2}$ is related to the measure-theoretic entropies $h(T_G) = \frac{\pi^2}{6 \log 2}$ and $h(T_{10}) = \log 10$ of the transformations T_G and $T_{10}(x) = 10x \pmod 1$ that generate regular continued fractions and decimal expansions, respectively. See also [23, 110] for similar results for other types of expansions. Later Lochs' theorem was extended and generalized in several directions, e.g., in [97] for random dynamical systems and in [17] to include systems with zero entropy.

Let us consider now the case of multidimensional continued fraction algorithms. For a general description of their ergodic properties, see [138]. A first step in the ergodic study of a continued fraction algorithm consists of proving the existence of an a.c.i.m. (absolutely continuous invariant measure), playing a similar role as the Gauss measure defined in (11). Note that a very efficient way to find a.c.i.m.'s for continued fraction transformations is via the natural extensions; see, e.g., [6, 7, 13, 55, 93].

Let us take a closer look at the convergence properties from a dynamical viewpoint. The a.e. exponential (strong) convergence of the Brun [60, 113, 139] and Jacobi–Perron algorithm [14] (see also [105, 137]) holds in dimension $d = 2$: there exists $\delta > 0$ s.t. for a.e. (α_1, α_2), there exists $n_0 = n_0(\alpha, \beta)$ s.t. for all $n \geq n_0$

$$|\alpha_1 - p_1^{(n)}/q^{(n)}| < \frac{1}{(q^{(n)})^{1+\delta}}, \quad |\alpha_2 - p_2^{(n)}/q^{(n)}| < \frac{1}{(q^{(n)})^{1+\delta}},$$

where $p_1^{(n)}, p_2^{(n)}, q^{(n)}$ are given by the Brun (resp. Jacobi–Perron) algorithm.

We have seen already in Sect. 3.3 that the quality of rational approximations provided by continued fraction algorithms obtained dynamically by an iteration of a map can be expressed in terms of the two first Lyapunov exponents λ_2 and λ_1: the coefficient δ corresponds to $-\lambda_2/\lambda_1$. There is, in fact, a strong link with the uniform approximation exponent $\eta_A^*(\boldsymbol{\alpha})$ (whose definition is recalled below) as shown in [105]; see also [70, Theorem 1], [15, Proposition 4] and [38, Section 2]. One considers as in (2) the maps $T : X \to X$, $A : X \to \mathrm{GL}(d+1, \mathbb{Z})$. For fixed $\boldsymbol{\alpha} \in X$ and $i \in \{1, \ldots, d+1\}$, we have

$$\eta_A^*(\boldsymbol{\alpha}, i) =$$

$$\sup \left\{ \delta > 0 \ : \ \exists n_0 = n_0(\boldsymbol{\alpha}, i, \delta) \in \mathbb{N} \text{ s. t. } \forall n \geq n_0, \ \left\| \boldsymbol{\alpha} - \frac{\mathbf{p}_i^{(n)}}{q_i^{(n)}} \right\| < (q_i^{(n)})^{-\delta} \right\},$$

where $\|\cdot\|$ is an arbitrary norm in $\mathbb{R}^d$. The quantity

$$\eta_A^*(\boldsymbol{\alpha}) = \min_{1 \leq i \leq d+1} \eta_A^*(\boldsymbol{\alpha}, i)$$

is called the *uniform approximation exponent* for $\boldsymbol{\alpha}$ using the algorithm A. Now, following [105, Theorem 4.1], we consider a d-dimensional multidimensional continued fraction algorithm A satisfying some mild ergodic conditions (called (H1) to (H5) in [105]). Let η_A^* be the uniform approximation exponent of A. We have $\lambda_1(A) > \lambda_2(A)$ and

$$\eta_A^*(\boldsymbol{\alpha}) = 1 - \frac{\lambda_2(A)}{\lambda_1(A)}$$

holds for almost all $\boldsymbol{\alpha} \in X$. In particular, if $\lambda_2(A) < 0$ then A is a.e. strongly convergent. Lyapunov exponents of classical algorithms such as the Jacobi–Perron algorithm or the Brun algorithm have been thoroughly studied in [14, 32]; see also [63] for results on the simplicity of the Lyapunov spectrum.

The computation of Lyapunov exponents is a challenging problem. For numerical results, see [38]. In practice, ergodic theorems provide efficient ways of estimating Lyapunov exponents numerically by following trajectories and then taking averages over truncated trajectories. This has been developed, e.g., in [29, 72] for continued fractions or in [157] for interval exchanges. Transfer operators are efficient ways to reach them, such as developed, e.g., in [81, 128], by getting in some cases exact computations for the top Lyapunov exponents of random products of matrices using transfer operators. Indeed, transfer operators come with the analogue of Perron–Frobenius theory for non-negative matrices. They are well suited to provide approximations by working in finite dimension, for instance by truncating Taylor expansions of analytic functions [49, 107]. See also [144] for lower and upper bounds for the Lyapunov exponents for random products of square matrices with determinant 1 having real and non-negative entries.

4.5 Classical Examples

With this section we want to illustrate the variety of continued fractions algorithms defined according to the formalism of Sect. 4.4, as well as to focus on two classical algorithms, namely the Brun and the Jacobi–Perron algorithms.

Consider algorithms based on elementary matrices, i.e., on subtractions, such as described in Sect. 4.4. In order to stress the simple rules that govern them, we express them in dimension $d + 1 = 3$. We thus start with parameters $\boldsymbol{\ell} = (\ell_1, \ell_2, \ell_3) \in \mathbb{R}_+^3$. We have to decide which number has to be subtracted, and with respect to which number it has to be done. Usually numbers ℓ_1, ℓ_2, ℓ_3 are sorted in increasing (or decreasing) order. We stress the subtraction rule but it is usually preceded and followed by a sorting operation.

For Jacobi–Perron, we subtract the second entry from the other ones, with the entries being ordered in such a way that the first entry is the largest one (but with no further order restriction on the two other entries). For Brun, we subtract the second largest from the largest one. For Poincaré, we subtract the second largest entry from the largest one, the third largest from the second largest, etc. For Selmer, we subtract the smallest (positive) entry from the largest one. For the fully subtractive algorithm, we subtract the smallest (positive entry) from all the larger ones.

We also recall that given an algorithm described at the level of a $d + 1$-dimensional space, there exist several possible projectivizations. This is the case of the various forms taken by the Brun algorithm. In particular, the Brun algorithm is also called modified Jacobi–Perron algorithm: the modified Jacobi–Perron algorithm introduced by Podsypanin in [127] is a two-point extension of the Brun algorithm.

Consider now the Jacobi–Perron algorithm (that will be handled in more details with respect to the nearest integer part in Sect. 7). The linear form of the Jacobi–Perron algorithm is defined on $\{(y_0, y_1, y_2) \in \mathbb{R}^3 \setminus \{\mathbf{0}\} : \ y_0 \geq y_1, y_2 \geq 0\}$ by

$$(y_0, y_1, y_2) \mapsto (y_1, y_2 - \lfloor y_2/y_1 \rfloor y_1, y_0 - \lfloor y_0/y_1 \rfloor y_1).$$

If we set

$$x_1 := y_1/y_0, \ x_2 := y_2/y_0,$$

we recover its projective version defined on $[0, 1]^2$ as

$$(x_1, x_2) \mapsto (\{x_2/x_1\}, \{1/x_1\}) .$$

Let us stress some differences between Brun's and Jacobi–Perron's rule. The Brun algorithm is a space-ordering algorithm according to the terminology introduced in [71]. (Note that it is called ordered Jacobi–Perron in [70].) Furthermore, each step of the Brun algorithm produces only one digit. This helps in computing the natural extension and the invariant measure of the Brun algorithm (see, e.g., [7]). Contrary to the Brun algorithm, the role played by y_1 and y_2 (in the description above) is not determined by a comparison between both parameters in the Jacobi–Perron case; this might explain the fact that an explicit expression of the natural extension of this algorithm is still not known. Nevertheless, the framework of S-expansions and the so-called techniques of Insertion and Singularization (see [75]) allow one to relate both algorithms as shown in [140]; see also [141]. Both algorithms (Brun and Jacobi–Perron) are known to have an invariant ergodic probability measure equivalent to the Lebesgue measure (an a.c.i.m.); see for instance [135] and [138]. However, this measure is not known explicitly for Jacobi–Perron (the density of the measure is shown to be a piecewise analytic function in [32]), whereas it is known explicitly for Brun [7, 60]. Note that the Brun algorithm can be considered as an additive algorithm [27], where it is proved that partial quotients tend to be equal to 1. Lastly, let us quote [119] for Borel-Bernstein type theorems on the growth of partial quotients.

5 Lattice Reduction and Rational Approximations

We now focus on the second approach discussed in Sect. 3.2 based on lattice reduction. Lattice reduction methods induce indeed a particularly fruitful way of exhibiting good simultaneous approximations, or else small values for linear forms. Algorithms based on lattice reduction theory are based on the following idea: lattice reduction algorithms do not produce a priori the smallest vector of a lattice, but a reasonably small vector. That is, a vector that is small enough for guarantying Diophantine approximation properties that can be compared with Dirichlet's quality up to an approximation factor exponential in the dimension. We can thus consider these algorithms as providing effective versions of Dirichlet's theorem, yielding a satisfying compromise between efficient computation and sharpness of the obtained bounds, that is, between algorithmic issues and Diophantine quality.

The range of applications of lattice reduction is quite wide for the following reasons. Firstly, lattice reduction algorithms play a central algorithmic role in cryptology, computer algebra, integer linear programming and algorithmic number theory. Secondly, they are particularly versatile in terms both of existing variants and algebraic contexts where they can be developed, see, e.g., [40, 62, 117]. Thirdly, they are efficient: LLL has a polynomial runtime with respect to the dimension. In particular, in the present context, they produce efficient gcd algorithms (see, e.g., [73]), and there exist promising attempts in order to devise continued fractions upon them (see [21, 37, 106]). However, there remains much to understand concerning their executions and the geometry of the outputs [120, 155].

Lattice reduction is based on the following elementary basis transformations: they can be described in terms of size reduction (the vector $\mathbf{b}_i$ of the basis $(\mathbf{b}_1, \ldots, \mathbf{b}_{d+1})$ is replaced by $\mathbf{b}_i - \lambda \mathbf{b}_j$ with $1 \leq j < i$), and of exchange steps, also called swaps (one exchanges $\mathbf{b}_i$ and $\mathbf{b}_{i+1}$). These operations are decided with respect to the Gram–Schmidt orthogonalization of the basis $\mathbf{b}$.

More precisely, let $(\mathbf{b}_i^*)$ stand for the basis obtained via the Gram–Schmidt orthogonalization from the basis $(\mathbf{b}_1, \ldots, \mathbf{b}_{d+1})$, i.e., $\mathbf{b}_i^*$ is the orthogonal projection of $\mathbf{b}_i$ on the orthogonal of the space generated by $\mathbf{b}_i, \cdots, \mathbf{b}_{i-1}$. One writes $\mathbf{b}_i^* = \mathbf{b}_i - \sum_{k=1}^{i-1} \mu_{ik} \mathbf{b}_k^*$ with $\mu_{ik} = \frac{\langle \mathbf{b_i}, \mathbf{b}_k^* \rangle}{\langle \mathbf{b_k^*}, \mathbf{b}_k^* \rangle}$ for $k \leq i-1$. A basis $(\mathbf{b}_1, \cdots, \mathbf{b}_d)$ is said to be LLL-reduced if

- $|\mu_{ik}| \leq 1/2$ for all i, k with $1 \leq i \leq d+1$ and $k \leq i-1$ (the basis is said *proper*).
- $3/4\|\mathbf{b}_i^*\| \leq \|\mu_{i+1,i}\mathbf{b}_i^* + \mathbf{b}_{i+1}^*\|$ for all i (this condition is called *Lovasz' condition*).

The factor $3/4$ can be replaced by a parameter t with $3/4 < t < 1$. The LLL algorithm consists of two steps.

- First, make the basis proper by replacing $\mathbf{b}_i$ by $\mathbf{b}_i - \|\mu_{ij}\|\mathbf{b}_j$, for $j < i$, where $\|\mu_{ij}\|$ stands for the distance to the nearest integer.
- If for some i, Lovasz' condition is not satisfied, then swap $\mathbf{b}_i$ and $\mathbf{b}_{i+1}$ and go the previous step.

Recall that we have sketched the basic strategy underlying the use of lattice reduction in this framework in Sect. 3.2. One starts with the lattice Λ_t generated by the columns of the matrix

$$M_t := \begin{bmatrix} 1 & 0 & \cdots & 0 & -\alpha_1 \\ 0 & 1 & \cdots & 0 & -\alpha_2 \\ \vdots & \vdots & \ddots & \vdots & \vdots \\ 0 & 0 & \cdots & 1 & -\alpha_d \\ 0 & 0 & \cdots & 0 & t \end{bmatrix}.$$

Note that $\det(M_t) = t$, hence, the lattice Λ_t changes at each step of the algorithm. One lets t tend to 0. Let us stress the fact that this strategy differs from the one discussed in Sect. 4.5 where one worked with bases of the fixed lattice $\mathbb{Z}^{d+1}$.

Lattice reduction algorithms such as LLL then perform a succession of permutations and subtractions on the matrix M_t, i.e., they multiply the matrix M_t by elementary matrices and permutation matrices. This is a common feature between unimodular continued fraction algorithms and algorithms based on lattice reduction, namely that they are made of a succession of permutations and subtractions. The decisions are taken for classical unimodular continued fractions by comparing entries, whereas lattice reduction involves quadratic comparisons in the sense that they depend on the Gram–Schmidt orthogonalization.

Lattice reduction provides best approximations of a real number, see [121, p.226,267], and for a survey on the overall strategy for getting constructive type results in Diophantine approximation based on LLL, see [121, p. 222]. Nevertheless, note that even in dimension 2, when using the Gauss algorithm whose efficiency has been largely proved, one has "little control on the convergent which is returned; in particular, this is *not* the largest convergent with denominator less than $2\sqrt{C/3}$", as quoted in [121, p.226 Example 1]; here the bound $2\sqrt{C/3}$ comes from Theorem 7 of [121, Chapter 6].

As an illustration of the efficiency of lattice reduction in algorithmic number theory, see, e.g., [129] where algorithms for solving the dual problems of approximating linear forms and of simultaneous approximation in number fields are developed by applying the LLL algorithm. A system of independent units of full rank is generated in any algebraic number field, which also leads to an algorithmic generalization of Lagrange's theorem.

Several attempts already exist in order to use lattice reduction to get simultaneous approximations. Let us quote [58, 59, 82–84, 103, 104] for strongly convergent algorithms; however they do not present the same advantages as more classical memory-less algorithms. Let us quote also [104] and [106] based on Minkowski lattice reduction; this approach is not effective, but it produces best approximations. This study is also extended in [66], [46] and [92].

Some approaches are built specifically upon LLL; see [21, 37]. In [37] an iterated LLL algorithm is designed that is obtained by decreasing the parameter t from the lattice Λ_t, by dividing it by a given constant (including also experimental data). In [21], the conditions that occur in LLL are proved to be linear in the parameter t (tending to 0). The idea at step k is to consider the smallest parameter t_k for which Λ_{t_k} is reduced, and then perform a reduction with $t_k - \varepsilon$.

An efficient way to input some dynamics with this reduction viewpoint is to rely on homogeneous dynamics with the (left) action of the diagonal flow (g_t) defined by

$$\begin{bmatrix} e^t I_d & 0 \\ 0 & e^{-dt} \end{bmatrix}$$

on the space of unimodular lattices, i.e., on the homogeneous space $SL(d+1, \mathbb{R})/SL(d+1, \mathbb{Z})$. This is a very fruitful way to combine lattice reduction with the strength of dynamical methods such as highlighted in the survey [46]. This amounts to changing the parameter t. See in particular [92] for a strongly convergent multidimensional continued fraction algorithm with effective hyperbolicity estimates.

This is particularly relevant for continued fractions defined in terms of best approximations. In [42, 43], Levy's result about the almost sure growth rate of the denominators of the convergents from (12) is extended to the best Diophantine approximations (see Definition 2). The value of the limit is given by an integral over a codimension one submanifold in the space of lattices $SL(d+1, \mathbb{R})/SL(d+1, \mathbb{Z})$. An analogue of the Doeblin-Lenstra discussed in Sect. 4.4 is also given in [43]. As highlighted in [43], the section is provided by lattices whose first two minima of lattices are equal and the first return map of the geodesic flow in the transversal plays the role of an invertible extension of the missing Gauss map. The main idea is to relate shortest and so-called minimal vectors of the lattice Λ_t with best approximations (see [46, Lemma 8]) together with a suitable choice of a norm (see [46, Section 3.2] and [45]).

6 Some Applications of Continued Fractions

In this section we give a diverse range of applications of continued fractions in arithmetics, cryptography and symbolic dynamics. We then propose possible hints for improvements of dynamical unimodular continued fraction algorithms.

We start with arithmetical applications. Continued fractions play an important role in the **arithmetic of algebraic curves**. The relation between the geometry of the plane quartic model of an elliptic curve, given by the equation $y^2 = x^4 - 6ax^2 - 8bx + c$, and the continued fraction of y with respect to the x^{-1}-adic valuation is given in [8] and [1]. The given elliptic curve has two rational points at infinity, call them P and O, such that O is the origin of the Mordell-Weil group of the elliptic curve. Then the order of the image of P in the Jacobian (called the infinity divisor of the curve) is finite if and only if the continued fraction of y is periodic. Moreover, the order depends on the period of the continued fraction. This result has been generalized for the hyperelliptic curves in [126]. The torsion order of the infinity divisor is given as the sum of the degrees of all partial fractions from 0 to the period of the continued fraction of y in [8, 130, 150]. Furthermore, thanks to the periodic continued fraction of y, for a given even degree hyperelliptic curve

$y^2 = f(x)$, the Pell equation for f is proved to have a non-trivial solution, i.e., there exist p, q in $k[x]$, p not a constant, such that $p^2 - q^2 f = 1$, in [8, 126].

Cryptography is another field where continued fractions occur in various places. First of all, continued fractions can be used in cryptanalysis, e.g., to attack the RSA cryptosystem. This cryptosystem is based on the mathematical problem of factoring a natural number that is a product of two large prime numbers. It is an *asymmetric cryptosystem*, i.e., a publicly known key is used for encryption and a secret key is used for decryption. If the secret key is chosen too small, one can use continued fraction expansions to efficiently compute the private key and break the cryptosystem. This attack is known as Wiener's attack (see [152]).

Continued fractions can be used to directly factorize integers. The continued fraction factorization method (CFRAC) and Shanks's square forms factorization (SQUFOF) are two such algorithms. For example, the main idea behind the CFRAC algorithm is to compute the continued fraction expansion of $\sqrt{n}$, where n is the number we want to factorize. This continued fraction expansion is then used to obtain a factor dividing n. However, these factorization algorithms are not efficient enough to factorize numbers in the magnitude of RSA parameters.

Apart from cryptanalysis, there is a connection between stream ciphers and continued fraction expansions. Stream ciphers are *symmetric cryptosystems*, i.e., the same key is used for encryption and decryption. A stream cipher generates a pseudorandom bit string. To encrypt a message, each bit of the pseudorandom bit string is combined with a bit of the secret message (e.g., with the XOR operation). Niederreiter uses continued fraction expansions of generating functions to analyze the randomness of pseudorandom sequences (see [118]). In [85], Kane constructs a stream cipher using continued fractions. The cryptosystem KronCrypt is a block cipher that is based on a constructive proof of Kronecker's approximation theorem (see [56]). Block ciphers encrypt blocks of the secret message, instead of generating a bit string that is combined with the secret message bitwise. KronCrypt has the structure of a Feistel cipher with key-dependent S-boxes. The continued fraction expansion is used for the key generation.

A rather indirect connection to cryptography is the shared interest in lattice reduction algorithms in the setting of post-quantum cryptography. As described in Sect. 5, these algorithms compute relatively short vectors of lattices. Various cryptosystems such as Frodo, Dilithium, and Kyber are based on the problem of finding short vectors in lattices. Hence, the study of lattice reduction algorithms such as the LLL algorithm is important for the security analysis of lattice-based cryptosystems.

We also note that due to its simplicity, the Brun algorithm appears in various application fields in cryptography and beyond. See [54], where efficient exponentiation using addition chains is used. Note that continued fractions were already used for addition chains; see, e.g., [16]. See also the survey [154] for application of lattice reduction and of the Brun algorithm in wireless communications and statistical signal processing.

We now consider some applications in **(symbolic) dynamics** with the seminal example of Sturmian dynamical systems, introduced by Morse and Hedlund in [114]. There is an impressive literature devoted to their study and to their possible generalizations in word combinatorics, and in digital geometry [133]. This is due to several factors. They provide symbolic codings for the simplest arithmetic systems, namely the irrational translations on the circle, they also code discrete lines and are unidimensional models of quasicrystals [22]. Moreover, the scale invariance of Sturmian dynamical systems allows their description using a renormalization scheme governed by usual continued fractions via the geodesic flow acting on the modular surface. More generally, renormalization schemes can often be interpreted as continued fractions [156]; a striking illustration comes from the study of interval exchanges in relation with the Teichmüller flow, through the work, among others, of Veech, Masur, Yoccoz, and Avila.

Similarly as for continued fractions, there is no canonical generalization of Sturmian words. Episturmian words, also called Arnoux-Rauzy words, have attracted a lot of attention in this direction. See in particular [4, 5, 44] for the study of associated continued fractions on the Rauzy gasket. More generally, it has been a long-standing problem to find good symbolic codings for translations on the d-dimensional torus that enjoy the beautiful properties of Sturmian sequences. This is the object of [39] which defines symbolic codings of toral translations based on exponentially convergent multidimensional continued fraction algorithms, leading to renormalization schemes. However these results rely on the positiveness of the second Lyapunov exponent, and they thus hold only in small dimensions.

We end this section by discussing possible ways to improve existing continued fractions algorithms. Indeed, in view of the non-positivity of the second Lyapunov exponent, there is a need to design strongly convergent continued fractions algorithms in high dimensions. In fact, positivity of the second Lyapunov exponent has only been proved in dimension $d = 2$ for classical algorithms, and $d = 3$ for the Brun algorithm. One natural approach for the design of continued fractions consists in trying to derive continued fraction algorithms from lattice reduction algorithms using the fact that they compute short vectors and that they reach Dirichlet's bound (up to a constant depending exponentially on the dimension). The design of continued fractions will go along with the dynamical modeling of lattice reduction algorithms (and more specifically LLL) for their probabilistic analysis. There is also a need to improve existing algorithms of a dynamical nature by taking advantage of their dynamical properties.

For other types of number expansions, in particular for β-expansions, it turned out to be very fruitful to introduce a random dynamical system to produce the expansions. Where a deterministic system is defined as a pair (X, T), a *random dynamical system* makes use of a family of transformations $(T_i : X \to X)_{i \in I}$ (for some index set I) all defined on the same domain X, where each map is chosen with a certain probability. One then studies the compositions of the form

$$T_{i_n} \circ \cdots \circ T_{i_1}(x), \quad i_j \in I,$$

instead of $T^n(x)$. For β-expansions, i.e., expansions of real numbers x of the form $\sum_{k\geq 1} \frac{b_k}{\beta^k}$, with $\beta > 1$ a non-integer and each b_i an integer between 0 and β, a corresponding random dynamical system was first introduced in [52]. Applications of random β-expansions with respect to rational approximations of real numbers were then described by Daubechies et al. [47], see also [50, 90], in relation to analog-to-digital conversion and in [79, 80] in relation to random number generation. Random one-dimensional continued fraction algorithms and their invariant measures were studied in [33, 53, 91, 94, 147]. Often, in terms of dynamics and the corresponding approximation properties, such a random system performs comparably to the best performing deterministic system present in the family $(T_i : X \to X)_{i\in I}$, but with the added flexibility that real numbers now have many different expansions assigned to them. Even though it is not quite clear at the moment whether placing multidimensional continued fraction algorithms in a random framework could yield similar advantages, it might be of interest to study this further.

7 Improving Jacobi–Perron Algorithm

In 1981, both Ito and Tanaka in [148] and Nakada in [116] introduced the notion of α-continued fractions for $\alpha \in [1/2, 1]$. This was done by replacing $\lfloor \frac{1}{x} \rfloor$ in the definition of the Gauss map T_G by $\lfloor \frac{1}{x} + 1 - \alpha \rfloor$ in [148] and by $\lfloor \frac{1}{|x|} + 1 - \alpha \rfloor$ in [116]. For the setup of Nakada, the sequence of the convergents of the corresponding α-continued fraction of a point form a subsequence of its regular sequence of convergents, and hence provide better approximations. In particular, for $\alpha \in [\frac{1}{2}, \frac{\sqrt{5}-1}{2}]$, all the corresponding α-continued fraction algorithms are isomorphic and provide better approximations than the ones given by the regular continued fractions. See in particular [96] for results on the entropy. One can use a similar idea to improve the convergence properties of the Jacobi–Perron algorithm. To keep the exposition simple, we will consider the nearest integer case, corresponding to $\alpha = \frac{1}{2}$, and we will not take absolute values inside the floor function (so the digits generated could be negative).

It is well known that when it comes to convergence speed, in one dimension the nearest integer algorithm performs best, see, e.g., [19]. For further results on nearest integer continued fractions see [3, 76, 153]. Dynamically the algorithm is given by the map $T(x) = \frac{1}{x} - |||\frac{1}{x}|||$, where $||| \cdot |||$ denotes the distance to the nearest integer as before. The map is well defined on the interval $[-\frac{1}{2}, \frac{1}{2}]$. The possible continued fraction digits for this algorithm are all integers n with $|n| \geq 2$. In higher dimensions a nearest integer version of the Jacobi–Perron algorithm has attracted some attention. See Sect. 7.2 for numerical values for the nearest integer Jacobi–Perron algorithm from [145]. We also recall that the Markov conditions on the digits produced by the classical Jacobi–Perron algorithm have a very simple form and that the piecewise analyticity of the density of its invariant measure has been proved in [32].

7.1 First Dynamical Properties of the Algorithm

Let $d \geq 2$. The definition of the *d-dimensional nearest integer Jacobi–Perron map (NIJP)* on $C = \left[-\frac{1}{2}, \frac{1}{2} \right]^d$ is given by

$$T_0(x_1, \ldots, x_d) = \left(\frac{x_2}{x_1} - \left\lfloor \frac{x_2}{x_1} + \frac{1}{2} \right\rfloor, \ldots, \frac{x_d}{x_1} - \left\lfloor \frac{x_d}{x_1} + \frac{1}{2} \right\rfloor, \frac{1}{x_1} - \left\lfloor \frac{1}{x_1} + \frac{1}{2} \right\rfloor \right)$$

and can be used to create for each d-tuple $(x_1, \ldots, x_d) \in C$ a sequence of continued fraction approximations with the same denominator.

The matrix version of T_0 is therefore defined on

$$\Lambda = \{(y_0, y_1, \ldots, y_d) \in \mathbb{R}^{d+1} \setminus \{\mathbf{0}\} : \, y_0 \geq y_1, \cdots, y_d \geq 0\}$$

as

$$(y_0, y_1, \ldots, y_d) \mapsto \left(y_1, y_2 - \left\lfloor \frac{y_2}{y_1} + \frac{1}{2} \right\rfloor y_1, \ldots, y_d - \left\lfloor \frac{y_d}{y_1} + \frac{1}{2} \right\rfloor y_1, y_0 - \left\lfloor \frac{y_0}{y_1} + \frac{1}{2} \right\rfloor y_1 \right),$$

and we have

$$\begin{aligned} &{}^t(y_0, y_1, \ldots, y_d) = {}^t A_0(y_0, y_1, \ldots, y_d) \cdot \\ &{}^t\left(y_1, y_2 - \left\lfloor \frac{y_2}{y_1} + \frac{1}{2} \right\rfloor y_1, \ldots, y_d - \left\lfloor \frac{y_d}{y_1} + \frac{1}{2} \right\rfloor y_1, y_0 - \left\lfloor \frac{y_0}{y_1} + \frac{1}{2} \right\rfloor y_1 \right) \end{aligned}$$

with

$${}^t A_0(y_0, y_1, \ldots, y_d) = \begin{bmatrix} \lfloor \frac{1}{y_1} + \frac{1}{2} \rfloor & 1 & \lfloor \frac{y_2}{y_1} + \frac{1}{2} \rfloor & \cdots & \lfloor \frac{y_d}{y_1} + \frac{1}{2} \rfloor \\ 0 & 0 & 1 & \cdots & 0 \\ \vdots & \vdots & \ddots & \ddots & \vdots \\ 0 & 0 & \cdots & \cdots & 1 \\ 1 & 0 & \cdots & \cdots & 0 \end{bmatrix}.$$

Iterates of T_0 then produce a matrix

$$\begin{bmatrix} q_0^{(n)} & p_{0,1}^{(n)} & \cdots & p_{0,d}^{(n)} \\ q_1^{(n)} & p_{1,1}^{(n)} & \cdots & p_{1,d}^{(n)} \\ \vdots & \vdots & \ddots & \vdots \\ q_d^{(n)} & p_{d,1}^{(n)} & \cdots & p_{d,d}^{(n)} \end{bmatrix}.$$

Note that this is not exactly the same matrix as in (3). This is mainly due to the fact that we use a different normalization for technical reasons: the largest coordinate is the first one (namely y_0), and not the last one, as before when working with vectors of the form $(\boldsymbol{\alpha}, 1)$. The $q_i^{(n)}$ are thus given here in the first column.

For $\mathbf{x} = (x_1, \ldots, x_d) \in C = \left[-\frac{1}{2}, \frac{1}{2}\right]^d$, set

$$a = a(\mathbf{x}) = \left|\left|\left|\frac{1}{x_1}\right|\right|\right|, \quad b^{(i)} = b^{(i)}(\mathbf{x}) = \left|\left|\left|\frac{x_i}{x_1}\right|\right|\right|, \; 2 \le i \le d.$$

The functions a and $b^{(i)}$ are piecewise constant on C. To be precise, for each $\mathbf{x} \in C$, it holds that $a(\mathbf{x}) = k$, $k \in \mathbb{Z}$, if and only if $\frac{2}{2k+1} < x_1 \le \frac{2}{2k-1}$, and, for each $2 \le i \le d$, it holds that $b^{(i)}(\mathbf{x}) = k$ if and only if

$$\begin{cases} x_1\left(k - \frac{1}{2}\right) \le x_2 < x_1\left(k + \frac{1}{2}\right), & \text{if } x_1 > 0, \\ x_1\left(k + \frac{1}{2}\right) < x_2 \le x_1\left(k - \frac{1}{2}\right), & \text{if } x_1 < 0. \end{cases}$$

Hence, if we let

$$C_{a,b_2,\ldots,b_d} = \left\{\mathbf{x} \in C \,:\, a(\mathbf{x}) = a,\; b^{(i)}(\mathbf{x}) = b_i,\; 2 \le i \le d\right\},$$

then the collection $\mathcal{C} = \{C_{a,b_2,\ldots,b_d}\}$ yields a partition of C, the elements of which are called the *cylinder sets* of T_0. Figure 1 shows the partition $\mathcal{C}$ of C for $d = 2$.

Note that $C_{a,b_2,\ldots,b_d} \neq \emptyset$ if and only if $|a| \ge 2$ and $0 \le |b_i| \le \lceil \frac{a}{2} \rceil$ for each $2 \le i \le d$. This implies that for the linear version of the NIJP algorithm not all matrices of the form

$$A_0 = \begin{bmatrix} a & 0 & 0 & \cdots & 0 & 1 \\ 1 & 0 & 0 & \cdots & 0 & 0 \\ b_2 & 1 & 0 & \cdots & 0 & 0 \\ b_3 & 0 & 1 & \cdots & 0 & 0 \\ \vdots & \vdots & \vdots & \ddots & \vdots & \vdots \\ b_d & 0 & 0 & \cdots & 1 & 0 \end{bmatrix} \tag{16}$$

are allowed. It needs to hold that $|a| \ge 2$ and $0 \le |b_i| \le \lceil \frac{a}{2} \rceil$ for each $2 \le i \le d$, but there are more restrictions. Below we describe these restrictions in detail for the case $d = 2$ by showing that in this case T_0 admits a *Markov partition*, i.e., there

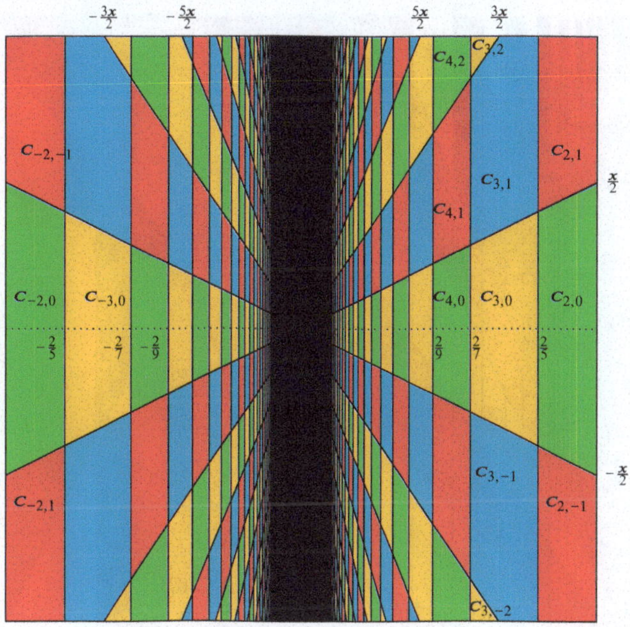

Fig. 1 The cylinder sets of the NIJP. The colors just serve to distinguish the different cylinder sets

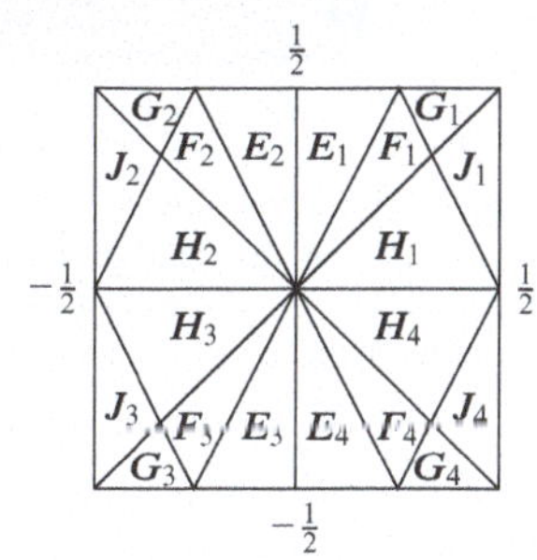

Fig. 2 The Markov partition for the NIJP map for $d = 2$

exists a finite collection $\mathcal{P}$ of polygonal subsets of $[-\frac{1}{2}, \frac{1}{2}]^2$ that have the property that for any set $P \cap A$ with $P \in \mathcal{P}$ and $A \in C$, there exist $P_1, \ldots, P_N \in \mathcal{P}$ such that $T_0(P \cap A) = \bigcup_{i=1}^{N} P_i$ up to sets of zero Lebesgue measure. Figure 2 shows the 20 sets that are in $\mathcal{P}$ for the map T_0.

Theorem 7 *Let $d = 2$. Let $\mathcal{P}$ be the collection of disjoint subsets of C bounded by the ten lines $x_1 = 0$, $x_2 = 0$, $x_2 = \pm x_1$, $x_2 = \pm 2x_1$, $x_2 = \pm 1 \pm 2x_1$. Then $\mathcal{P}$ is a Markov partition for T_0.*

Proof To show that $\mathcal{P}$ from Fig. 2 is a Markov partition, we need to consider the image under T_0 of all sets $P \cap C_{a,b}$ for $P \in \mathcal{P}$ and the cylinder set $C_{a,b} \in C$. Due to symmetry, it is enough to only consider the sets in the first quadrant. We label

Fig. 3 The black lines indicate the boundaries of cylinder sets. The blue lines indicate the boundaries of the Markov partition elements. The colors of the polygons indicate the different types of sets $C_{a,b,P}$ according to their images under the NIJP map. Sets with the same color have the same type

the sets of $\mathcal{P}$ by E_i, F_i, G_i, H_i, J_i, $1 \le i \le 4$, as shown in Fig. 2 and use $C_{a,b,P}$ to denote the set $P \cap C_{a,b}$ with $P \in \mathcal{P}$ and $C_{a,b} \in C$. Figure 3 shows several of these sets in the first quadrant. Based on the images of the sets $C_{a,b,P}$ we distinguish 15 types, indicated by the different colors in Fig. 3. Hence, to prove that T_0 admits a Markov partition for $d = 2$ we need to compute the image under T_0 of each of these types of sets. We describe what happens to the set $C_{2,0,H_1}$.

On $C_{2,0,H_1}$ the map T_0 is given by $T_0(x_1, x_2) = (\frac{x_2}{x_1}, \frac{1}{x_1} - 2)$. The boundary of $C_{2,0,H_1}$ is the union of three sets (we take all intervals open, because the endpoints of the intervals have no impact on the Lebesgue measure of the sets):

Table 1 Images of the sets $C_{a,b,P}$ for $a, b \geq 0$ and $P \in \mathcal{P}$ under T_0

	$C_{a,b,P}$	$T(C_{a,b,P})$
Type 1	$C_{2,0,H_1}$, $C_{3,1,F_1}$	$E_1 \cup F_1 \cup G_1$
Type 2	$C_{2,0,J_1}$	$H_1 \cup J_1$
Type 3	$C_{2,1,J_1}$	$E_2 \cup F_2 \cup G_2 \cup H_2 \cup J_2$
Type 4	$C_{2,1,G_1}$, $C_{4,2,E_1}$	E_1
Type 5	$C_{a,0,H_1}$ for $a \geq 3$, $C_{a,1,F_1}$ for $a \geq 4$, $C_{a,2,E_1}$ for $a \geq 6$	$\bigcup_{i=1,4}(E_i \cup F_i \cup G_i \cup H_i \cup J_i)$
Type 6	$C_{3,1,H_1}$, $C_{4,2,F_1}$	$E_2 \cup F_2 \cup G_2 \cup H_2 \cup J_2 \cup H_3 \cup J_3$
Type 7	$C_{3,1,J_1}$	$E_3 \cup F_3 \cup G_3$
Type 8	$C_{3,1,G_1}$	$H_1 \cup J_1 \cup E_4 \cup F_4 \cup H_4$
Type 9	$C_{3,2,G_1}$, C_{2k-1,k,E_1} for $k \geq 3$	$G_2 \cup J_2$
Type 10	$C_{a,1,H_1}$ for $a \geq 4$, $C_{a,2,F_1}$ for $a \geq 5$	$\bigcup_{i=2,3}(E_i \cup F_i \cup G_i \cup H_i \cup J_i)$
Type 11	$C_{4,2,G_1}$	$F_3 \cup G_3$
Type 12	$C_{5,2,E_1}$	$E_1 \cup F_1 \cup G_1 \cup H_1 \cup J_1 \cup E_4 \cup F_4 \cup H_4$
Type 13	C_{2k,k,E_1} for $k \geq 3$	$E_1 \cup E_2 \cup F_2 \cup G_2 \cup H_2 \cup J_2 \cup F_3 \cup G_3 \cup H_3 \cup J_3$
Type 14	C_{2k+1,k,E_1} for $k \geq 3$	$C \setminus (G_4 \cup J_4)$
Type 15	C_{a,b,E_1} for $b \geq 3$ and $a \geq 2b+2$	C

$$\partial_1 = \left\{(x_1, x_2) \in C \ : \ x_2 = 0,\ x_1 \in \left(\frac{2}{5}, \frac{1}{2}\right)\right\},$$

$$\partial_2 = \left\{(x_1, x_2) \in C \ : \ x_1 = \frac{2}{5},\ x_2 \in \left(0, \frac{1}{5}\right)\right\},$$

$$\partial_3 = \left\{(x_1, x_2) \in C \ : \ x_1 \in \left(\frac{2}{5}, \frac{1}{2}\right),\ x_2 = 1 - 2x_1\right\}.$$

Then

$$T_0(\partial_1) = \left\{(x_1, x_2) \in C \ : \ x_1 = 0,\ x_2 \in \left(0, \frac{1}{2}\right)\right\},$$

$$T_0(\partial_2) = \left\{(x_1, x_2) \in C \ : \ x_2 = \frac{1}{2},\ x_1 \in \left(0, \frac{1}{2}\right)\right\},$$

$$T_0(\partial_3) = \left\{(x_1, x_2) \in C \ : \ x_2 = x_1,\ x_1 \in \left(0, \frac{1}{2}\right)\right\}.$$

From this we can conclude that $T_0(C_{2,0,H_1}) = E_1 \cup F_1 \cup G_1$. A similar computation can be done for each of the sets $C_{a,b,P}$. Table 1 lists the images of each type of set $C_{a,b,P}$ in the first quadrant. See also Fig. 4. By symmetry, similar results are obtained for sets $C_{a,b,P}$ in the other quadrants, from which we can deduce that the collection $\mathcal{P}$ is a Markov partition for T_0 with $d = 2$.

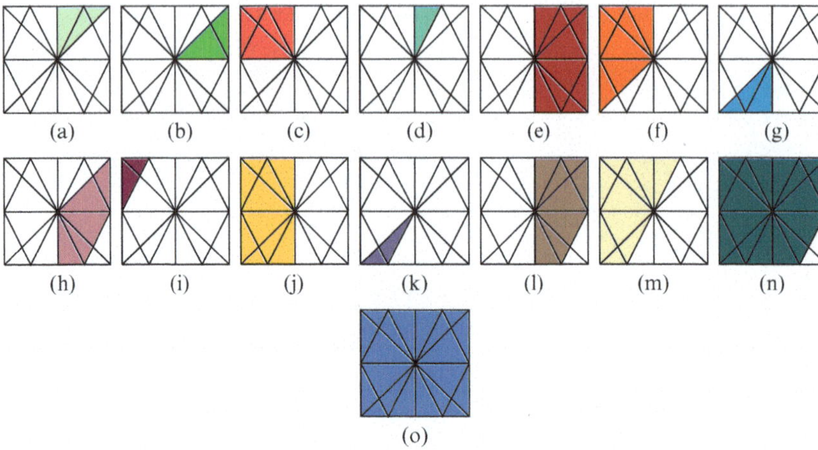

Fig. 4 The images of the sets from Fig. 3. (**a**) type 1 (**b**) type 2. (**c**) type 3. (**d**) type 4. (**e**) type 5. (**f**) type 6. (**g**) type 7. (**h**) type 8. (**i**) type 9. (**j**) type 10. (**k**) type 11. (**l**) type 12. (**m**) type 13. (**n**) type 14. (**o**) type 15

From Table 1 we can deduce the restrictions that apply to applications of the matrices from (16). For example, from the first two lines we read that the digit $(2, 0)$ can only be followed by a digit (a, b) with $a, b \geq 0$. Some restrictions are more complicated to describe and carry further. For example, the digit $(3, 2)$ can be followed by the digits $(-2, 0)$, $(-2, -1)$, $(-3, -1)$, $(-3, -2)$ and $(-4, -2)$. The digit $(-2, 0)$ can in principle be followed by any digit (a, b) with $a, b \leq 0$, but if one sees $(3, 2)$ followed by $(-2, 0)$, then this can only be followed by those digit (a, b) with $a, b \leq 0$ such that $C_{a,b} \in H_3 \cup J_3$. Knowing which sequences of digits are allowed tells us which matrix products involving matrices of the form (16) we have to analyze in order to obtaining numerical information about the approximation properties of the nearest integer Jacobi–Perron algorithm. Giving a fuller description of the allowed digit sequences from the results in Table 1 would be a first step in this direction.

Having a Markov partition can also help in finding invariant measures for the dynamical system given by T_0. Here we describe a different approach that might lead to an invariant measure that is absolutely continuous with respect to the Lebesgue measure. To prove the existence of an a.c.i.m., we follow the book of Schweiger [138, Chapter 4], where he analyzed the ergodic properties of the Jacobi–Perron algorithm (see also [14, 112]). The analysis is similar; we will be brief. First notice that only cylinders of type 15 are full, i.e., $T_0([a, b)]) = C$ up to sets of zero Lebesgue measure. All other types are non-full. However, after at most three iterations, any non-full cylinder is mapped to a region containing at least a fixed positive proportion, say q, of full cylinders. This implies that every non-full cylinder can be written as a countable union of disjoint full cylinders of higher rank (i.e., where more digits are specified), so that the collection of full cylinders generate the Borel σ-algebra. This then allows one to define a jump transformation with full cylinders and satisfying the conditions of Rényi [132] (see also Theorem 8 in [138]). From this one concludes that the jump transformation admits an absolutely continuous invariant ergodic measure. Then, Theorem 11 (see also Theorem 18) of [138] implies that T_0 admits a finite absolutely continuous invariant ergodic measure.

7.2 Experimental Data

We now provide some experimental data due to Steiner [145] indicating a better behavior of the nearest integer Jacobi–Perron algorithm in terms of Lyapunov exponents than of its usual version.

For the nearest integer Jacobi–Perron algorithm (on the left) and the usual Jacobi–Perron algorithm (on the right), one gets the following experimental data: the usual Jacobi-Perron algorithm stops being strongly convergent in dimension 10, while the nearest integer Jacobi-Perron algorithm is strongly convergent up to dimension 13.

d	λ_1	λ_2	$1 - \frac{\lambda_2}{\lambda_1}$
d = 2	1.72241	−0.691444	1.40144
d = 3	1.72394	−0.388217	1.22520
d = 4	1.72400	−0.24779	1.14372
d = 5	1.72408	−0.16873	1.09786
d = 6	1.72417	−0.11892	1.06897
d = 7	1.72413	−0.08522	1.049430
d = 8	1.72417	−0.06122	1.03551
d = 9	1.72409	−0.04347	1.02522
d = 10	1.72414	−0.02995	1.01737
d = 11	1.72413	−0.01939	1.01125
d = 12	1.72414	−0.01102	1.00640
d = 13	1.72409	−0.00425	1.00247
d = 14	1.72414	0.001304	0.99924

d	λ_1	λ_2	$1 - \frac{\lambda_2}{\lambda_1}$
d = 2	1.20052	−0.448404	1.37351
d = 3	1.18560	−0.227877	1.19220
d = 4	1.17295	−0.13064	1.11138
d = 5	1.16579	−0.07882	1.067614
d = 6	1.16224	−0.04798	1.041279
d = 7	1.16068	−0.028202	1.02430
d = 8	1.15992	−0.014708	1.01268
d = 9	1.15956	−0.00505	1.004358
d = 10	1.159476	0.00217	0.99813
d = 11	1.159360	0.00776	0.993308
d = 12	1.159364	0.01221	0.98946
d = 13	1.159401	0.01586	0.986320
d = 14	1.15930	0.01889	0.983705

Acknowledgments We are greatly indebted to Wolfgang Steiner for the numerical data he provided us in Sect. 7.2. We would like to thank Women in Numbers Europe-4 (WINE4) for bringing us together and for giving us the opportunity to work on this project. The research of A.T. was supported by the DFG-project MO 1884/2-1 and the Collaborative Research Centre TRR 326 "Geometry and Arithmetic of Uniformized Structures". The research of V.B. was supported by the Agence Nationale de la Recherche through the project CODYS (ANR-18-CE40-0007)

References

1. N.H. Abel, Über die Integration der Differential-Formel $\frac{\varrho d.x}{\sqrt{R}}$, wenn R und ϱ ganze Functionen sind. J. Reine Angew. Math. **1826**(1), 185–221 (1826)
2. P. Arnoux, V. Berthé, M. Minervino, W. Steiner, J. Thuswaldner, *Nonstationary Markov partitions and multidimensional continued fractions*. Preprint (2023)
3. W.W. Adams, On a relationship between the convergents of the nearest integer and regular continued fractions. Math. Comp. **33**(148), 1321–1331 (1979)
4. A. Avila, P. Hubert, A. Skripchenko, Diffusion for chaotic plane sections of 3-periodic surfaces. Invent. Math. **206**(1), 109–146 (2016)
5. A. Avila, P. Hubert, A. Skripchenko, On the Hausdorff dimension of the Rauzy gasket. Bull. Soc. Math. France **144**, 539–568 (2016)
6. P. Arnoux, S. Labbé, On some symmetric multidimensional continued fraction algorithms. Ergodic Theory Dynam. Syst. **38**(5), 1601–1626 (2018)
7. P. Arnoux, A. Nogueira, Mesures de Gauss pour des algorithmes de fractions continues multidimensionnelles. Ann. Sci. École Norm. Sup. **26**(6), 645–664 (1993)
8. W.W. Adams, M.J. Razar, Multiples of points on elliptic curves and continued fractions. Proc. Lond. Math. Soc. **s3-41**(3), 481–498 (1980)
9. V.I. Arnold, A-graded algebras and continued fractions. Commun. Pure Appl. Math. **42**(7), 993–1000 (1989)
10. L. Arnold, *Random Dynamical Systems*. Dynamical systems (Montecatini Terme, 1994). Lecture Notes in Math., vol. 1609 (Springer, Berlin, 1995), pp. 1–43
11. L. Arnold, *Random Dynamical Systems*. Springer Monographs in Mathematics (Springer, Berlin, 1998)
12. V.I. Arnold, Higher-dimensional continued fractions. Regul. Chaotic Dyn. **3**(3), 10–17 (1998)
13. P. Arnoux, T.A. Schmidt, Natural extensions and Gauss measures for piecewise homographic continued fractions. Bull. Soc. Math. France **147**(3), 515–544 (2019)
14. A. Broise-Alamichel, Y. Guivarc'h, Exposants caractéristiques de l'algorithme de Jacobi-Perron et de la transformation associée. Ann. Inst. Fourier (Grenoble) **51**(3), 565–686 (2001)
15. P.R. Baldwin, A convergence exponent for multidimensional continued-fraction algorithms. J. Statist. Phys. **66**(5-6), 1507–1526 (1992)
16. F. Bergeron, J. Berstel, S. Brlek, C. Duboc, Addition chains using continued fractions. J. Algorithms **10**(3), 403–412 (1989)
17. V. Berthé, E. Cesaratto, P. Rotondo, M.D. Safe, Lochs-type theorems beyond positive entropy. Monatshefte für Mathematik **200**, 737–779 (2023)
18. W. Bosma, K. Dajani, C. Kraaikamp, *Entropy and Counting Correct Digits*. Tech. Report 9925, University of Nijmegen, 1999. http://www-math.sci.kun.nl/math/onderzoek/reports/reports1999.html
19. J. Bourdon, B. Daireaux, B. Vallée, *Dynamical Analysis of α-Euclidean Algorithms*, vol. 44, 2002. Analysis of Algorithms, pp. 246–285
20. L. Bernstein, *The Jacobi-Perron Algorithm—Its Theory and Application*. Lecture Notes in Mathematics, vol. 207 (Springer, Berlin, New York, 1971)
21. F. Beukers, Geodesic continued fractions and LLL. Indag. Math. (N.S.) **25**, 632–645 (2014)
22. M. Baake, U. Grimm, *Aperiodic Order, vol. 1*. Encyclopedia of Mathematics and Its Applications, vol. 149 (Cambridge University Press, Cambridge, 2013)
23. L. Barreira, G. Iommi, Partial quotients of continued fractions and β-expansions. Nonlinearity **21**(10), 2211–2219 (2008)
24. P. Billingsley, *Ergodic Theory and Information* (Robert E. Krieger Publishing Co., Huntington, NY, 1978). Reprint of the 1965 original
25. L. Babai, B. Just, F. Meyer auf der Heide, On the limits of computations with the floor function. Inform. Comput. **78**(2), 99–107 (1988)
26. W. Bosma, H. Jager, F. Wiedijk, Some metrical observations on the approximation by continued fractions. Nederl. Akad. Wetensch. Indag. Math. **45**(3), 281–299 (1983)

27. V. Berthé, L. Lhote, B. Vallée, The Brun gcd algorithm in high dimensions is almost always subtractive. J. Symbolic Comput. **85**, 72–107 (2018)
28. F. Bagemihl, J.R. McLaughlin, Generalization of some classical theorems concerning triples of consecutive convergents to simple continued fractions. J. Reine Angew. Math. **221**, 146–149 (1966)
29. V. Baladi, A. Nogueira, Lyapunov exponents for non-classical multidimensional continued fraction algorithms. Nonlinearity **9**, 1529–1546 (1996)
30. É. Borel, Contribution à l'analyse arithmétique du continu, J. Math. Pures et Appl. (5) **9**, 329–375 (1903)
31. A.J. Brentjes, *Multidimensional Continued Fraction Algorithms* (Mathematisch Centrum, Amsterdam, 1981)
32. A. Broise, Fractions continues multidimensionnelles et lois stables. Bull. Soc. Math. France **124**(1), 97–139 (1996)
33. W. Bahsoun, M. Ruziboev, B. Saussol, Linear response for random dynamical systems. Adv. Math. **364**, 107011, 44 (2020)
34. V. Brun, En generalisation av kjedebrøken I. Skr. Vidensk.-Selsk. Christiana Math.-Nat. Kl. **6**, 1–29(1919)
35. V. Brun, En generalisation av kjedebrøken II. Skr. Vidensk.-Selsk. Christiana Math.-Nat. Kl. **6**, 1–24 (1920)
36. V. Brun, *Algorithmes euclidiens pour trois et quatre nombres*, Treizième congrès des mathèmaticiens scandinaves, tenu à Helsinki 18-23 août 1957 (Mercators Tryckeri, Helsinki, 1958), pp. 45–64
37. W. Bosma, I. Smeets, *Finding Simultaneous Diophantine Approximations with Prescribed Quality*, in *ANTS X—Proceedings of the Tenth Algorithmic Number Theory Symposium*. Open Book Ser., vol. 1 (Math. Sci. Publ., Berkeley, CA, 2013), pp. 167–185
38. V. Berthé, W. Steiner, J.M. Thuswaldner, On the second Lyapunov exponent of some multidimensional continued fraction algorithms. Math. Comput. **90**(328), 883–905 (2021)
39. V. Berthé, W. Steiner, J.M. Thuswaldner, Multidimensional continued fractions and symbolic codings of toral translations. J. Eur. Math. Soc. (JEMS) **25**, 4997–5057 (2023)
40. T. Camus, *Méthodes algorithmiques pour les réseaux algébriques. (algorithmic methods for algebraic lattices)*, Ph.D. thesis, Grenoble Alpes University, France, 2017
41. J.W.S. Cassels, *An Introduction to Diophantine Approximation* (Hafner Publishing Co., New York, 1972). Facsimile reprint of the 1957 edition, Cambridge Tracts in Mathematics and Mathematical Physics, No. 45.
42. Y. Cheung, N. Chevallier, *About the value of the two dimensional levy's constant*. Preprint (2021). https://arxiv.org/pdf/2107.01907.pdf
43. Y. Cheung, N. Chevallier, *Levy-khintchin theorem for best simultaneous diophantine approximations*. Preprint (2019). arXiv:1906.11173
44. J. Cassaigne, S. Ferenczi, A. Messaoudi, Weak mixing and eigenvalues for Arnoux-Rauzy sequences. Ann. Inst. Fourier (Grenoble) **58**(6), 1983–2005 (2008)
45. Y. Cheung, Hausdorff dimension of the set of singular pairs. Ann. Math. (2) **173**(1), 127–167 (2011)
46. N. Chevallier, Best simultaneous Diophantine approximations and multidimensional continued fraction expansions. Mosc. J. Comb. Number Theory **3**, 3–56 (2013)
47. I. Daubechies, R.A. DeVore, C.S. Güntürk, V.A. Vaishampayan, A/D conversion with imperfect quantizers. IEEE Trans. Inform. Theory **52**(3), 874–885 (2006)
48. K. Dajani, A. Fieldsteel, Equipartition of interval partitions and an application to number theory. Proc. Am. Math. Soc. **129**(12), 3453–3460 (2001)
49. H. Daudé, P. Flajolet, B. Vallée, An average-case analysis of the Gaussian algorithm for lattice reduction. Combin. Probab. Comput. **6**, 397–433 (1997)
50. I. Daubechies, S. Güntürk, Y. Wang, Ö. Yı lmaz, The golden ratio encoder. IEEE Trans. Inform. Theory **56**(10), 5097–5110 (2010)
51. K. Dajani, C. Kraaikamp, *Ergodic Theory of Numbers*. Carus Mathematical Monographs, vol. 29 (Mathematical Association of America, Washington, DC, 2002)

52. K. Dajani, C. Kraaikamp, Random β-expansions. Ergodic Theory Dynam. Syst. **23**(2), 461–479 (2003)
53. K. Dajani, C. Kalle, M. Maggioni, Matching for random systems with an application to minimal weight expansions. Nonlinearity **34**(6), 3676–3708 (2021)
54. P. de Rooij, *Efficient Exponentiation Using Precomputation and Vector Addition Chains*. Advances in Cryptology—EUROCRYPT '94 (Perugia). Lecture Notes in Comput. Sci., vol. 950 (Springer, Berlin, 1995), pp. 389–399
55. H. Ei, S. Ito, H. Nakada, R. Natsui, On the construction of the natural extension of the Hurwitz complex continued fraction map. Monatsh. Math. **188**(1), 37–86 (2019)
56. C. Elsner, M. Schmidt, *Kroncrypt—a new symmetric cryptosystem based on kronecker's approximation theorem*. Cryptology ePrint Archive. Paper 2009/416, 2009. https://eprint.iacr.org/2009/416
57. M. Einsiedler, T. Ward, *Ergodic Theory with a View Towards Number Theory*. Graduate Texts in Mathematics, vol. 259 (Springer, London, 2011)
58. H.R.P. Ferguson, A noninductive $\mathrm{GL}(n, \mathbf{Z})$ algorithm that constructs integral linear relations for n $\mathbf{Z}$-linearly dependent real numbers. J. Algorithms **8**(1), 131–145 (1987)
59. H.R.P. Ferguson, R.W. Forcade, Generalization of the Euclidean algorithm for real numbers to all dimensions higher than two. Bull. Am. Math. Soc. (N.S.) **1**(6), 912–914 (1979)
60. T. Fujita, S. Ito, M. Keane, M. Ohtsuki, On almost everywhere exponential convergence of the modified Jacobi-Perron algorithm: a corrected proof. Ergodic Theory Dynam. Syst. **16**(6), 1345–1352 (1996)
61. H. Furstenberg, H. Kesten, Products of random matrices. Ann. Math. Statist. **31**, 457–469 (1960)
62. C. Fieker, D. Stehlé, *Short Bases of Lattices over Number Fields*. Algorithmic Number Theory. Lecture Notes in Comput. Sci., vol. 6197 (Springer, Berlin, 2010), pp. 157–173
63. C. Fougeron, A. Skripchenko, Simplicity of spectra for certain multidimensional continued fraction algorithms. Monatsh. Math. **194**(4), 767–787 (2021)
64. H. Furstenberg, *Stationary Processes and Prediction Theory*. Annals of Mathematics Studies, vol. 44 (Princeton University Press, Princeton, NJ, 1960)
65. H. Furstenberg, Noncommuting random products. Trans. Am. Math. Soc. **108**, 377–428 (1963)
66. D.J. Grabiner, J.C. Lagarias, Cutting sequences for geodesic flow on the modular surface and continued fractions. Monatsh. Math. **133**(4), 295–339 (2001)
67. D.J. Grabiner, Farey nets and multidimensional continued fractions. Monatsh. Math. **114**(1), 35–61 (1992)
68. E. Heine, C.G.J. Jacobi, Allgemeine Theorie der kettenbruchähnlichen Algorithmen, in welchen jede Zahl aus drei vorhergehenden gebildet wird. J. Reine Angew. Math. **69**, 29–64 (1868)
69. J. Håstad, B. Just, J.C. Lagarias, C.-P. Schnorr, Polynomial time algorithms for finding integer relations among real numbers. SIAM J. Comput. **18**(5), 859–881 (1989)
70. D.M. Hardcastle, K. Khanin, On almost everywhere strong convergence of multi-dimensional continued fraction algorithms. Ergodic Theory Dynam. Syst. **20**(6), 1711–1733 (2000)
71. D.M. Hardcastle, K. Khanin, Continued fractions and the d-dimensional Gauss transformation. Commun. Math. Phys. **215**(3), 487–515 (2001)
72. D.M. Hardcastle, K. Khanin, The d-dimensional Gauss transformation : strong convergence and Lyapunov exponents. Exp. Math. **11**, 119–129 (2002)
73. G. Havas, B.S. Majewski, K.R. Matthews, Extended GCD and Hermite normal form algorithms via lattice basis reduction. Exp. Math. **7**, 125–136 (1998)
74. G.H. Hardy, E.M. Wright, *An Introduction to the Theory of Numbers* (Oxford Science Publications, Oxford, 1979)
75. M. Iosifescu, C. Kraaikamp, *Metrical Theory of Continued Fractions* (Dordrecht Kluwer Academic Publishers, 2002)
76. H. Jager, Metrical results for the nearest integer continued fraction. Nederl. Akad. Wetensch. Indag. Math. **47**(4), 417–427 (1985)

77. H. Jager, The distribution of certain sequences connected with the continued fraction. Nederl. Akad. Wetensch. Indag. Math. **48**(1), 61–69 (1986)
78. H. Jager, C. Kraaikamp, On the approximation by continued fractions. Nederl. Akad. Wetensch. Indag. Math. **51**(3), 289–307 (1989)
79. Y. Jitsumatsu, K. Matsumura, A β-ary to binary conversion for random number generation using a β encoder. IEICE (NOLTA), 38–55 (2016)
80. Y. Jitsumatsu, K. Matsumura, T. Kohda, K. Aihara, *Pseudo-random number generator using beta-encoder cmos circuit*, in *The 3rd Int. Symp. Innovative Mathematical Modelling* (2013), p. 107
81. O. Jenkinson, M. Pollicott, P. Vytnova, Rigorous computation of diffusion coefficients for expanding maps. J. Stat. Phys. **170**, 221–253 (2018)
82. B. Just, *Integer Relations Among Algebraic Numbers*. Mathematical Foundations of Computer Science, 1989 (Porąbka-Kozubnik, 1989). Lecture Notes in Comput. Sci., vol. 379 (Springer, Berlin, 1989), pp. 314–320
83. B. Just, Integer relations among algebraic numbers. Math. Comput. **54**(189), 467–477 (1990)
84. B. Just, Generalizing the continued fraction algorithm to arbitrary dimensions. SIAM J. Comput. **21**(5), 909–926 (1992)
85. A.M. Kane, On the use of continued fractions for stream ciphers. IACR Cryptol. **2013**, 319 (2009). ePrint Arch.
86. O.N. Karpenkov, Constructing multidimensional periodic continued fractions in the sense of Klein. Math. Comput. **78**(267), 1687–1711 (2009)
87. O.N. Karpenkov, *Geometry of Continued Fractions*. Algorithms and Computation in Mathematics, vol. 26 (Springer, Heidelberg, 2013)
88. O.N. Karpenkov, On Hermite's problem, Jacobi-Perron type algorithms, and Dirichlet groups. Acta Arith. **203**(1), 27–48 (2022)
89. A.Ya. Khintchine, *Continued Fractions*. Translated by Peter Wynn (P. Noordhoff Ltd., Groningen, 1963)
90. T. Kohda, Y. Horio, Y. Takahashi, K. Aihara, Beta encoders: symbolic dynamics and electronic implementation. Int. J. Bifur. Chaos Appl. Sci. Eng. **22**(9), 1230031, 55 (2012)
91. C. Kalle, T. Kempton, E. Verbitskiy, The random continued fraction transformation. Nonlinearity **30**(3), 1182–1203 (2017)
92. K. Khanin, J. Lopes Dias, J. Marklof, Multidimensional continued fractions, dynamical renormalization and KAM theory. Commun. Math. Phys. **270**(1), 197–231 (2007)
93. C. Kalle, N. Langeveld, M. Maggioni, S. Munday, Matching for a family of infinite measure continued fraction transformations. Discrete Contin. Dyn. Syst. **40**(11), 6309–6330 (2020)
94. C. Kalle, V. Matache, M. Tsujii, E. Verbitskiy, Invariant densities for random continued fractions. J. Math. Anal. Appl. **512**(2), Paper No. 126163, 28 (2022)
95. E. Korkina, La périodicité des fractions continues multidimensionnelles. C. R. Acad. Sci. Paris Sér. I Math. **319**(8), 777–780 (1994)
96. C. Kraaikamp, T.A. Schmidt, W. Steiner, Natural extensions and entropy of α-continued fractions. Nonlinearity **25**(8), 2207–2243 (2012)
97. C. Kalle, E. Verbitskiy, B. Zeegers, Random Lochs' theorem. Studia Math. **267**(2), 201–239 (2022)
98. G. Lachaud, Polyèdre d'Arnol' d et voile d'un cône simplicial: analogues du théorème de Lagrange. C. R. Acad. Sci. Paris Sér. I Math. **317**(8), 711–716 (1993)
99. G. Lachaud, *Klein Polygons and Geometric Diagrams*. Number theory (Tiruchirapalli, 1996). Contemp. Math., vol. 210 (Amer. Math. Soc., Providence, RI, 1998), pp. 365–372
100. G. Lachaud, *Sails and Klein Polyhedra*. Number theory (Tiruchirapalli, 1996). Contemp. Math., vol. 210 (Amer. Math. Soc., Providence, RI, 1998), pp. 373–385
101. J.C. Lagarias, Best simultaneous Diophantine approximations. I. Growth rates of best approximation denominators. Trans. Am. Math. Soc. **272**, 545–554 (1982)
102. J.C. Lagarias, Best simultaneous Diophantine approximations. II. Behavior of consecutive best approximations. Pacific J. Math. **102**, 61–88 (1982)

103. J.C. Lagarias, *The computational complexity of simultaneous Diophantine approximation problems*, in *23rd Annual Symposium on Foundations of Computer Science* (Chicago, Ill., 1982) (IEEE, New York, 1982), pp. 32–39
104. J.C. Lagarias, The computational complexity of simultaneous Diophantine approximation problems. SIAM J. Comput. **14**, 196–209 (1985)
105. J.C. Lagarias, The quality of the Diophantine approximations found by the Jacobi-Perron algorithm and related algorithms. Monatsh. Math. **115**(4), 299–328 (1993)
106. J.C. Lagarias, Geodesic multidimensional continued fractions. Proc. Lond. Math. Soc. (3) **69**, 464–488 (1994)
107. L. Lhote, *Computation of a Class of Continued Fraction Constants*. Analytic Algorithmics and Combinatorics ANALCO 2014 (SIAM, 2004), pp. 199–210
108. A.K. Lenstra, H.W. Lenstra, Jr., L. Lovász, Factoring polynomials with rational coefficients. Math. Ann. **261**(4), 515–534 (1982)
109. G. Lochs, Vergleich der Genauigkeit von Dezimalbruch und Kettenbruch. Abh. Math. Sem. Univ. Hamburg **27**, 142–144 (1964)
110. B. Li, J. Wu, Beta-expansion and continued fraction expansion. J. Math. Anal. Appl. **339**(2), 1322–1331 (2008)
111. R. Mönkemeyer, Über Fareynetze in n Dimensionen. Math. Nachr. **11**, 321–344 (1954)
112. D.H. Mayer, Approach to equilibrium for locally expanding maps in $\mathbf{R}^k$. Commun. Math. Phys. **95**(1), 1–15 (1984)
113. R. Meester, A simple proof of the exponential convergence of the modified Jacobi-Perron algorithm. Ergodic Theory Dynam. Syst. **19**(4), 1077–1083 (1999)
114. M. Morse, G.A. Hedlund, Symbolic dynamics II. Sturmian trajectories. Am. J. Math. **62**, 1–42 (1940)
115. N.G. Moshchevitin, *Best Diophantine Approximations: The Phenomenon of Degenerate Dimension*. Surveys in geometry and number theory: reports on contemporary Russian mathematics. London Math. Soc. Lecture Note Ser., vol. 338 (Cambridge University Press, Cambridge, 2007), pp. 158–182
116. H. Nakada, Metrical theory for a class of continued fraction transformations and their natural extensions. Tokyo J. Math. **4**(2), 399–426 (1981)
117. H. Napias, A generalization of the LLL-algorithm over Euclidean rings or orders. J. Théor. Nombres Bordeaux **8**, 387–396 (1996)
118. H. Niederreiter, *Sequences with Almost Perfect Linear Complexity Profile*, in *Advances in Cryptology—EUROCRYPT' 87* (Berlin, Heidelberg), ed. by D. Chaum, W.L. Price (Springer, Berlin, Heidelberg, 1988), pp. 37–51
119. A. Nogueira, The Borel-Bernstein theorem for multidimensional continued fractions. J. Anal. Math. **85**, 1–41 (2001)
120. P.Q. Nguyen, D. Stehlé, *LLL on the Average*. Algorithmic number theory. Lecture Notes in Comput. Sci., vol. 4076 (Springer, Berlin, 2006), pp. 238–256
121. P.Q. Nguyen, B. Vallée (ed.) *The LLL Algorithm*. Survey and Applications. Information Security and Cryptography (Springer, Dordrecht, 2010)
122. N. Obrechkoff, Sur l'approximation des nombres irrationnels par des nombres rationnels. C. R. Acad. Bulgare Sci. **3**(1), 1–4 (1950,1951)
123. G. Panti, Multidimensional continued fractions and a Minkowski function. Monatsh. Math. **154**(3), 247–264 (2008)
124. O. Perron, Grundlagen für eine Theorie des Jacobischen Kettenbruchalgorithmus. Math. Ann. **64**(1), 1–76 (1907)
125. O. Perron, Über diophantische approximationen. Math. Annalen **83**, 77–84 (1921)
126. V. Platonov, G. Fedorov, On the periodicity of continued fractions in hyperelliptic fields. Doklady Math. **95**, 254–258 (2017)
127. E.V. Podsypanin, A generalization of the continued fraction algorithm that is related to the Viggo Brun algorithm. Zap. Naučn. Sem. Leningrad. Otdel. Mat. Inst. Steklov. (LOMI) **67**, 184–194, 227 (1977). Studies in number theory (LOMI), 4

128. M. Pollicott, Maximal Lyapunov exponents for random matrix products. Invent. Math. **181**(1), 209–226 (2010)
129. A. Pethő, M.E. Pohst, C. Bertók, On multidimensional Diophantine approximation of algebraic numbers. J. Number Theory **171**, 422–448 (2017)
130. F. Pappalardi, A.J. Van Der Poorten, Pseudo-elliptic integrals, units, and torsion. J. Aust. Math. Soc. **79**(3), 335–347 (2005). Cited by: 4; All Open Access, Bronze Open Access, Green Open Access
131. N. Pytheas Fogg, in *Substitutions in Dynamics, Arithmetics and Combinatorics*, ed. by V. Berthé, S. Ferenczi, C. Mauduit, A. Siegel. Lecture Notes in Mathematics, vol. 1794 (Springer, 2002)
132. A. Rényi, Representations for real numbers and their ergodic properties. Acta Math. Acad. Sci. Hungar. **8**, 477–493 (1957)
133. A. Rosenfeld, R. Klette, Digital straightness. Electron. Notes Theor. Comput. Sci. **46**, 1–32 (2001). IWCIA 2001, 8th International Workshop on Combinatorial Image Analysis
134. F. Schweiger, *The Metrical Theory of Jacobi-Perron Algorithm*. Lecture Notes in Mathematics, vol. 334 (Springer, Berlin, New York, 1973)
135. F. Schweiger, On the invariant measure for Jacobi-Perron algorithm. Math. Pannon. **1**(2), 91–106 (1990)
136. W.S. Schmidt, *Diophantine Approximation*. Lecture Notes in Mathematics, vol. 785 (Springer, 1996)
137. F. Schweiger, *The Exponent of Convergence for the 2-Dimensional Jacobi-Perron Algorithm*, in *Proceedings of the Conference on Analytic and Elementary Number Theory* (Vienna), ed. by W.G. Nowak, J. Schoissengeier (1996), pp. 207–213
138. F. Schweiger, *Multidimensional Continued Fractions*. Oxford Science Publications (Oxford University Press, Oxford, 2000)
139. B. R. Schratzberger, The quality of approximation of Brun's algorithm in three dimensions. Monatshefte für Mathematik **134**, 143–157 (2001)
140. B. Schratzberger, On the singularization of the two-dimensional Jacobi-Perron algorithm. Exp. Math. **16**(4), 441–454 (2007) (English)
141. B.R. Schratzberger, A conversion algorithm based on the technique of singularization. Theor. Comput. Sci. **391**(1-2), 138–149 (2008) (English)
142. E.S. Selmer, Continued fractions in several dimensions. Nordisk Nat. Tidskr. **9**, 37–43, 95 (1961)
143. B. Sendov, Der Zahlensche Satz über die singulären Kettenbrüche und die Kettenbrüche nach nächsten Ganzen. Annuaire Univ. Sofia Fac. Sci. Phys. Math. Livre 1 Math. **54**, 251–258 (1959/1960)
144. R. Sturman, J.-L. Thiffeault, Lyapunov exponents for the random product of two shears. J. Nonlinear Sci. **29**(2), 593–620 (2019)
145. W. Steiner, Personnal communication
146. G. Szekeres, Multidimensional continued fractions. Ann. Univ. Sci. Budapest. Eötvös Sect. Math. **13**, 113–140 (1970/1971)
147. T. Taylor-Crush, On the regularity and approximation of invariant densities for random continued fractions. Dyn. Syst. **36**(1), 1–18 (2021)
148. S. Tanaka, S. Ito, On a family of continued-fraction transformations and their ergodic properties. Tokyo J. Math. **4**(1), 153–175 (1981)
149. J.C. Tong, The conjugate property of the Borel theorem on Diophantine approximation. Math. Z. **184**(2), 151–153 (1983)
150. A.J. Van der Poorten, X.C. Tran, Quasi-elliptic integrals and periodic continued fractions. Monatshefte für Mathematik **131**(2), 155–169 (2000)
151. M. Viana, *Lectures on Lyapunov Exponents*. Cambridge Studies in Advanced Mathematics, vol. 145 (Cambridge University Press, Cambridge, 2014)
152. M. Wiener, *Cryptanalysis of Short RSA Secret Exponents*, in *Advances in Cryptology—EUROCRYPT '89* (Berlin, Heidelberg), ed. by J.-J. Quisquater, J. Vandewalle (Springer, Berlin, Heidelberg, 1990), pp. 372–372

153. H.C. Williams, Some results concerning the nearest integer continued fraction expansion of $\sqrt{D}$. J. Reine Angew. Math. **315**, 1–15 (1980)
154. D. Wübben, D. Seethaler, J. Jaldén, G. Matz, Lattice reduction. IEEE Signal Process. Mag. **28**(3), 70–91 (2011)
155. Y. Yu, L. Ducas, *Second Order Statistical Behavior of LLL and BKZ*. Selected areas in cryptography—SAC 2017. Lecture Notes in Comput. Sci., vol. 10719 (Springer, 2018), pp. 3–22
156. J.-C. Yoccoz, *Continued Fraction Algorithms for Interval Exchange Maps: an Introduction*. Frontiers in Number Theory, Physics, and Geometry. I (Springer, Berlin, 2006), pp. 401–435
157. A. Zorich, Deviation for interval exchange transformations. Ergodic Theory Dynam. Syst. **17**, 1477–1499 (1997)

Power Values of Power Sums: A Survey

Nirvana Coppola, Mar Curcó-Iranzo, Maleeha Khawaja, Vandita Patel, and Özge Ülkem

1 Introduction

Euler, in 1770, noted the remarkable identity

$$3^3 + 4^3 + 5^3 = 6^3$$

in his famous *Vollständige Anleitung zur Algebra* [73, art. 249]. He further asked if there were other instances of sums of consecutive cubes that are perfect powers. Dickson's encyclopaedic *History of the Theory of Numbers* [68, pp. 582 – 588] extensively surveys this problem citing contributions from many 19th century renowned mathematicians such as Cunningham, Catalan, Genocchi and Lucas.

Consecutive cubes that sum to a square have the well-known parametric family of solutions

N. Coppola
Vrije Universiteit Amsterdam, Amsterdam, The Netherlands

M. Curcó-Iranzo
Hans-Freudental Gebouw, Utrecht University, Utrecht, The Netherlands
e-mail: m.curcoiranzo@uu.nl

M. Khawaja
School of Mathematics and Statistics, University of Sheffield, Sheffield, UK
e-mail: mkhawaja2@sheffield.ac.uk

V. Patel (✉)
School of Mathematics, University of Manchester, Manchester, UK
e-mail: vandita.patel@manchester.ac.uk

Ö. Ülkem
Galatasaray University, Istanbul, Turkey

R. Abdellatif et al. (eds.), *Women in Numbers Europe IV*, Association for Women in Mathematics Series 32, https://doi.org/10.1007/978-3-031-52163-8_6

$$\sum_{i=1}^{d} i^3 = \left(\frac{d(d+1)}{2}\right)^2.$$

The question of existence of solutions beyond this family poses a more interesting and challenging problem. Further studies began with Catalan, Cantor and Richaud, who produced incomplete results by generating solutions via Pell's equation [68, pp. 586].

Lucas proceeded to boldly make the following three claims ([119, pg. 92]).

(a) For $d = 2$: "the only square that is the sum of two consecutive positive cubes is $3^2 = 1^3 + 2^3$".
(b) For $d = 3$: "the sum of 3 consecutive cubes is never a square, except for the solution $1^3 + 2^3 + 3^3 = 6^2$".
(c) For $d = 5$: "the sum of 5 consecutive cubes is never a square, except for in the cases where the middle cube is either 2, 3, 27, 98 and 120".

Claim (b) turned out to be faulty: Cassels [50] and Uchiyama [166] independently prove that the complete list of solutions are actually

$$\begin{aligned} 0 &= (-1)^3 + 0^3 + 1^3, \\ 3^2 &= 0^3 + 1^3 + 2^3, \\ 6^2 &= 1^3 + 2^3 + 3^3 = 6^2, \\ 204^2 &= 23^3 + 24^3 + 25^3. \end{aligned}$$

On the other hand, Stroeker [163] proves that claims (a) and (c) are indeed correct! In fact, Stroeker studies the problem of squares that are sums of at most 50 consecutive cubes using a (by now) standard method based on linear forms in elliptic logarithms. Nowadays, the problem of finding squares that are sums of d consecutive cubes for a fixed integer d is more commonly viewed as a problem of determining all integral points on certain genus 1 curves [26].

Consecutive cubes that sum to a cube possess the lesser-known parametric family of solutions

$$\left(\frac{m^5 + m^3 - 2m}{6}\right)^3 = \sum_{i=1}^{m^3} \left(\frac{m^4 - 3m^3 - 2m^2 - 2}{6} + i\right)^3$$

where $m \equiv 2 \text{ or } 4 \pmod 6$ to guarantee that the cubes are integral. The construction was given in 1829 by Pagliani, whose apparent motivation was to answer a challenge posed presumably by the journal editor Gergonne: to find 1000 consecutive cubes whose sum is a perfect cube. Notable mathematicians have contributed to providing parametric families of solutions, including Lenhart and Matthiessen, or proving that in certain instances there are no solutions, such as Lebesgue, Evans and Escott.

Once more, finding solutions outside of Pagliani's family is the more compelling question.

The literature surrounding the general problem of expressing perfect powers as a sum of (a small number of) consecutive cubes (or higher powers) is far more sparse, most likely due to the fact that modern tools and techniques are necessary, even in the case of partial resolution of problems. Since we aim to find all integer solutions to certain exponential Diophantine equations, we can naturally view this as a subset of all of the rational solutions of the same equation. In other words, we are interested in the set of rational points of the algebraic curve C defined by $f(x, y) \in \mathbb{Z}[x, y]$ obtained from the given equation. If C is smooth, Faltings' theorem [74] (which awarded him the Fields Medal in 1986) determines the cardinality of the set of rational points according to the genus of C.

- If the genus of C is 0, then C is isomorphic to $\mathbb{P}^1$ and $C(\mathbb{Q})$ is either empty or has infinitely many points parameterised by $\mathbb{P}^1(\mathbb{Q})$.
- If the genus of C is 1, and C has a rational point, then C is an elliptic curve and $C(\mathbb{Q})$ is isomorphic to a finitely generated Abelian group $\mathbb{Z}^r \oplus T$, where T is finite.
- If the genus of C is greater than or equal to 2, then the set $C(\mathbb{Q})$ is finite.

In particular, note that fixing the genus (equivalent to fixing certain exponents) of an exponential Diophantine equation enables us to conclude, via Faltings' theorem, that there are finitely many integral solutions. However, we encounter two significant drawbacks: firstly, Faltings' proof is ineffective, thus we are unable to explicitly resolve problems such as those posed by Euler and Gergonne, and secondly, exponential Diophantine equations usually do not have a fixed genus and so we are unable to conclude finiteness results. An effective version of Faltings' proof would indeed be momentous and certainly pave the way for significant breakthroughs in the explicit resolution of Diophantine equations!

In this survey, we aim to give a comprehensive overview of results towards the following families of exponential Diophantine equations:

$$x^k + (x+r)^k + \cdots + (x+(d-1)r)^k = y^n, \qquad d, k, n, r, x, y, \in \mathbb{Z}, \tag{1}$$

with parameters (k, d, r, n);

$$s(1^k + 2^k + \cdots + x^k) + r(x) = y^n, \qquad k, n, x, y, s \in \mathbb{Z}, r(x) \in \mathbb{Z}[x], \tag{2}$$

with parameters $(k, n, s, r(x))$.

We begin by highlighting some of the key pioneering advances in number theory that have enabled explicit resolutions of difficult multi-parametric exponential Diophantine equations. One such strategy used to reach a full explicit resolution is to first bound the exponent(s). This reduces the number of unknowns in the Diophantine equation. The next step is to eliminate equations by ruling out the existence of integral solutions for certain combinations of variables. The remaining handful of equations are then resolved using ad hoc techniques.

1.1 Bounding the Exponent

We present techniques used to bound exponents in certain exponential Diophantine equations.

1.2 Diophantine Approximation via Linear Forms in Logarithms

Let $\alpha_1, \alpha_2, \ldots, \alpha_n$ and $\beta_1, \beta_2, \ldots, \beta_n$ be algebraic numbers. A linear form in logarithms is an expression of the form

$$\beta_1 \log \alpha_1 + \cdots + \beta_n \log \alpha_n. \tag{3}$$

The motivation behind (see [53, Section 12.1.2]) obtaining a bound on (3) follows from the following idea. After successfully associating a linear form in logarithms to a putative solution to a given Diophantine equation D, the successive step is to then obtain a bound B_D for the α_i, β_i. This provides an upper bound for the solutions to D. If B_D is small enough then one can attempt to determine all solutions to D. We note that for Diophantine applications, it is essential that $\beta_i \in \mathbb{Z}$. Moreover, if the α_i are known, then one should be able to find a smaller bound on β_i than B_D.

The first effective bound was obtained by Baker [7] in 1966, awarding him the Fields medal. We note that this is the most general result as it holds for arbitrary n, see [9, pg. 257].

Theorem 1 (Baker) *Let K be a number field. Let $\alpha_1, \ldots, \alpha_n \in K^*$ and $\beta_1, \ldots, \beta_n \in K$. For any constant ν, define*

$$\begin{aligned}\tau(\nu) : &= \tau(\nu; \alpha_1, \ldots, \alpha_n; \beta_1, \ldots, \beta_n) \\ &= h([1, \beta_1, \ldots, \beta_n]) h([1, \alpha_1, \ldots, \alpha_n])^{\nu}\end{aligned}$$

where h is a logarithmic height function. Fix an embedding $K \subset \mathbb{C}$ and let $|\cdot|$ be the corresponding absolute value. Assume that

$$\beta_1 \log \alpha_1 + \cdots + \beta_n \log \alpha_n \neq 0.$$

Then there are effectively computable constants $C, \nu > 0$ which depend only on n and $[K : \mathbb{Q}]$ such that

$$|\beta_1 \log \alpha_1 + \cdots + \beta_n \log \alpha_n| > C^{-\tau(\nu)}.$$

Baker's theory, albeit effective, usually produces astronomical bounds. Substantial work has been undertaken to reduce such bounds, with the majority of the focus over the last century residing within the realm of linear forms in *two* logarithms. One such result is due to Mignotte [126], which uses linear forms in two logarithms to bound the integral coefficients of Diophantine equations of the form $ax^m - by^m = c$. Novel advances in the general reduction of bounds arising from linear forms in *three* logarithms have only appeared extremely recently [127], demonstrating computational finesse in [28] and [29].

1.3 Primitive Divisors of Lucas–Lehmer Sequences

Let (α, β) be a pair of algebraic integers. We say (α, β) is a Lehmer pair if $(\alpha+\beta)^2$ and $\alpha\beta$ are non-zero coprime rational integers with α/β not a root of unity. The Lehmer sequence associated to the pair (α, β) is then defined as

$$\tilde{u}_n(\alpha, \beta) = \begin{cases} \frac{\alpha^n - \beta^n}{\alpha - \beta} & \text{if } n \text{ is odd} \\ \frac{\alpha^n - \beta^n}{\alpha^2 - \beta^2} & \text{if } n \text{ is even.} \end{cases}$$

Similarly, (α, β) is a Lucas pair if $\alpha + \beta$ and $\alpha\beta$ are non-zero coprime rational integers and α/β is not a root of unity. The corresponding Lucas sequence is defined as

$$u_n(\alpha, \beta) = \frac{\alpha^n - \beta^n}{\alpha - \beta} \qquad (n = 0, 1, 2, \dots)$$

which can be rewritten as

$$u_n(\alpha, \beta) = \begin{cases} \tilde{u}_n(\alpha, \beta) & \text{if } n \text{ is odd} \\ (\alpha + \beta)\tilde{u}_n(\alpha, \beta) & \text{if } n \text{ is even,} \end{cases}$$

showing that every Lucas pair is also a Lehmer pair.

A primitive divisor of $\tilde{u}_n$ is a prime divisor p of $\tilde{u}_n$ not dividing $(\alpha^2 - \beta^2)^2 \tilde{u}_1 \cdots \tilde{u}_{n-1}$ (analogously for u_n). The celebrated theorem of Bilu, Hanrot and Voutier [36] is not just impressive in its own right but lends a hand to many Diophantine applications (see [5, 55, 81, 141, 144]).

Theorem 2 (Bilu, Hanrot and Voutier) *Any Lucas or Lehmer sequence has a primitive divisor for all* $n > 30$ *and for all primes* $n > 13$.

For example, the following theorem due to Patel [141] is proved using the aforementioned theorem of Bilu, Hanrot and Voutier in combination with a factorisation argument over imaginary quadratic fields, displaying how well exponents in exponential Diophantine equations can be bounded.

Theorem 3 *Let $C_1 \geq 1$ be a square-free integer and C_2 a positive integer. Assume that $C_1C_2 \not\equiv 7 \pmod 8$. Let p be an odd prime for which*

$$C_1x^2 + C_2 = y^p, \qquad \gcd(C_1x^2, C_2, y^p) = 1 \tag{4}$$

has a solution $(x, y) \in \mathbb{N}^2$ such that $xy \neq 0$. Writing $C_1C_2 = cd^2$ with c square-free, then either

- *$p \leq 5$ or*
- *$p = 7$ and $y = 3, 5$ or 9 or*
- *p divides the class number of $\mathbb{Q}(\sqrt{-c})$ or*
- *p divides $\left(q - \left(\frac{-c}{q}\right)\right)$, where q is a prime $q \mid d$ and $q \nmid 2c$.*

1.4 Eliminating Equations

Suppose we have bounded the exponent of the exponential Diophantine equation that we are interested in solving. We are now interested in concretely solving as many possibilities as we can. This gives us finitely many equations to resolve. We are now interested in finding the full set of solutions to a given equation or proving that it has no solutions (in other words, eliminating the equation). In this section we summarise two of the most powerful tools used to eliminate equations: Sophie Germain's empty set criterion and Wiles' Modularity Theorem.

1.5 Sophie Germain's Empty Set Criterion

Sophie Germain (1776–1831) made significant contributions to number theory despite being born into an era in which it was socially improper for women to practice mathematics. In pursuit of her dream, having adopted the identity of the former student *Monsieur Auguste Antoine LeBlanc*, she enrolled at the newly opened École Polytechnique in Paris in 1794. Sophie began corresponding with Gauss after her interest in number theory piqued through a serious study of Legendre's *Théorie des nombres* and Gauss' *Disquisitiones Arithmeticae*. She soon became the first to provide evidence for Fermat's Last Theorem through studying residue classes of the exponent. Although her work was left unrecognised during her lifetime, her contributions to mathematics did receive recognition after her death. In 1882, the street *Rue Sophie-Germain* was named in her honour (located in the 14th *Arrondissement* of Paris). Significantly later on, in 2003, the French Academy of Sciences officiated an annual prize to mark her legacy.

In one of her 14 letters to Gauss, Sophie writes (taken from [63, pp. 358–359]):

> Prenons pour exemple l'équation même de Fermat qui est la plus simple de toutes celles dont il s'agit ici.
>
> Soit donc, p étant un nombre premier, $z^p = x^p + y^p$. Je dis que si cette équation est possible, tout nombre premier de la forme $2Np + 1$ (N tant un entier quelconque) pour lequel il n'y aura pas deux résidus $p^{iéme}$ puissance placés de suite dans la série des nombres naturels divisera nécessairement l'un des nombres x, y et z.
>
> Cela est évident, car l'équation $z^p = x^p + y^p$ donne la congruance [congruence] $1 \equiv r^{sp} + r^{tp}$ dans laquelle r représente une racine primitive et s et t des entiers.
>
> On sait que l'équation a une infinité de solutions lorsque $p = 2$. Et en effet tous les nombres, exceptés 3 et 5 ont au moins deux résidus quarrés dont la différence est l'unité. Aussi dans ce cas la forme connue savoir $h^2 + f^2, 2fh, h^2 - f^2$ des nombres $z[x]$, y et z montre-t-elle que l'un de ces nombres est multiple de 3 et aussi que l'un des mêmes nombres est multiple de 5.
>
> Il est aisé de voir que si un nombre quelconque k est résidu puissance $p^{iéme}$ mod. $2Np + 1$ et qu'il y ait deux résidus puissance $p^{iéme}$ méme mod. dont la différence soit l'unité, il y aura aussi deux résidus puissance $p^{iéme}$ dont la différence sera k.
>
> Mais il peut arriver qu'on ait deux résidus $p^{iéme}$ dont la différence soit k, sans que k soit résidu $p^{iéme}$.
>
> Cela posé voici l'équation générale dont la solution me semble dépendre comme celle de Fermat de l'ordre des résidus: $kz^p = x^p \pm y^p$ car d'aprés ce que vient d'être dit on voit que tout nombre premier de la forme $2Np + 1$ pour lequel deux résidus $p^{iéme}$ n'ont pas le nombre k pour différence divise le nombre z [l'un des nombres x, y, z]. Il suit delé que s'il y avait un nombre infini de tels nombres l'équation serait impossible.

In the above correspondence, Germain essentially provides a proof for the following theorem.

Theorem 4 (Germain) *Let $p \geq 3$ be a prime. Let $q = 2Np + 1$ be an auxiliary prime, where N is an integer not divisible by 3. Suppose the following hold:*

(i) If $x^p + y^p + z^p \equiv 0 \pmod q$ then $xyz \equiv 0 \pmod q$.
(ii) p is not a pth power residue modulo q.

Then there are no rational solutions to the Fermat equation with exponent p when $p \nmid xyz$.

The following criterion due to Patel [139], inspired by the above work of Sophie Germain, is a computational method of eliminating Fermat-type equations of signature $(p, 2p, 2)$. Since all computations are performed over finite fields this method is computationally highly efficient.

Theorem 5 *Let $p \geq 3$ be a prime. Let a, b and c be positive integers such that* $\gcd(a, b, c) = 1$. *Let $q = 2kp + 1$ be a prime that does not divide a. Define*

$$\mu(p, q) = \{\eta^{2p} \ : \ \eta \in \mathbb{F}_q\} = \{0\} \cup \{\zeta \in \mathbb{F}_q^* \ : \ \zeta^k = 1\}$$

and

$$B(p, q) = \left\{\zeta \in \mu(p, q) \ : \ ((b\zeta + c)/a)^{2k} \in \{0, 1\}\right\}.$$

If $B(p, q) = \emptyset$, then there are no solutions to $aw_2^p - bw_1^{2p} = c$.

We demonstrate the computational power of this result through the following example.

Example 1 In [55], the authors resolve the equation

$$(x - 4r)^3 + (x - 3r)^3 + \cdots + x^3 + \cdots + (x + 3r)^3 + (x + 4r)^3 = y^p$$

under the constraints $x, r, y, p \in \mathbb{Z}$, p a prime, $\gcd(x, r) = 1$ and $0 < r \le 10^6$. Descent, and using the divisibility conditions $2 \nmid x$, $3 \nmid x$, and $5 \mid x$, lead to the ternary equation

$$3^{p-2} \cdot w_2^p - 5^{2p-3} w_1^{2p} = 4r^2. \tag{5}$$

Once the exponent p has been bounded, there are still 1, 951, 465, 447 equations to be resolved in unknown variables x and y. Germain's methodology results in only 2, 821 equations surviving after applying Theorem 5 to (5).

1.6 Wiles' Modularity Theorem

Perhaps the most fashionable (albeit computationally challenging) method of studying exponential Diophantine equations is the one pioneered by Wiles and his collaborators during the resolution of the Fermat equation (over the rational field) during the late 20th century. We give a brief outline of the proof here.

Suppose we have a non-trivial rational solution (a, b, c) to the Fermat equation of degree p

$$x^p + y^p = z^p, \tag{6}$$

for prime $p \ge 5$. We can associate a semistable elliptic curve E (called the *Frey-Hellegouarch* curve) to this putative solution

$$E : y^2 = x(x - a^p)(x + b^p),$$

where E has conductor N. A key non-trivial ingredient in this method is the modularity of E. The modularity of all semistable elliptic curves over $\mathbb{Q}$ was established by Wiles [177] and Taylor–Wiles [164]. A few years later, the modularity of all elliptic curves over $\mathbb{Q}$ was established by Breuil-Conrad-Diamond-Taylor [43].

The absolute Galois group $G_{\mathbb{Q}}$ of $\mathbb{Q}$ acts on the group of p–torsion points on E. This induces the group representation of $G_{\mathbb{Q}}$

$$\bar{\rho}_{E,p} : G_{\mathbb{Q}} \to \mathrm{GL}_2(\mathbb{F}_p)$$

known as the mod p Galois representation $\bar{\rho}_{E,p}$ associated to E. By work of Mazur [123], we know that $\bar{\rho}_{E,p}$ is irreducible. By the modularity of E, there is a newform of level N associated to $\bar{\rho}_{E,p}$. The final step is to apply Ribet's level-lowering theorem [149] which yields a newform of level 2. Since there are no newforms of level 2, this contradicts the existence of the putative solution.

This initiated a movement of establishing the modularity of elliptic curves over other families of totally real number fields (see [41, 65, 75, 97]). Subsequently, the Fermat equation has been studied extensively over totally real number fields using an extension [77, Theorem 3] of the modular method outlined above (see [77, 98, 105, 125]). A particular highlight is the trailblazing work in [105], which is the very first instance of a full resolution to the Fermat equation over a field with degree greater than 2 (real biquadratic field specifically).

Comparatively, less progress has been made in the direction of establishing the modularity of elliptic curves over number fields with at least one pair of complex embeddings. However, under the assumption of Serre's modularity conjecture (see [56, Conjecture 3.1]), the Fermat equation has been resolved over some small imaginary quadratic fields by Ţurcaş [57, 58]. In very recent work, Caraiani and Newton [49] have established the modularity of all elliptic curves over imaginary quadratic fields F such the Mordell-Weil group $X_0(15)(F)$ is finite, opening up the road for unconditional results in the direction of the resolution of Diophantine equations over imaginary quadratic fields.

Wiles' Theorem (for which he was subsequently awarded the Abel Prize (2016)) paved the way for further advances in the explicit resolution of (ternary exponential) Diophantine equations. Some of the novelties established to date include Kraus [107], Bennett and Skinner [30] and Bennett, Vatsal and Yazdani [31] which have generalised the construction of an elliptic curve out of hypothetical solutions of generalised Fermat equations of signatures (p, p, p), $(p, p, 2)$ and $(p, p, 3)$ respectively. For these three cases, the association process is very detailed. Moreover, thanks to the well-written and explicit survey [156], using Wiles' modular method is more accessible and the enthusiastic Diophantine solver may employ these tactics to arrive at a contradiction.

1.7 Determining All Integral Solutions

Ideally, we would like to determine all of the integral points of a certain Diophantine equation from the offset. However, current tools to achieve this are either not applicable in general cases or otherwise computationally expensive and infeasible. Once we manage to sieve out impossibilities, we should hopefully be left with a handful of equations that in the end require a resolution. In this section, we distinguish between the common types of equations encountered, once we have finished sieving.

1.8 Integral Points on Genus 1 Curves

In 1929, Siegel [155] announced the following groundbreaking result.

Theorem 6 (Siegel) *Let C be a smooth affine algebraic curve of genus g defined over a number field K. Then there are only finitely many K-integral points on C, provided that g is strictly positive.*

Siegel's Theorem is the first major result within Diophantine equations that depends only on the genus of the curve and not directly on the form of the equations. Setting $g = 1$ and $K = \mathbb{Q}$ in Siegel's Theorem tells us that there are finitely many integral points on a given elliptic curve. Siegel used Thue's methods from Diophantine approximation to prove this result, however, since Thue's methods are ineffective, so is the proof of Siegel's Theorem. Not all is lost, as Theorem 1 (due to Baker) is applicable in this case, which yields effective results that are absolutely key to perform explicit computations. We demonstrate this with some examples.

Let us consider the elliptic curve $E/\mathbb{Q}$ with Cremona label `11a1` with minimal Weierstrass equation $E : y^2 + y = x^3 - x^2 - 10x - 20$. The Mordell-Weil group of $E/\mathbb{Q}$ is $E(\mathbb{Q}) = \mathbb{Z}/5\mathbb{Z} = \{0_E, (5, 5), (5, -6), (16, 60), (16, -61)\}$. Since E has finitely many rational points, we automatically know that E has finitely many integral points. Let us now consider the elliptic curve $E'/\mathbb{Q}$ with Cremona label `37a1` with minimal Weierstrass equation $E' : y^2 + y = x^3 - x$. The Mordell-Weil group of $E'/\mathbb{Q}$ is given by $E'(\mathbb{Q}) = \mathbb{Z} = \langle (0, 0) \rangle$. In other words, E' has infinitely many rational points. From this information, we cannot deduce the state of the integral points on E' straight away. The determination of the Mordell-Weil group of an elliptic curve over $\mathbb{Q}$ is generally a difficult problem, but once this has been completed, then the set of integral points can be computed with ease. Suppose an elliptic curve $E/\mathbb{Q}$ has rank r. We can write any rational (and indeed integral) point as $P = Q + a_1 R_1 + \cdots + a_r R_r$ where $a_i \in \mathbb{Z}$, Q is torsion point on E and $R_1, \cdots, R_r$ generate $E(\mathbb{Q})$ modulo torsion. If P is an integral point then by Theorem 6 there are only finitely many possibilities for P. Effective bounds on $|a_i|$ is due to Baker, which can be reduced significantly further by lattice reduction algorithms (see [52, Section 2.3.5, Section 8.7] for details). Once a reasonable bound on the $|a_i|$ is obtained, it becomes feasible to completely determine the set of integral points on E. We note that this method has been well-implemented computationally. For example, computing the integral points on the elliptic curve E' on the online `MAGMA` calculator (http://magma.maths.usyd.edu.au/calc/) via the `IntegralPoints` command [40] took 0.520 seconds.

1.9 Integral Points on Higher Genus Curves

In many Diophantine problems it is common to run into hyperelliptic curves, for example, after performing a descent step. It is possible to solve some of these equations by using the method of Chabauty via a MAGMA implementation [40]. The aim of Chabauty's method is to recover the set of rational points $C(\mathbb{Q})$ of a smooth projective curve $C/\mathbb{Q}$ from its Jacobian $J(C)$ through the Abel-Jacobi map when the rank of $J(C)$ is not "too big". We refer the reader to [157] or [138] for a comprehensive survey of the methodology and we proceed to summarise the main ideas below.

Let $C/\mathbb{Q}$ be a curve of genus g and let $J(C)$ denote its Jacobian. Assume that we have a known rational point $P_0 \in C(\mathbb{Q})$; the Abel-Jacobi embedding can be written as

$$\iota : C \to J(C), \qquad P \mapsto [P - P_0].$$

Consider $\Omega_C := H^0(C, \Omega_1)$ the g-dimensional vector space of regular differentials of the curve C. Given a prime $p \in \mathbb{N}$, we have the pairing

$$\Omega_C \times J(\mathbb{Q}_p) \to \mathbb{Q}_p$$

defined by

$$\left\langle \omega, \left[\sum_i P_i - Q_i\right]\right\rangle := \sum \int_{Q_i}^{P_i} \omega.$$

If the divisor $\left[\sum_i P_i - Q_i\right]$ is a torsion element of $J(\mathbb{Q}_p)$, then we have $\sum \int_{Q_i}^{P_i} \omega = 0$. Thus $J(\mathbb{Q})_{\text{tor}} \subset J(\mathbb{Q}_p)_{\text{tor}}$ lies in the right kernel of the pairing $\langle , \rangle$. Thus, if we have $D_1, \ldots, D_r$ a basis for the free part of $J(\mathbb{Q}_p)$, then the space $J(\mathbb{Q})^{\perp}$ of differentials ω annihilated by all elements D in $J(\mathbb{Q})$ is determined by the ones annihilated by $D_1, \ldots, D_r$. That is, $J(\mathbb{Q})^{\perp}$ is given by the equations

$$\int_{D_i} \omega = 0, \qquad i = 1, \ldots, r.$$

It follows then that $\dim J(\mathbb{Q})^{\perp} > g - r$. If the condition $r < g$ (known as the Chabauty condition) holds, then $\dim J(\mathbb{Q})^{\perp} \geq 1$ and there exists a differential $\omega_0 \in \dim J(\mathbb{Q})^{\perp}$. Then, the points $C(\mathbb{Q})$ are determined by the equation

$$\int_{\iota P} \omega_0 = 0.$$

This equation with a finite number of solutions can be solved by using p-adic power series. Finally, the set of rational points $C(\mathbb{Q})$ can be determined using a Mordell-Weil sieve. The combination of these techniques is currently implemented in MAGMA [40] for hyperelliptic curves of genus 2. The implementation is often capable of determining $C(\mathbb{Q})$.

Example 2 In [55], the authors resolve the equation

$$(x-4r)^3+(x-3r)^3+\cdots+x^3+\cdots+(x+3r)^3+(x+4r)^3=y^p$$

under the constraints $x, r, y, p \in \mathbb{Z}$, p a prime, $\gcd(x, r) = 1$ and $0 < r \leq 10^6$. For $p = 5$, descent and the divisibility conditions $3 \nmid x$ and $4 \mid x$ and $5 \nmid x$ lead to the ternary equation $3^3 w_2^5 - 2^4 w_1^{10} = 5r^2$. Letting $X = w_2/w_1^2$ and $Y = 5r$ yields the genus 2 curve

$$C:\ Y^2 = 3^3 \cdot 5 \cdot X^5 - 2^4 \cdot 5.$$

The Jacobian of C has rank 1 over $\mathbb{Q}$. Thus, the Chabauty condition is satisfied and the Chabauty implementation in MAGMA [40] is used to find that

$$C(\mathbb{Q}) = \{\infty\}.$$

1.10 Irreducible Bivariate Forms

A Thue equation of degree $n \geq 3$ is a Diophantine equation of the form

$$f(x, y) = C \tag{7}$$

where f is an irreducible degree n polynomial with integer coefficients and C is a known integer. It is well-known, due to a non-effective proof of Thue [165], that there are finitely many integral solutions to (7).

Effectivity was first achieved via Baker's theory of linear forms in logarithms. There are now readily available algorithms (implemented in MAGMA [40]) that solve (7) for small values of n. For a detailed survey, we refer the reader to [173].

Example 3 In [55], the authors resolve the equation

$$(x-4r)^3+(x-3r)^3+\cdots+x^3+\cdots+(x+3r)^3+(x+4r)^3=y^p$$

under the constraints $x, r, y, p \in \mathbb{Z}$, p a prime, $\gcd(x, r) = 1$ and $0 < r \leq 10^6$. Within each descent case, there are a small number of equations that remain once we have sieved out possibilities. These are usually dealt with using a Thue equation solver. Following previous approaches taken in [3, 4] and [5] proved to be futile as the authors encountered Thue equations with very large exponents and coefficients. These were (miraculously) eliminated via an application of the methods presented in Sect. 1.3. Here is an example of a Thue equation that was bypassed:

$$X^{19} - 4853698686832378443796187639236450195 3125(Y^2)^{19} = 274877906944.$$

We note that this particular Thue equation can (just about) be solved using MAGMA's [40] inbuilt Thue solver, under GRH assumption.

2 Perfect Powers as Sums of Like Powers

In this section, we give an overview of the results to date on various subfamilies of (1);

$$x^k+(x+r)^k+\cdots+(x+(d-1)r)^k=y^n \quad d,k,n,x,y,\in\mathbb{Z}, \quad r\geq 1; \quad d,k,n\geq 2$$

with parameters (k,d,r,n). We present tabular summaries at the end of each subsection (Tables 1, 2, 3).

2.1 *Sums of Squares*

Let us consider the problem of determining perfect powers that are sums of d consecutive squares, i.e., we consider the following subfamily of (1):

$$x^2+(x+1)^2+\cdots+(x+d-1)^2=y^n \qquad d,n,x,y,\in\mathbb{Z}, \quad d,n\geq 2. \tag{8}$$

The case $d=2$ was first studied by Cohn [54] in 1996 using elementary number theory techniques in conjunction with the Fibonacci sequence. It is well-known that (8) with exponent $n=2$ possesses infinitely many solutions. For example, (8) in the case $n=2$ and $d=2$ admits the following infinite family of solutions ([142]),

$$2x+2+y\sqrt{2}=\pm(1+\sqrt{2})^{2a+1}, \ a\in\mathbb{Z}.$$

Table 1 Progress towards Eq. (1) with $k=2$

(k,d,r,n)	Reference(s)	Method(s) used
$(2,2,1,n\geq 3)$	Cohn [54]	Elementary
$(2,d\geq 2,1,2)$ for $1\leq x\leq 100$	Bremner, Stroeker and Tzanakis [42]	Sects. 1.2 and 1.8
$(2,d=x,1,\text{ odd } n>2)$	Zhang and Bai [181]	Applied [16], Sect. 1.8
$(2,3,1,n\geq 2)$	Zhang [178]	Algebraic number theory
$(2,2\leq d\leq 10,1,n\geq 3)$	Patel [140]	Elementary, Sect. 1.3
$(2,3,1\leq r\leq 10^6,n\geq 2)$	Koutsianas and Patel [106]	Sect. 1.3
$(2,2\leq d\leq 10,1\leq r\leq 10^6,n\geq 2)$	Kundu and Patel [111]	Sect. 1.3

Table 2 Progress towards Eq. (1) with $k = 3$

(k, d, r, n)	Reference(s)	Method(s) used
$(3, 3, 1, 2)$	Cassels [50], Uchiyama [166]	Elementary
$(3, 3, 1, n \geq 2)$	Zhang [178]	Applied [23, 50] and [31]
$(3, 2 \leq d \leq 50, 1, 2)$	Stroeker [163]	Sect. 1.2
$(3, 2 \leq d \leq 50, 1, n \geq 2)$	Bennett, Patel and Siksek [27]	Sects. 1.2, 1.8, 1.5, 1.6, and 1.10
$(3, 3, 1 \leq r \leq 10^6$, prime $p \geq 2)$	Argáez-García and Patel [4]	Sects. 1.2, 1.8, 1.5, 1.6, 1.9, and 1.10
$(3, 3, r \geq 1$, prime $p \geq 7)$	Edis [70]	Sect. 1.6
$(3, 5, 1 \leq r \leq 10^6$, prime $p \geq 5)$	Argáez-García [3]	Sects. 1.2, 1.3, 1.5, 1.9, and 1.10
$(3, 7, 1 \leq r \leq 10^6$, prime $p \geq 5)$	Argáez-García and Patel [5]	Sect. 1.2, 1.3, 1.5, 1.9, and 1.10
$(3, 9, 1 \leq r \leq 10^6$, prime $p \geq 2)$	Coppola, Curcó-Iranzo, Khawaja, Patel and Ülkem [55]	Sects. 1.2, 1.8, 1.3, 1.5, 1.9, and 1.10

Table 3 Progress towards Eq. (1) with $k \geq 4$

(k, d, r, n)	Reference(s)	Method(s) used
$(4, 3, 1, n \geq 2)$	Zhang [178]	Sects. 1.8, 1.6, and 1.10
$(4, 3, 1 \leq r \leq 6, n \geq 2)$	Zhang [179]	Sects. 1.8, 1.6, and 1.10
$(4, 3, r \geq 1, n \geq 2)$	van Langen [170]	Sect. 1.6, analysis of $\mathbb{Q}$-curves
$(5 \leq k \leq 6, 3, 1, n \geq 2)$	Bennett, Patel and Siksek [26]	Sects. 1.6, 1.10, and 1.8, Hecke polynomials
$(5, 3, r \geq 1, n \geq 2)$	Bennett and Koutsianas [25]	Sects. 1.6, and 1.9
"For even k, and $n \geq 2$, Eq. (1) has no solutions for almost all $d \geq 2$"	Patel and Siksek [142]	Bernoulli polynomial, and its Galois group and Newton polygon

In [42], Bremner, Stroeker and Tzanakis study (8) also with $n = 2$, however, they fix values of x and let d vary. In this manner, they are able to re-frame the problem to one of determining integral points on certain elliptic (genus 1) curves. Letting $1 \leq x \leq 100$, the authors found that the principal difficulty arose in determining the torsion-free part of the curve. The authors subsequently used the method of linear forms in logarithms to obtain a sensible bound on the heights of integral points on the curve, allowing them to feasibly search for all such points.

After the passing of a couple of decades, Zhang and Bai [181] provided a resolution of (8) when $d = x$ using work of Bazsó, Bérczes, Győry and Pintér [16] on the equation $Ax^n - By^n = C$. A year later, Zhang [178], using algebraic number theory, provides a complete resolution to the problem of determining when sums of three consecutive squares forms a perfect power; he proves that this cannot happen.

In an extension to this problem, one may consider (8), where the terms are taken in (bounded) arithmetic progression. Application of work of Bilu, Hanrot and Voutier (see Sect. 1.3) leads to the resolution of further general subfamilies, see [106, 111, 113, 140].

The first glimmer of density results can be found in [181]. Theorem 1.2 in [181] states that there are infinitely many values of d such that the sum of d consecutive squares will never be a perfect power. A straightforward application of Dirichlet's Theorem gives the desired density result. We further explore density approaches in Sect. 2.3.

? Open Problems

1. Solve (8) for all values of d and $n \geq 3$.
2. Solve (8) with fixed value of d, and unrestricted arithmetic progression for $n \geq 3$.
3. As of now, the problem of perfect powers that are sums of squares has not been particularly amenable to the modular approach (as well as generally being ungovernable by the approaches outlined in Sects. 1.2 and 1.5). Could a new take on the modular approach hope for some resolution?

2.2 *Sums of Cubes*

Whilst one can easily make some progress on finding perfect powers that are sums of squares via some elementary algebraic number theory, the question of finding perfect powers as sums of cubes enters a completely different realm of number theory.

$$x^3 + (x+1)^3 + \cdots (x+d-1)^3 = y^n, \qquad x, y, d, n \in \mathbb{Z}, \quad d, n \geq 2. \tag{9}$$

Finding squares and cubes that are a sums of a fixed number of consecutive cubes ([50, 163, 166]) nowadays boils down to finding integral points on certain genus 1 curves [27].

Finding perfect powers (beyond squares and cubes) that are sums of consecutive cubes has only been touched on recently, due to contemporary techniques required in order to reach a full resolution. Zhang [178], building upon Cassels' work [50] by a clever application of the innovative and impressive work on *the generalised Fermat equation* presented in [23] and [31], determines *all* perfect powers that arise as a sum of three consecutive cubes. Zhang recovers Euler's identity, in addition to the solutions provided by Cassels, and concludes that there are no further solutions for all $n > 1$.

The first approach unifying many theories in the world of explicit number theory was given by Bennett, Patel and Siksek [27], who provide a complete list of solutions for perfect powers that are sums of at most 50 consecutive cubes. The methods employed include those summarised in Sects. 1.2, 1.8, 1.5, 1.6, and 1.10. In theory it would be possible, via existing techniques, to reach satisfactory conclusions for *any* value of d, however, current computational techniques used would need to be enhanced in order for the computations to be completed within a reasonable time.

Extending [27] to the study of perfect powers that arise from cubes of an arithmetic progression, we note the following papers: [3–5] and [55]. The exact number of cubes required is specified, and the length of the arithmetic progression is bounded (usually by 10^6). As the number of cubes increases, complexities arise due to the synthesis of vastly differing theories. Whilst usually this is the biggest hindrance, the authors in [55] have managed to synergise various theories to drastically increase efficiency and find all perfect powers that are sums of nine cubes with bounded arithmetic progression. Disappointingly, for $p \geq 5$, no solutions were found! In order to reach contradictions for such a large volume of equations, refining existing methodologies (Sects. 1.2, 1.5, and 1.3) was crucial.

We reveal the state of the art here (noting that it is natural to take $n = p$ to be a prime, as we can absorb excess powers into y, see [55] for full details):

Theorem 7 (Coppola, Curcó-Iranzo, Khawaja, Patel and Ülkem) *For a prime $p \geq 5$, there are no non-trivial integer solutions to*

$$(x-4r)^3 + (x-3r)^3 + \cdots + x^3 + \cdots + (x+3r)^3 + (x+4r)^3 = y^p \qquad (10)$$

for coprime x, r and $1 \leq r \leq 10^6$, where the trivial solutions satisfy $xy = 0$.

Ideally, one would like to resolve (10) for any arithmetic progression. We note that current modularity methods are unable to achieve this and completely new ideas are required.

As of now, there exists one complete resolution of Eq. (10) with unrestricted arithmetic progression. Edis [70] solves the problem of perfect powers that are sums of three cubes in *any* arithmetic progression, proving that no prime power greater than 7 equals the sum of three cubes in an arithmetic progression, thus engulfing the results presented in [4]. Edis uses modularity techniques and by twisting the Frey-Hellegouarch curves, he is able to elegantly reach a contradiction.

? Open Problems

1. Study families of (or sporadic) solutions for (10) for any d and any arithmetic progression with $n = 3$.

2. Solve (10) for all values of d and $n \geq 5$.
3. Solve (10) with fixed value of d, and unrestricted arithmetic progression for $n \geq 5$.
4. Density results? Or an average statement?

2.3 *Beyond Sums of Squares or Cubes*

A natural continuation of the preceding subsections is to consider (1) with exponents k greater than 3. As one can imagine, the literature now begins to rapidly thin out! We begin with the work of Zhang once more. In [178], it is shown that a sum of three consecutive fourth powers is never a perfect power. In this instance, the author employs the modular method (by hand!) following the recipe provided in [30] which attaches Frey-Hellegouarch curves to generalised Fermat-type equations of signature $(p, p, 2)$, along with considering integral points on certain elliptic curves, as well as solving a handful of Thue equations to reach a full resolution (Sects. 1.6, 1.8, and 1.10).

Bennett, Patel and Siksek [26] further this work and show that sums of three consecutive fifth or sums of three consecutive sixth powers is never a perfect power. In order to reach such a resolution, the modular method is significantly improved upon to handle larger dimensional newform spaces, notably via a clever consideration of their Hecke polynomials. The authors employ a *Multi-Frey-Hellegouarch* approach, pioneered by Bugeaud, Luca, Mignotte and Siksek [47]. The crux of the idea is to associate two (or more) different curves to the Diophantine equation, which results in performing computations within two different newform spaces that must still be compatible via modularity of the associated Galois representations. This increases the chances of reaching contradictions. We emphasise that the results in [26] for exponent $k = 6$ is only possible as the authors attach Frey-Hellegouarch curves to certain binary cubic forms, as pioneered in [22].

The echoing question of higher powers in arithmetic progression is once again considered first by Zhang [179] with partial results when considering fourth powers that depend upon the arithmetic progression. Remarkably, van Langen [170] provides a full resolution to the question of perfect powers that are sums of three fourth powers in *any* arithmetic progression. A modular approach is undertaken; this time with a careful analysis of certain $\mathbb{Q}$-curves.

Building upon [26], Bennett and Koutsianas [25] completely solve the question of perfect powers expressible as sums of three fifth powers in arithmetic progression (superseding earlier work [60]). They dealt with small exponents using the method of Chabauty over number fields.

To date, only a single result exists towards studying the *entire* Diophantine family (1). Patel and Siksek [142] prove that in the case of even exponents k, and any arithmetic progression r, solutions are rare.

Theorem 8 (Patel and Siksek) *Let $k \geq 2$ be an even integer, $r \in \mathbb{Z}_{>0}$. Let $\mathcal{A}_{k,r}$ denote the set of integers $d \geq 2$ such that there is a solution to* (1) *for $n \geq 2$. Then the set $\mathcal{A}_{k,r}$ has natural density 0.*

The proof relies on the study of the Galois group and the Newton polygon of Bernoulli polynomials. Bernoulli polynomials are conjectured to be an irreducible family of polynomials, however, at this moment, we seem to be quite far away from a resolution to this long-standing conjecture. We further remark that we can not hope for such a strong density result for odd values of k. This is due to the existence of the solution $(x, y, n) = (r(1-d)/2, 0, n)$, where d is any odd integer.

2.4 The Asymptotic Approach: A Way Forward?

In 2014, Freitas and Siksek initiated the study of viewing the Fermat equation asymptotically. As in [78], we say *Asymptotic Fermat's Last Theorem* holds over a (totally real) number field K if there is a constant B_K such that for all $p > B_K$ there are no non-trivial solutions to the Fermat equation over the ring of integers of K, namely, $\mathcal{O}_K$.

The asymptotic approach begins, in the same way as the modular approach, by attaching a Frey curve E to a putative non-trivial solution to the Fermat equation. If E is modular over K, then E corresponds to some modular form. Level-lowering produces a Hilbert newform $\mathfrak{f}$ of level $\mathcal{N}_p$ such that $\bar{\rho}_{E,p} \sim \bar{\rho}_{\mathfrak{f},\bar{\omega}}$. In a diversion to the modular approach, Freitas and Siksek then use an Eichler-Shimura type assumption to construct an elliptic curve $E_1 : y^2 = (x-e_1)(x-e_2)(x-e_3)$ of conductor $\mathcal{N}_p$, corresponding to $\mathfrak{f}$. Let S denote the set of primes of bad reduction for E_1. Writing $\lambda = (e_1 - e_2)/(e_1 - e_3)$, $\mu = (e_2 - e_3)/(e_1 - e_3)$ gives a solution to the S-unit equation $\lambda + \mu = 1$, where S is the set of primes dividing $2\mathcal{N}_p$. The idea is then to study solutions to the Fermat equation over K by studying solutions to the S-unit equation over K. This method has since been extended to study other Fermat-type equations (see [56, 62, 76, 94, 95, 100, 108, 130] and [129]).

? Open Problems

1. Can we find all solutions to (1) for some fixed d and r with $n > 3$?
2. Solve (1) for higher exponents $k \geq 4$ and larger length d of consecutive terms.
3. Solve (1) for higher exponents $k \geq 4$ with fixed d and unrestricted arithmetic progression.
4. Consider perfect powers that are sums of differing powers as studied in [180].

5. Following [180], is there any hope to solve this for differing powers taken in an arithmetic progression? What about *all* arithmetic progressions?
6. Density results for odd exponents, modulo trivial solutions? Or an average statement?
7. Proving irreducibility of the entire family of Bernoulli polynomials with even degree.
8. Proving irreducibility of the entire family of Bernoulli polynomials modulo the factors arising from the well-known rational roots for all odd degree.

---

3 Perfect Powers as Sums of Consecutive Integers Starting from 1

In this section, we present an overview of the results to date on various subfamilies of (2):

$$s(1^k + 2^k + \cdots + x^k) + r(x) = y^n \qquad k, n, x, y, s \in \mathbb{Z},\ r(x) \in \mathbb{Z}[x], \quad k, n \geq 2$$

with parameters $(k, n, s, r(x))$. Notice that when $s = 1$ and $r(x) = 0$ this equation gives a subfamily of Eq. (1). A note of great importance is that the methodology and results obtained for this subfamily differ vastly from those given in the previous section. We summarise key contributions at the end of each subsection in Tables 4 and 5, as well as listing open problems.

This story begins with Lucas' Problem 1180 [117]:

> Une pile de boulets à base carée ne contient un nombre de boulets égal au carré d'un nombre entier que lorsqu'elle en contient vingt-quatre sur le côté de la base.

In this assertion, E. Lucas claimed that the only (positive integer) solutions to the equation

$$1^2 + \cdots + x^2 = y^2$$

are $(x, y) \in \{(1, 1), (24, 70)\}$. Attempts to prove this claim by himself [118] and Moret-Blanc [136] turned out to be both defective, as they contained errors in the argumentation. It was only in 1918 that a correct proof materialised.

Watson [175] was the first to do so, but many others have given alternative proofs of the theorem; see [2, 59, 116, 120, 121] and [21].

3.1 Erdős–Moser Conjecture

In the early 1950's, Erdős and Moser studied the more general equation

$$S_k(m) := 1^k + 2^k + \cdots + (m-1)^k = m^k. \tag{11}$$

It is easy to see that for $k = 1$ we can rewrite the left hand side of (11) as $m(m-1)/2$ and find the unique trivial solution $m = 2$. However, as Erdős noticed in a letter to Moser, taking $k \geq 2$, Eq. (11) seems to have not a single solution.

Most Diophantine problems (of interest) seem to be easy to state, yet notoriously difficult to solve; naturally this is the case for Eq. (11), as it still remains conjectural.

Conjecture 1 (Erdős–Moser) The Eq. (11) only has one integral solution when $k = 1$ and $m = 2$ and no others.

There exist (partial) results that should convince the reader to stay positive about Conjecture 1 being true. For odd k, it is easy to see that the conjecture is indeed true. By induction, one can see that for such k the sum $S_k(m)$ needs to be a multiple of the integer $m(m-1)/2$; provided that $m > 2$, if there were a solution to Conjecture (1) this would contradict the coprimality of m and $m-1$. Hence k has to be even. Moser himself [137], assuming that a solution exists to (11), with $k \geq 2$ even, ingeniously provided a proof for a large lower bound on m.

Table 4 Achievements towards the resolution of Eqs. (11), (12) and (13)

(a, α, k, m)	Reference(s)	Method(s) used
$(1, 1, k, m \leq 10^{10^6})$	Moser [137]	Elementary
$(1, 1, k, m \leq 10^{10^6})$	Moree [133]	Four mathemagical rabbits pulled from a tophat
$(1, 1, k \text{ odd}, m \geq 3)$	Moser [137], Moree [133]	Elementary
$(1, 1, k, m \leq 1.485 \cdot 10^{9321155})$	Butske, Jaje and Mayernick [48]	Elementary
$(1, 1, k, m \leq 10^{10^9})$	Gallot, Moree and Zudilin [79]	Continued fractions, builds upon [35, 135]
$(a, 1, k, m < \max(10^{9 \cdot 10^6}))$	Moree [134]	Elementary, builds upon [132]
$(1, 2 \leq \alpha \leq 1500, k, m < 3)$	Baoulina and Moree [13]	Elementary, uses regular primes and Bernoulli numbers
"For even α or divisible by regular primes, and $m > 3$, Eq. (13) has no solutions"	Baoulina and Moree [13]	Elementary, uses regular primes and Bernoulli numbers
"For α square-free, Eq. (13) has no solutions with $m, k \leq 3.44 \times 10^{82}$"	Baoulina and Moree [13]	Elementary, builds upon [137]

Table 5 Progress towards Schäffer's Conjecture 4

$(k, n, s, r(x))$	Reference(s)	Method(s) used
(2, 2, 1, 0)	Watson [175]	Elliptic functions
(2, 2, 1, 0)	Ljunggren [116]	Algebraic proof
(2, 2, 1, 0)	Ma [121], Cucurezeanu [59], Anglin [2]	Elementary
(2, 2, 1, 0)	Bennett [21]	Sects. 1.8, and 1.2
For $(k, n) \in S$, $(s, r(x)) = (1, 0)$ (2) has infinitely many solutions.	Schäffer [152]	Elementary
For $(k, n) \notin S$, $(s, r(x)) = (1, 0)$, (2) has finitely many solutions (ineffective).	Schäffer [152]	Sects. 1.10 and 1.8
Determination of all solutions for certain small pairs $(k, n) \leq (11, 641)$; see [82]	Schäffer [152]	Sect. 1.10
For $k \notin \{3, 5\}$ and $s = 1$ (2) has finitely many solutions (ineffective).	Voorhoeve, Győry and Tijdeman [172]	Diophantine approximation
For $k \notin \{3, 5\}$, $\square \nmid s$ and $r(x) \in \mathbb{Z}$, (2) has finitely many solutions (effective).	Voorhoeve, Győry and Tijdeman [83]	Diophantine approximation, Sect. 1.2
Gives effective commutable bounds for x, y and n solutions of [172] in terms of k and s.	Brindza [44]	Diophantine approximation, Sect. 1.2
If $n \notin \{1, 2, 4\}$, $s = 1$ and $r(x) = 0$, (2) has at most e^{7k} solutions	Brindza [45]	Diophantine approximation, Sect. 1.2
$(k \leq 58, 2, 1, 0)$	Jacobson, Pintér and Walsh [96]	Sects. 1.2 and 1.8
$(1 \leq k \leq 11, 2 \leq n, 1, 0)$	Bennett, Győry and Pintér [24]	Sects. 1.2, 1.10, and 1.6
$(1 \leq k \leq 170$ odd, even, 1, 0)	Pintér [146]	Methods in [96] with Sects. 1.2 and 1.6
Solved Eq. (2) for a positive proportion of x and constructed bounds on n in terms of x and k.	Hajdu [84]	Elementary
Solve (2) with $s = 1$ and $r(x) = 0$ for $x < 25$ and $n \geq 3$	Bérczes, Hajdu, Miyazaki and Pink [32]	Methods in [84] and Sect. 1.2

Theorem 9 (Moser, 1953) *If* (m, k) *is a solution to* (11)*, with* $k \geq 2$*, then* $m > 10^{10^6}$.

Moser's proof unfolds from four simple identities (or mathemagical rabbits) that putative solutions of (11) need to satisfy. Moree [133] gave a top hat to these magical identities by noticing that they can be pulled up from a theorem on harmonic series *à la* von Staudt-Clausen's Theorem. By doing some juggling with the four mathemagical rabbits of Moser, one can arrive to the conclusion that any solution m

would need to satisfy

$$\sum_{p|M} \frac{1}{p} + \frac{1}{m-1} + \frac{2}{m+1} + \frac{2}{2m-1} + \frac{4}{2m+1} > 3.16$$

where M is a square-free integer that can be constructed from m. Now the lower bound on m follows from looking for the smallest such M such that the value of prime harmonic series – which is a divergent series – is big enough. Moser estimated this value, thus obtaining an "estimated value" for the lower bound on m.

With the arrival of the new millennia and the emergence of modern techniques, Butske, Jaje and Mayernik [48] managed to refine Moser's lower bound by explicitly computing the value of m derived from the original outlined proof. The improvements situated m just above $1.485 \cdot 10^{9321155}$.

Theorem 10 (Butske, Jaje and Mayernik, 2000) *If* (m, k) *is a solution to* (11), *with* $k \geq 2$, *then* $m > 10^{9 \cdot 10^6}$.

Notice that Butske, Jaje and Mayernik [48] are just shy of the 10^{10^7} mark! The authors remark "...the authors hope that new insights will eventually make it possible to reach the more natural benchmark 10^{10^7}..."

Readers, fear not, as the change of millennia also brought with it completely novel methods: Gallot, Moree and Zudilin used continued fractions digits calculation of $\log 2/N$ for an integer N and managed to obtain significant improvements on the lower bound m.

Theorem 11 (Gallot, Moree and Zudilin, 2011) *If* (m, k) *is a solution to* (11), *with* $k \geq 2$, *then* $m > 10^{10^9}$.

Taking the methods of Gallot, Moree and Zudilin as in [79] further seems infeasible with current technology, as computing sufficiently many continued fractions digits is the bottleneck. However, if it were possible to write down as many digits as one wanted, the methods in [79] lead us to believe that a lower bound for m as big as $10^{10^{400}}$ could be achieved. Readers are encouraged to consult the excellent survey paper by Moree [134] for more details surrounding the mysterious Erdős–Moser conjecture.

Like every other notorious conjecture in mathematics, it is also of great interest to study generalisations of the Erdős–Moser Conjecture. Krzysztofek [110] was the first to introduce an additional integer variable to right hand side of Eq. (11).

Conjecture 2 (generalised Erdős–Moser, 1966) There are no integral solutions (a, k, m) to the Diophantine equation

$$S_k(m) = am^k \tag{12}$$

with $m \geq 2$, $k \geq 2$ and $a \geq 1$.

Towards the generalised Erdős–Moser conjecture, one can use Moser's original method to show that Eq. (12) has no integer solutions (a, m, k) with

$$k > 1, \qquad m < \max(10^{9 \cdot 10^6}, a \cdot 10^{28})$$

This result is the best known lower bound to date (see [134], which improved upon bounds in [132]) since we face the same obstructions as in previous works on the Erdős–Moser Conjecture 1; to improve on the current known lower bounds, it seems that we are required to find a completely new approach.

The generalised Erdős–Moser Conjecture is not the only generalisation of Conjecture 1. In another related direction, we shed light on the so called Kellner–Erdős–Moser Conjecture [101], which makes speculations on the ratios of consecutive power sums.

Conjecture 3 (Kellner–Erdős–Moser) Let k and m be positive integers with $m \geq 3$, then the quotient $S_k(m+1)/S_k(m)$ is an integer if and only if $(m, k) = (3, 1)$ or $(m, k) = (3, 3)$.

In particular, the Kellner–Erdős–Moser Conjecture is equivalent to solving the Diophantine equation

$$\alpha S_k(m) = m^k, \tag{13}$$

in the integral variables (α, k, m), where it is now apparent that we have indeed generalised Erdős' original stipulation.

To contrast with Eq. (12), Eq. (13) is a bit more tame and allows one to obtain stronger results. Using a refinement of the Voronoi congruence, the notion of "regular" and "irregular" primes as well as previous techniques which relate power sums to Bernoulli Numbers, in [13] the authors prove the following:

Theorem 12 (Baoulina and Moree, 2016) *If α is even or α has a regular prime divisor or $2 \leq \alpha \leq 1500$, then Eq. 13 has no solutions (m, k) with $m > 3$.*

The study of "regular" and "irregular" primes are currently of great interest to number theorists as their behaviour seems to underpin many infamous open conjectures in number theory. Some well-known examples are the results of Kummer relating the class number of cyclotomic extensions with regular primes, or his proof of Fermat's Last theorem for regular prime exponents. We expect irregular primes to behave in the following manner: as $x \to \infty$, the number of irregular primes $p \leq x$ should be bounded above by $\delta x/(\log x)$, where δ is a real constant $\delta < 1$. Conjecturally, Siegel [154] claimed that δ should be 0.3935, which has the implication that approximately 60% of all primes are regular. Today, Siegel's claim still remains an open problem. In fact, we do not even know if the number of regular primes is infinite!

Despite the study of regular and irregular primes being highly conjectural, when pairing Theorem 12 with expected behaviour of irregular primes, we obtain much stronger results; it then follows that, for a set of integers α of density 1, the Eq. (13) has no solution with $m \geq 4$. In addition to these much stronger density-type results, Baoulina and Moree [13] also provide explicit results for square-free integers α whereby they give astronomical lower bounds on k and m for Eq. (13) having solutions.

Theorem 13 (Baoulina and Moree, 2016) *If $m \geq 4$ and α is square-free, then both k and m exceed* $3.44 \cdot 10^{82}$.

The generalisations of equations of Erdős–Moser type does not end here; there is a whole world of exploration unfolding from Eq. (11). To the reader willing to dive down this mathemagical rabbit hole, we recommend to consult [12, 122, 132, 134, 160]. Notable milestones towards the Erdős–Moser Conjecture and its generalisations are summarised in Table 4. We end this section with some open problems.

? Open Problems

1. Develop new techniques to study (11).
2. Fully resolve the infamous Erdős–Moser Conjecture 1.
3. Reach the natural benchmark of $m \geq 10^{10^7}$ in Conjecture 2.
4. Density results for Conjecture 2.
5. Generalise and study Eq. (11) over number fields?
6. Prove Siegel's Conjecture on the behaviour of regular/irregular primes.

3.2 Schäffer's Conjecture

Schäffer, in 1956, taking inspiration from the speculation of Erdős and Moser, considered the general equation

$$S_k(x) = 1^k + 2^k + \cdots + x^k = y^n. \tag{14}$$

Schäffer himself, [152] gives all the (infinitely many) solutions of Eq. (14) for (n, k) in the set $S := \{(1, 2), (2, 3), (3, 4), (5, 2)\}$ and proves that for all (n, k) not in this set, Eq. (2) can only have a finite number of solutions. His proof was non-effective. However, that did not stop him from making the following conjecture.

Conjecture 4 (Schäffer) For $k \geq 1$ and $n \geq 2$ with (k, n) not in the set S, Eq. (2) has only one non-trivial solution, namely $(k, n, x, y) = (2, 2, 24, 70)$.

In 1979, Voorhoeve, Győry, and Tijdeman [172] study Eq. (2) for $s = 1$ and $r(x)$ any polynomial with integer coefficients and prove that for $k \geq 2$ such equation has finitely many solutions.

Theorem 14 (Voorhoeve, Győry, and Tijdeman) *Let $s = 1$ and $r(x) \in \mathbb{Z}[x]$, for $k \geq 2$, $k \notin \{3, 5\}$, Eq.* (2) *has finitely many integer solutions and n can be effectively bounded.*

In further results [83], the same authors restrict $r(x) \in \mathbb{Z}$ to be an integer and present similar results for s any square-free integer. However, these are now effective.

Theorem 15 (Voorhoeve, Győry, and Tijdeman) *Let s be a square-free integer and $r(x) \in \mathbb{Z}$. For $k \geq 2$, $k \notin \{3, 5\}$, Eq.* (2) *has finitely many integer solutions and they can be made effective.*

In particular, for $s = 1$ and $r = 0$ they give an effective proof of Schäffer's Theorem whenever $k \neq 3, 5$. Their methods use a theorem of LeVeque [114], based on Siegel's method [155] for finding an asymptotic estimate for the frequency of solutions of a superelliptic curve.

The effective results in [83] rely on the fact that for $n \geq 2$, and $r(x) \in \mathbb{Z}$, Eq. (2) gives a superelliptic equation where methods of Baker [8] based on linear forms in logarithms can be combined with LeVeque's Theorem to bound n, but not x and y.

After this initial leap, some authors worked on generalisations of the results. In [44], Brindza considers a more general equation involving (2) and, with an effective version of LeVeque's Theorem, proves an effective version of [172]; in [45], the same author uses similar techniques to find effective bounds for x and y in [83]; in [99], Kano considers more general s whenever $k \not\equiv 1 \pmod 4$; and Dilcher in [69] and Urbanowicz in [167, 168] and [169], introduce characters in the equation.

However effective, none of these results are explicit. We can find the first explicit bound for n in work of Pintér [145]. On the other hand, Brindza [45] proves that for any given $n \neq 3$ or 4, the number of integer solutions of (2) cannot be more than e^{7k}. Brindza and Pintér [46] further show that for $(k, n) \neq (3, 4)$, Eq. (2) has at most $\max\{C, e^{3k}\}$ positive integer solutions when $n > 2$, where C is an effectively computable absolute constant (they also give analogous results for $n = 2$).

In order to tackle Schäffer's Conjecture, we need to completely solve equations in the family given by (2). Schäffer himself was able to confirm his conjecture for a certain finite number of pairs (k, n) with $k \leq 11$, $n \geq 2$ (see [152] or [82] for details). When $n = 2$, Jacobson, Pintér and Walsh [96] confirmed Conjecture 4 for another handful of k's, namely for all even numbers below 58 (and below 70 assuming GRH). Bennett, Győry and Pintér [24] were the first ones to prove Schäffer's Conjecture for an infinite subfamily of (2): they completely solved the problem for all $1 \leq k \leq 11$. It took all of the tools in the box to do so: from classical local and elementary methods to more powerful modularity techniques for ternary equations, passing by linear forms in logarithms and in elliptic logarithms, and finally appealing to computational methods such as Thue and Siegel. Pintér

[146] improved on their results by showing that the conjecture still holds for $n > 4$ even and $1 \leq k < 170$.

In another direction towards tackling Schäffer's Conjecture, by imposing certain conditions on x, Hajdu [84] recently obtained further results. Precisely, by studying the valuation of the primes 2 and 3 of the polynomial $S_k(x)$, Hajdu constructs upper bounds for n and proves the following.

Theorem 16 (Hajdu) *Assume that* $x \equiv 3, 4 \pmod 8$. *Then* (2) *has no solutions with* $k = 1$ *or* k *even. Furthermore, Schäffer's Conjecture holds if* x *satisfies one of the following congruence conditions*

$$x \equiv h_i \pmod{m_i}, \quad \text{with } h_i \in H_i, \quad i = 1, 2, 3, 4$$

where $H_1 = \{2\}$, $H_2 = \{5, 7\}$, $H_3 = \{2, 7, 9, 14\}$, $H_4 = \{18, 22\}$, *and* $m_1 = 5$, $m_2 = 13$, $m_3 = 17$, $m_4 = 41$.

The theorem can be paraphrased as "the only putative solutions to (14) that satisfy certain congruence conditions for x are exactly the ones given by the conjectured equations".

Following up on this line of thought, Bérczes, Hajdu, Miyazaki and Pink [32] managed to extend Hajdu's results and show that for all $x < 25$, all solutions of equation (14) with $n \geq 3$ are given by $(x, k, y, n) = (1, k, 1, n), (8, 3, 6, 4)$, recovering Schäffer's observed solutions.

The authors treat the cases where x is not covered by Hajdu's Theorem [84] via bounds from linear forms in two logarithms and polynomial-exponential congruences.

Taking the ideas developed in [32] further, and combining with Baker's theory, results are obtained towards a *generalised Schäffer problem*, namely

$$(x + 1)^k + \cdots + (\ell x)^k = y^n \quad k, \ell, n, x, y, s \in \mathbb{Z} \tag{15}$$

In [33], the authors consider $\ell = 2$ in (15) and for $2 \leq x \leq 13$, $k \geq 1$, $y \geq 2$, and $n \geq 3$, prove that there are no solutions to (15), as well as obtaining upper bounds for the exponent n in terms of 2-adic and 3-adic valuations of some functions depending on x and k.

In [161] and [14], the authors continue studies on (15) and prove results that are in a similar vein to Schäffer's original problem. The authors prove the effective finiteness of the solutions of (14) when taking fixed $k \geq 2$ and $\ell \geq 2$. In other words, they show that $\max\{x, y, n\} < C$, where C is an effectively computable constant depending only on k and ℓ. Other accounts of generalisations on Schäffer's problem are considered in [19, 147].

Another trend on generalisations of Schäffer's problem was established in [17], where the authors take on the endeavour of studying the polynomial

$$S_{a,b}^k(x) := b^k + (a + b)^k + (2a + b)^k + \ldots + (a(x - 1) + b)^k.$$

Several results on the number of solution of Diophantine equations constructed upon the polynomial $S^k_{a,b}(x)$ are given in [15, 18].

This closes the story of all known results on Schäffer's Conjecture and generalisations, for now. For a thorough early history and comprehensive study, please consult [153].

? Open Problems

1. Solve Schäffer's Conjecture 4.
2. Study and solve generalisations of Schäffer's Conjecture.

4 To the Power 1: Integers That are Sums of Like Powers

A problem that may seem similar to that treated in this survey, but is actually extremely different in terms of the methodology used to tackle it, is to determine which integers are sums of like powers. We give a brief overview of this related problem for completeness.

We begin by stating Fermat's infamous Two Square Theorem.

Theorem 17 (Fermat) *An odd prime can be represented as the sum of two integral squares if and only if the prime is congruent to 1 modulo 4.*

Although the first observation (not necessarily restricted to primes) of this identity was in 1625 by Girard, Fermat meticulously expanded upon this observation in a letter to Mersenne dated 25 December 1640: for this reason, the theorem is sometimes known as *Fermat's Christmas Theorem.*

The story then gains momentum once again in 1770, with Lagrange announcing his proof of the Four Square Theorem:

Theorem 18 (Lagrange) *Every natural number can be represented as a sum of four non-negative integral squares.*

Bachet's 1621 Latin translation of Diophantus' *Arithmetica* stated this theorem, however, lacked a proof. Around the same time, Waring stated general observations in his Meditationes Algebraicae (pg.203–204):

> Omnis integer numerus vel est cubus, vel e duobus, tribus, 4, 5, 6, 7, 8, vel novem cubus compositus: est etiam quadratoquadratus; vel e duobus, tribus &c. usque ad novemdecim compositus &sic deinceps.

In a later edition, Waring adds (pg.349)

> ...consimilia etiam affirmari possunt (exceptis excipiendis) do oedem numero quantitatum earundem dimensionum.

Waring's assertions/problem is more familiarly stated as

> Can every positive integer be represented as the sum of at most $g(k)$ kth powers of positive integers, where $g(k)$ depends only on k and not the number being represented?

Results so far show that every positive integer can be represented with at most 4 squares (Lagrange, 1770); at most 9 positive cubes [102, 176]; at most 19 fourth powers [10, 11]; at most 37 positive fifth powers [51] and at most 73 sixth powers [143].

During the 1920's, Hardy and Littlewood introduced in a series of papers ([85–90]) what we may now consider to be a more fundamental number than $g(k)$; the number $G(k)$, which is defined to be the least positive integer such that all sufficiently large integers may be represented as a sum of at most $G(k)$ kth powers of **positive** integers. This can be seen as a natural generalisation to Waring's Problem, or perhaps even a strengthening to Waring's Problem.

Since Wieferich has already shown that every positive integer can be written as a sum of at most 9 positive cubes [176], it seems natural to study the number $G(3)$, i.e., representations of integers as sums of less than 9 cubes.

Landau [112] showed that every sufficiently large integer is the sum of at most 8 positive cubes, however, Landau's result was not effective. Effectivity came in 1913, only two years later, where Baer [6] showed that every integer greater than 2.26×10^{15} is the sum of at most 8 positive cubes. Dickson [67] refined this to the best result possible; he proves that every positive integer except 23 and 239 can be written as the sum of at most 8 positive cubes.

As one can imagine, the problem very steeply increases with difficulty when only allowing for smaller and smaller number of positive cubes! Linnik [115] was the first, in 1943, to show that every sufficiently large integer is the sum of seven positive cubes, again, an ineffective result. A simpler, albeit still ineffective, proof was given by Watson in 1951 [174]. It was not until 1984 that we have effective results, the first given by McCurley [124] which states that any integer greater than exp(exp 13.94) can be written as the sum of at most 7 positive cubes. Ramaré, [148] improved this to $3.72. \times 10^{227}$. Partial results followed; see [34, 37, 66, 72], but none settled the question once and for all.

The question of integers represented by 7 positive cubes was finally settled in 2016 [158].

Theorem 19 (Siksek) *Every positive integer except*

$$15, 22, 23, 50, 114, 167, 175, 186, 212, 231, 238, 239, 303, 364, 420, 428, 454$$

is the sum of seven cubes.

Currently, we know that $4 \leq G(3) \leq 7$. It is conjectured that $G(3) = 4$, however, we are yet to determine whether every sufficiently large integer can be represented as the sum of 6 positive cubes, let alone specific results along the lines of Theorem 19.

In Table 6 we summarise the known bounds on $G(k)$. An excellent survey article on Waring's Problem and its generalisations is [171] and the references therein, where there are more results on the function $G(k)$.

We now shift gears slightly, to talk about *explicit* representations of certain integers as the sum of a specified number of kth powers - we are able to speak of such things as we no longer specify that the kth powers need to be positive. In this direction of study, Lagrange's results were extended by Legendre in 1797–1798, who considered integers that are sums of three squares:

Theorem 20 (Legendre) *A natural number can be represented as the sum of three squares if and only if it is not of the form* $4^a(8b+7)$ *for some integers a and b.*

In determining integers that are sums of a small number of (integer) cubes, the problem becomes significantly more difficult. We begin gently with the problem of expressing integers as a sum of two integer cubes, in other words, let k be an arbitrary integer, then, is it possible to write

$$k = x^3 + y^3, \qquad \text{for some } x, y \in \mathbb{Z}?$$

This problem is actually not too hard to solve, since we have $x^3 + y^3 = (x + y)(x^2 - xy + y^2) = rs$, with $r = x + y$, and from this we obtain $s = 3x^2 - 3rx + r^2$, which has integer solutions if and only if $12s - 3r^2$ is a square, say t^2, and in particular the solutions are $x = (3r + t)/6,\ y = (3r - t)/6$.

It is worth noticing that, if instead of integer x and y we allow for rational numbers, the problem of solving $k = x^3 + y^3$ for a given k transforms to that of finding rational points on an elliptic curve. The curve $x^3 + y^3 = k$ has Weierstrass form $Y^2 = X^3 - 432k^2$, it has non-trivial torsion if and only if k is a cube or twice a cube, and otherwise we have a solution if and only if it has positive rank. There are some partial results towards this, which we list below.

- In [1], it is proven that a positive proportion of k yield a rank zero elliptic curve and a positive proportion of k yield a positive rank elliptic curve.
- In [61, 71, 109, 151] there are results concerning the case where k is a prime number, namely if $k \equiv 2, 5 \pmod 9$ then it yields a rank zero elliptic curve,

Table 6 Known bounds on $G(k)$

		Known Bounds		
$1 = G(1) = 1$	$6 \le G(5) \le 17$	$13 \le G(9) \le 50$	$14 \le G(13) \le 84$	$18 \le G(17) \le 117$
$4 = G(2) = 4$	$9 \le G(6) \le 24$	$12 \le G(10) \le 59$	$15 \le G(14) \le 92$	$27 \le G(18) \le 125$
$4 \le G(3) \le 7$	$8 \le G(7) \le 33$	$12 \le G(11) \le 67$	$16 \le G(15) \le 100$	$20 \le G(19) \le 134$
$16 = G(4) = 16$	$32 \le G(8) \le 42$	$16 \le G(12) \le 76$	$64 \le G(16) \le 109$	$25 \le G(20) \le 142$

while if $k \equiv 4, 7, 8 \pmod 9$ it yields a rank 1 elliptic curve. For $k \equiv 1 \pmod 9$, work of [150] shows that the underlying elliptic curve may have rank either 0 or 2.

- In [104], the cases where k is twice a prime p or twice the square of a prime p satisfying $p \equiv 2, 5 \pmod 9$ are dealt with by using Iwasawa theory. For a more informative but less technical reading on the subject, it is worth consulting [103] (in Japanese).

A well-known result concerns the case of the sum of five or more cubes: indeed every integer has infinitely many representations as the sum of five cubes, since the following identity holds:

$$6m = (m+1)^3 + (m-1)^3 - 2m^3$$

and if we write $k = 6n + r$, we can apply this identity to $m = m(n) = (k - (6n + r)^3)/6$, obtaining:

$$k = (6n+r)^3 + (m(n)+1)^3 + (m(n)-1)^3 - (m(n))^3 - (m(n))^3.$$

In [64] it is shown, with a similar approach, that every integer $k \not\equiv \pm 4 \pmod 9$ can be represented as a sum of four cubes in infinitely many ways, and it is conjectured (see [162]) that in fact every integer k can be written as a sum of four cubes in infinitely many ways.

Finally, let us consider the problem of whether a given integer can be written as a sum of three cubes. First of all, a simple congruence condition modulo 9 shows that if $k \equiv \pm 4 \pmod 9$ then it is not the sum of three cubes. Moreover, it is easy to show that if k is a cube or twice a cube then again there are infinitely many ways to write it as a sum of three cubes – it is sufficient to multiply by a suitable cube one of the following identities:

$$\begin{aligned} 0 &= n^3 - n^3 + 0^3; \\ 1 &= (9n^4)^3 + (3n - 9n^4)^3 + (1 - 9n^3)^3; \\ 2 &= (1 + 6n^3)^3 + (1 - 6n^3)^3 + (-6n^2)^3 \end{aligned}$$

It remains to understand whether integers $k \not\equiv \pm 4 \pmod 9$ and not being a cube or twice a cube can be written as the sum of three cubes, and if so, whether there are infinitely many ways of doing so. In 1956 [131], Mordell wrote: *I do not know anything about the integer solutions of* $X^3 + Y^3 + Z^3 = 3$ *beyond the existence of* $(1, 1, 1)$, $(4, 4, -5)$, and indeed this was such a hard problem that only over 60 years later a less trivial solution was found [39] by Booker and Sutherland, namely:

$$\begin{aligned} 3 = {} & 569936821221962380720^3 \\ & + (-569936821113563493509)^3 \end{aligned}$$

$$+ (-472715493453327032)^3$$

They also treat rigorously and computationally the challenge of writing, more generally, an integer k as the sum of three cubes, other than the cases that are already understood. Although it may seem an innocuous computational task, it is incredibly hard to compute numbers with so many digits by a naïve brute force algorithm. Indeed, already in 1954, Miller and Woollett [128] programmed the EDSAC (Electronic Delay Storage Automatic Calculator) in Cambridge to conduct the search for solutions to Mordell's equation with $0 \leq k \leq 100$ and $|Z| \leq |Y| \leq |X| \leq 3164$, and succeeded in finding solutions for $k \not\equiv \pm 4 \pmod 9$ in this range, with the exceptions of

$$k = 30, 33, 39, 42, 52, 74, 75, 84, 87.$$

Later, in 1964, a more ambitious and sophisticated search [80] involved all $0 \leq k \leq 1000$ and $|Z| \leq |Y| \leq |X| \leq 2^{16}$, but was only able to produce one extra solution, for $k = 87$.

In 1992, Heath-Brown [91] finally made a more precise conjecture, namely that every integer $k \not\equiv \pm 4 \pmod 9$ has infinitely many writings as sum of three cubes. He also conjectured an asymptotic formula for this number of solutions, as follows: if $N(B)$ is the number of solutions in the box $\max\{|X|, |Y|, |Z|\} \leq B$, then

$$N(B) \sim \frac{1}{9}\frac{\Gamma(1/3)^2}{\Gamma(2/3)}\left(\prod_{p \text{ prime}} \sigma_p\right) \log B$$

where σ_p is defined, for example, in [159].

Due to the above partial results and significant progress towards the conjectures, there has been strong motivation towards finding more optimised algorithms to search for solutions to Mordell's equation (also where 3 is replaced by an integer k). Following [159], we illustrate here the most significant algorithm currently known, which was used, among others, by Booker and Sutherland to find the result mentioned above, as well as to write every number below 100 (with the usual restrictions) as a sum of three cubes. This algorithm is based on the ideas and work of [20, 38, 39, 92, 93] and it is somewhat similar to the factorisation algorithm mentioned above for the case of two cubes.

Fix k, suppose there is a solution (X, Y, Z) to $X^3 + Y^3 + Z^3 = k$ and let $d = X + Y$. Then, taking $X = d - Y$ we obtain a quadratic equation for Y in terms of d, Z, such that its discriminant is a square. Therefore, we have two constraints on Z, namely:

$$\begin{aligned} Z^3 &\equiv k \pmod d, \\ 3d(4(k - Z^3) - d^3) &\text{ is a square} \end{aligned}$$

Table 7 Last remaining numbers to be written as a sum of three cubes

k	X, Y, Z
87	4271, −4126, −1972
39	134476, 117367, −159380
84	41639611, −41531726, −8241191
75	4381159, 435203083, −435203231
30	2220422932, −283059965, −2218888517
52	60702901317, 23961292454, −61922712865
74	−284650292555885, 66229832190556, 283450105697727
33	8866128975287528, −8778405442862239, −2736111468807040
42	−80538738812075974, 80435758145817515, 12602123297335631

At this point, Booker [38] computes cube roots of k modulo d for d varying over a large range of values, ordered by prime factorisation (rather than running the algorithm on consecutive d's). This way, it is possible to find solutions in the box $\max\{|X|, |Y|, |Z|\} \leq B$ in $O(B \log\log B \log\log\log B)$ arithmetic operations and table lookups.

To end this section, we include a table with the solutions to $X^3 + Y^3 + Z^3 = k$ for $k \in \{30, 33, 39, 42, 52, 74, 75, 84, 87\}$. For references about who the first authors to find such solutions were, see [159, Table 1]. We list them here in order of "discovery" (Table 7).

? Open Problems

1. Partial, or even full progress on Waring's Problem.
2. Can every sufficiently large integer be represented as the sum of 6 positive cubes?
3. Can every sufficiently large integer be represented as the sum of 5 positive cubes?
4. Explicit determination of integers that are a small number of cubes.

Acknowledgments This collaboration grew from the WiNE-4 workshop that was hosted at Universiteit Utrecht in August 2022. The authors would like to express their sincere gratitude to the organisers of the workshop; Ramla Abdellatif, Valentijn Karemaker, Ariane Mézard and Nirvana Coppola, for creating a welcoming and hospitable environment that resulted in a fruitful and thoroughly enjoyable meeting. The authors would like to thank the anonymous referee for carefully reading the paper and suggesting several improvements. The authors would like to extend their thanks to Pieter Moree, and Ákos Pintér for pointing out key references.

Competing Interests

N. Coppola is supported by the NWO Vidi grant No. 639.032.613, New Diophantine Directions.

M. Khawaja is supported by an EPSRC studentship from the University of Sheffield (EPSRC grant no. EP/T517835/1).

Ö. Ülkem is supported by TÜBITAK project no. 119F405.

The authors have no conflicts of interest to declare that are relevant to the content of this chapter.

References

1. L. Alpöge, M. Bhargava, A. Shnidman, *Integers expressible as the sum of two rational cubes (with an appendix by A. Burungale and C. Skinner)* (2022). arXiv: 2210.10730
2. W.S. Anglin, The square pyramid puzzle. Am. Math. Monthly **97**, 120–124 (1990)
3. A. Argáez-García, On perfect powers that are sums of cubes of a five term arithmetic progression. J. Number Theory **201**, 460–472 (2019)
4. A. Argáez-García, V. Patel, On perfect powers that are sums of cubes of a three term arithmetic progression. J. Comb. Number Theory **10**(3), 147–160 (2019)
5. A. Argáez-García, V. Patel, On perfect powers that are sums of cubes of a seven term arithmetic progression. J. Number Theory **214**, 440–451 (2020)
6. W.S. Baer, Über die Zerlegung der ganzen Zahlen in sieben Kuben. Math. Ann. **74**, 511–514 (1913)
7. A. Baker, Linear forms in logarithms of algebraic numbers. Mathematika **13**, 204–216 (1966)
8. A. Baker, Bounds for the solutions of the hyperelliptic equation. Proc. Camb. Philos. Soc. **65**, 439–444 (1969)
9. A. Baker, *Transcendental Number Theory*, 2nd edn. Cambridge Mathematical Library (Cambridge University Press, Cambridge, 1990)
10. R. Balasubramanian, J.-M. Deshouillers, F. Dress, Problème de Waring pour les bicarrés. I. Schéma de la solution. C. R. Acad. Sci. Paris Sér. I Math. **303**, 85–88 (1986)
11. R. Balasubramanian, J.-M. Deshouillers, F. Dress, Problème de Waring pour les bicarrés. II. Résultats auxiliaires pour le théorème asymptotique. C. R. Acad. Sci. Paris Sér. I Math. **303**, 161–163 (1986)
12. I.N. Baoulina, On the unsolvability of certain equations of Erdős-Moser type. Elem. Math. **74**, 1–9 (2019)
13. I.N. Baoulina, P. Moree, *Forbidden Integer Ratios of Consecutive Power Sums*, in *From Arithmetic to Zeta-Functions* (Springer, Cham, 2016), pp. 1–30
14. D. Bartoli, G. Soydan, The Diophantine equation $(x+1)^k + (x+2)^k + \cdots + (\ell x)^k = y^n$ revisited. Publ. Math. Debrecen **96**, 111–120 (2020)
15. A. Bazsó, *Effective results for polynomial values of (alternating) power sums of arithmetic progressions*. Preprint (2020)
16. A. Bazsó, A. Bérczes, K. Győry, A. Pintér, On the resolution of equations $Ax^n - By^n = C$ in integers x, y and $n \geq 3$. II. Publ. Math. Debrecen **76**, 227–250 (2010)
17. A. Bazsó, D. Kreso, F. Luca, Á. Pintér, On equal values of power sums of arithmetic progressions. Glasnik matematički **47**, 253–263 (2012)
18. A. Bazsó, D. Kreso, F. Luca, A. Pintér, Cs. Rakaczki, *On equal values of products and power sums of consecutive elements in an arithmetic progression*. Preprint (2023)
19. A. Bazsó, A. Pintér, H.M. Srivastava, A refinement of Faulhaber's theorem concerning sums of powers of natural numbers. Appl. Math. Lett. **25**, 486–489 (2012)
20. M. Beck, E. Pine, W. Tarrant, K. Yarbrough Jensen, New integer representations as the sum of three cubes. Math. Comp. **76**, 1683–1690 (2007)
21. M.A. Bennett, Lucas' square pyramid problem revisited. Acta Arith. **105**, 341–347 (2002)
22. M.A. Bennett, S.R. Dahmen, Klein forms and the generalized superelliptic equation. Ann. Math. (2) **177**, 171–239 (2013)

23. M.A. Bennett, K. Győry, M. Mignotte, A. Pintér, Binomial Thue equations and polynomial powers. Compos. Math. **142**, 1103–1121 (2006)
24. M.A. Bennett, K. Győry, A. Pintér, On the Diophantine equation $1^k + 2^k + \cdots + x^k = y^n$. Compos. Math. **140**, 1417–1431 (2004)
25. M.A. Bennett, A. Koutsianas, The equation $(x-d)^5 + x^5 + (x+d)^5 = y^n$. Acta Arith. **198**, 387–399 (2021)
26. M.A. Bennett, V. Patel, S. Siksek, Superelliptic equations arising from sums of consecutive powers. Acta Arith. **172**, 377–393 (2016)
27. M.A. Bennett, V. Patel, S. Siksek, Perfect powers that are sums of consecutive cubes. Mathematika **63**, 230–249 (2017)
28. M.A. Bennett, S. Siksek, Differences between perfect powers : prime power gaps. Algebra Number Theory **17**, 1789–1846 (2023)
29. M.A. Bennett, S. Siksek, Differences between perfect powers: the Lebesgue-Nagell equation. Trans. Am. Math. Soc. **376**, 335–370 (2023)
30. M.A. Bennett, C.M. Skinner, Ternary Diophantine equations via Galois representations and modular forms. Canad. J. Math. **56**, 23–54 (2004)
31. M.A. Bennett, V. Vatsal, S. Yazdani, Ternary Diophantine equations of signature $(p, p, 3)$. Compos. Math. **140**, 1399–1416 (2004)
32. A. Bérczes, L. Hajdu, T. Miyazaki, I. Pink, On the equation $1^k + 2^k + \cdots + x^k = y^n$ for fixed x. J. Number Theory **163**, 43–60 (2016)
33. A. Bérczes, I. Pink, G. Savaş, G. Soydan, On the Diophantine equation $(x+1)^k + (x+2)^k + \cdots + (2x)^k = y^n$. J. Number Theory **183**, 326–351 (2018)
34. F. Bertault, O. Ramaré, P. Zimmermann, On sums of seven cubes. Math. Comp. **68**, 1303–1310 (1999)
35. M.R. Best, H.J.J. te Riele, *On a conjecture of Erdős concerning sums of powers of integers*. Stichting Mathematisch Centrum. Numerieke Wiskunde (1976), p. Report NW 23/76
36. Yu. Bilu, G. Hanrot, P.M. Voutier, Existence of primitive divisors of Lucas and Lehmer numbers (with an appendix by M. Mignotte). J. Reine Angew. Math. **539**, 75–122 (2001)
37. K.D. Boklan, N.D. Elkies, *Every multiple of* 4 *except* 212, 364, 420, *and* 428 *is the sum of seven cubes* (2009). arXiv: 0903.4503
38. A.R. Booker, Cracking the problem with 33. Res. Number Theory **5**, 1–6 (2019)
39. A.R. Booker, A.V. Sutherland, On a question of Mordell. Proc. Natl. Acad. Sci. USA **118**, e2022377118 (2021)
40. W. Bosma, J. Cannon, C. Playoust, The Magma algebra system. I. The user language. J. Symbolic Comput. **24**, 235–265 (1997). Computational algebra and number theory (London, 1993)
41. J. Box, Elliptic curves over totally real quartic fields not containing $\sqrt{5}$ are modular. Trans. Am. Math. Soc. **375**, 3129–3172 (2022)
42. A. Bremner, R.J. Stroeker, N. Tzanakis, On sums of consecutive squares. J. Number Theory **62**, 39–70 (1997)
43. C. Breuil, B. Conrad, F. Diamond, R. Taylor, On the Modularity of Elliptic Curves over $\mathbb{Q}$: *W*ild 3-Adic Exercises. J. Am. Math. Soc. **14**, 843–939 (2001)
44. B. Brindza, On some generalizations of the Diophantine equation $1^k + 2^k + \cdots + x^k = y^z$. Acta Arith. **44**, 99–107 (1984)
45. B. Brindza, *Power Values of Sums* $1^k + 2^k + \cdots + x^k$, in *Number Theory*, vol. II (Budapest, 1987), vol. 51 of Colloq. Math. Soc. János Bolyai (North-Holland, Amsterdam, 1990), pp. 595–611
46. B. Brindza, A. Pintér, On the number of solutions of the equation $1^k + 2^k + \cdots + (x-1)^k = y^z$. Publ. Math. Debrecen. **56**, 271–277 (2000)
47. Y. Bugeaud, M. Mignotte, S. Siksek, A multi-Frey approach to some multi-parameter families of Diophantine equations. Canad. J. Math. **60**, 491–519 (2008)
48. W. Butske, L.M. Jaje, D.R. Mayernik, On the equation $\sum_{p|N}(1/p) + (1/N) = 1$, pseudoperfect numbers, and perfectly weighted graphs. Math. Comp. **69**, 407–420 (2000)

49. A. Caraiani, J. Newton, *On the modularity of elliptic curves over imaginary quadratic fields* (2023). arXiv: 2301.10509
50. J.W.S. Cassels, A Diophantine equation. Glasgow Math. J. **27**, 11–18 (1985)
51. J. Chen, Waring's problem for $g(5) = 37$. Sci. Sinica, **13**, 1547–1568 (1964)
52. H. Cohen, *Number Theory: Volume I: Tools and Diophantine Equations*, vol. 239 of Grad. Texts in Math. (Springer, New York, 2007)
53. H. Cohen, *Number Theory: Volume II: Analytic and Modern Tools*, vol. 240 of Grad. Texts in Math. (Springer, New York, 2007)
54. J.H.E. Cohn, Perfect Pell Powers. Glasg. Math. J. **38**, 19–20 (1996)
55. N. Coppola, M. Curcó-Iranzo, M. Khawaja, V. Patel, O. Ülkem, *On perfect powers that are sums of cubes of a nine term arithmetic progression* (2023). arXiv: 2307.01815
56. M.H. Şengün, S. Siksek, On the asymptotic Fermat's last theorem over number fields. Comment. Math. Helv. **93**, 359–375 (2018)
57. G.C. Ţurcaş, On Fermat's equation over some quadratic imaginary number fields. Res. Number Theory **4**, Paper No. 24, 16 (2018)
58. G.C. Ţurcaş, On Serre's modularity conjecture and Fermat's equation over quadratic imaginary fields of class number one. J. Number Theory **209**, 516–530 (2020)
59. I. Cucurezeanu, An elementary solution of Lucas' problem. J. Number Theory **44**, 9–12 (1993)
60. P. Das, P.K. Dey, A. Koutsianas, N. Tzanakis, Perfect powers in sum of three fifth powers. J. Number Theory, **236**, 443–462 (2022)
61. S. Dasgupta, J. Voight, Sylvester's problem and mock Heegner points. Proc. Am. Math. Soc. **146**, 3257–3273 (2018)
62. H. Deconinck, On the generalized Fermat equation over totally real fields. Acta Arith. **173**, 225–237 (2016)
63. A. Del Centina, Unpublished manuscripts of Sophie Germain and a revaluation of her work on Fermat's Last Theorem. Arch. Hist. Exact Sci. **62**, 349–392 (2008)
64. V.A. Dem'janenko, Sums of four cubes. Izv. Vysš. Učebn. Zaved. Matematika, 64–69 (1966)
65. M. Derickx, F. Najman, S. Siksek, Elliptic curves over totally real cubic fields are modular. Algebra Number Theory **14**, 1791–1800 (2020)
66. J.-M. Deshouillers, F. Hennecart, B. Landreau, 7 373 170 279 850. Math. Comp. **69**, 421–439 (2000)
67. L.E. Dickson, All integers except 23 and 239 are sums of eight cubes. Bull. Am. Math. Soc. **45**, 588–591 (1939)
68. L.E. Dickson, *History of the Theory of Numbers*, vol. 2 (Diophantine Analysis, Chelsea, New York, 1971)
69. K. Dilcher, On a Diophantine equation involving quadratic characters. Compos. Math. **57**, 383–403 (1986)
70. S. Edis, *On Arithmetic Progressions and Perfect Powers*, PhD thesis, University of Sheffield, 2019
71. N.D. Elkies, *Heegner Point Computations*, in *Algorithmic Number Theory* (Springer, Berlin, Heidelberg, 1994), pp. 122–133
72. N.D. Elkies, *Every even number greater than 454 is the sum of seven cubes* (2010). arXiv: 1009.3983
73. L. Euler, *Vollständige Anleitung zur Algebra* (Reclam-Verlag, Stuttgart, 1959)
74. G. Faltings, Diophantine approximation on abelian varieties. Ann. Math. **133**, 549–576 (1991)
75. N. Freitas, B.V.L. Hung, S. Siksek, Elliptic curves over real quadratic fields are modular. Invent. Math. **201**, 159–206 (2015)
76. N. Freitas, A. Kraus, S. Siksek, Class field theory, Diophantine analysis and the asymptotic Fermat's last theorem. Adv. Math. **363**, 106964, 37 (2020)
77. N. Freitas, S. Siksek, Fermat's Last Theorem over some small real quadratic fields. Algebra Number Theory **9**, 875–895 (2014)
78. N. Freitas, S. Siksek, The asymptotic Fermat's Last Theorem for five-sixths of real quadratic fields. Compos. Math. **151**, 1395–1415 (2015)

79. Y. Gallot, P. Moree, W. Zudilin, The Erdős-Moser equation $1^k + 2^k + \cdots + (m-1)^k = m^k$ revisited using continued fractions. Math. Comp. **80**, 1221–1237 (2011)
80. V. Gardiner, R.B. Lazarus, P.R. Stein, Solutions of the Diophantine equation $x^3+y^3 = z^3-d$. Math. Comp. **18**, 408–413 (1964)
81. S. Gou, T. Wang, The Diophantine equation $x^2 + 2a \cdot 17b = y^n$. Czechoslovak Math. J., **62**, 645–654 (2012)
82. K. Győry, A. Pintér, On the equation $1^k + 2^k + \cdots + x^k = y^n$. Publ. Math. Debrecen **62**, 403–414 (2003)
83. K. Győry, R. Tijdeman, M. Voorhoeve, On the equation $1^k + 2^k + \cdots + x^k = y^z$. Acta Arith. **37**, 233–240 (1980)
84. L. Hajdu, On a conjecture of Schäffer concerning the equation $1^k + \cdots + x^k = y^n$. J. Number Theory **155**, 129–138 (2015)
85. G.H. Hardy, J.E. Littlewood, Some problems of "partitio numerorum": II. Proof that every large number is the sum of at most 21 biquadrates. Math. Z. **9**, 14–27 (1921)
86. G.H. Hardy, J.E. Littlewood, Some problems of 'Partitio Numerorum': IV. The singular series in Waring's Problem and the value of the number $G(k)$. Math. Z. **12**, 161–188 (1922)
87. G.H. Hardy, J.E. Littlewood, Some problems of 'Partitio numerorum'; III: On the expression of a number as a sum of primes. Acta Math. **44**, 1–70 (1923)
88. G.H. Hardy, J.E. Littlewood, Some problems of 'Partitio Numerorum'(V): A further contribution to the study of Goldbach's problem. Proc. Lond. Math. Soc. (2), **22**, 46–56 (1924)
89. G.H. Hardy, J.E. Littlewood, Some problems of 'Partitio numerorum' (VI): Further researches in Waring's Problem. Math. Z. **23**, 1–37 (1925)
90. G.H. Hardy, J.E. Littlewood, Some problems of 'Partitio Numerorum' (VIII): The number Gamma(k) in Waring's problem. Proc. Lond. Math. Soc. (2) **28**, 518–542 (1928)
91. D.R. Heath-Brown, The Density of Zeros of Forms for which Weak Approximation Fails. Math. Comp. **59**, 613–623 (1992)
92. D.R. Heath-Brown, *Searching for Solutions of $x^3 + y^3 + z^3 = k$*, in *Séminaire de Théorie des Nombres*, Paris, 1989–90, vol. 102 of Progr. Math. (Birkhäuser Boston, Boston, MA, 1992), pp. 71–76
93. D.R. Heath-Brown, W.M. Lioen, H.J.J.T. Riele, On solving the Diophantine equation $x^3 + y^3 + z^3 = k$ on a vector computer. Math. Comp. **61**, 235–244 (1993)
94. E. Isik, Y. Kara, E. Ozman, On ternary Diophantine equations of signature $(p, p, 2)$ over number fields. Turkish J. Math., **44**, 1197–1211 (2020)
95. E. Isik, Y. Kara, E. Ozman, On ternary Diophantine equations of signature $(p, p, 3)$ over number fields. Canad. J. Math. **75**(4), 1293–1313 (2023)
96. M.J. Jacobson, Jr., A. Pintér, P.G. Walsh, A computational approach for solving $y^2 = 1^k + 2^k + \cdots + x^k$. Math. Comp. **72**, 2099–2110 (2003)
97. F. Jarvis, J. Manoharmayum, On the modularity of supersingular elliptic curves over certain totally real number fields. J. Number Theory **128**, 589–618 (2008)
98. F. Jarvis, P. Meekin, The Fermat equation over $\mathbb{Q}(\sqrt{2})$. J. Number Theory **109**, 182–196 (2004)
99. H. Kano, On the equation $s(1^k + 2^k + \cdots + x^k) + r = by^z$. Tokyo J. Math. **13**, 441–448 (1990)
100. Y. Kara, E. Ozman, Asymptotic generalized Fermat's last theorem over number fields. Int. J. Number Theory **16**, 907–924 (2020)
101. B.C. Kellner, On stronger conjectures that imply the Erdős-Moser conjecture. J. Number Theory **131**, 1054–1061 (2011)
102. A. Kempner, Bemerkungen zum Waringschen problem. Math. Ann. **72**, 387–399 (1912)
103. Y. Kezuka, *Which Numbers Are Sums of Two Cubes? (in japanese)*, in *Proceedings of the Waseda Number Theory Workshop 2022* (2022)
104. Y. Kezuka, Y. Li, A classical family of elliptic curves having rank one and the 2-primary part of their Tate-Shafarevich group non-trivial. Doc. Math. **25**, 2115–2147 (2020)
105. M. Khawaja, F. Jarvis, *Fermat's Last Theorem over $\mathbb{Q}(\sqrt{2}, \sqrt{3})$* (2022). arXiv: 2210.03744

106. A. Koutsianas, V. Patel, Perfect powers that are sums of squares in a three term arithmetic progression. Int. J. Number Theory **14**, 2729–2735 (2018)
107. A. Kraus, Majorations effectives pour l'équation de Fermat généralisée. Canad. J. Math. **49**, 1139–1161 (1997)
108. A. Kraus, Le théorème de Fermat sur certains corps de nombres totalement réels. Algebra Number Theory **13**, 301–332 (2019)
109. D. Kriz, *Supersingular main conjectures, Sylvester's conjecture and Goldfeld's conjecture.* Preprint (2020)
110. B. Krzysztofek, The equation $1^n + 2^n + \cdots + m^n = (m+1)^n \cdot k$ (in polish). Wyż. Szkoł. Ped. w Katowicach—Zeszyty Nauk. Sekc. Mat. **5**, 47–54 (1966)
111. D. Kundu, V. Patel, Perfect powers that are sums of squares of an arithmetic progression. Rocky Mountain J. Math. **51**, 933–949 (2021)
112. E. Landau, Über die Zerlegung positiver ganzer Zahlen in positive Kuben. Arch. der Math. u. Phys. **3**, 248–252 (1911)
113. M. Le, G. Soydan, On the power values of the sum of three squares in arithmetic progression. Math. Commun. **27**, 137–150 (2022)
114. W.J. LeVeque, On the equation $y^m = f(x)$. Acta Arith. **9**, 209–219 (1964)
115. U.V. Linnik, On the representation of large numbers as sums of seven cubes. Rec. Math. [Mat. Sbornik] N.S., 218–224 (1943)
116. W. Ljunggren, New solution of a problem proposed by E. Lucas. Norsk Mat. Tidsskr. **34**, 65–72 (1952)
117. E. Lucas, Problem 1180. Nouvelle Ann. Math., **14**, 336 (1875)
118. E. Lucas, *Solution to Problem* 1180. Nouvelle Ann. Math. **16**, 429–432 (1877)
119. E. Lucas, *Recherches sur l'analyse indéterminée et l'Arithmétique de Diophante* (Librairie Scientifique et Technique Albert Blanchard, Paris, 1961)
120. D.G. Ma, An elementary proof of the solutions to the Diophantine equation $6y^2 = x(x+1)(2x+1)$. Sichuan Daxue Xuebao, 107–116 (1985)
121. D.G. Ma, On the Diophantine equation $6Y^2 = X(X+1)(2X+1)$. Kexue Tongbao (English Ed.) **30**, 1266 (1985)
122. K. MacMillan, J. Sondow, Divisibility of power sums and the generalized Erdős-Moser equation. Elem. Math., **67**, 182–186 (2012)
123. B. Mazur, Rational isogenies of prime degree. Invent. Math. **44**, 129–162 (1978)
124. K.S. McCurley, An effective seven cube theorem. J. Number Theory **19**, 176–183 (1984)
125. P. Michaud-Jacobs, Fermat's Last Theorem and modular curves over real quadratic fields. Acta Arith. **203**, 319–351 (2022)
126. M. Mignotte, A note on the equation $ax^n - by^n = c$. Acta Arith. **75**, 287–295 (1996)
127. M. Mignotte, P.M. Voutier, *A kit for linear forms in three logarithms (with an appendix by M. Laurent)* (2022). arXiv: 2205.08899
128. J.C.P. Miller, M.F.C. Woollett, Solutions of the Diophantine equation: $x^3 + y^3 + z^3 = k$. J. Lond. Math. Soc. **s1-30**, 101–110 (1955)
129. D. Mocanu, Asymptotic Fermat for signatures $(p, p, 2)$ *and* $(p, p, 3)$ over totally real fields. Mathematika **68**, 1233–1257 (2022)
130. D. Mocanu, *Asymptotic Fermat for signatures* (r, r, p) *using the modular approach* (2022). arXiv: 2212.10627
131. L.J. Mordell, On the Integer Solutions of the Equation $x^2 + y^2 + z^2 + 2xyz = n$. J. Lond. Math. Soc., **s1-28**, 500–510 (1953)
132. P. Moree, Diophantine equations of Erdős-Moser type. Bull. Aust. Math. Soc. **53**, 281–292 (1996)
133. P. Moree, A top hat for Moser's four mathemagical rabbits. Am. Math. Monthly **118**, 364–370 (2011)
134. P. Moree, Moser's mathemagical work on the equation $1^k + 2^k + \cdots + (m-1)^k = m^k$. Rocky Mountain J. Math. **43**, 1707–1737 (2013)
135. P. Moree, H.J.J. te Riele, J. Urbanowicz, Divisibility properties of integers x, k *satisfying* $1^k + \cdots + (x-1)^k = x^k$. Math. Comp. **63**, 799–815 (1994)

136. Moret-Blanc, [Solution to] question 1180. Nouvelle Ann. Math. **15**, 46–48 (1876)
137. L. Moser, On the diophantine equation $1^n + 2^n + 3^n + \cdots + (m-1)^n = m^n$. Scripta Math. **19**, 84–88 (1953)
138. J.S. Müller, *Rational points on curves* (2019). Online Lecture Notes
139. V. Patel, *Perfect powers that are sums of consecutive like powers*, PhD thesis, University of Warwick, 2017
140. V. Patel, Perfect powers that are sums of consecutive squares, Comptes Rendus Math. **40**, 33–38 (2018)
141. V. Patel, A Lucas–Lehmer approach to generalised Lebesgue–Ramanujan–Nagell equations. Ramanujan J. **56**, 585–596 (2021)
142. V. Patel, S. Siksek, On powers that are sums of consecutive like powers. Res. Number Theory **3**, 1–7 (2017)
143. S.S. Pillai, On Waring's problem $g(6) = 73$. Proc. Indian Acad. Sci. Sect. A. **12**, 30–40 (1940)
144. I. Pink, Z. Rábai, On the Diophantine equation $x^2 + 5^k 17^l = y^n$. Commun. Math. **19**, 1–9 (2011).
145. A. Pintér, A note on the equation $1^k + 2^k + \cdots + (x-1)^k = y^m$. Indag. Math. (N.S.) **8**, 119–123 (1997)
146. Á. Pintér, On the power values of power sums. J. Number Theory **125**, 412–423 (2007)
147. Cs. Rakaczki, On some generalizations of the Diophantine equation $s(1^k + 2^k + \cdots + x^k) + r = dy^n$. Acta Arith. **151**, 201–216 (2012)
148. O. Ramaré, An explicit result of the sum of seven cubes. Manuscripta Math. **124**, 59–75 (2007)
149. K. Ribet, On modular representations of $Gal(\bar{\mathbb{Q}}/\mathbb{Q})$ arising from modular forms. Invent. Math. **100**, 431–476 (1990)
150. F. Rodríguez Villegas, D. Zagier, *Which primes are sums of two cubes?*, in *Number Theory* (Halifax, NS, 1994), vol. 15 of CMS Conf. (Proc., Amer. Math. Soc., Providence, RI, 1995), pp. 295–306
151. P. Satge, Groupes de Selmer et corps cubiques. J. Number Theory **23**, 294–317 (1986)
152. J.J. Schäffer, The equation $1^p + 2^p + 3^p + \cdots + n^p = m^q$. Acta Math. **95**, 155–189 (1956)
153. T.N. Shorey, R. Tijdeman, *Exponential Diophantine Equations*, vol. 87 of Cambridge Tracts in Mathematics (Cambridge University Press, Cambridge, 1986)
154. C.L. Siegel, Zu zwei Bemerkungen Kummers. Nachr. Akad. Wiss. Göttingen Math.-Phys. Kl. II , 51–57 (1964)
155. C.L. Siegel, *Über einige Anwendungen diophantischer Approximationen*, in *On Some Applications of Diophantine Approximations*, vol. 2 of Quad./Monogr., Ed. (Norm., Pisa, 2014), pp. 81–138
156. S. Siksek, *The Modular Approach to Diophantine Equations*, vol. 36 of *Panor. Synthéses* (Soc. Math., France, Paris, 2012), pp. 151–179
157. S. Siksek, *Chabauty and the Mordell-Weil sieve*, vol. 41 of NATO Sci. Peace Secur. Ser. D Inf. Commun. Secur. (IOS, Amsterdam, 2015), pp. 194–224
158. S. Siksek, Every integer greater than 454 is the sum of at most seven positive cubes. Algebra Number Theory **10**, 2093–2119 (2016)
159. S. Siksek, Sums of integer cubes. Proc. Natl. Acad. Sci. USA **118**, e2103697118 (2021)
160. J. Sondow, K. MacMillan, Reducing the Erdős-Moser equation $1^n + 2^n + \cdots + k^n = (k+1)^n$ modulo k and k^2. Integers **11**, A34, 8 (2011)
161. G. Soydan, On the Diophantine equation $(x+1)^k + (x+2)^k + \cdots + (lx)^k = y^n$. Publ. Math. Debrecen **91**, 369–382 (2017)
162. W. Spierpiński, On some unsolved problems of arithmetics. Manuscripta Math. **25**, 125–136 (1960)
163. R.J. Stroeker, On the sum of consecutive cubes being a perfect square. Compos. Math. **97**, 295–307 (1995)
164. R. Taylor, A. Wiles, Ring-theoretic properties of certain Hecke algebras. Ann. Math. **141**, 553–572 (1995)

165. A. Thue, Über Annäherungswerte algebraischer Zahlen. J. Reine Angew. Math. **135**, 284–305 (1909)
166. S. Uchiyama, On a Diophantine equation. Proc. Jpn. Acad. Ser. A Math. Sci. **55**, 367–369 (1979)
167. J. Urbanowicz, Remarks on the equation $1^k + 2^k + \cdots + (x-1)^k = x^k$. Nederl. Akad. Wetensch. Indag. Math. **50**, 343–348 (1988)
168. J. Urbanowicz, On Diophantine equations involving sums of powers with quadratic characters as coefficients. I. Compos. Math. **92**, 249–271 (1994)
169. J. Urbanowicz, On Diophantine equations involving sums of powers with quadratic characters as coefficients. II. Compos. Math. **102**, 125–140 (1996)
170. J.M. van Langen, On the sum of fourth powers in arithmetic progression. Int. J. Number Theory **17**, 191–221 (2021)
171. R.C. Vaughan, T.D. Wooley, *Waring's Problem: A Survey*, in *Number Theory for the Millennium, III* (Urbana, IL, 2000) (A K Peters, Natick, MA, 2002), pp. 301–340
172. M. Voorhoeve, K. Győry, R. Tijdeman, On the Diophantine equation $1^k + 2^k + \cdots + x^k + R(x) = y^z$. Acta Math. **143**, 1–8 (1979)
173. M. Waldschmidt, *Thue Diophantine equations – A Survey* (Springer, Singapore, 2020), pp. 25–41
174. G.L. Watson, A proof of the seven cube theorem. J. Lond. Math. Soc. **26**, 153–156 (1951)
175. G.N. Watson, The problem of the square pyramid. Messenger Math. **48**, 1–22 (1918)
176. A. Wieferich, Beweis des Satzes, daßsich eine jede ganze Zahl als Summe von höchstens neun positiven Kuben darstellen läßt. Math. Ann. **66**, 95–101 (1908)
177. A. Wiles, Modular elliptic curves and Fermat's Last Theorem. Ann. Math. **142**, 443–551 (1995)
178. Z. Zhang, On the Diophantine equation $(x-1)^k + x^k + (x+1)^k = y^n$. Publ. Math. Debrecen **85**, 93–100 (2014)
179. Z. Zhang, On the Diophantine equation $(x-d)^4 + x^4 + (x+d)^4 = y^n$. Int. J. Number Theory **13**, 2229–2243 (2017)
180. Z. Zhang, On the Diophantine equations $(x-1)^3 + x^5 + (x+1)^3 = y^n$ and $(x-1)^5 + x^3 + (x+1)^5 = y^n$. Publ. Math. Debrecen **91**, 383–390 (2017)
181. Z. Zhang, M. Bai, On the Diophantine equation $(x+1)^2 + (x+2)^2 + \cdots + (x+d)^2 = y^n$. Funct. Approx. Comment. Math. **49**, 73–77 (2013)

Transcendence Measure of $e^{1/n}$

Marta Dujella, Anne-Maria Ernvall-Hytönen, Linda Frey, and Bidisha Roy

1 Introduction

Let ξ be a transcendental number. Then $P(\xi) \neq 0$ for any polynomial P with integer coefficients.

Let $k, H \geq 1$ and $\omega(\xi, k, H)$ is the infimum of the numbers $r > 0$ satisfying the estimate

$$\left|\lambda_0 + \lambda_1\xi + \lambda_2\xi^2 + \ldots + \lambda_k\xi^k\right| > \frac{1}{H^r}, \tag{1}$$

for all $\overline{\lambda} = (\lambda_0, \ldots, \lambda_k)^T \in \mathbb{Z}^{k+1} \setminus \{\overline{0}\}$ with $\max_{1\leq i\leq k}\{|\lambda_i|\} \leq H$. Any function greater than or equal to $\omega(\xi, k, H)$ is a *transcendence measure of* ξ.

Bounding transcendence measures of different constants is a classical problem in number theory. It is widely investigated in particular in the context of the Napier's constant e. In 1873, Hermite proved e to be transcendental [5]. The work with the transcendence measure of e has a long history, and it has been summarized in [4]

M. Dujella
University of Basel, Basel, Switzerland
e-mail: marta.dujella@unibas.ch

A.-M. Ernvall-Hytönen (✉)
University of Helsinki, Helsinki, Finland
e-mail: anne-maria.ernvall-hytonen@helsinki.fi

L. Frey
Universität Göttingen, Göttingen, Germany
e-mail: lindafre89@gmail.com

B. Roy
Scuola Normale Superiore di Pisa, Pisa, Italy
e-mail: bidisha.roy@sns.it

R. Abdellatif et al. (eds.), *Women in Numbers Europe IV*, Association for Women in Mathematics Series 32, https://doi.org/10.1007/978-3-031-52163-8_7

and [3], but let us also go through it here. In 1899, Borel [1] proved the bound $c \log\log H$ for some constant c depending on k. Popken [9, 10] improved the bound to $k + \frac{c}{\log\log H}$, where c depends on k in 1928–1929. Mahler [8] was the first to move the result into a more explicit direction. He derived the bound $k + \frac{ck^2 \log(k+1)}{\log\log H}$, where c is an absolute constant. In 1991, Khassa and Srinivasan [6] showed that $c = 98$ works for $\log\log H \geq d(k+1)^{6k}$ for some constant $d > e^{950}$. This result was improved by Hata in 1995 [4] who showed that one can choose $c = 1$ and also considerably improved the lower bound for H. This was further improved by Ernvall-Hytönen, Matala-aho, and Seppälä in 2018 [3].

In this paper, we concentrate on the expression

$$\left|\lambda_0 + \lambda_1 e^{1/n} + \ldots + \lambda_k e^{k/n}\right|,$$

where $k \geq n \geq 2$. This gives a transcendence measure for roots of e. Since e is transcendental, all its roots and powers are also transcendental. Ernvall-Hytönen, Matala-aho, and Seppälä considered also sparse polynomials in the context of the transcendence measure of e. As a corollary, they also derived a transcendence measure for integer powers of e. However, we were not able to find any transcendence measures tailored for *roots of e* in the literature. There are some general results, though. One such result is due to Shidlovskii [11] (and its English translation [12]), proving for a very large class of transcendence measures among degree k polynomials, that the main term must be k. However, the other term is not bounded, and the result is not explicit. Also Mahler [7] has considered a more general setting and proved a bound which can be used to derive an explicit transcendence measure in the current setting. Finally, the generalized transcendence measure by Ernvall-Hytönen, Leppälä, and Matala-aho can be used to derive a bound. Our bound will be compared to these bounds in Sect. 2.

In this article, we prove the following bound:

Theorem 1 *Assume $k \geq n \geq 2$. We have*

$$\omega(e^{1/n}, k, H) \leq k + \frac{k^2 \log k}{\log\log H} d(k)$$

for each H with $\log H \geq s(n,k)e^{s(n,k)}$, where $s(n,k) = (k+n)(\log(k+n))^2$ and

$$d(k) = \begin{cases} 1.596 \text{ for } k = 2 \\ 1.069 \text{ for } k = 3 \\ 1.051 \text{ for } k = 4 \\ \left(1 + \frac{0.69}{\log k - 1}\right) \text{ for } k \geq 5. \end{cases}$$

We follow the approach used in [3, 4].

2 Earlier Results and Comparisons to Our Bound

The following result can be obtained as a corollary of a much more general result presented by Mahler in 1975.

Theorem 2 (Theorem 1 in [7]) *Take $a_i = i$ for $i = 0, \ldots, k$ (with $k \geq 2$) and $a = n$ a positive integer. Let $\lambda_0, \ldots, \lambda_k$ be integers not all zero (in Mahler's paper $x_0, \ldots, x_k$), $C(r) = (k+1)^2 r\sqrt{(\log(n+k)\log r}$ and T be the product of the non-zero λ_i. Then for r the smallest integer for which*

$$\frac{(r-1)!}{e^{2C(r-1)}} \leq \max\{|\lambda_i|\} < \frac{r!}{e^{2C(r)}}$$

we have

$$|T(\lambda_0 + \lambda_1 e^{\frac{1}{n}} + \ldots + \lambda_k e^{\frac{k}{n}})| > \frac{\max\{|\lambda_i|\}}{e^{(2(k+1)-\frac{1}{4})C(r)}}. \tag{2}$$

The following result is due to Ernvall-Hytönen, Leppälä, and Matala-aho and can be obtained as a corollary of a much more general result presented in [2].

Theorem 3 (Corollary 3.5 of [2]) *Let $k \geq 1$. Denote $\alpha_j = \frac{x_j}{y_j}$ are $k+1$ different numbers such that $x_j \in \mathbb{Z}$, $y_j \in \mathbb{Z}^+$ and $\gcd(x_j, y_j) = 1$, for $j = 0, 1, \ldots, k$. Also consider $\overline{\alpha} = (\alpha_0, \alpha_1, \ldots, \alpha_k)$. Then, we have*

$$\left|\lambda_0 + \lambda_1 e^{\alpha_1} + \cdots + \lambda_k e^{\alpha_k}\right| > \frac{M^{1-\hat{\delta}(M)}}{h_0 h_1 \ldots h_k},$$

where $M = \max_{0 \leq i \leq k}\{|\lambda_i|\}$, $\hat{\delta}(M) \leq c_k k^2 \sqrt{\log(g_1(\overline{\alpha})(1+g_3(\overline{\alpha})))}/\sqrt{\log\log M}$ such that $g_1(\overline{\alpha}) = lcm(y_0, \ldots, y_n)$, $g_3(\overline{\alpha}) = \max_{0 \leq j \leq n} |\alpha_j|$ and $h_i = \max\{1, |\lambda_i|\}$, for $i = 0, \ldots, k$. Moreover, $c_k = 13$ if $k < 3$ and 12 otherwise.

In particular, for $\overline{\alpha} = (0, \frac{1}{n}, \ldots, \frac{k}{n})$ we have $g_1(\overline{\alpha}) = n$ and $g_3(\overline{\alpha}) = \frac{k}{n}$. Therefore,

$$\left|\lambda_0 + \lambda_1 e^{1/n} + \cdots + \lambda_k e^{k/n}\right| > \frac{M}{h_0 \cdots h_k M^{\frac{c_k k^2 \sqrt{\log(n(1+k/n))}}{\sqrt{\log\log M}}}}.$$

Let us now compare these results with our bound.

Example 1 Let us look at the family of polynomials with $\frac{H}{2} \leq |\lambda_i| \leq H$ for all coefficients λ_i when $0 \leq i \leq k$ and compare our result with results in [2] and [7]. Assume $k \geq 5$. Our result gives the bound

$$\left|\lambda_0 + \lambda_1 e^{1/n} + \cdots + \lambda_k e^{k/n}\right| > H^{-k - \frac{k^2 \log k}{\log\log H}\left(1 + \frac{0.69}{\log k - 1}\right)}.$$

First we compare this to the bound by Ernvall-Hytönen, Matala-aho, and Leppälä:

$$\left|\lambda_0 + \lambda_1 e^{1/n} + \cdots + \lambda_k e^{k/n}\right| > \frac{M}{h_0 \cdots h_k M^{\frac{c_k k^2 \sqrt{\log(n+k)}}{\sqrt{\log\log M}}}}.$$

We want to show that their bound is worse (i.e., smaller) than ours in this family. To do that, we look at the case, when their bound is at its best. By the definition of M and h_i-s in Theorem 3, we have that $M = h_{i_0}$ for some $i_0 \in \{0, \ldots, k\}$. Moreover, by the assumptions on $|\lambda_i|$-s, we have that $h_i \geq \frac{H}{2}$ for all i. This implies

$$\frac{M}{h_0 \cdots h_k M^{\frac{c_k k^2 \sqrt{\log(n(1+k/n))}}{\sqrt{\log\log M}}}} \leq \left(\frac{H}{2}\right)^{-k} M^{\frac{-c_k k^2 \sqrt{\log(n(1+k/n))}}{\sqrt{\log\log M}}}.$$

Since $\frac{H}{2} \leq M \leq H$, this is at most

$$\left(\frac{H}{2}\right)^{-k-\frac{12k^2\sqrt{\log(n(1+k/n))}}{\sqrt{\log\log H}}} = H^{-k-\frac{12k^2\sqrt{\log(n+k)}}{\sqrt{\log\log H}}+\frac{\log 2}{\log H}(k+\frac{12k^2\sqrt{\log(n+k)}}{\sqrt{\log\log H}})}.$$

This is weaker than our bound for large values of H because for large $\sqrt{\log\log H}$ grows slower than $\log\log H$.

Next, we compare our bound to the result of Mahler from 1975, which states that

$$|\lambda_0 + \lambda_1 e^{\frac{1}{n}} + \ldots + \lambda_k e^{\frac{k}{n}}| > \frac{\max\{|\lambda_i|\}}{T e^{(2(k+1)-\frac{1}{4})C(r)}},$$

where T is the product of all non-zero λ_i's. With a similar argument as above, we have that

$$\frac{\max\{|\lambda_i|\}}{T e^{(2(k+1)-\frac{1}{4})C(r)}} \leq \frac{1}{\left(\frac{H}{2}\right)^k e^{(2(k+1)-1/4)C(r)}}. \tag{3}$$

Mahler gives the bound $\frac{\log x}{\log\log x} < r < \frac{6\log x}{\log\log x}$, where in his notation x is the maximum of the absolute values of the coefficients of the polynomial. In our setting, this inequality would approximately translate to

$$\frac{\log H}{\log\log H} < r < \frac{6\log H}{\log\log H}.$$

We are losing some accuracy here because we only expected the coefficients of the polynomial to be on the interval $[\frac{H}{2}, H]$. So actually, in the inequality above, we should have the actual height of the polynomial. However, for the current purposes, this is not an issue.

The denominator of (3) can now be written as

$$\left(\frac{H}{2}\right)^k e^{(2(k+1)-1/4)C(r)} = H^{k-\frac{k\log 2-(2(k+1)-1/4)C(r)}{\log H}} = H^{k-\frac{k\log 2}{\log H}+\frac{(2(k+1)-1/4)C(r)}{\log H}}.$$

Let us now look at the expression $\frac{(2(k+1)-1/4)C(r)}{\log H}$. Let us use the expression for $C(r)$ and the bound for r:

$$\begin{aligned}C(r) &= (k+1)^2 r\sqrt{\log(n+k)\log r}\\ &\approx (k+1)^2\frac{\log H}{\log\log H}\sqrt{\log(n+k)\log\frac{\log H}{\log\log H}}\\ &\approx (k+1)^2\frac{\log H\sqrt{\log k}}{\sqrt{\log\log H}},\end{aligned}$$

where we used the bound $n \le k$ to estimate $\log(n+k) \approx \log k$, and that for large H, $\log\frac{\log H}{\log\log H} \approx \log\log H$. In the above lines $\approx$ means that for large values of H, the difference between two sides tends to be zero. Hence,

$$\begin{aligned}\frac{(2(k+1)-1/4)C(r)}{\log H} &\approx \frac{(2(k+1)-1/4)}{\log H}\cdot(k+1)^2\frac{\log H\sqrt{\log k}}{\sqrt{\log\log H}}\\ &\approx \frac{2(k+1)^3\sqrt{\log k}}{\sqrt{\log\log H}}.\end{aligned}$$

Hence, our bound is also better than Mahler's bound, because $\sqrt{\log\log H}$ grows slower than $\log\log H$, and the numerator is bigger (dependence on k^3 instead of k^2).

3 Preliminaries and the Outline of the Method

Ernvall-Hytönen, Seppälä, and Matala-aho [3] used the following common approach:

Assume that there is a sequence of simultaneous approximations

$$L_{m,j}(h) = B_{m,0}(h)\Theta_j + B_{m,j}(h)$$

with $m = 0, 1, \ldots, k$ and $j = 1, 2, \ldots, k$. Further assume $B_{m,j}(h) \in \mathbb{Z}$ for all $m, j \in \{0, 1, \ldots, k\}$. Assume further that the coefficients $B_{m,j}$ satisfy the following determinant condition:

$$\begin{vmatrix} B_{0,0} & B_{0,1} & \cdots & B_{0,k} \\ B_{1,0} & B_{1,1} & \cdots & B_{1,k} \\ \vdots & \vdots & \ddots & \vdots \\ B_{k,0} & B_{k,1} & \cdots & B_{k,k} \end{vmatrix} \neq 0.$$

This assumption is by far not trivial but can be proved the same way as in [3]. Also, this can be found as Proposition 2.9 in [13].

Pick the functions $Q(h)$, $q(h)$, $R(h)$, and $r(h)$ to be such that they satisfy the following inequalities:

$$B_{m,0}(h) \leq Q(h) = e^{q(h)}$$

and

$$\sum_{j=1}^{k} |L_{m,j}(h)| \leq R(h) = e^{-r(h)},$$

for all $h \geq h_0$, for a constant h_0, such that the functions are of the form

$$q(h) = ah \log h + bh$$

and

$$-r(h) = -ch \log h + dh.$$

Assume that $z(y)$ is the inverse function of the function $y(z) = z \log z$. Further, denote

$$\begin{aligned} B &= b + \frac{ad}{c}, \\ C &= a, \\ D &= a + b + e^{-s(m)}, \\ F^{-1} &= 2e^{D}, \\ v &= c - \frac{d}{s(m)}, \\ h_1 &= \max\{h_0, e, e^{s(m)}\}. \end{aligned}$$

Our choice will be $s(n,k) = (n+k)(\log(n+k))^2$, and we will actually have $h_1 = e^{s(n,k)}$. Under the assumptions above, they proved the following lemma:

Lemma 1 ([3]) *Let $m \geq 1$ and $\log(2H) \geq vh_1 \log h_1$. Then under the assumptions above*

$$|\lambda_0 + \lambda_1\Theta_1 + \cdots + \lambda_m\Theta_m| > F(2H)^{-\frac{a}{c}-\epsilon(H)}, \tag{4}$$

where

$$\epsilon(H)\log(2H) = Bz\left(\frac{\log(2H)}{v}\right) + C\log\left(z\left(\frac{\log(2H)}{v}\right)\right).$$

Furthermore, they gave the following construction for the approximations in the case of e^{α_j}: Write $\overline{\alpha} = (\alpha_0, \ldots, \alpha_k)$ and set

$$\Omega(x, \bar{\ell}, \overline{\alpha}) = \prod_{j=0}^{m}(\alpha_j - x)^{\ell_j} = \sum_{i=0}^{L}\sigma_i(\overline{\ell}, \overline{\alpha})x^i, \tag{5}$$

where $L = \ell_0 + \ell_1 + \cdots + \ell_m$ and $\sigma_i = \sigma_i(\overline{\ell}, \overline{\alpha})$. Then choosing

$$A_{\bar{\ell},0}(t) = \sum_{i=\ell_0}^{L} t^{L-i} i!\sigma_i(\overline{\ell}, \overline{\alpha}),$$

we get

$$e^{\alpha_j t}A_{\bar{\ell},0}(t) - A_{\bar{\ell},j}(t) = R_{\bar{\ell},j}(t), \tag{6}$$

where $A_j(t)$ is a polynomial with integer coefficients and

$$\begin{cases} \deg A_{\bar{\ell},0}(t) = L - \ell_0 \\ \deg A_{\bar{\ell},j}(t) = L - \ell_j \\ \operatorname{ord}_{t=0} R_{\bar{\ell},j}(t) \geq L + 1, \end{cases}$$

Notice that the polynomials depend on the values of $\ell_0, \ell_1, \ldots, \ell_m$ and on $\overline{\alpha}$. We will explicitly describe A_j and R_j in the following chapter.

In the following, we will be choosing $\Theta_j = e^{j/n}$ for some $n \geq 2$. We will then proceed in the same fashion as in [3] to construct the explicit polynomials used in the simultaneous approximations and to bound them. Finally, we simplify the estimate given by (4).

4 Explicit Polynomial Construction

We start by constructing the simultaneous approximations of the powers of the roots of e. For estimating the required term, we set $\overline{\alpha} = (\alpha_0, \ldots, \alpha_k)$ with $\alpha_s = s/n$, for $s = 0, 1, \ldots k$. Let $\overline{\ell} = (\ell_0, \ldots, \ell_k) \in \mathbb{Z}_{\geq 1}^{k+1}$ and $L = \ell_0 + \ldots + \ell_k$. As explained in the previous chapter, we get the following approximation formulas for $j = 1, \ldots, k$

$$e^{\alpha_j t} A_{\bar{\ell},0}(t) - A_{\bar{\ell},j}(t) = L_{\bar{\ell},j}(t),$$

where

$$A_{\bar{\ell},0}(t) = \sum_{i=\ell_0}^{L} t^{L-i} i! \sigma_i(\overline{\ell}, \overline{\alpha}).$$

With a direct computation (similarly as in [3]), we obtain

$$\begin{aligned}
\sigma_i(\overline{\ell}, \overline{\alpha}) &= (-1)^i \sum_{\ell_0+i_1+\ldots i_k=i} \binom{\ell_1}{i_1} \cdots \binom{\ell_k}{i_k} \left(\frac{1}{n}\right)^{\ell_1-i_1} \cdots \left(\frac{k}{n}\right)^{\ell_k-i_k} \\
&= (-1)^i \sum_{\ell_0+i_1+\ldots i_k=i} \binom{\ell_1}{i_1} \cdots \binom{\ell_k}{i_k} n^{-L+i} 2^{\ell_2-i_2} \ldots k^{\ell_k-i_k}.
\end{aligned}$$

Furthermore, $\sigma_i(\overline{\ell}, \overline{\alpha}) = 0$ when $0 \le i < \ell_0$ and so

$$A_{\bar{\ell},0}(t) = \sum_{i=0}^{L} t^{L-i} i! \sigma_i(\overline{\ell}, \overline{\alpha}).$$

We wish to now bound the polynomials. We use a similar approach using Laplace transforms as for example in [3, 4], because Laplace transform gives us a tool to switch from sums to integrals, which is helpful in estimates. Since $\frac{i!\sigma_i(\overline{\ell},\overline{\alpha})}{t^{i+1}} = \mathcal{L}(\sigma_i(\overline{\ell}, \overline{\alpha})x^i)(t)$ (where $\mathcal{L}$ denotes the Laplace transform), we have

$$\begin{aligned}
A_{\bar{\ell},0}(t) &= \sum_{i=0}^{L} t^{L-i} i! \sigma_i(\overline{\ell}, \overline{\alpha}) = t^{L+1} \sum_{i=0}^{L} \mathcal{L}(\sigma_i(\overline{\ell}, \overline{\alpha})x^i)(t) \\
&= t^{L+1} \int_0^\infty e^{-xt} \Omega(x, \bar{\ell}, \overline{\alpha}) dx,
\end{aligned}$$

where $\Omega(x, \bar{\ell}, \overline{\alpha})$ is given by (5). Now for any α_j, we have

$$\begin{aligned}
e^{\alpha_j t} A_{\bar{\ell},0}(t) &= t^{L+1} \int_0^\infty e^{(\alpha_j - x)t} \Omega(x, \bar{\ell}, \overline{\alpha}) dx \\
&= t^{L+1} \left(\int_0^{\alpha_j} + \int_{\alpha_j}^\infty \right) e^{(\alpha_j - x)t} \Omega(x, \bar{\ell}, \overline{\alpha}) dx.
\end{aligned}$$

Changing the variable in the second integral: $y = x - \alpha_j$ gives us:

$$e^{\alpha_j t} A_{\bar{\ell},0}(t) = t^{L+1} \int_0^{\alpha_j} e^{(\alpha_j - x)t} \Omega(x, \bar{\ell}, \overline{\alpha}) dx + t^{L+1} \int_0^{\infty} e^{-yt} \Omega(y + \alpha_j, \bar{\ell}, \overline{\alpha}) dy.$$

Hence we get

$$A_{\bar{\ell},j}(t) = t^{L+1} \int_0^{\infty} e^{-yt} \Omega(y + \alpha_j, \bar{\ell}, \overline{\alpha}) dy$$

and

$$L_{\bar{\ell},j}(t) = t^{L+1} \int_0^{\alpha_j} e^{(\alpha_j - x)t} \Omega(x, \bar{\ell}, \overline{\alpha}) dx \tag{7}$$

for $j = 1, \ldots, k$.

We can make the result stronger if the terms in (6) are as small as possible. But at the same time, we want to keep the coefficients in $A_{\bar{\ell},j}(t)$ as integers. Therefore, we try to find as large common factors in coefficients as possible.

To do that, we proceed as in [3]. We start by picking very specific values of $\ell_0, \ell_1, \ldots, \ell_k$ in relation to each other.

For any u with $0 \leq u \leq k$, we take $\ell_s^{(u)} = \begin{cases} \ell - 1 & \text{if } s = u \\ \ell & \text{otherwise} \end{cases}$ and $\overline{\ell}^{(u)} = (\ell_0^{(u)}, \ldots, \ell_k^{(u)})$. For these values of $\overline{\ell}$, we denote $A_{\overline{\ell},j}(t)$ by $A_{u,j}(t)$ and $L_{\overline{\ell},j}(t)$ by $L_{u,j}(t)$.

$$A_{u,0}(t) = \sum_{i=\ell_0}^{L} t^{L-i} i!(-1)^i \sum_{\ell_0^{(u)} + i_1 + \ldots i_k = i} \binom{\ell_1^{(u)}}{i_1} \cdots \binom{\ell_k^{(u)}}{i_k} n^{-L+i} 2^{\ell_2^{(u)} - i_2} \ldots k^{\ell_k^{(u)} - i_k}.$$

For our chosen $\overline{\ell}$-s we always have $\ell_0 \in \{\ell, \ell - 1\}$, so we can see that

$$\frac{n^{L-\ell+1}}{(\ell - 1)!} A_{u,0}(t) \in \mathbb{Z}[t].$$

Similarly, we can look at polynomials $A_{u,j}(t)$ for $j = 1, \ldots, k$. We have

$$A_{u,j}(t) = \int_0^{\infty} e^{-yt} \Omega(y + \alpha_j) dy = t^{L+1} \sum_{i=0}^{L} \mathcal{L}(\sigma_i(\overline{\ell}, \overline{\beta^{(j)}})) = \sum_{i=0}^{L} t^{L-i} i! \sigma_i(\overline{\ell}, \overline{\beta^{(j)}}),$$

where $\overline{\beta^{(j)}} = (\alpha_0 - \alpha_j, \ldots, \alpha_k - \alpha_j)$ for each j. Since the coefficients $\sigma_i(\overline{\ell}, \overline{\beta^{(j)}})$ are defined using the polynomial $\Omega(\overline{\ell}, \overline{\beta^{(j)}})$, and in the product representation of Ω, the term $(t - \beta_j^{(j)})^{\ell_j} = (t - (\alpha_j - \alpha_j))^{\ell_j} = t^{\alpha_j}$ will define the lowest degree of terms occurring in $\Omega(\overline{\ell}, \overline{\beta})$, and thereby in $A_{u,j}$, we have $\sigma_i(\overline{\ell}, \overline{\beta}) = 0$ unless $i \geq \ell_j$. Hence, all the coefficients in the representation of $A_{u,j}(t)$ have the factor $\ell_j!$, which is again either ℓ or $\ell - 1$. On the other hand, these terms have the denominator n^{L-i}, so again

$$\frac{n^{L-\ell+1}}{(\ell-1)!}A_{u,j}(t) \in \mathbb{Z}[t].$$

5 Estimation of $A_{u,0}(1)$

Next, we would like to estimate the term $A_{u,0}(t)$, for which its representation as an integral will be useful. It suffices to bound these for $t = 1$, because that way we can bound the coefficients of the linear forms.

We have fixed $\overline{\alpha} = (0, 1/n, \dots, k/n)$ and so $\Omega(x, \bar{\ell}, \overline{\alpha}) = \prod_{j=0}^{k}(j/n - x)^{\ell_j}$. Furthermore, for the choices of $\overline{\ell}$ as in the previous section we have $L+1 = (k+1)\ell$. Therefore, $A_{u,0}(1)$ looks like

$$\begin{aligned} A_{u,0}(1) &= \int_0^\infty e^{-x}\prod_{j=0}^{k}\left(\frac{j}{n}-x\right)^{\ell_j^{(u)}}dx \\ &= \int_0^\infty e^{-x}(-x)^\ell\left(\frac{1}{n}-x\right)^{\ell}\cdots\left(\frac{u}{n}-x\right)^{\ell-1}\cdots\left(\frac{k}{n}-x\right)^{\ell}dx. \end{aligned}$$

Note that $\left|\frac{x^\ell(x-1/n)^\ell\cdots(x-k/n)^\ell}{(x-u/n)}\right| \leq x^{(k+1)\ell-1} \leq x^{(k+1)\ell}$, for $x > \frac{k}{n}$. This gives us an idea of how the function inside the integral behaves: while $0 \leq x \leq \frac{k}{n}$, the function stays relatively small, and it touches zero at points $0, \frac{1}{n}, \dots, \frac{k}{n}$. However, when $x \geq \frac{k}{n}$, it starts behaving roughly as $x^{(k+1)\ell-1}e^{-xt}$.

Therefore, we split the above integral in the following way

$$\begin{aligned} &\int_0^\infty e^{-xt}\prod_{j=0}^{k}\left(\frac{j}{n}-x\right)^{\ell_s}dx \\ &= \left(\int_0^{k/n}+\int_{k/n}^{2(k+1)\ell}+\int_{2(k+1)\ell}^{\infty}\right)e^{-xt}\prod_{j=0}^{k}\left(\frac{j}{n}-x\right)^{\ell_j^{(u)}}dx. \qquad (8) \end{aligned}$$

Now,

$$\left|\left(\int_{k/n}^{2(k+1)\ell}+\int_{2(k+1)\ell}^{\infty}\right)e^{-x}\frac{\prod_{s=0}^{k}\left(\frac{s}{n}-x\right)^{\ell}}{\left(\frac{u}{n}-x\right)}dx\right|$$

$$\leq \left(\int_{k/n}^{2(k+1)\ell} + \int_{2(k+1)\ell}^{\infty} \right) e^{-x} x^{(k+1)\ell-1} dx := I_1 + I_2.$$

For estimating the integrals above, we first consider I_1. In that case,

$$|I_1| = \left| \int_{k/n}^{2(k+1)\ell} e^{-x} x^{(k+1)\ell-1} dx \right| < 2(k+1)\ell e^{-(k+1)\ell+1} \left((k+1)\ell - 1\right)^{(k+1)\ell-1}$$

because the expression $e^{-x} x^{(k+1)\ell-1}$ is maximal when $x = (k+1)\ell - 1$.

Let us now move to I_2.

Lemma 2 *Let $c > 1$ be a constant. We have*

$$\int_{c\ell(k+1)}^{\infty} e^{-x} \frac{\left(\prod_{j=0}^{k} \left|\frac{j}{n} - x\right|\right)^{\ell}}{\left|\frac{k'}{n} - x\right|} \leq \frac{c}{c-1} e^{-c(k+1)^{\ell}} (c(k+1)\ell)^{\ell(k+1)-1}$$

for any $0 \leq k' \leq k$.

Proof We can use partial integration:

$$\int e^{-x} x^t dx = [-e^{-x} x^t] + \int e^{-x} t x^{t-1},$$

which gives us the series expansion for the integral I_2:

$$\begin{aligned}
&\int_{c\ell(k+1)}^{\infty} e^{-x} x^{\ell(k+1)-1} dx \\
&= e^{-c(k+1)\ell} (c(k+1)\ell)^{\ell(k+1)-1} + (\ell(k+1) - 1) e^{-c(k+1)\ell} (c(k+1)\ell)^{\ell(k+1)-2} \\
&\quad + (\ell(k+1) - 1)(\ell(k+1) - 2) e^{-c(k+1)\ell} (c(k+1)\ell)^{\ell(k+1)-3} + \dots \\
&\leq e^{-c(k+1)l} (c(k+1)\ell)^{\ell(k+1)-1} \\
&\quad \times \left(1 + \frac{(k+1)\ell - 1}{c(k+1)\ell} + \frac{(k+1)\ell - 1}{c(k+1)\ell} \cdot \frac{(k+1)\ell - 2}{c(k+1)\ell} \dots \right) \\
&\leq \frac{c}{c-1} e^{-c(k+1)\ell} (c(k+1)\ell)^{\ell(k+1)-1}.
\end{aligned}$$

□

Notice that if we pick $c = 2$, we get the following corollary:

Corollary 1 *We have the following estimate*

$$|I_2| = \int_{2\ell(k+1)}^{\infty} e^{-x} x^{\ell(k+1)-1} dx \leq 2e^{-2(k+1)\ell} (2(k+1)\ell)^{\ell(k+1)-1}.$$

Next, it remains to get a bound for

$$\int_0^{k/n} e^{-y} \frac{\prod_{s=0}^{k} \left(\frac{s}{n} - y\right)^{\ell}}{\left(\frac{u}{n} - y\right)} dy.$$

The following lemma will be useful while proving Lemma 7 (in the next section). This lemma is stated here because the idea of the proof is repeatedly used in other proofs, also within this section.

Lemma 3 *Let $k \geq 2$ and $n \geq 2$. Then*

$$\max_{0<x<k/n} \left| x \left(\frac{1}{n} - x\right) \left(\frac{2}{n} - x\right) \cdot \ldots \cdot \left(\frac{k}{n} - x\right) \right| \leq \frac{k!}{c_k n^{k+1}},$$

where $c_k = 6$ for $k \geq 3$, and $c_2 = 3\sqrt{3}$.

Proof By doing a change of variable $y = nx$ the expression on the left becomes

$$\max_{0<x<k/n} \left| x \left(\frac{1}{n} - x\right) \cdot \ldots \cdot \left(\frac{k}{n} - x\right) \right| = \frac{1}{n^{k+1}} \max_{0<y<k} |y(1-y) \cdot \ldots \cdot (k-y)|.$$

By analyzing the function $|y(1-y)(2-y) \cdot \ldots \cdot (k-y)|$ we see that its maximum on the interval $(0, k)$ is attained for the first time already on the interval $(0, 1)$. If $k \geq 3$ we have the following

$$\begin{aligned} &\max\nolimits_{0<y<k} |y(1-y)\ldots(k-y)| \\ &\leq \max\nolimits_{0<y<1} |(4-y)\ldots(k-y)| \cdot \max\nolimits_{0<y<1} |y(1-y)(2-y)(3-y)| \\ &\leq \tfrac{k!}{3!} \max\nolimits_{0<y<1} y(1-y)(2-y)(3-y). \end{aligned}$$

Using Wolfram alpha, we get that the function $y(y-1)(y-2)(y-3)$ achieves its maximum 1 for $y = (3 \pm \sqrt{5})/2$, which finally implies that

$$\max_{0<x<k/n} \left| x \left(\frac{1}{n} - x\right) \cdot \ldots \cdot \left(\frac{k}{n} - x\right) \right| \leq \frac{k!}{6n^{k+1}}.$$

Similarly for $k = 2$ we need to analyze the function $y(y-1)(y-2)$, whose maximum $\frac{2}{3\sqrt{3}}$ is achieved for $y = 1 \pm \frac{1}{\sqrt{3}}$, from which the claim follows. □

Lemma 4 *Assume $k \geq 5$. Now*

$$\int_0^{k/n} e^{-y} \frac{\prod_{s=0}^{k} \left(\frac{s}{n} - y\right)^{\ell}}{\left(\frac{u}{n} - y\right)} dy \leq \frac{(k!)^{\ell}}{120^{\ell-1} n^{k\ell-5(\ell-1)}} c(n)^{\ell-1},$$

where $c(n) = \max_{0\le y\le 1} \prod_{s=0}^{5} \left|\frac{s}{n} - y\right|^{\ell-1}$. *Furthermore,* $|c(n)| \le 1$.

Proof Observe that $\max_{v\le ny\le v+1} \prod_{s=0}^{k} \left|\frac{s}{n} - y\right|^{\ell-1} \le \max_{0\le ny\le 1} \prod_{s=0}^{k} \left|\frac{s}{n} - y\right|^{\ell-1}$, for positive integer v with $0 \le v \le k-1$. Therefore, with the same technique as in Lemma 3

$$\left|\frac{\prod_{s=0}^{k}\left(\frac{s}{n} - y\right)^{\ell}}{\left(\frac{u}{n} - y\right)}\right| \le \frac{k!}{n^k} \max_{0\le y\le 1} \prod_{s=0}^{k} \left|\frac{s}{n} - y\right|^{\ell-1}.$$

Hence

$$\begin{aligned}
&\left|\int_0^{k/n} e^{-y} \frac{\prod_{s=0}^{k}\left(\frac{s}{n} - y\right)^{\ell}}{\left(\frac{u}{n} - y\right)} dy\right| \\
&\le \int_0^{k/n} e^{-y} \frac{k!}{n^k} \max_{0\le z\le 1} \prod_{s=0}^{k} \left|\frac{s}{n} - z\right|^{\ell-1} dy \\
&\le \frac{k!}{n^k} \frac{(k!)^{\ell-1}}{(5!)^{\ell-1} n^{(k-5)(\ell-1)}} \int_0^{k/n} e^{-y} \max_{0\le z\le 1} \prod_{s=0}^{5} \left|\frac{s}{n} - z\right|^{\ell-1} dy \\
&\le \frac{(k!)^{\ell}}{120^{\ell-1} n^{k\ell-5(\ell-1)}} c(n)^{\ell-1},
\end{aligned}$$

where $c(n) := \max_{0<z<1} \prod_{s=0}^{5} \left|\frac{s}{n} - z\right|$.

By checking small values individually, and bounding

$$\prod_{s=0}^{5} \left|\frac{s}{n} - y\right| \le 1$$

for $n \ge 5$, we obtain $|c(n)| \le 1$. □

Lemma 5 *Assume* $k \le 4$. *Now*

$$\int_0^{k/n} e^{-y} \frac{\prod_{s=0}^{k}\left(\frac{s}{n} - y\right)^{\ell}}{\left(\frac{u}{n} - y\right)} dy \le \frac{k!}{n^k} c(n,k)^{\ell-1},$$

where

$$c(n,k)=\begin{cases}0.049 & \text{for } (n,k)=(2,2)\\ \frac{1}{16} & \text{for } (n,k)=(2,3)\\ \frac{1}{81} & \text{for } (n,k)=(3,3)\\ 0.114 & \text{for } (n,k)=(2,4)\\ 0.015 & \text{for } (n,k)=(3,4)\\ 0.004 & \text{for } (n,k)=(4,4)\end{cases}.$$

Proof We simply bound

$$\int_0^{k/n} e^{-y}\frac{\prod_{s=0}^{k}\left(\frac{s}{n}-y\right)^{\ell}}{\left(\frac{u}{n}-y\right)}dy \leq \frac{k!}{n^k}\max_{0\leq y\leq 1}\prod_{s=0}^{k}\left|y-\frac{s}{n}\right|^{\ell-1}\int_0^{k/n} e^{-y}dy,$$

where the integral can be bounded to be at most 1, and the individual maxima can be determined using WolframAlpha. □

Lemma 6 *Assume $k \geq 2$. We have*

$$\int_0^{\infty} e^{-y}\frac{\prod_{s=0}^{k}(\frac{s}{n}-y)^{\ell}}{(\frac{u}{n}-y)}dy$$
$$\leq \exp\left(\log 4+(k+1)\ell\log 2+(\ell(k+1)-1)\log((k+1)\ell)-\ell(k+1)+1\right).$$

Proof Assume first $k \geq 5$. Now taking the above three estimations into account, we obtain

$$\begin{aligned}
\left|A_{u,0}(1)\right| &= \left|\int_0^{\infty} e^{-y}(-y)^{\ell}\left(\frac{1}{n}-y\right)^{\ell}\cdots\left(\frac{u}{n}-y\right)^{\ell-1}\cdots\left(\frac{k}{n}-y\right)^{\ell}dy\right|\\
&\leq \frac{(k!)^{\ell}}{120^{\ell-1}n^{k\ell-5(\ell-1)}}c(n)^{\ell-1}+2(k+1)\ell e^{-(k+1)\ell+1}((k+1)\ell-1)^{(k+1)\ell-1}\\
&\quad +2e^{-2(k+1)l}(2(k+1)\ell)^{\ell(k+1)-1}\\
&\leq \frac{(k!)^{\ell}}{n^{k\ell-5(\ell-1)}}\left(\frac{c(n)}{120}\right)^{\ell-1}+\left(2(k+1)\ell+2\cdot 2^{\ell(k+1)-1}\right)\\
&\quad \times e^{-(k+1)\ell+1}((k+1)\ell)^{\ell(k+1)-1}\\
&\leq \frac{(k!)^{\ell}}{n^{k\ell-5(\ell-1)}}\left(\frac{c(n)}{120}\right)^{\ell-1}+3\cdot 2^{(k+1)\ell}e^{-(k+1)\ell+1}((k+1)\ell)^{\ell(k+1)-1}\\
&\leq 4\cdot 2^{(k+1)\ell}e^{-(k+1)\ell+1}((k+1)\ell)^{\ell(k+1)-1}\\
&\leq \exp\left(\log 4+(k+1)\ell\log 2+(\ell(k+1)-1)\log((k+1)\ell)-\ell(k+1)+1\right).
\end{aligned}$$

For $k \in \{2, 3, 4\}$ the only thing that changes is the first term, but because $c(n, k) < 1$ for all choices of k and n that interest us, we have

$$\frac{k!}{n^k} c(n, k)^{\ell-1} \leq 2^{(k+1)\ell} e^{-(k+1)\ell+1} ((k+1)\ell)^{\ell(k+1)-1},$$

so the final bound is also valid for these values. □

Finally, we actually need to get a bound for $A^{\star}_{u,0}(1)$. The following corollary gives us the desired bound.

Corollary 2 *For* $k \geq 3$ *and* $\ell \geq \exp(s(n, k))$, *we have*

$$\left|\frac{n^{L-\ell+1}}{(\ell-1)!} A_{u,0}(1)\right| \leq \exp\left(\ell k \log \ell + \ell \left[k \log k + k \log n + 0.72k + 0.000003\right]\right).$$

For $k = 2$, *we have* $\left|\frac{n^{L-\ell+1}}{(\ell-1)!} A_{u,0}(1)\right| \leq \exp(2\ell \log \ell + 3.377257\ell + 2\ell \log 2)$.

Proof Applying the previous lemma and Stirling's formula we have

$$\begin{aligned} &\log \left|A_{u,0}(1)\right| - \log((l-1)!) \\ &\leq \ell k \log \ell + \ell\left((k+1)\log(k+1) + \log \ell - \log(\ell-1) - k + (k+1)\log 2\right) \\ &\quad - \log \ell - \log(k+1) + \tfrac{1}{2}\log(\ell-1) + \log 4 - \log \sqrt{2\pi}. \end{aligned}$$

We first deal with the case when $k \geq 3$. With this assumption, because $\log \ell \geq s(n, k) \geq s(2, 3)$ we have that

$$\log(\ell) - \log(\ell-1) = \int_{\ell-1}^{\ell} \frac{dx}{x} \leq \frac{1}{\ell-1} < 0.000003.$$

Furthermore,

$$(k+1)\log(k+1) - k\log(k) = k\log(1+1/k) + \log(k+1) \leq 1 + \log(k+1)$$

and

$$-k + (k+1)\log 2 + 1 + \log(k+1) < 0.72k.$$

Additionally, for all $k \geq 2$

$$\frac{-\log \ell - \log(k+1) + \frac{1}{2}\log(\ell-1) + \log 4 - \log \sqrt{2\pi}}{\ell} < 0$$

and goes to 0 as ℓ grows.

Thus, multiplying by $n^{L-\ell+1} = n^{k\ell}$ we get

$$\left|\frac{n^{L-\ell+1}}{(\ell-1)!}A_{u,0}(1)\right| \le \exp\left(\ell k \log \ell + \ell\left[k\log k + k\log n + 0.72k + 0.000003\right]\right).$$

Let us now move to the case $k = 2$. We have

$$\log(\ell) - \log(\ell-1) = \int_{\ell-1}^{\ell} \frac{dx}{x} \le \frac{1}{\ell-1} < 0.00046.$$

For $k = 2$ see that

$$\begin{aligned}
&\log\left|A_{u,0}(1)\right| - \log\left((l-1)!\right) \\
&\quad\le (\log 4 - 3\ell + 3\ell\log(6\ell)) - (\ell-1)\log(\ell-1) \\
&\qquad + (\ell-1) - \frac{1}{2}\log(\ell-1) - \log\sqrt{2\pi} \\
&\quad\le 3\ell\log\ell + \ell\left(3\log 6 - 3 + \frac{\log 4 - 1}{\ell} - \frac{\ell\log(\ell-1)}{\ell} + 1\right. \\
&\qquad\qquad \left. + \frac{\log(\ell-1)}{2\ell} - \frac{\log\sqrt{2\pi}}{\ell}\right) \\
&\quad\le 2\ell\log\ell + \ell\left(3\log 6 - 2 + \frac{\log 4/\sqrt{2\pi} - 1}{\ell} + \log\left(\frac{\ell}{\ell-1}\right) + \frac{\log(\ell-1)}{2\ell}\right) \\
&\quad\le 2\ell\log\ell + 3.377\ell.
\end{aligned}$$

Therefore, in this case, $\left|\frac{n^{L-\ell+1}}{(\ell-1)!}A_{u,0}(1)\right| \le \exp(2\ell\log\ell + 3.377\ell + 2\ell\log 2)$. □

6 Integrals Corresponding to Terms $L_{u,j}n^{L-\ell+1}/(\ell-1)!$

At first, we denote $L_{u,j}n^{L-\ell+1}/(\ell-1)!$ as $L^*_{u,j}$ (recall its formula (7) in section 4). Next we need to get a suitable bound on terms $L^*_{u,j}$, or more precisely their sum $\sum_{j=1}^{k}\left|L^*_{u,j}\right|$.

Lemma 7 *Let $k \ge 2$ and $n \ge 2$. Then*

$$|L^*_{u,j}(1)| \le n^{L-\ell+1}\frac{(e^{\frac{j}{n}} - 1)(k!)^{\ell}}{(\ell-1)!(c_k n^{k+1})^{\ell-1}n^k},$$

where $c_k = 6$ *for* $k \geq 3$ *and* $c_2 = 3\sqrt{3}$.

Proof Let $j \in \{1, \ldots, k\}$. By the definition of $|L^*_{u,j}(1)|$ we have

$$\begin{aligned}|L^*_{u,j}(1)|(\ell-1)! &= n^{L-\ell+1} e^{\frac{j}{n}} \int_0^{\frac{j}{n}} e^{-x} \frac{\prod_{r=0}^{k} |\frac{r}{n} - x|^\ell}{|\frac{u}{n} - x|} dx \\ &= n^{L-\ell+1} e^{\frac{j}{n}} \int_0^{\frac{j}{n}} e^{-x} \prod_{r=0}^{k} \left|\frac{r}{n} - x\right|^{\ell-1} \frac{\prod_{r=0}^{k} \left|\frac{r}{n} - x\right|}{\left|\frac{u}{n} - x\right|} dx \\ &\leq n^{L-\ell+1} e^{\frac{j}{n}} \int_0^{\frac{j}{n}} e^{-x} \prod_{r=0}^{k} \left|\frac{r}{n} - x\right|^{\ell-1} \frac{k!}{n^k} dx.\end{aligned}$$

Because $j \leq k$, we have that $\max_{0<x<j/n} \prod_{r=0}^{k} |\frac{r}{n} - x| \leq \max_{0<x<k/n} \prod_{r=0}^{k} |\frac{r}{n} - x|$ which is at most $\frac{k!}{c_k n^{k+1}}$ due to Lemma 3. So we further have

$$\begin{aligned}|L^*_{u,j}(1)|(\ell-1)! &\leq n^{L-\ell+1} e^{\frac{j}{n}} \int_0^{\frac{j}{n}} e^{-x} \left(\frac{k!}{c_k n^{k+1}}\right)^{\ell-1} \frac{k!}{n^k} dx \\ &\leq n^{L-\ell+1} \frac{(k!)^l}{(c_k n^{k+1})^{\ell-1} n^k} e^{\frac{j}{n}} \int_0^{\frac{j}{n}} e^{-x} dx \\ &\leq n^{L-\ell+1} \frac{(k!)^\ell}{(c_k n^{k+1})^{\ell-1} n^k} (e^{\frac{j}{n}} - 1).\end{aligned}$$

□

Lemma 8 *Let* $k \geq 2$ *and* c_k *as in the previous lemma. We have*

$$\sum_{j=1}^{k} |L^*_{u,j}| \leq \frac{(k!)^\ell}{c_k^{\ell-1}(\ell-1)!} n^{2-\ell} e^{(k+1)/n}.$$

Proof We have the following:

$$\sum_{j=1}^{k} (e^{j/n} - 1) < \sum_{j=1}^{k} e^{j/n} = \frac{e^{(k+1)/n} - e^{1/n}}{e^{1/n} - 1} < \frac{e^{(k+1)/n}}{e^{1/n} - 1}.$$

Since

$$e^{1/n} - 1 = \int_0^{1/n} e^x dx > \frac{1}{n},$$

this can be further estimated to

$$\sum_{j=1}^{k}(e^{j/n}-1) < ne^{(k+1)/n}.$$

By summing up the above estimation for $j = 1, \dots, k$ we get

$$\sum_{j=1}^{k}|L^*_{u,j}| \le n^{L-\ell+1}\frac{(k!)^\ell}{(c_k n^{k+1})^{\ell-1}n^k(\ell-1)!}ne^{(k+1)/n}.$$

We can further simplify the above expression, by noticing that

$$\frac{n^{L-\ell+1}n}{(n^{k+1})^{\ell-1}n^k} = n^{2-\ell},$$

which follows from $L = (k+1)\ell - 1$. This finally gives us the bound

$$\sum_{j=1}^{k}|L^*_{u,j}| \le \frac{(k!)^\ell}{c_k^{\ell-1}(\ell-1)!}n^{2-\ell}e^{(k+1)/n}.$$

□

To make this bound suitable for application in $r(\ell) = \exp(R(\ell))$ we need to simplify it further.

Lemma 9 *Let $k \ge 3$ and $\ell \ge e^{s(k,n)} = e^{(k+n)(\log(k+n))^2}$. We have*

$$\sum_{j=1}^{k}|L^*_{u,j}| \le \exp\left(-\ell\log\ell + \ell(k\log k - 0.81k - \log n + 0.17)\right).$$

For $k = 2$, we have

$$\sum_{j=1}^{2}|L^*_{u,j}| \le \exp\left(-\ell\log\ell - 0.64\ell\right).$$

Proof First let $k \ge 3$. We need to simplify the following expression (since $c_k = 6$ for $k \ge 3$):

$$\log\left(\frac{(k!)^\ell}{6^{\ell-1}(\ell-1)!}n^{2-\ell}e^{(k+1)/n}\right) = \ell\log(k!) - (\ell-1)\log 6 - \log(\ell-1)!$$
$$+(2-\ell)\log n + \frac{k+1}{n}.$$

We have $\log(\ell-1)! = \log \ell! - \log \ell$. Further, we can use Stirling's formula to bound the factorials:

$$\sqrt{2\pi\ell}\left(\frac{\ell}{e}\right)^{\ell} e^{1/(12\ell+1)} < \ell! < \sqrt{2\pi\ell}\left(\frac{\ell}{e}\right)^{\ell} e^{1/(12\ell)}.$$

Hence

$$\log \ell! > \frac{1}{2}\log(2\pi) + \frac{1}{2}\log \ell + \ell \log \ell - \ell + \frac{1}{12\ell+1}$$

and similarly for $k!$:

$$\log k! < \frac{1}{2}\log(2\pi) + \frac{1}{2}\log k + k \log k - k + \frac{1}{12k}.$$

Hence, we have

$$\begin{aligned}
&\log\left(\frac{(k!)^{\ell}}{6^{\ell-1}(\ell-1)!}n^{2-\ell}e^{(k+1)/n}\right) \\
&< \ell\left(\frac{1}{2}\log(2\pi) + \frac{1}{2}\log k + k\log k - k + \frac{1}{12k}\right) - (\ell-1)\log 6 + \log \ell \\
&- \log \ell! + (2-\ell)\log n + \frac{k+1}{n} \\
&< \ell\left(\frac{1}{2}\log(2\pi) + \frac{1}{2}\log k + k\log k - k + \frac{1}{12k}\right) - (\ell-1)\log 6 + \log \ell \\
&- \left(\frac{1}{2}\log(2\pi) + \frac{1}{2}\log \ell + \ell\log \ell - \ell + \frac{1}{12\ell+1}\right) + (2-\ell)\log n + \frac{k+1}{n} \\
&< -\ell\log\ell + \ell(k\log k - 0.81k - \log n + 0.16) + \frac{1}{2}\log\ell + 2\log n + \frac{k+1}{n} + 0.88,
\end{aligned}$$

because $\log 6 - \frac{1}{2}\log(2\pi) - \frac{1}{12\ell+1} < 0.88$ and $\frac{1}{2}\log(2\pi) - \log 6 + \frac{1}{12k} + 1 < 0.16$ and $\frac{1}{2}\log k - k < -0.81k$. We can further simplify by using the inequality

$$\left(\frac{1}{2}\log \ell + 2\log n + \frac{k+1}{n} + 0.88\right) < 0.00004\ell$$

to obtain

$$\log\left(\frac{(k!)^{\ell}}{6^{\ell-1}(\ell-1)!}n^{2-\ell}e^{(k+1)/n}\right) < -\ell\log\ell + \ell(k\log k - 0.81k - \log n + 0.17).$$

Let us now look at $k = 2$. Now also $n = 2$ and $k! = 2$, so we do not need the Stirling formula for that term. Otherwise, the calculation is similar. We obtain:

$$\begin{aligned}
\log\left(\sum_{j=1}^{2}|L^*_{u,j}|\right) &\le \ell \log\left(\frac{(2!)^\ell}{c_2^{\ell-1}(\ell-1)!}n^{2-\ell}e^{3/n}\right)\\
&= \ell\log 2 + (2-\ell)\log 2 + \frac{3}{2} - (\ell-1)\log(3\sqrt{3}) - \log \ell! + \log \ell\\
&\le 2\log 2 + \frac{3}{2} - (\ell-1)\log(3\sqrt{3})\\
&\quad - \left(\frac{1}{2}\log(2\pi) + \frac{1}{2}\log\ell + \ell\log\ell - \ell + \frac{1}{12\ell+1}\right) + \log\ell\\
&\le -\ell\log\ell - 0.64\ell.
\end{aligned}$$

□

7 Transcendence Measure for $e^{1/n}$

First we treat the case when $k \ge 3$. We are now ready to put together the bounds

$$q(\ell) = \ell k \log\ell + \ell\left(k\log k + k\log n + 0.72k + 0.000003\right),$$

which is true for $k \ge 3$ and for $k = 2$

$$q(\ell) = 2\ell\log\ell + \ell(3.377257 + 2\log 2).$$

Estimating the sum of $L^*_{u,j}$, for $k \ge 3$, we obtained

$$-r(\ell) = -\ell\log\ell + \ell(k\log k - 0.81k - \log n + 0.17)$$

and for $k = 2$, it is

$$-r(\ell) = -\ell\log\ell - 0.64\ell.$$

Using notation from Section 3 in [3], equations (8) and (9), we have

$$\begin{aligned}
s(n,k) &= (k+n)(\log(k+n))^2. \quad \text{This function is in place of } s(m) \text{ in [3].}\\
a &= k,\\
b &= k\log k + k\log n + 0.72k + 0.000003,\\
c &= 1,
\end{aligned}$$

$$d = k\log k - 0.81k - \log n + 0.17.$$

Now with the notation from Section 3 in [3], equation (10), we have

$$\begin{aligned}
B &= b + \frac{ad}{c}\\
&= k\log k + k\log n + 0.72k + 0.000003 + k(k\log k - 0.81k - \log n + 0.17)\\
&= k\log k + 0.89k + 0.000003 + k^2\log k - 0.81k^2,\\
C &= a = k,\\
D &= a + b + ae^{-s(k,n)}\\
&= k + k\log k + k\log n + 0.72k + 0.000003 + \frac{k}{e^{(k+n)(\log(k+n))^2}},\\
F^{-1} &= 2e^D,\\
v &= 1 - \frac{k\log k - 0.81k - \log n + 0.17}{(k+n)(\log(k+n))^2},\\
n_1 &= e^{(n+k)(\log(n+k))^2}.
\end{aligned}$$

Now we have

$$|\Lambda| := |\lambda_0 + \lambda_1 e^{1/n} + \cdots + \lambda_k e^{k/n}| > F(2H)^{-a/c-\epsilon(H)},$$

where

$$\epsilon(H) = \frac{1}{\log(2H)}\left(Bz\left(\frac{\log(2H)}{v}\right) + C\log\left(z\left(\frac{\log(2H)}{v}\right)\right)\right).$$

The term $H^{-a/c} = H^{-k}$ will form the main term, and everything else will be put together into the second term in the exponent (of H). This second term will be formed of the terms

$$F2^{-a/c}(2H)^{-\epsilon(H)}.$$

Therefore, $|\Lambda| > H^{-k-Y}$, where

$$\begin{aligned}
Y &:= \frac{1}{\log H}\left(Bz\left(\frac{\log(2H)}{v}\right) + C\log\left(z\left(\frac{\log(2H)}{v}\right)\right) + D + (k+1)\log 2\right)\\
&\le \frac{1}{\log H}\left(\frac{uB}{v}\frac{\log(2H)}{\log\log(2H)} + k\log\left(\frac{u}{v}\frac{\log(2H)}{\log\log(2H)}\right) + D + (k+1)\log 2\right)
\end{aligned}$$

and

$$u = 1 + \frac{\log(s(n,k))}{s(n,k)}.$$

We use the fact that $\log H \geq s(n,k)e^{s(n,k)}$. Now, we need to estimate the term Y.

$$\begin{aligned}
D + (k+1)\log 2 &= k + k\log k + k\log n \\
&\quad +0.72k + 0.000003 + \frac{k}{e^{(k+n)(\log(k+n))^2}} + (k+1)\log 2 \\
&= k + k\log k + k\log n \\
&\quad +k\left(0.72 + \frac{0.000003}{k} + \frac{1}{e^{(k+n)(\log(k+n))^2}} + \log 2 + \frac{\log 2}{k}\right) \\
&\leq k\log k + k\log n + 3.4k.
\end{aligned}$$

For the next term, we observe

$$k\log\left(\frac{u}{v}\frac{\log(2H)}{\log\log(2H)}\right) \leq k\log(u\log(2H)).$$

Therefore, we get

$$\begin{aligned}
Y &\leq \frac{1}{\log H}\left(\frac{uB}{v}\frac{\log(2H)}{\log\log(2H)} + k\log(u\log(2H)) + k\log k + k\log n + 3.4k\right) \\
&\leq \frac{1}{\log H}\left(\frac{uB}{v}\frac{\log(2H)}{\log\log(2H)} + k\left(\log(2\log H) + \log\log 4 + \log k + \log n + 3.4\right)\right) \\
&\leq \frac{1}{\log H}\left(\frac{uB}{v}\frac{\log(2H)}{\log\log(2H)} + k\left(\log(2\log H) + \frac{1}{2} + \log k + \log n + 3.4\right)\right) \\
&\leq \frac{1}{\log H}\left(\frac{uB}{v}\frac{\log(2H)}{\log\log(2H)} + \frac{69}{10}k\log(2\log H)\right) \\
&\leq \frac{1}{\log\log H}\left(\frac{\log(2H)}{\log H}\cdot\frac{uB}{v} + \frac{6.9\log\log H\cdot k\log(2\log H)}{\log H}\right) \\
&= \frac{u}{v\log\log H}\left(B + \frac{1}{\log H}\left(\log(2)B + \frac{v\log\log H\cdot 6.9k\log(2\log H)}{u}\right)\right).
\end{aligned}$$

Let us write

$$\theta = \frac{1}{\log H}\left(\log(2)B + \frac{v\log\log H\cdot 6.9k\log(2\log H)}{u}\right),$$

so $Y \leq \frac{u}{v\log\log H}(B+\theta)$.

Let us now look at $k \geq 3$. We have $v \leq 1$ and $u \geq 1$. Therefore,

$$\theta \leq \frac{1}{\log H} \left(\log(2) B + \log\log H \cdot 6.9k \log(2 \log H)\right),$$

which is decreasing in H for any given value of k, meaning that $n = 2$ yields the largest bound. Furthermore,

$$B = k^2 \log k + k \log k + 0.89k + 0.000003 - 0.81k^2 \leq 2k^2 \log k$$

and

$$6.9k \log(2 \log H) \leq 13.8k \log\log H.$$

Therefore, we need to bound the expression

$$\frac{1}{\log H} \left(2 \log(2) k^2 \log k + 13.8k (\log\log H)^2\right)$$

with $\log H \geq (k+2)(\log(2+k))^2 e^{(k+2)(\log(k+2))^2}$. Now

$$\begin{aligned}
&\frac{1}{\log H} \left(2 \log(2) k^2 \log k + 13.8k (\log\log H)^2\right) \\
&\leq \frac{2k \log 2}{e^{(k+2)(\log(k+2))^2}} + \frac{13.8k(2(k+2)(\log(k+2))^2)^2}{(k+2)(\log(2+k))^2 e^{(k+2)(\log(k+2))^2}} \\
&= \frac{2k \log 2}{e^{(k+2)(\log(k+2))^2}} + \frac{55.2k(k+2)(\log(k+2))^2}{e^{(k+2)(\log(k+2))^2}} \\
&< \frac{2k \log 2}{e^{(k+2)(\log(k+2))^2}} + \frac{55.2(k+2)^2(\log(k+2))^4}{e^{(k+2)(\log(k+2))^2}}.
\end{aligned}$$

Both of the terms containing k are decreasing, so it suffices to evaluate them for $k = 3, 4, 5$. For $k = 3$, we obtain:

$$\theta \leq \frac{2k \log 2}{e^{(k+2)(\log(k+2))^2}} + \frac{55.2(k+2)^2(\log(k+2))^4}{e^{(k+2)(\log(k+2))^2}} < 0.022.$$

For $k = 4$, we get $\theta \leq 0.00009$ and for $k \geq 5$, we have $\theta \leq 2 \cdot 10^{-7}$.

We need to do a bit of casework: We treat $k = 2$ separately at the end because it is a completely different case. For $k \geq 3$, we do some preparational work first, then compute for $k = 3$ and $k = 4$, and then bound all the other values together.

Let us now look at the case $k \geq 3$. For simplicity's sake, write

$$Y \leq \frac{u}{v \log\log H}(B+\theta),$$

with θ as before. Define

$$\begin{aligned} f(k,n) &= \frac{u}{vk^2 \log k}(B+\theta) \\ &= \frac{\left(1+\frac{\log((k+n)(\log(k+n))^2)}{(k+n)(\log(k+n))^2}\right)\left(1+\frac{1}{k}+\frac{0.89}{k\log k}+\frac{0.000003+\theta}{k^2\log k}-\frac{0.81}{\log k}\right)}{1-\frac{k\log k-0.81k-\log n+0.17}{(k+n)(\log(k+n))^2}}. \end{aligned}$$

For $k=3$, we obtain $f(2,3)<1.069$ and $f(3,3)<0.999$. Since $f(2,3)$ is the largest one, it will yield the final bound in this case. Similarly for $k=4$, we have $f(2,4)<1.051$ and $f(3,4)<0.980$ and $f(4,4)<0.941$, so $f(2,4)$ yields the largest bound.

This expression can be further simplified to

$$f(k,n)=\frac{\left(1+\frac{1}{(k+n)\log(k+n)}+\frac{2\log\log(k+n)}{(k+n)(\log(k+n))^2}\right)\left(1+\frac{1}{k}+\frac{0.89}{k\log k}+\frac{0.000003+\theta}{k^2\log k}-\frac{0.81}{\log k}\right)}{1-\frac{k\log k-0.81k-\log n+0.17}{(k+n)(\log(k+n))^2}}.$$

Let us look at the numerator. We have

$$\begin{aligned} &\left(1+\frac{1}{(k+n)\log(k+n)}+\frac{2\log\log(k+n)}{(k+n)(\log(k+n))^2}\right) \\ &\cdot\left(1+\frac{1}{k}+\frac{0.89}{k\log k}+\frac{0.000003+\theta}{k^2\log k}-\frac{0.81}{\log k}\right) \\ &=1+\frac{1}{k}+\frac{0.89}{k\log k}+\frac{0.000003+\theta}{k^2\log k}-\frac{0.81k}{\log k}+\frac{1}{(k+n)\log(k+n)} \\ &+\frac{1}{k(k+n)\log(k+n)}+\frac{0.89}{k(k+n)\log k\log(k+n)}+\frac{0.000003+\theta}{k^2\log k(k+n)\log(k+n)} \\ &-\frac{0.81}{(k+n)\log(k+n)\log k}+\frac{2\log\log(k+n)}{(k+n)(\log(k+n))^2}+\frac{2\log\log(k+n)}{k(k+n)(\log(k+n))^2} \\ &+\frac{2\cdot 0.89\log\log(k+n)}{k(k+n)\log k(\log(k+n))^2}+\frac{2\cdot(0.000003+\theta)\log\log(k+n)}{k^2(k+n)\log k(\log(k+n))^2} \\ &-\frac{2\cdot 0.81\log\log(k+n)}{(k+n)(\log(k+n))^2\log k}. \end{aligned}$$

We can now verify using WolframAlpha that

$$\frac{1}{k(k+n)\log(k+n)}+\frac{0.89}{k(k+n)\log k\log(k+n)}+\frac{0.000003+\theta}{k^2\log k(k+n)\log(k+n)}-\frac{0.81}{(k+n)\log(k+n)\log k}<0$$

and

$$\frac{2\log\log(k+n)}{k(k+n)(\log(k+n))^2}+\frac{2\cdot 0.89\log\log(k+n)}{k(k+n)\log k(\log(k+n))^2}+\frac{2\cdot(0.000003+\theta)\log\log(k+n)}{k^2(k+n)\log k(\log(k+n))^2}-\frac{2\cdot 0.81\log\log(k+n)}{(k+n)(\log(k+n))^2\log k}<0$$

for $k\geq 3$. Furthermore, the denominator can be written as

$$\begin{aligned}1&-\frac{k\log k}{(k+n)(\log(k+n))^2}+\frac{0.81k}{(k+n)(\log(k+n))^2}\\&+\frac{\log n}{(k+n)(\log(k+n))^2}-\frac{0.17}{(k+n)(\log(k+n))^2}\\>1&-\frac{k\log k}{(k+n)(\log(k+n))^2}+\frac{0.81k}{(k+n)(\log(k+n))^2},\end{aligned}$$

since $\frac{\log n}{(k+n)(\log(k+n))^2}-\frac{0.17}{(k+n)(\log(k+n))^2}>0$. We can thus estimate

$$f(n,k)<\frac{1+\frac{1}{k}+\frac{0.89}{k\log k}+\frac{0.000003+\theta}{k^2\log k}-\frac{0.81}{\log k}+\frac{1}{(k+n)\log(k+n)}+\frac{2\log\log(k+n)}{(k+n)(\log(k+n))^2}}{1-\frac{k\log k}{(k+n)(\log(k+n))^2}+\frac{0.81k}{(k+n)(\log(k+n))^2}}.$$

Now we need to derive a bound on $f(n,k)$ when $k\geq 5$. Estimating

$$\frac{1}{(k+n)\log(k+n)}+\frac{2\log\log(k+n)}{(k+n)(\log(k+n))^2}<\frac{0.81k}{(k+n)(\log(k+n))^2}$$

(which is based on the fact that after multiplying the inequality with $(k+n)\log(k+n)$, the second term on the left side is at most $1+\frac{2}{e}<1.74$ and the term on the right side is at least >1.75 if $k\geq 5$ and $n\leq k$), and

$$\frac{k\log k}{(k+n)(\log(k+n))^2}<\frac{1}{\log k},$$

we can further simplify the expression:

$$f(n,k) < \frac{1 + \frac{1}{k} + \frac{0.89}{k\log k} + \frac{0.000003+\theta}{k^2\log k} - \frac{0.81}{\log k} + \frac{0.81k}{(k+n)(\log(k+n))^2}}{1 - \frac{1}{\log k} + \frac{0.81k}{(k+n)(\log(k+n))^2}}.$$

Now

$$\frac{1}{k} + \frac{0.89}{k\log k} + \frac{0.000003+\theta}{k^2\log k} = \frac{1}{\log k}\left(\frac{\log k}{k} + \frac{0.89}{k} + \frac{0.000003+\theta}{k^2}\right)$$
$$< \frac{1}{\log k}\left(\frac{\log 5}{5} + \frac{0.89}{5} + \frac{0.0000003+\theta}{25}\right) < \frac{0.5}{\log k},$$

where the terms are estimated using $k \geq 5$. Therefore,

$$f(n,k) < \frac{1 - \frac{0.31}{\log k} + \frac{0.81k}{(k+n)(\log(k+n))^2}}{1 - \frac{1}{\log k} + \frac{0.81k}{(k+n)(\log(k+n))^2}} = 1 + \frac{0.69}{\log k - 1 + \frac{0.81k\log k}{(k+n)(\log(k+n))^2}}$$
$$< 1 + \frac{0.69}{\log k - 1}.$$

Hence, we get the desired estimate

$$Y \leq \frac{k^2\log k}{\log\log H} f(n,k) < \frac{k^2\log k}{\log\log H}\left(1 + \frac{0.69}{\log k - 1}\right).$$

We finish by considering the case $k = 2$, for which the only possibility is $n = 2$. We had the bounds

$$q(\ell) = 2\ell\log\ell + \ell(3.377257 + 2\log 2)$$

and

$$-r(\ell) = -\ell\log\ell - 0.64\ell.$$

This gives us the values

$$a = 2, \quad b = 3.377 + 2\log 2,$$
$$c = 1, \quad d = -0.64.$$

The same formulas as in the general case imply

$$B \leq 3.48356, \quad D = 6.76447 \quad \text{and} \quad u \leq 1.26532.$$

We also have $\log H \geq s(2,2)e^{s(2,2)} = 7.68 \times e^{7.68}$. For $k = 2$, we have $v \geq 1$. Hence,

$$
\begin{aligned}
Y &= \frac{1}{\log H}\left(Bz\left(\frac{\log(2H)}{v}\right) + 2\log\left(z\left(\frac{\log(2H)}{v}\right)\right) + D + 3\log 2\right) \\
&\leq \frac{1}{\log H}\left(Bz\left(\log(2H)\right) + 2\log\left(z\left(\log(2H)\right)\right) + D + 3\log 2\right) \\
&\leq \frac{1}{\log H}\left(Bu\frac{\log(2H)}{\log\log(2H)} + 2\log\left(u\frac{\log(2H)}{\log\log(2H)}\right) + D + 3\log 2\right),
\end{aligned}
$$

where the last inequality follows from [3] Lemma 3.3. We may further simplify

$$
\begin{aligned}
Y &\leq \frac{1}{\log H}\left(Bu\frac{\log(2H)}{\log\log(H)} + 2\log\left(u\log(2H)\right) + D + 3\log 2\right) \\
&\leq \frac{2^2\log(2)}{\log\log H}\left(Bu\frac{\log(2H)}{2^2\log(2)\log(H)} + 2\frac{\log\log H\log\left(u\log(2H)\right)}{2^2\log(2)\log(H)}\right. \\
&\quad \left. + \frac{(D + 3\log 2)\log\log H}{2^2\log(2)\log H}\right).
\end{aligned}
$$

The first and the last terms are clearly decreasing. The second term can be thought in two parts, the most difficult one being $\frac{\log\log(H)\log\log(2H)}{\log(H)}$, but also this is decreasing when $H \geq 596$ (checked with WolframAlpha). Therefore, we obtain

$$
Y \leq \frac{2^2\log(2)}{\log\log H} \cdot 1.596.
$$

Acknowledgments We thank the organizers of the conference Women in Numbers Europe 4, especially but not exclusively Valentijn Karemaker and Nirvana Coppola. This conference set the foundation for this article. Furthermore, we thank the Universiteit Utrecht for granting unlimited coffee. We would also like to thank the anonymous referees for helpful comments. The research of Ernvall-Hytönen is supported by the Emil Aaltonen foundation. Research of Marta Dujella was supported by the Swiss National Science Foundation grant "Diophantine Equations: Special Points, Integrality, and Beyond" (no.200020 184623).

References

1. É. Borel, Sur la nature arithmétique du nombre *e*. C. R. Acad. Sci. Paris. **128**, 596–599 (1899)
2. A.-M. Ernvall-Hytönen, K. Leppälä, T. Matala-aho, An explicit Baker-type lower bound of exponential values. Proc. Roy. Soc. Edinburgh Sect. A Math. (2015). https://doi.org/10.1017/S0308210515000049
3. A.-M. Ernvall-Hytönen, T. Matala-aho, L. Seppälä, On Mahler's transcendence measure for e. Constr. Approx. (2019). https://doi.org/10.1007/s00365-018-9429-3
4. M. Hata, Remarks on Mahler's transcendence measure for e. J. Number Theory **54**, 81–92 (1995)

5. Ch. Hermite, Sur la fonction exponentielle. C. R. Acad. Sci. **77**, 18–24, 74–79, 226–233, 285–293 (1873)
6. D.S. Khassa, S.A. Srinivasan, Transcendence measure for e. J. Indian Math. Soc. **56**, 145–152 (1991)
7. K. Mahler, On a paper by A. Baker on the approximation of rational powers of e. Acta Arithmetica **27**, 61–87 (1975). https://doi.org/10.4064/aa-27-1-61-87
8. K. Mahler, Zur Approximation der Exponentialfunktion und des Logarithmus. Teil I. J. Reine Angew. Math. **166**, 118–136 (1931)
9. J. Popken, Sur la nature arithmétique du nombre e. C. R. Acad. Sci. Paris **186**, 1505–1507 (1928)
10. J. Popken, Zur Transzendenz von e. Math. Z. **29**, 525–541 (1929)
11. A.B. Shidlovskii, Transcendence measure estimates for the values of E-functions (in Russian). Mat. Zametki **2:1**, 33–44 (1967); Math. Notes **2:1**, 508–513 (1967)
12. A.B. Shidlovskii, Transcendence measure estimates for the values of E-functions. Math. Notes **2**, 508–513 (1967)
13. M. Waldschmidt, Introduction To Diophantine methods: irrationality and transcendence (2019). https://webusers.imj-prg.fr/~michel.waldschmidt/articles/pdf/IntroductionDiophantineMethods.pdf

Variants of the Li-Type Criteria for the Generalized Riemann Hypothesis

A.-M. Ernvall-Hytönen, A. Odžak, L. Smajlović, and M. Zubača

1 Introduction

The Riemann hypothesis, which claims that all zeros of the Riemann zeta function lie on the line $\mathrm{Re}(s) = \frac{1}{2}$, is one of the most significant conjectures in mathematics, as its confirmation would have many important consequences, e.g., it would yield that the prime numbers are distributed as regularly as possible. For this reason, many different attempts have been made in a quest to prove (or disprove) it, resulting in many statements equivalent to the Riemann hypothesis; the most extensive collection of equivalent statements (and many related topics) up to 2017 is given in [7] and [8].

1.1 Li Criterion for the Riemann Hypothesis

A particularly simple sequence and a criterion for the Riemann hypothesis appeared in the work of X.-J. Li in 1997 [30]. More precisely, the Li criterion proved in [30] states that positivity of the sequence of coefficients

The research of Ernvall–Hytönen is supported by the Emil Aaltonen Foundation.

A.-M. Ernvall-Hytönen
University of Helsinki, Helsingin yliopisto, Finland
e-mail: anne-maria.ernvall-hytonen@helsinki.fi

A. Odžak (✉) · L. Smajlović · M. Zubača
Department of Mathematics, University of Sarajevo, Sarajevo, Bosnia and Herzegovina
e-mail: almasa.odzak@pmf.unsa.ba; lejlas@pmf.unsa.ba; medina.zubaca@pmf.unsa.ba

R. Abdellatif et al. (eds.), *Women in Numbers Europe IV*, Association for Women in Mathematics Series 32, https://doi.org/10.1007/978-3-031-52163-8_8

$$\lambda_n = \sum_{\rho\in Z(\zeta)}{}^{*}\left(1-\left(1-\frac{1}{\rho}\right)^n\right) = \sum_{\rho\in Z(\zeta)}{}^{*}\left(1-\left(\frac{\rho}{\rho-1}\right)^n\right), \tag{1}$$

for $n \in \mathbb{N}$, is a necessary and sufficient condition for the Riemann hypothesis. The sums in (1) are taken over the set $Z(\zeta)$ of all non-trivial zeros of the Riemann zeta function, counted with multiplicities, and * indicates that they need to be interpreted as the limit as $T \to \infty$ of the sum over zeros ρ such that $|\mathrm{Im}(\rho)| \leq T$. The equality in (1) follows from the symmetry $\rho \leftrightarrow 1-\rho$ of zeros of the Riemann zeta function.

The coefficients λ_n appeared first in the work of Keiper [28, formula (34) on p. 769] who proved that if the Riemann hypothesis is true, then $\lambda_n \geq 0$ for all positive integers n. For this reason, numbers λ_n are sometimes called the Keiper–Li coefficients.[1]

Bombieri and Lagarias [6] generalized the work [30] of Li showing that the Li-type criterion can be derived for functions F such that their set of zeros $Z(F)$ counting multiplicities satisfies the properties that $1 \notin Z(F)$ and that

$$\sum_{\rho\in Z(F)} \frac{1+\mathrm{Re}(\rho)}{(1+|\rho|)^2} < \infty. \tag{2}$$

They also deduced an arithmetic formula for the numbers λ_n and proved that establishing one-sided tempered growth of λ_n as $n \to \infty$ is sufficient to prove the Riemann hypothesis.

Going further, Voros [61] proved there are only two possible asymptotic behaviors of the Li coefficients λ_n, as $n \to \infty$: Either they grow as $\frac{1}{2}n\log n$ (which is equivalent to the Riemann hypothesis being true) or they oscillate with exponentially growing amplitude (which is equivalent to the Riemann hypothesis being false). Voros' result regarding to the asymptotic behavior of λ_n as $n \to \infty$ under the Riemann hypothesis was further refined by Lagarias in [29], who proved that $\lambda_n = \frac{1}{2}n\log n + (\gamma - \log(2\pi) - 1)\,n + O(\sqrt{n}\log n)$, where γ is the Euler constant. Moreover, in [1] the validity of the Riemann hypothesis was proven to be equivalent to the statement $\{\lambda_n/n - \frac{1}{2}(\log n + \gamma - \log(2\pi) - 1)\} \in \ell^2$.

Further study of the Li criterion for the Riemann zeta function has also been conducted by Coffey in [13–15]. The Li coefficients are used in the paper [21] to deduce an interesting link between analytic and transcendental number theory, under the Riemann hypothesis.

1.2 Li-Type Criterion for the Generalized Riemann Hypothesis

The generalized Riemann hypothesis (GRH) in the context of this chapter is the statement that all non-trivial zeros of a certain zeta or L-function of number-

[1] Keiper introduces the scaling factor of $\frac{1}{n}$, meaning that Keiper's nth coefficient equals $\frac{1}{n}\lambda_n$, where λ_n is defined by (1).

theoretic interest are located on the critical line. The critical line is the vertical line that is in the middle of the critical strip in the complex plane, i.e., in the middle of the strip that contains all non-trivial zeros of that function. The critical strip is determined by combining the zero-free half-plane stemming from some type of Euler product representation and (sometimes conjectural) functional equation. For example, in the case when functions belong to the modified Selberg class $\mathcal{S}^{\sharp\flat}$ defined below, the critical strip is always the strip $0 \leq \mathrm{Re}(s) \leq 1$, and the critical line is then the line $\mathrm{Re}(s) = 1/2$.

Assume that F is a zeta or L-function with the set of non-trivial zeros $Z(F)$ (including multiplicities) such that the bound (2) is true and such that $\rho \in Z(F)$ implies $1 - \overline{\rho} \in Z(F)$ with the same multiplicities. Assume further that $0 \notin Z(F)$. Then, the generalized Li coefficients attached to such a function F are well defined for all integers $n \neq 0$ through the series

$$\lambda_F(n) = \sum_{\rho \in Z(F)}{}^{*} \left(1 - \left(1 - \frac{1}{\rho}\right)^n\right), \tag{3}$$

where the sum is to be interpreted in the sense of the limit as $|\mathrm{Im}(\rho)| \to \infty$. The symmetry of zeros yields that $\overline{\lambda_F(n)} = \lambda_F(-n)$, and the Li-type positivity criterion, in this broad setting, is the statement that zeros of F are located on the critical line $\mathrm{Re}(s) = 1/2$ if and only if $\mathrm{Re}(\lambda_F(n)) \geq 0$ for all positive integers n.

Many classes of functions satisfy the assumptions of Bombieri and Lagarias [6]. Hence, the Li-type criteria for the GRH were formulated, and the generalized Li coefficients were studied for the following types of functions: automorphic L-functions attached to irreducible unitary automorphic representations in [29], functions in the Selberg class in [50, 51], and [52], functions in the modified Selberg class $\mathcal{S}^{\sharp\flat}$ in [60] and [39], Rankin–Selberg L-functions in [43], Dirichlet L-functions in [9, 31] and [36], Hecke L-functions in [31–33, 54], and [53], Epstein zeta function in [10] and [47], zeta functions on function fields in [40, 48, 49], and [3], and Ihara zeta function in [22].

1.3 Li-Type Criterion for Zero-Free Half-planes

The problem of finding zero-free half-planes for the Riemann zeta function turned out to be almost as difficult as the Riemann hypothesis, since the only unconditional zero-free half-plane of the form $\mathrm{Re}(s) > \tau/2$ for some $\tau \in [1, 2]$ for the Riemann zeta function is known to be $\mathrm{Re}(s) > 1$. For functions in the Selberg class, even less is known, e.g., non-vanishing on the line $\mathrm{Re}(s) = 1$ is proved only under an additional "normality" conjecture, see [27].

In 2006, Freitas [20] noticed that a Li-type criterion can be formulated for zero-free half-planes of the Riemann zeta function. Namely, for $\tau \in [1, 2)$ and nonzero integers n, he studied the τ-Li coefficients $\lambda_\zeta(n, \tau)$ associated to the Riemann zeta function, defined by Eq. (4) below with $F = \zeta$, and proved that $\lambda_\zeta(n, \tau) \geq 0$ if and only if all non-trivial zeros ρ of $\zeta(s)$ lie on the strip $1 - \tau/2 \leq \mathrm{Re}(\rho) \leq \tau/2$. In other words, given the symmetry properties of zeros, the inequality $\lambda_\zeta(n, \tau) \geq 0$ is true if and only if $\zeta(s)$ is non-vanishing on the half-plane $\mathrm{Re}(s) > \tau/2$.

For a real number $\tau \geq 1$, the $\tau/2$-generalized Riemann hypothesis ($\tau/2$-GRH, for short) is the statement that a zeta or L-function F of number-theoretical interest possesses no zeros on the half-plane $\mathrm{Re}(s) > \tau/2$.

Following the work of Freitas, for $\tau \geq 1$, Droll defined in his thesis [16] the generalized τ-Li coefficients associated to a zeta or L-function F with the set of zeros $Z(F)$ (including multiplicities) such that $0, \tau \notin Z(F)$, the bound (2) holds and such that $\rho \in Z(F)$ implies $1 - \overline{\rho} \in Z(F)$ with the same multiplicities. Those coefficients are defined for $\tau \in [1, 2)$ as

$$\lambda_F(n, \tau) = \sum_{\rho \in Z(F)}{}^{*} \left(1 - \left(\frac{\rho}{\rho - \tau}\right)^n\right). \tag{4}$$

We notice that $\lambda_F(n, 1) = \lambda_F(-n) = \overline{\lambda_F(n)}$; hence, $\mathrm{Re}(\lambda_F(n, 1)) = \mathrm{Re}(\lambda_F(n))$. Therefore, the Li-type criterion for the generalized Riemann hypothesis for F satisfying the conditions above follows from the symmetry of zeros and the Li-type criterion for $\tau/2$-GRH when $\tau = 1$.

The generalized τ-Li coefficients and the Li-type criteria for $\tau/2$-GRH have been studied for many classes of functions that include: the Rankin–Selberg L-functions in [11], Dirichlet series with real coefficients in [12], a very broad class $\mathcal{S}^{\sharp\flat}(\sigma_0, \sigma_1)$ of Dirichlet series containing the Selberg class in [18, 42] and [19], and automorphic L-functions in [37].

1.4 *Variants of the Li Coefficients and the Li Criteria*

Different variants of the Li and τ-Li coefficients, generalizing (4), have also been studied. For example, Sekatskii [58] generalized the results of [6] by introducing a real parameter $a \neq 1/2$ and for positive integers n defined coefficients

$$k(n, a) = \sum_{\rho \in Z(\zeta)}{}^{*} \left(1 - \left(\frac{\rho - a}{\rho + a - 1}\right)^n\right).$$

He proved that the Riemann hypothesis is equivalent to non-negativity of $k(n, a)$ for all positive n and any $a < 1/2$. The study of those coefficients has been carried out for the Selberg class in [38].

The so-called *base point* for the Li coefficients is the point $s = 1$, since the Li coefficients can be expressed in terms of the coefficients in the Taylor series expansion of the completed function at $s = 1$, see, e.g., Theorem 1. The base point for the coefficients $k(n, a)$ is the point $s = 1 - a \neq 1/2$, see [58, theorem 5].

The *centered variant* of the Li coefficients with the base point $1/2$ instead of 1 or $1-a \neq 1/2$ has been proposed by Voros in [62], and it was further studied in [4] in the context of function fields.

In [62], Voros also proposed the study of the discretized Li coefficients associated to the Riemann zeta function, which are more amenable for numerical computations, and which are related to the Riemann hypothesis via a similar asymptotic criterion. Namely, Voros defined a new set of coefficients such that calculating those coefficients requires knowledge of the values of the completed zeta function at even positive integers only, which is convenient given that these values can be explicitly expressed in terms of Bernoulli numbers.

The discretized τ-Li coefficients for a very broad class $\mathcal{S}^{\sharp\flat}(\sigma_0, \sigma_1)$ of Dirichlet series, described in Sect. 2.3, have been introduced in [46]. For $F \in \mathcal{S}^{\sharp\flat}$ and $\tau \in [1, 2]$, the discretized τ-Li coefficient associated to F is defined by

$$\Lambda_F(n, \tau) = \frac{n}{\tau^{n-1}} \frac{1}{2\pi i} \int_{C_n} \log \tilde{\Phi}_F(u) f_n(u) du, \tag{5}$$

where $\tilde{\Phi}_F$ is the completed function associated to F,

$$f_n(u) = \frac{\tau^n}{u(u-\tau)} \prod_{j=1}^{n} \frac{u + 2j\tau - \tau}{u - 2j\tau}, \tag{6}$$

and the contour C_n encircles the real interval $[\tau, 2n\tau]$ positively, leaving all singularities of the function $\log \tilde{\Phi}_F(u)$ outside the contour.

The coefficients $\Lambda_F(n, \tau)$ can be expressed as a finite linear combination of values of $\log \tilde{\Phi}_F$ evaluated at $2m\tau$, $m = 1, \ldots, n$, which is very well-suited for numerical computations; see formula (22).

In [46], it is proved that the existence of a zero $\rho \in Z(F)$ outside the strip $1 - \tau/2 \leq \text{Re}(s) \leq \tau/2$ yields oscillations of $\Lambda_F(n, \tau)$ with power-like growing amplitudes. The class $\mathcal{S}^{\sharp\flat}(\sigma_0, \sigma_1)$ contains certain functions for which it is expected that the GRH does not hold true. Hence, numerical computations of $\Lambda_F(n, \tau)$ might be very helpful in finding values of τ for which the half-plane $\text{Re}(s) > \tau/2$ is not zero-free.

1.5 The Purpose of the Paper

The Li-type criteria for the (G)RH and the $\tau/2$-(G)RH have generated a lot of interest in the past 25 years. Different expressions for the Li and τ-Li coefficients have been deduced in many papers cited above relating the (generalized) Li coefficients with the prime number theorem, the Euler–Stieltjes constants, and other arithmetic data. The τ-Li coefficients have also been related to explicit zero-free regions of L-functions (e.g., [9] or [55]).

The simplicity of expressions for the Li and τ-Li coefficients in terms of derivatives of the completed L-function (see Sects. 3.1 and 3.2) is also suitable for numerical experiments in order to establish asymptotic behavior of those coefficients, see, e.g., computations of Keiper [28] and Maslanka [34] for the Riemann zeta function and more recent work on the generalized Li and τ-Li coefficients in [12, 18, 19, 42], and [62].

The purpose of this chapter is to present an overview of results related to the analytic and numerical study of the generalized Li-type coefficients that are present in the vast literature on this topic. Many of those results have been conceived during the first Women in Numbers Europe (WINE)[2] workshop held in Luminy in 2013 and further developed by members of the WINE group with various coauthors during the past ten years, with the first published paper being [11].

More precisely, after introducing the classes of functions related to the Selberg class in Sect. 2, we will present a survey of results on generalized Li coefficients and Li-type criteria for the modified Selberg class $\mathcal{S}^{\sharp\flat}$ defined below. Most of the results hold true for more general classes of functions, but to ease the notation we state almost all results for functions in the class $\mathcal{S}^{\sharp\flat}$.

In Sect. 3, we list many equivalent formulas for computation of the generalized Li and τ-Li coefficients and the discretized Li coefficients. The generalized Li and τ-Li coefficients are represented as a sum of two terms: the Archimedean and the non-Archimedean term; the asymptotic behavior of both terms is also discussed in Sect. 3.

Variants of the Li-type positivity criteria for the GRH and $\tau/2-$GRH are discussed in Sect. 4. The study of the discretized Li coefficients for a very large class of functions containing the class $\mathcal{S}^{\sharp\flat}$ is described in Sects. 3.3 and 4.3. An overview of various numerical results related to zero-free regions of some selected L-functions, which illustrates theoretical results described in Sects. 3 and 4, is given in Sect. 5. We conclude the paper with the table summarizing the bibliography related to the Li coefficients and the Li-type criteria for different classes of functions. This table also contains the location of the sections in which the corresponding results are further elaborated.

2 The Selberg Class of Functions and Its Extensions

In this section, we define the Selberg class of functions and certain classes that appear naturally in the study of zeta and L-functions and that contain the Selberg class.

---

[2] Women in Numbers Europe 2013 Workshop website: https://mathweb.ucsd.edu/~alina/luminy/.

2.1 The Selberg Class

The Selberg class of functions $\mathcal{S}$, introduced in [59], is a general class of Dirichlet series F that possesses a meromorphic continuation to the whole complex plane, satisfies a functional equation of the Riemann type and the Ramanujan conjecture, and can be written as an Euler product. More precisely, the Selberg class is the class of functions F satisfying the following axioms:

(i) (Dirichlet series) The function F has a Dirichlet series representation

$$F(s) = \sum_{n=1}^{\infty} \frac{a_F(n)}{n^s}$$

that converges absolutely for $\mathrm{Re}(s) > 1$.

(ii) (Meromorphic continuation) There exists an integer $m \geq 0$ such that $(s-1)^m F(s)$ is an entire function of finite order. The smallest such number is denoted by m_F and called the polar order of F.

(iii) (Functional equation) The function F satisfies the functional equation

$$\Phi_F(s) = w\overline{\Phi_F(1-\bar{s})}, \tag{7}$$

where

$$\Phi_F(s) = F(s) Q_F^s \prod_{j=1}^{r} \Gamma(\lambda_j s + \mu_j), \tag{8}$$

with $Q_F > 0$, $r \geq 0$, $\lambda_j > 0$, $|w| = 1$, $\mathrm{Re}(\mu_j) \geq 0$, $j = 1, \ldots, r$. When $r = 0$, the empty product is defined to be identically one.

(iv) (Ramanujan conjecture) For every $\epsilon > 0$, $a_F(n) \ll n^{\epsilon}$.

(v) (Euler product) The logarithm of the function F possesses a Dirichlet series representation

$$\log F(s) = \sum_{n=1}^{\infty} \frac{b_F(n)}{n^s},$$

converging absolutely for $\mathrm{Re}(s) > 1$. Here $b_F(n) = 0$ unless n is a prime or a prime power and $b_F(n) \ll n^{\theta}$ for some $\theta < 1/2$.

The axiom (v) is called the Euler product axiom because it implies that the coefficients $a_F(n)$ are multiplicative; hence, $F(s)$ can be expressed as a product over primes (see, e.g., discussion on p. 28 of [56] where the rationale behind the assumption $\theta < 1/2$ is also explained).

The extended Selberg class $\mathcal{S}^\sharp$ is the class of functions F satisfying axioms (i), (ii), and (iii). We refer the interested reader to the surveys [26], [56], and [57] where properties of the Selberg class and the extended Selberg class are nicely summarized.

2.2 *The Modified Selberg Class* $\mathcal{S}^{\sharp\flat}$

The Selberg class S is assumed to contain all automorphic L-functions. However, some required properties of automorphic L-functions, such as the Ramanujan conjecture, have not been proved (yet). For this reason, different classes of functions that are related to the Selberg class have been introduced with the aim of deducing unconditional results for different types of automorphic L-functions or Rankin–Selberg L-functions.

The class of functions we will focus on in this chapter is the modified Selberg class $\mathcal{S}^{\sharp\flat}$ defined in [60], where it is proved that $\mathcal{S}^{\sharp\flat}$ contains both the Selberg class and the class of all automorphic L-functions attached to irreducible unitary automorphic representations. The modified Selberg class $\mathcal{S}^{\sharp\flat}$ is the class of functions F satisfying the (i) and (ii) above and the following axioms:

(iii') (Functional equation) The function F satisfies the functional equation (7), where $\Phi_F(s)$ is defined by (8) with $Q_F > 0$, $r \geq 0$, $\lambda_j > 0$, $|w| = 1$, $\mathrm{Re}(\mu_j) > -\frac{1}{4}$, $\mathrm{Re}(\lambda_j + 2\mu_j) > 0$, $j = 1, \ldots, r$.

(v') (Euler sum) The logarithmic derivative of the function F possesses a Dirichlet series representation

$$\frac{F'}{F}(s) = -\sum_{n=2}^{\infty} \frac{c_F(n)}{n^s},$$

converging absolutely for $\mathrm{Re}(s) > 1$.

The axiom (v') is called the *Euler sum* axiom, to emphasize that, as opposed to the Euler product axiom (v), it does not imply that coefficients $a_F(n)$ are multiplicative. This is consistent with the definition of the fundamental class of functions in [25, pp. 45–46].

Properties of the class $\mathcal{S}^{\sharp\flat}$, which are of interest for the investigation of generalized Li coefficients and their modifications, can be found in [16, 44, 60] and [38]. For the purposes of this chapter, it is relevant to mention that every $F \in \mathcal{S}^{\sharp\flat}$ is a meromorphic function of order one with a possible pole at $s = 1$ only. The set $Z(F)$ of non-trivial zeros of F is a subset of the critical strip $0 \leq \mathrm{Re}(s) \leq 1$. Also, under assumption that $0 \notin Z(F)$, we have

$$\frac{\tilde{\Phi}'_F}{\tilde{\Phi}_F}(s) = \sum_{\rho \in Z(F)}^{*} \frac{1}{s-\rho},$$

for all complex numbers $s \notin Z(F)$, where

$$\tilde{\Phi}_F(s) = s^{m_F}(s-1)^{m_F}\Phi_F(s) \tag{9}$$

is the completed function. The function $\tilde{\Phi}_F(s)$ is an entire function of order one and satisfies the functional equation $\tilde{\Phi}_F(s) = \overline{\tilde{\Phi}_F(1-\overline{s})}$.

From the functional equation axiom (iii'), it is obvious that $\rho \in Z(F)$ if and only if $1-\overline{\rho} \in Z(F)$ with the same multiplicity. These properties of $F \in \mathcal{S}^{\sharp\flat}$ ensure that, for all integers $n \neq 0$, the generalized Li coefficients $\lambda_F(n)$ given by (3) are well defined and that the τ-Li coefficients given by (4) are also well defined for all $F \in \mathcal{S}^{\sharp\flat}$ such that $\tau \notin Z(F)$.

2.3 *Other Classes of Functions Related to* $\mathcal{S}$

The class $\mathcal{S}^{\sharp\flat}$, though broad enough to unconditionally contain all L-functions attached to irreducible unitary automorphic representations, does not contain some zeta and L-functions of number-theoretic interest. For example, the Rankin–Selberg L-functions might have a pole at a point on the line $\mathrm{Re}(s) = 1$ that is different from $s = 1$, thus violating axiom (ii). Some L-functions are given as a Dirichlet series such that the smallest number σ_0 for which this Dirichlet series converges absolutely in the half-plane $\mathrm{Re}(s) > \sigma_0$ is bigger than one, thus violating axiom (i). Moreover, some L-functions satisfy a functional equation relating s to $\sigma_1 - s$, for $\sigma_1 \neq 1$, thus violating axiom (iii'). (For example, Hecke L-functions associated to un-normalized Fourier coefficients of certain even weight $k > 2$ automorphic newforms violate axioms (i) and (iii'), see [26, Section 1.4.].) For all those functions, it is of interest to deduce properties of the generalized Li and τ-Li coefficients.

For this reason, a very broad class of functions, denoted by $\mathcal{S}^{\sharp\flat}(\sigma_0, \sigma_1)$, is introduced in [18]. It consists of all Dirichlet series that converge in some half-plane $\mathrm{Re}(s) > \sigma_0 \geq 1$ and have a continuation to the complex plane $\mathbb{C}$, and this continuation is a meromorphic function of finite order with at most finitely many poles. Also, $F \in \mathcal{S}^{\sharp\flat}(\sigma_0, \sigma_1)$ is assumed to satisfy a functional equation of the Riemann type (i.e., with factors being products of gamma functions) relating values $F(s)$ with $\overline{F(\sigma_1 - \overline{s})}$, for some $\sigma_1 \leq 2\sigma_0$, and, moreover, the logarithmic derivative of F is assumed to possess a Dirichlet series representation converging absolutely for $\mathrm{Re}(s) > \sigma_0$. An interested reader may consult [24, Section 7] and [25, pp. 43–46] for examples of functions in a so-called *fundamental class* that contains $\mathcal{S}^{\sharp\flat}(\sigma_0, \sigma_1)$ as well as for examples of zeta and L-functions with a functional equation that is not of the Riemann type.

The class $\mathcal{S}^{\sharp\flat}(\sigma_0, \sigma_1)$ contains products of suitable shifts of L-functions from $\mathcal{S}^{\sharp\flat}$, as well as products of shifts of certain L-functions possessing an Euler product representation that are not necessarily in $\mathcal{S}^{\sharp\flat}$ (such as the Rankin–Selberg L-functions that belong to $\mathcal{S}^{\sharp\flat}(1, 1)$). Some properties of the class $\mathcal{S}^{\sharp\flat}(\sigma_0, \sigma_1)$ are derived in [18, 19], and [46].

The Dirichlet series representation of the logarithmic derivative $\frac{F'}{F}$ implies that $F \in \mathcal{S}^{\sharp\flat}(\sigma_0, \sigma_1)$ is non-vanishing in the half-plane $\mathrm{Re}(s) > \sigma_0$. This, together with the functional equation axiom, yields that the natural range of the parameter τ is $[\sigma_1, 2\sigma_0]$ for the study of the τ-Li coefficients in the class $\mathcal{S}^{\sharp\flat}(\sigma_0, \sigma_1)$.

It is proved in [12, p. 260] for functions F in the extended Selberg class $\mathcal{S}^{\sharp}$ such that $0, \tau \notin Z(F)$ that the series (3) and (4) defining generalized Li and τ-Li coefficients for $\tau \in [1, 2]$, respectively, are convergent. Application of results of Bombieri and Lagarias [6] yields that the non-negativity of $\mathrm{Re}(\lambda_F(n))$ for all positive integers n is equivalent to the GRH for F and that the non-negativity of $\mathrm{Re}(\lambda_F(n, \tau))$ for all positive integers n is equivalent to the $\tau/2$-GRH for F.

However, with the existing methods (which make extensive use of the Euler sum axiom (v')), it is not possible to relate those coefficients to the Laurent/Taylor series expansion of $\log \tilde{\Phi}_F(s)$ or $\frac{\tilde{\Phi}'_F}{\tilde{\Phi}_F}(s)$, where $\tilde{\Phi}_F$ is defined by (9). This makes it very difficult to deduce any further arithmetic information regarding the Li or the τ-Li coefficients for a general $F \in \mathcal{S}^{\sharp}$. Moreover, it also makes it very difficult to numerically evaluate those coefficients without an explicit knowledge of the zeros of $F \in \mathcal{S}^{\sharp}$.

For this reason, in [12], the authors studied the class $\mathcal{S}^{\sharp}_{\mathbb{R}}$ that consists of Dirichlet series from $\mathcal{S}^{\sharp}$ with real coefficients $a_F(n)$ and real spectral parameters μ_j, $j = 1, \ldots, r$, in axiom (iii). Some properties of the class $\mathcal{S}^{\sharp}_{\mathbb{R}}$ are derived in [12], where the authors have also studied the asymptotic behavior of the τ-Li coefficients.

Some interesting arithmetic properties of the generalized Li and τ-Li coefficients can be deduced for certain subclasses of the Selberg class. For example, the class $\mathcal{S}^{\flat}$ that consists of functions $F \in \mathcal{S}$ with an error term in the prime number theorem of the form $O(x/\log^{\ell} x)$ for any positive ℓ, as $x \to \infty$, was studied in [60] where an arithmetic expression for the generalized Li coefficients attached to $F \in \mathcal{S}^{\flat}$ was deduced.

Another subclass of the Selberg class, in which the coefficients of the Dirichlet series are periodic, was studied in [17], where a necessary and sufficient condition for a Dirichlet series to belong to this class was derived.

3 Generalized Li and τ-Li Coefficients

The expressions (3) and (4) defining the generalized Li and τ-Li coefficients, respectively, are suitable for deducing the relation of the coefficients to the GRH and τ-GRH. However, those expressions are not well-suited for numerical studies especially when the lists of zeros are not known. Moreover, the arithmetic information contained in (3) and (4) is well-hidden in the zeros, and it is not so clear how those coefficients are related to arithmetic data attached to a function F. Therefore, it is of interest to derive other equivalent expressions for the generalized Li and τ-Li coefficients as well as arithmetic formulas for those coefficients. Those expressions are described and further analyzed in the first two subsections of this section.

On the other hand, from the definition (5) of the discretized Li coefficients $\Lambda_F(n,\tau)$, it is not clear how those are related to non-trivial zeros of F. In Sect. 3.3, we describe this relationship through Eq. (23) and also give Eq. (22) expressing those coefficients as a finite sum that is easy to compute in numerical experiments.

3.1 Generalized Li Coefficients

Many authors (starting with Li) have deduced different formulas for the generalized Li coefficients in [6, 13, 16, 50, 51, 60], to name a few. We summarize those formulas, for $F \in \mathcal{S}^{\sharp\flat}$ in the following theorem which we quote from [60], Appendix A.

Theorem 1 *Let $F \in \mathcal{S}^{\sharp\flat}$ be such that $0 \notin Z(F)$ and let $\tilde{\Phi}_F$ denote its completed function. The Li coefficients $\lambda_F(n)$ defined by (3) can be expressed as*

$$\lambda_F(n) = \frac{1}{(n-1)!}\left[\frac{d^n}{ds^n}(s^{n-1}\log\tilde{\Phi}_F(s))\right]_{s=1}. \tag{10}$$

They are also the coefficients in the Taylor series expansion of the logarithmic derivative of $\tilde{\Phi}_F$ evaluated at a suitable point, namely,

$$\frac{d}{dz}\log\tilde{\Phi}_F\left(\frac{1}{1-z}\right) = \sum_{n=0}^{\infty}\lambda_F(n+1)z^n. \tag{11}$$

In the case of the Riemann zeta function, formulas (10) and (11) were proved by Li [30]. For example, the Li coefficient λ_n given by (1) can be expressed as

$$\lambda_n = \frac{1}{(n-1)!}\left[\frac{d^n}{ds^n}(s^{n-1}\log\xi(s))\right]_{s=1},$$

where $\xi(s) = s(s-1)\pi^{-s/2}\Gamma(s/2)\zeta(s)$ is the completed zeta function.

From the previous representations, one can prove that

$$\lambda_F(n) = \frac{n}{2\pi i}\int_C \frac{\log\tilde{\Phi}_F\left(\frac{1}{1-s}\right)}{s^n}ds,$$

where C is a positively oriented closed contour enveloping the origin as the only singularity of the integrand.

The following arithmetic formula for the coefficients was deduced in [60].

Theorem 2 *Assume $F \in \mathcal{S}^{\sharp\flat}$ is such that $0 \notin Z(F)$. For all positive integers n, the generalized Li coefficient $\lambda_F(-n)$ can be written as*

$$\lambda_F(-n) = S_{A,F}(n) + S_{NA,F}(n), \tag{12}$$

where

$$S_{A,F}(n)=m_F+n\log Q_F+\sum_{\ell=1}^{n}\binom{n}{\ell}\eta_F(\ell-1) \quad \text{and} \quad S_{NA,F}(n)=\sum_{j=1}^{n}\binom{n}{\ell}\gamma_F(\ell-1)$$

are the Archimedean and the non-Archimedean parts, respectively. Here

$$\eta_F(0) = \sum_{j=1}^{r}\lambda_j\psi(\lambda_j+\mu_j) \quad \text{and} \quad \eta_F(\ell-1) = \sum_{j=1}^{r}(-\lambda_j)^{\ell}\sum_{k=0}^{\infty}\frac{1}{(\lambda_j+\mu_j+k)^{\ell}},$$

where $\psi(s) = \frac{\Gamma'}{\Gamma}(s)$ is the digamma function. The coefficients γ_F arise from the Laurent series of the logarithmic derivative of F at 1, namely

$$\frac{F'}{F}(s) = -\frac{m_F}{s-1} + \sum_{n=0}^{\infty}\gamma_F(n)(s-1)^n. \tag{13}$$

In the setting of the Riemann zeta function, Bombieri and Lagarias [6] were the first to observe that the Li coefficients λ_n can be distributed into parts based on which term in the completed function representation gave rise to these terms. Namely, the Archimedean part contains contributions from the gamma factors of the functional equation, while the non-Archimedean part can be expressed in terms of the function F alone. It is assumed that the Archimedean part, as $n \to \infty$ contributes to the growth of λ_n as $\frac{1}{2}n\log n + (\gamma - \log(2\pi) - 1)\,n + const$, while the non-Archimedean part is assumed to oscillate with a small amplitude; more details are given in Sect. 4.1.

This separation into Archimedean $S_{A,F}$ and non-Archimedean $S_{NA,F}$ parts of the Li coefficients for wider classes of functions was done by Lagarias [29], who investigated automorphic functions. Those contributions were further studied in [45] and [44] for the class $\mathcal{S}^{\sharp\flat}$. In [45], a very precise asymptotic expansion of the Archimedean contribution to $\lambda_F(-n)$, for $F \in \mathcal{S}^{\sharp\flat}$, as $n \to \infty$ was deduced. We cite it as the following theorem.

Theorem 3 *Let $F \in \mathcal{S}^{\sharp\flat}$ be such that $0 \notin Z(F)$. The Archimedean contribution to $\lambda_F(-n)$ for all positive integers n can be approximated, for any positive integer K as*

$$S_{A,F}(n) = \frac{d_F}{2} n \log n + C_F n + \sum_{j=1}^{r} (\mu_j - 1/2) + m_F + \frac{d_F}{4}$$

$$- \frac{d_F}{2} \sum_{k=1}^{K} \frac{B_{2k}}{2k} n^{1-2k} + O_{F,K}(n^{-2K}),$$

as $n \to \infty$*, where* B_{2k} *are the Bernoulli numbers,* $d_F = 2 \sum_{j=1}^{r} \lambda_j$ *is the degree of* F*, and*

$$C_F = \log Q_F + \frac{d_F}{2}(\gamma - 1) + \sum_{j=1}^{r} \lambda_j \log \lambda_j. \tag{14}$$

Sometimes it is useful to define the incomplete Li coefficients up to height T:

$$\lambda_F(n, T) = \sum_{\substack{\rho \in Z(F) \\ |\mathrm{Im}(\rho)| < T}} \left(1 - \left(1 - \frac{1}{\rho}\right)^n\right).$$

The behavior of the non-Archimedean part is close to the behavior of the incomplete coefficient with a suitable cutting point. Namely, for the incomplete coefficients, Odžak and Smajlović [45] proved the following asymptotic result:

Proposition 1 *Let* $F \in \mathcal{S}^{\sharp\flat}$ *be such that* $0 \notin Z(F)$*. As* $n \to \infty$*, we have the asymptotic formula*

$$S_{NA,F}(n) = \lambda_F(-n, \sqrt{n}) + O(\sqrt{n} \log n). \tag{15}$$

3.2 Generalized τ-Li Coefficients

Alternative representations for the τ-Li coefficients for functions from $\mathcal{S}^{\sharp\flat}$, as well as functions from the class $\mathcal{S}^{\sharp\flat}(\sigma_0, \sigma_1)$, which are convenient for numerical calculations, experiments, and study of the asymptotic behavior using the Arb package [23], were derived in [16, 18, 46].

Let $d_F(n, s_0)$ be the power series coefficients in the expansion of the logarithmic derivative of $\tilde{\Phi}_F\left(\frac{1}{1-s}\right)$ around the point $s_0 \neq 1$ that is not a zero of $\tilde{\Phi}_F\left(\frac{1}{1-s}\right)$. We have the following theorem, which we quote from [16, 18, 46] and state for the class $\mathcal{S}^{\sharp\flat}$.

Theorem 4 *Let* $F \in \mathcal{S}^{\sharp\flat}$ *and let* $\tau \in [1, 2]$ *be an arbitrary fixed real number such that* $0, \tau \notin Z(F)$*. For every positive integer* n*, we have*

$$\lambda_F(n,\tau) = \frac{\tau}{(n-1)!}\left[\frac{d^n}{ds^n}\left(s^{n-1}\log\tilde{\Phi}_F(s)\right)\right]_{s=\tau} = \frac{1}{\tau^n}d_F\left(n-1, 1-\frac{1}{\tau}\right), \tag{16}$$

where $\tilde{\Phi}_F$ is the completed function defined by (9).

Using the calculus of residues starting with (16), one can obtain an integral representation of τ-Li coefficients given by

$$\lambda_F(n,\tau) = \frac{n}{\tau^n}\frac{1}{2\pi i}\int_{C_\tau}\frac{\log\tilde{\Phi}_F\left(\frac{1}{1-s}\right)}{\left(s-1+\frac{1}{\tau}\right)^{n+1}}ds,$$

where C_τ is a positively oriented closed contour chosen so that it contains $s = 1-\frac{1}{\tau}$ as the only singularity of the integrand.

The following theorem, proved in [16], gives an arithmetic formula for the computation of the τ-Li coefficients.

Theorem 5 *Let n be a positive integer and $F \in \mathcal{S}^{\sharp\flat}$. Then whenever $\tau \geq 1$, if $0, \tau \notin Z(F)$, we have*

$$\lambda_F(n,\tau) = m_F + n\tau\log Q_F + \sum_{k=1}^{n}\binom{n}{k}\frac{\tau^k}{(k-1)!}\left[\frac{d^{k-1}}{ds^{k-1}}\left(\frac{F'}{F}(s) + \frac{m_F}{s-1}\right)\right]_{s=\tau}$$
$$+\sum_{k=1}^{n}\binom{n}{k}\frac{\tau^k}{(k-1)!}\sum_{j=1}^{r}\lambda_j^k\psi^{(k-1)}(\lambda_j\tau+\mu_j). \tag{17}$$

Moreover, for $\tau > 1$, we have

$$\lambda_F(n,\tau) = n\tau\log Q_F + m_F\left[2 + (-1)^{n+1}\left(\frac{1}{\tau-1}\right)^n\right]$$
$$+\sum_{k=1}^{n}\binom{n}{k}\frac{\tau^k}{(k-1)!}\left[\frac{d^{k-1}}{ds^{k-1}}\left(\frac{F'}{F}(s)\right)\right]_{s=\tau}$$
$$+\sum_{k=1}^{n}\binom{n}{k}\frac{\tau^k}{(k-1)!}\sum_{j=1}^{r}\lambda_j^k\psi^{(k-1)}(\lambda_j\tau+\mu_j).$$

The expression (17) enables us to write τ-Li coefficients $\lambda_F(n,\tau)$ as a sum of two parts: the Archimedean part stemming from the factor of the functional equation and the non-Archimedean part given in terms of the coefficients in the Laurent series expansion of $\frac{F'}{F}$ at $s = \tau$. More precisely,

$$\lambda_F(n,\tau) = S_{A,F}(n,\tau) + S_{NA,F}(n,\tau),$$

where

$$S_{A,F}(n,\tau) = \sum_{k=1}^{n} \binom{n}{k} \frac{\tau^k}{(k-1)!} \sum_{j=1}^{r} \lambda_j^k \psi^{(k-1)}(\lambda_j \tau + \mu_j) + n\tau \log Q_F$$

and

$$S_{NA,F}(n,\tau) = m_F + \sum_{k=1}^{n} \binom{n}{k} \frac{\tau^k}{(k-1)!} \left[\frac{d^{k-1}}{ds^{k-1}} \left(\frac{F'}{F}(s) + \frac{m_F}{s-1} \right) \right]_{s=\tau}. \tag{18}$$

Asymptotic expansions of $S_{A,F}(n,\tau)$ and $S_{NA,F}(n,\tau)$, for $F \in \mathcal{S}^{\sharp\flat}(\sigma_0,\sigma_1)$, are derived in [19]. We state the obtained results for τ-Li coefficients attached to $F \in \mathcal{S}^{\sharp\flat}$ in the following two theorems.

Theorem 6 *Let* $\tau \in (1,2]$ *and* $F \in \mathcal{S}^{\sharp\flat}$ *such that* $0, \tau \notin Z(F)$. *Then*

$$S_{A,F}(n,\tau) = \tau \frac{d_F}{2} n \log n + \left(C_F + \frac{d_F}{2} \log \tau \right) n\tau + \frac{\tau}{4} d_F \tag{19}$$

$$+ \sum_{j=1}^{r} (\mu_j - \frac{1}{2}) + \tau \frac{d_F}{2} \sum_{k=1}^{K} \frac{B_{2k}}{2k} n^{-2k+1} + O_K(n^{-2K}),$$

as $n \to \infty$, *where* B_{2k} *are the Bernoulli numbers,* $K \in \mathbb{N}$, *and* C_F *is given by* (14).

The incomplete τ-Li coefficient up to height $T > 0$, the *cutoff parameter*, is defined by

$$\lambda_F(n,\tau;T) = \sum_{\substack{\rho \in Z(F) \\ |\mathrm{Im}(\rho)| \le T}} \left(1 - \left(\frac{\rho}{\rho - \tau} \right)^n \right). \tag{20}$$

For large n, the non-Archimedean contribution $S_{NA,F}(n,\tau)$ can be approximated with the incomplete τ-Li coefficient up to height $\sqrt{n}$, as shown by the following theorem.

Theorem 7 *Let* $\tau \in (1,2]$ *and* $F \in \mathcal{S}^{\sharp\flat}$ *such that* $0, \tau \notin Z(F)$. *As* $n \to \infty$, *we have the asymptotic formula*

$$S_{NA,F}(n,\tau) = \lambda_F(n,\tau;\sqrt{n}) + O(\sqrt{n}\log n). \tag{21}$$

3.3 *Discretized τ-Li Coefficients*

The expression of the discretized τ-Li coefficients in terms of a finite sum is given in the following proposition, proved in [46] for the class $\mathcal{S}^{\sharp\flat}(\sigma_0,\sigma_1)$.

Proposition 2 *Let $\tau \in [1, 2]$ and $F \in \mathcal{S}^{\sharp\flat}$ such that $\tau \notin Z(F)$. Then, for every positive integer n*

$$\Lambda_F(n,\tau) = n(-1)^n \left(\frac{-\log \tilde{\Phi}_F(\tau)}{A^{\tau}_{n,0}} + \sum_{m=1}^{n} (-1)^m A^{\tau}_{n,m} \log \tilde{\Phi}_F(2m\tau) \right), \tag{22}$$

where the coefficients $A^{\tau}_{n,m}$ are defined as

$$A^{\tau}_{n,m} = \frac{\prod_{k=1}^{n}(2m+2k-1)}{2^n m!(n-m)!(2m-1)}, \quad m = 0, 1, 2, \ldots, n.$$

The relation between the discretized τ-Li coefficients $\Lambda_F(n,\tau)$ and the non-trivial zeros of F is described in the following proposition, proved in [46] for the class $\mathcal{S}^{\sharp\flat}(\sigma_0, \sigma_1)$.

Proposition 3 *Let $\tau \in [1, 2]$ and $F \in \mathcal{S}^{\sharp\flat}$ such that $0, \tau \notin Z(F)$. Then, for every positive integer n,*

$$\Lambda_F(n,\tau) = \frac{n}{\tau^{n-1}} \sum_{\rho \in Z(F)} F_n(\rho) + (-1)^{n+1} n A^{\tau}_{n,0} \log \tilde{\Phi}_F(0). \tag{23}$$

Here

$$F_n(z) = (-1)^n \tau^{n-1} \left(\frac{-1}{A^{\tau}_{n,0}} \log(z-\tau) + \sum_{m=0}^{n} (-1)^m A^{\tau}_{n,m} \log(z - 2m\tau) \right), \tag{24}$$

for $z \in \mathbb{C} \setminus U(n,\tau)$, where $U(n,\tau) = \bigcup_{m=0}^{n} [2m\tau - \delta_m, 2m\tau + \delta_m] \bigcup [\tau - \delta_\tau, \tau + \delta_\tau]$ and δ_m and δ_τ are arbitrary small numbers such that there is no non-trivial zero ρ of the function $F \in \mathcal{S}^{\sharp\flat}$ with $\rho \in U(n,\tau)$.

A finite sum representation for the discretized τ-Li coefficients given by (22) is suitable for numerical calculations. The representation in terms of zeros of $F \in \mathcal{S}^{\sharp\flat}$ given by (23) is convenient for asymptotic analysis, since the main interest in τ-Li coefficients and their discretization stems from their relation to the $\tau/2$-GRH.

4 Li-Type Criteria

In this section, we present variants of the Li-type criteria for the GRH and $\tau/2$-GRH, in terms of the asymptotic behavior of the generalized Li coefficients and τ-Li coefficients. We also present a criterion relating the behavior of the discretized Li coefficients $\Lambda_F(n,\tau)$ to the existence of a non-trivial zero outside the strip $1-\tau/2 \leq \mathrm{Re}(s) \leq \tau/2$.

4.1 Li Criterion and Its Variations

Originally, Li [30] stated that the Riemann hypothesis is equivalent to the coefficient λ_n being non-negative.

However, interestingly, the behavior of the coefficients is very different depending on whether the Riemann hypothesis is true. Odžak and Smajlović [44] proved that for functions in $\mathcal{S}^{\sharp\flat}$, the GRH is equivalent to the coefficients asymptotically behaving as

$$\lambda_F(-n) = \frac{d_F}{2} n \log n + C_F n + O(\sqrt{n} \log n),$$

when $n \to +\infty$. Here d_F is the degree of F, and the constant C_F is defined by (14).

Furthermore, in [44], it is proved that GRH being false is equivalent to having a positive constant δ such that $|\lambda_F(-n)| > e^{\delta n}$ for infinitely many $n \in \mathbb{N}$. As the Archimedean contribution has the lead term $\frac{d_F}{2} n \log n$ in the asymptotic expansion as $n \to \infty$, this means that if the GRH is false, then the non-Archimedean part very clearly takes over and decides the asymptotic behavior of $\lambda_F(-n)$, as $n \to \infty$.

4.2 Li-Type Criteria for τ-Li Coefficients

The Li-type criterion for the τ-Li coefficients was first formulated by Freitas [20] who proved that $\lambda_\zeta(n, \tau) \geq 0$ for all positive integers n if and only if the Riemann zeta function is zero-free in the region $\mathrm{Re}(s) > \tau/2$.

The Li-type criterion for zero-free region for the functions from the class $\mathcal{S}^{\sharp\flat}$, based on the Li criterion [6] for an arbitrary complex multiset, was proved in [16]. It states that a function $F \in \mathcal{S}^{\sharp\flat}$ such that $0, \tau \notin Z(F)$ satisfies the $\tau/2$-GRH for some $\tau \in [1, 2]$ if and only if $\mathrm{Re}\left(\lambda_F(n, \tau)\right) \geq 0$ for every positive integer n. Moreover, from [18] and [42], it follows that for any $\tau \in [1, 2]$ and $F \in \mathcal{S}^{\sharp\flat}$ such that $0, \tau \notin Z(F)$, the $\tau/2$-GRH is equivalent to the τ-Li coefficients behaving asymptotically as

$$\lambda_F(-n, \tau) = \tau \frac{d_F}{2} n \log n + \left(C_F + \frac{d_F}{2} \log \tau \right) \tau n + O(\sqrt{n} \log n),$$

as $n \to \infty$, where C_F is given by (14).

This is further generalized in [42] where the following theorem is proved for the class $\mathcal{S}^{\sharp\flat}(\sigma_0, \sigma_1)$.

Theorem 8 *Let $F \in \mathcal{S}^{\sharp\flat}(\sigma_0, \sigma_1)$ and let $\tilde{\Phi}_F$ denote its completed function. Let $\tau \in [\sigma_1, +\infty)$ be an arbitrary fixed real number such that $0, \tau \notin Z(F)$ and $\varepsilon \in (0, 1/\tau)$ is chosen such that the function $\frac{d}{dz} \log \tilde{\Phi}_F\left(\frac{1}{1-z}\right)$ is holomorphic inside the* *disc* $D_{\tau,\varepsilon} = D(1 - 1/\tau, 1/\tau - \varepsilon)$. *Let* $\rho_0 = \tau/2 - i\gamma_0$ *be a zero of the function* F *outside the strip* $\sigma_1 - \tau/2 \leq \mathrm{Re}(\rho) \leq \tau/2$.

Then, the following statements are true:

(i) *There are only finitely many other non-trivial zeros of the function* F *denoted by* $\rho_j = \tau/2 - i\gamma_j$, $j = 1, \ldots, j(\varepsilon)$ *such that they are images of the points from the disk* $D_{\tau,\varepsilon}$ *under the map* $m(s) : s \to 1/(1-s)$. *Moreover,*

$$\lambda_F(n,\tau) = -\mathrm{ord}\rho_0 \left(\frac{\gamma_0 - \frac{\tau}{2}i}{\gamma_0 + \frac{\tau}{2}i}\right)^{-n} - \sum_{j=1}^{j(\varepsilon)} \mathrm{ord}\rho_j \left(\frac{\gamma_j - \frac{\tau}{2}i}{\gamma_j + \frac{\tau}{2}i}\right)^{-n} + O(e^{n\tau\varepsilon}), \tag{25}$$

as $n \to \infty$, *where* $\mathrm{ord}\rho$ *denotes the order of zero* ρ.

(ii) *There exists a positive constant* δ *such that* $|\lambda_F(n,\tau)| > e^{\delta n}$ *for infinitely many* $n \in \mathbb{N}$.

Since $\mathcal{S}^{\sharp\flat} \subseteq \mathcal{S}^{\sharp\flat}(1,1)$, it is easy to obtain a corresponding asymptotic criterion for τ-Li coefficients attached to $F \in \mathcal{S}^{\sharp\flat}$.

4.3 Li-Type Criterion for Discretized Coefficients

The asymptotic behavior of the discretized Li coefficients is, as expected, related to the $\tau/2$-GRH. However, the relationship is slightly different than the one of the generalized Li and τ-Li coefficients described above.

Before we formulate the criterion, let us recall the definition (24) of the function F_n and define

$$\phi(z) := z \log\left(1 - \frac{1}{z^2}\right) + \log\frac{z+1}{z-1} \tag{26}$$

as the single-valued function on $\mathbb{C} \setminus [-1, 1]$ by taking $\log z = \log|z| + i\mathrm{Arg}(z)$ and $\mathrm{Arg}(z) \in (-\pi, \pi]$ and

$$\phi(z) \sim \frac{1}{z}, \quad \text{as} \quad |z| \to \infty.$$

The following theorem is a special case of Theorem 1 proved in [46].

Theorem 9 *Let* $F \in \mathcal{S}^{\sharp\flat}$ *and* $\tau \in [1, 2]$ *an arbitrary fixed real number such that* $0, \tau \notin Z(F)$. *Then*

$$\Lambda_F(n,\tau) = \frac{n}{\tau^{n-1}} \sum_{\mathrm{Re}\phi(z_\rho) > \phi_0} F_n(\rho) + (-1)^{n+1} n A^{\tau}_{n,0} \log \tilde{\Phi}_F(0) + o\left(nR^n\right) \tag{27}$$

for all $R > R_0 = e^{\phi_0}$ *as* $n \to \infty$, *where* ϕ_0 *is a fixed small number. The first sum on the right-hand side of* (27) *is taken over all* z_ρ *such that* $\rho = \frac{\tau}{2} + 2n\tau z_\rho \in Z(F)$.

A careful analysis of the asymptotic behavior of F_n conducted in [46] yields the following corollary.

Corollary 1 *Let F be a function from the class* $\mathcal{S}^{\sharp\flat}$ *and* $\tau \in [1, 2]$ *an arbitrary fixed real number such that* $0, \tau \notin Z(F)$. *Then, the existence of a zero* $\rho \in Z(F)$ *outside the strip* $1 - \tau/2 \leq \mathrm{Re}\rho \leq \tau/2$ *implies oscillations of the coefficients* $\Lambda_F(n, \tau)$ *with power-like growing amplitudes.*

5 Numerical Computations

A particular interest in the Li coefficients and their variations and generalizations is closely related to the possibility to use them for numerical computations and simulations.

Numerical evidence may be used in a deeper analysis of their properties, for deriving interesting conclusions and posing some conjectures. Of course, one should bear in mind that no matter how convincing numerical evidence is, it may be just illusory. Oesterlé's observation, stating that if all non-trivial zeros ρ of the Riemann zeta function with $|\mathrm{Im}(\rho)| < \sqrt{n}$ are on the critical line, then $\lambda_n \geq 0$ (see [2, page 441]), suggests that we need to be careful when deriving conclusions from numerical evidence. However, rapidly increasing capabilities of available computer systems, hardware and software, give us possibilities to obtain larger and larger and more precise sets of numerical results.

The main challenges in numerical calculations of the Li and related coefficients come from the fact that infinite sums appearing in these representations need to be truncated appropriately. Moreover, for some internal quantities such as the Riemann zeta zeros, the Euler–Stieltjes constants, the Bernoulli numbers, or values of the Gamma functions at specific points, only approximate values up to some number of correct digits are available. This makes it harder to control the error.

5.1 Numerical Results for the Li and Generalized Li Coefficients

Numerical calculation of the first few values of the Li coefficients attached to the Riemann zeta function is given in [2], while the first extensive research is presented in [28]. The numerical evidence obtained reveals a decomposition of the Li coefficients into two parts: a strictly growing trend and some tiny oscillations. In the notation of Theorem 2 with $F = \zeta$, the strictly growing trend stems from the Archimedean contribution $S_{A,\zeta}(n)$, and the tiny oscillations (as $n \to \infty$) stem from the non-Archimedean part $S_{NA,\zeta}(n)$. It suggests a reformulation of the Li criterion in terms of the requirement that the non-Archimedean part $S_{NA,\zeta}(n)$ is superimposed, as $n \to \infty$ by the lead term of the Archimedean contribution to coefficients; see also the discussion in Sect. 4.1.

The arithmetic formulas for the Li and the generalized Li coefficients associated to a function F presented in Sect. 3.1 reveal a deep connection between the generalized Li coefficients and the so-called Euler–Stieltjes constants $\gamma_F(n)$, i.e., the constants in the Laurent or Taylor series expansion (13) of $\frac{F'}{F}(s)$. The Euler–Stieltjes constants are, of course, related to the Stieltjes constants for F that are the constants in the Laurent or Taylor series expansion of $F(s)$ at $s = 1$, see, e.g., [13] in which a recurrence relation between the Stieltjes and the Euler–Stieltjes constants attached to the Riemann zeta function is given.

Calculations of the Li coefficients attached to the Riemann zeta function that are based on the Stieltjes constants for the Riemann zeta function $\zeta(s)$ are first conducted in [28]. The Stieltjes constants and the Euler–Stieltjes constants for the Riemann zeta function were used in calculations of the Li coefficients in [34], using Wolfram Mathematica 5 package. In the paper [13], the author tabulated numerical values for the first 100 Li coefficients; see also related results in [14] and [15].

Numerical calculations of the generalized Li coefficients associated with the Hecke L-functions for certain values of the conductor are presented in [54]. The authors used arithmetic formulas for the corresponding coefficients derived in [50] and [51] and Sage and PARI-GP as a tool for calculations. The presented numerical evidence supports Oesterlé's observation, recalled in the introduction of this section.

5.2 *Numerical Results for the Generalized τ-Li Coefficients in the Class $\mathcal{S}^{\sharp\flat}(\sigma_0, \sigma_1)$*

Numerical experiments and investigations with the generalized τ-Li coefficients are more interesting (and challenging) due to the presence of an extra parameter τ. In our opinion, especially interesting experiments can be carried out with functions that are not expected to have zeros accumulated on the single critical line, but which do belong to some natural extensions of the Selberg class.

Natural examples of such functions are products of the Riemann zeta functions with appropriately shifted arguments. These functions are members of the class $\mathcal{S}^{\sharp\flat}(\sigma_0, \sigma_1)$ described in Sect. 2.3. In [18], it was proved that

$$F(s) = \prod_{i=1}^{K} \zeta(s - \alpha_i)\zeta(s + \alpha_i) \in \mathcal{S}^{\sharp\flat}(\sigma_0, 1),$$

for arbitrary complex constants α_i, where $\sigma_0 = \max_{1 \leq i \leq K} \{|\mathrm{Re}\alpha_i| + 1\}$.

The results of numerical calculations of the τ-Li coefficients associated to function $F_4(s) = \prod_{j=1}^{4} \zeta(s-j)\zeta(s+j)$ are presented in [18]. Obtained numerical evidence is used to pose some conjectures relating the large-scale asymptotic behavior of generalized coefficients and $\tau/2$-GRH, see [18, Conjectures 26 and 27]. These are partially proved in [42].

The calculations in [42] use the definition (4) and tables of zeros of the Riemann zeta functions available at [41] as a starting point, but the error control is handled in a more general manner. For example, [42, Proposition 7] gives an upper bound for an error term when approximating $\lambda_F(n,\tau)$ attached to F with $\lambda_F(n,\tau;T)$ defined by (20) for any range of n, and any selected value of τ from the natural interval.

In [42], some additional properties concerning the growth of the oscillations of the coefficients have been observed. Namely, the change of τ implies the change of the strip in the formulation of the τ-Li criterion and thus the change in the proportion of the zeros of the corresponding L-function in that strip. It is observed that the proportion of zeros included in the strip is not linearly related to the order of the amplitudes of the oscillations. For example, let $\tau_1 < \tau_2$ be two parameters such that the proportion of zeros of F in the strip $1-\tau_1 < \mathrm{Re}(s) < \tau_1$ is one half of the proportion of zeros of F in the strip $1-\tau_2 < \mathrm{Re}(s) < \tau_2$, but F also has zeros outside the second strip. Then the generalized τ-Li coefficients $\lambda_F(n,\tau_1)$ and $\lambda_F(n,\tau_2)$ will oscillate with exponentially growing amplitudes. However, the size of the amplitude of $\lambda_F(n,\tau_1)$ will not be twice the size of the amplitude of $\lambda_F(n,\tau_2)$, in spite of the fact that a lot more zeros of F are contributing to the growth of $\lambda_F(n,\tau_2)$.

More precisely, it is noticed that σ_0 is the breaking point of the natural interval for τ when it comes to the behavior of the generalized τ-Li coefficients. The increase of the amplitudes of the oscillations of $\lambda_F(n,\tau)$ with the increase of τ is noticeable for τ in $[\sigma_1,\sigma_0)$, the interval that corresponds to values where the corresponding Dirichlet series is not absolutely convergent, while for τ in $[\sigma_0, 2\sigma_0]$ the gradual decrease of the amplitudes of the oscillations is observed. The interval $[\sigma_0, 2\sigma_0]$ corresponds to those values for which the Dirichlet series is absolutely convergent.

Further numerical calculations related to the τ-Li coefficients for products of the Riemann functions that are in the class $\mathcal{S}^{\sharp\flat}(\sigma_0,\sigma_1)$ are presented in [19]. They have been conducted using representation (16) of the τ-Li coefficients in terms of the nth derivative of the completed L-function combined with the power series representation of the corresponding L-function. The code is written using Arb, a C library for arbitrary-precision floating-point ball arithmetic, developed by F. Johansson [23]. The main advantage of using Arb as a tool in the computations is the fact that it supports efficient high-precision computation with power series and special functions over real and complex numbers with automatic error control.

Using this approach, it was possible to distinguish between different contributions, Archimedean and non-Archimedean, to τ-Li coefficients. The examples presented in [19] show that the main terms in asymptotic formulas for these contributions nicely describe their overall behavior, and they are considerably easier to calculate than the contributions themselves.

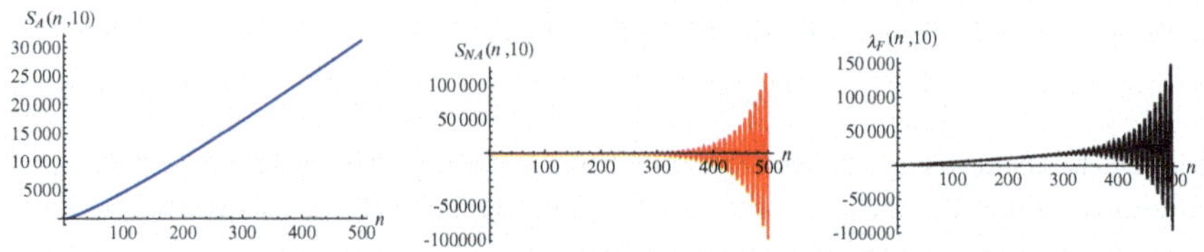

Fig. 1 Archimedean and non-Archimedean contributions, and the τ-Li coefficients attached to the function $F = F_5$ for $\tau = 10$

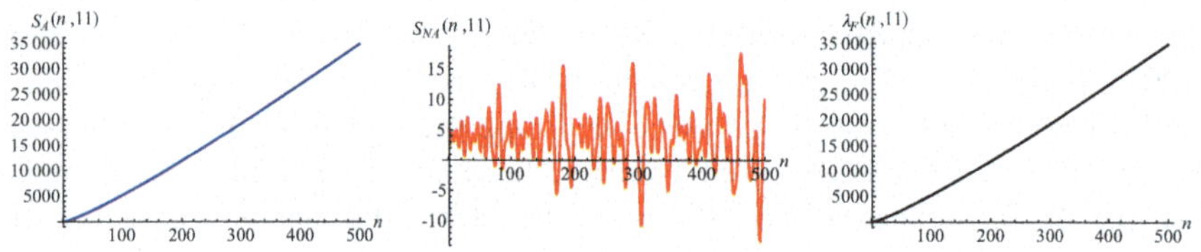

Fig. 2 Archimedean and non-Archimedean contributions, and the τ-Li coefficients attached to the function $F = F_5$ for $\tau = 11$

The results of numerical calculations of Archimedean and non-Archimedean contributions to the τ-Li coefficients associated to function $F_5(s) = \zeta(s-5)\zeta(s+5)$ are presented in [19]. Selected graphs are given in Figs. 1 and 2; for brevity, we omitted the index F in the notation for the contributions.

These graphs demonstrate two opposite situations. In both cases, the Archimedean contribution to the τ-Li coefficients grows with the lead term of order $n \log n$, but the non-Archimedean term exhibits oscillatory values with exponentially growing amplitudes when $\tau = 10$. However, the non-Archimedean term is considerably smaller than the corresponding values of the Archimedean term when $\tau = 11$. When $\tau = 10$, the Archimedean contribution to the τ-Li coefficients is completely dominant for n from 1 to 300, then the impact of the non-Archimedean term becomes visible, and finally, it produces a negative value of the τ-Li coefficients around $n = 430$. For $\tau = 11$, there is no visible impact of the non-Archimedean term to the τ-Li coefficients, values of $\lambda_{F_5}(n, 11)$ grow as n increases, and there are no oscillations and no negative values of τ-Li coefficients in this case.

Going further, in [19], the asymptotic expansions for the Archimedean and non-Archimedean contributions to the τ-Li coefficients are derived, and it is observed from the numerical examples that the main terms in the asymptotic formulas for these contributions nicely describe their overall behavior. The importance of this observation comes from the fact that the main terms are considerably easier to calculate than the contributions themselves.

5.3 Numerical Results for the Generalized τ-Li Coefficients in the Class $\mathcal{S}^{\sharp}_{\mathbb{R}}$

A very well-known example of an L-function with zeros not accumulated on a single critical line is the Davenport–Heilbronn function L_{DH}. This function served as a motivation to introduce a subclass $\mathcal{S}^{\sharp}_{\mathbb{R}}$ of the extended Selberg class containing functions that possess a Dirichlet series representation with real coefficients, as introduced in [12]. The class $\mathcal{S}^{\sharp}_{\mathbb{R}}$ contains the Bombieri–Ghosh-type functions considered in [5] under some additional conditions. More precisely, the Bombieri–Ghosh-type functions are defined for $\mathrm{Re}(s) \gg 0$ by the absolutely convergent Dirichlet series

$$f(s, \xi; \psi) = \sum_{n=1}^{\infty} \frac{a(n, \xi)\psi(n)}{n^s},$$

where ξ is an arbitrary complex number, ψ is a completely multiplicative function satisfying the condition $|\psi(p)| < p$ for all primes p, and coefficients $a(n, \xi)$ are defined as

$$a(n, \xi) = \begin{cases} 1 & n \equiv 1 \pmod 5 \\ \xi & n \equiv 2 \pmod 5 \\ -\xi & n \equiv 3 \pmod 5 \\ -1 & n \equiv 4 \pmod 5 \\ 0 & n \equiv 0 \pmod 5). \end{cases}$$

It is proved in [12] that for $\psi \equiv 1$, the Davenport–Heilbronn function $L_{DH}(s, \xi) = f(s, \xi; \psi)$ belongs to the class $\mathcal{S}^{\sharp}_{\mathbb{R}}$ if and only if $\xi = \xi_{\pm} = \phi \pm \sqrt{1 + \phi^2}$, where $\phi = (1 + \sqrt{5})/2$ is the golden ratio.

The class $\mathcal{S}^{\sharp}_{\mathbb{R}}$ contains Dirichlet series with coefficients that are periodic modulo arbitrary integer m, under certain assumptions posed on the first $m - 1$ defining coefficients. For more details on such series, we refer to [17].

A very nice property of the Dirichlet series with periodic coefficients is that it can be expressed as a linear combination of the Hurwitz zeta functions. This property was used in the numerical calculations of the τ-Li coefficients of Davenport–Heilbronn-type functions presented in [12]. The starting point for computations was the expression (16) in which the corresponding τ-Li coefficients associated to F are given in terms of coefficients in the Taylor (Laurent) series representation of the completed function $\tilde{\Phi}_F$. Given that F is a linear combination of the Hurwitz zeta functions, those coefficients can be expressed in terms of coefficients in the Taylor or Laurent series expansion of the Hurwitz zeta functions. The computation with such series is very nicely implemented in the interval arithmetic package Arb. Examples of functions with coefficients that are periodic with periods 5, 7, and 35

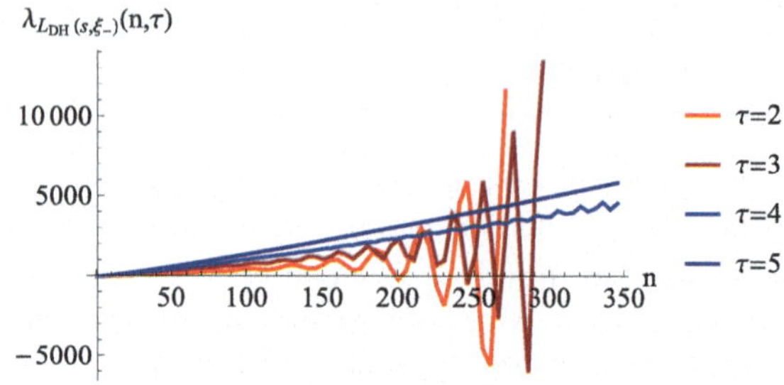

Fig. 3 τ-Li coefficients attached to $L_{DH}(s, \xi_-)$

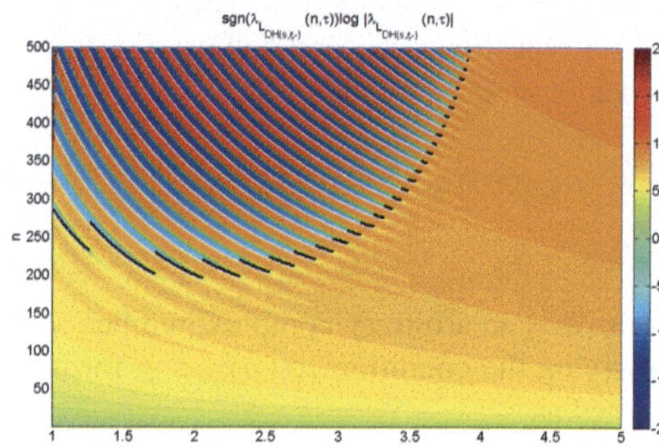

Fig. 4 τ-Li coefficients attached to $L_{DH}(s, \xi_-)$

are presented in [12], which enabled us to derive certain conclusions regarding the existence of zeros outside some strips.

In Fig. 3, some values of τ-Li coefficients attached to $L_{DH}(s, \xi_-)$ for $\tau \in \{2, 3, 4, 5\}$ are presented. The negative values of the τ-Li coefficients and their exponentially growing amplitudes for $\tau = 2$ and $\tau = 3$ imply the existence of zeros outside the strip $1 - \tau/2 \leq \text{Re}(s) \leq \tau/2$. This can also be concluded from the plot of zeros of $L_{DH}(s, \xi_-)$ given in [5], and it is in agreement with [12, Corollary 3.4]. When $\tau = 4$, small oscillations occur for $n \geq 300$, and their amplitude increases for larger values of n; this is further supported by additional numerical evidence presented in [12]. The case $\tau = 5$ suggests a complete absence of zeros in the right half-plane $\text{Re}(s) > 5/2$. Again, this agrees with the plot of zeros of $L_{DH}(s, \xi_-)$ given in [5].

Figure 4 gives a better insight into the behavior of $\lambda_{L_{DH}(s,\xi_-)}(n, \tau)$ for τ between 1 and 5 and for n from 1 to 500. Because of the existence of very small and very large values, a logarithmic scale while preserving the sign is used. Alternating blue and red sectors correspond to the ranges of n with negative and positive τ-Li coefficients, while a darker shade indicates the increase in amplitude. The black dots correspond to the smallest value of n for which $\lambda_{L_{DH}(s,\xi_-)}(n, \tau)$ is negative for a given τ.

Interestingly, in [12], it is noticed that the level of complexity of computations of the τ-Li coefficients $\lambda_F(n, \tau)$ is not a unique function of integer n and parameter τ, independent of the function F under consideration. Actually, it depends upon the overall distribution of the zeros of function F. More precisely, in the case when a large proportion of the zeros of $F(s)$ is concentrated close to the critical line, then one needs large values of n to notice oscillations in the τ-Li coefficients. An example of such a situation is the function $L_{DH}(s, \xi_+)$ and values of τ very close to 2. Similar conclusions were derived in [5] where the search for the zeros was performed using the secant method.

5.4 Numerical Calculations with the Discretized Li and τ-Li Coefficients

As described in the introduction and Sect. 3.3, the discretized variants of Li and generalized Li coefficients have been introduced with an aim to make numerical calculations easier and less challenging. The representation in terms of a finite sum (22) that does not exist in the case of the classical Li coefficients makes numerical computations with discretized Li coefficients a lot easier to handle from the aspect of error control.

Some values of discretized Li coefficients attached to the Riemann zeta function are listed in [62]. Also, plots of the values of the RH-true asymptotic form (that corresponds to the Archimedean contribution) and the reminder sequence (i.e., the difference between the discretized Li coefficient and the RH-true asymptotic form) are provided. These plots agree with results on sensitivity of the discretized Li coefficients to the Riemann hypothesis, see [62, section 3.1]. Plots of the discretized Li coefficients attached to the Davenport–Heilbronn functions $L_{DH}(s, \xi_\pm)$ are also presented in [62].

Numerical evidence related to the discretized τ-Li coefficients attached to functions from $\mathcal{S}^{\sharp\flat}(\sigma_0, \sigma_1)$ is obtained using the finite sum representation (22) and Arb as a tool for calculations. In Fig. 5, values for discretized τ-Li coefficients attached to function F_5, defined in 5.2, and $\tau \in \{1, 11\}$ are presented.

The existence of oscillations of $\Lambda_F(n, 1)$ is expected, since there are zeros of the function F_5 outside the strip $1 - \tau/2 \leq \mathrm{Re}(s) \leq \tau/2$ (which, in this case, is actually the line $\mathrm{Re}(s) = 1/2$), as noticed in Sect. 5.2. This is also illustrated in Fig. 5. When $\tau = 11$, under the Riemann hypothesis, all zeros of the function F_5 are contained

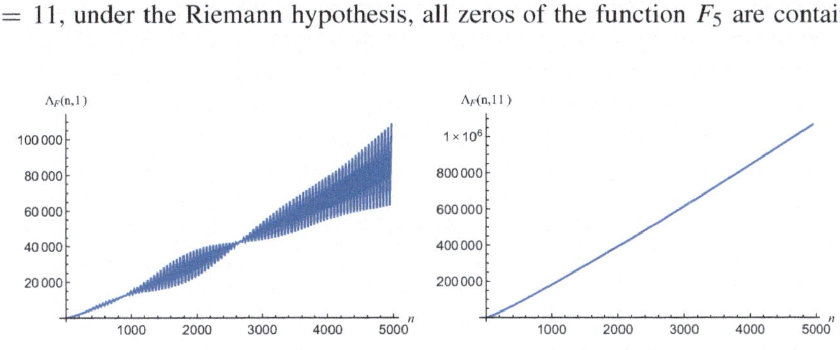

Fig. 5 Discretized τ-Li coefficients $\Lambda_F(n, \tau)$ attached to the function $F = F_5$ for $\tau = 1$ and $\tau = 11$

in the strip $-9/2 \leq \mathrm{Re}(s) \leq 11/2$, and no oscillations of the discretized τ-Li coefficients attached to F_5 are expected. The overall behavior (existence or absence of oscillations) of the discretized τ-Li coefficients is analogous to the behavior of τ-Li coefficients. However, in the case of the discretized coefficients, oscillations of the coefficients arise with power-like growing amplitudes, as proved in [46].

6 Summary of Results as a Table

Table 1 hopefully serves as a quick reference list for anybody looking for specific results about Li coefficients related to different classes of zeta and L-functions.

The first column "Function(s)" refers to the function or the class of functions under investigation, while the rest of Table 1 contains references and the numbers of sections from the current paper related to the specific topics given in the header row of the table. In the first column, some functions are grouped together, and listed in a hierarchical order, from specific examples to classes of functions with increased generality. More precisely, the Dirichlet L-functions belong to the Selberg class, which is a subclass of the modified Selberg class. The modified Selberg class also contains the class of all automorphic L-functions. For this reason, all results that are related to the modified Selberg class are true for functions in the classes listed above that class in the table. The class $\mathcal{S}^{\sharp\flat}(\sigma_0, \sigma_1)$ contains the modified Selberg class; hence, all results related to that class are true for the functions listed above.

The last three classes of functions listed in Table 1 have not been discussed in greater detail in this chapter, but we wanted to provide interested readers a list of references in which they can find further information related to the Li-type criteria for those functions. References/sections of the current paper dealing

Table 1 Reference list/sections from the current paper

	References/sections				
Function(s)	Li	τ-Li	Expressions	Numerics	Other
Riemann zeta	[1, 30, 61]	[20]	[6, 13, 14, 61]	[2, 28, 34]	[21, 58, 62]
	1.1	1.3		5.1	1.4
Dirichlet L	[31, 36]		[31, 36]		
Selberg class	[35, 50, 52, 60]		[39, 50, 52, 60]		[9, 38]
Automorphic L	[29, 43]	[11, 37]	[11, 29, 37, 43]		
Modified Selberg class	[45, 60]	[16]	[16, 45, 60]	[18]	[16, 44]
	1.2, 4.1	1.3, 4.2	3.1, 3.2		3.3, 4.3
$\mathcal{S}^{\sharp\flat}(\sigma_0, \sigma_1)$		[12, 18]	[12, 18]	[12, 18, 19, 42]	[46, 55]
	1.2			5.2, 5.3	1.4, 5.4
Hecke L	[31–33, 53]		[31–33]	[54]	
Function fields L	[3, 40, 48]		[3, 40, 48, 49]		[4]
Other zeta and L	[10, 22, 47]		[22]		

with the classical Li criterion are listed in the second column, titled "Li." The third column "τ-Li" refers to the results of the τ-Li criteria type. The fourth column "Expressions" refers to results giving different analytic and/or arithmetic expressions for the Li and the τ-Li coefficients. The fifth column titled "Numerics" refers to various numerical investigations around the Li coefficients and the τ-Li coefficients. In the last column "Other," we have collected other relevant results related to the Li-type coefficients and the Li-type criteria, for example, discretized coefficients, Stieltjes constants, Euler–Stieltjes constants, as well as results related to transcendence theory or zero-free regions.

Acknowledgments We would like to thank the anonymous referee for a careful reading of the manuscript and for their comments and recommendations that helped us improve the paper.

References

1. J. Arias de Reyna, Asymptotics of Keiper-Li coefficients. Funct. Approx. Comment. Math. **45**, part 1, 7–21 (2011)
2. P. Biane, P. Pitman, M. Yor, Probability laws related to the Jacobi theta and Riemann zeta functions, and Brownian excursions. Bull. Am. Math. Soc. **38**, 435–465 (2001)
3. K.H. Bllaca, K. Mazhouda, Explicit formula on function fields and application: Li coefficients. Ann. Mat. Pura Appl. (4) **200**(5), 1859–1869 (2021)
4. K.H. Bllaca, K. Mazhouda, Centralized variant of the Li criterion on functions fields. Finite Fields Appl. **72**, Paper No. 101800, 18 pp. (2021)
5. E. Bombieri, A. Ghosh, Around Davenport-Heilbronn function. Uspekhi Math. Nauk **66**, 15–66 (2011), translated in Russian Math. Surveys **66**, 221–270 (2011)
6. E. Bombieri, J.C. Lagarias, Complements to Li's criterion for the Riemann hypothesis. J. Number Theory **77**, 274–287 (1999)
7. K.A. Broughan, Equivalents of the Riemann hypothesis, vol. 1. Encyclopedia Math. Appl., vol. 164 (Cambridge Univ. Press, Cambridge, 2017)
8. K.A. Broughan, Equivalents of the Riemann hypothesis, vol. 2. Analytic equivalents. Encyclopedia Math. Appl., vol. 165 (Cambridge University Press, Cambridge, 2017)
9. F.C.S. Brown, Li's criterion and zero-free regions of L-functions. J. Number Theory **111**, 1–32 (2005)
10. F. Brown, S. Omar, Li's criterion for Epstein zeta functions, generalization of Kronecker's limit formula and the Gauss problem. J. Number Theory **158**, 90–103 (2016)
11. A. Bucur, A.-M. Ernvall-Hytönen, A. Odžak, E. Roditty-Gershon, L. Smajlović, On τ-Li coefficients for Rankin-Selberg L-functions, in *Women in Numbers Europe - WINE 2013*, ed. by M.J. Bertin, A. Bucur, B. Feigon, L. Schneps (Springer International Publishing Switzerland, 2015), pp. 167–190
12. A. Bucur, A.-M. Ernvall-Hytönen, A. Odžak, L. Smajlović, On a Li-type criterion for zero-free regions of certain Dirichlet series with real coefficients. LMS J. Comput. Math. **19**(1), 259–280 (2016)
13. M.W. Coffey, Toward verification of the Riemann hypothesis: Application of the Li criterion. Math. Phys. Anal. Geom. **8**, 211–255 (2005)
14. M.W. Coffey, New results concerning power series expansions of the Riemann xi function and the Li/Keiper constants. Proc. R. Soc. Lond. Ser. A Math. Phys. Eng. Sci. **464**(2091), 711–731 (2008)
15. M.W. Coffey, Summation properties of the η_j and Li constants. J. Comput. Appl. Math. **233**(3), 667–673 (2009)

16. A.D. Droll, Variations of Li's criterion for an extension of the Selberg class, PhD thesis, Queen's University Ontario, Canada, July 2012
17. A.-M. Ernvall-Hytönen, A. Odžak, L. Smajlović, On a class of periodic Dirichlet series with functional equation. Math. Commun. **25**(1), 35–47 (2020)
18. A.-M. Ernvall-Hytönen, A. Odžak, L. Smajlović, M. Sušić, On the modified Li criterion for a certain class of L-functions. J. Number Theory **156**, 340–367 (2015)
19. A.-M. Ernvall-Hytönen, A. Odžak, M. Sušić, On asymptotic behaviour of generalized Li coefficients. Taiwanese J. Math. **22**(6), 1321–1346 (2018)
20. P. Freitas, A Li-type criterion for zero-free half-planes of Riemann's zeta function. J. Lond. Math. Soc. **73**, 399–414 (2006)
21. S. Gun, M.R. Murty, P. Rath, Transcendental sums related to the zeros of zeta functions. Mathematika **64**(3), 875–897 (2018)
22. H.-W. Huang, Ihara zeta function, coefficients of Maclaurin series and Ramanujan graphs. Int. J. Math. **31**(10), 2050082, 10 pp. (2020)
23. F. Johansson, Arb: a C library for ball arithmetic. ACM Commun. Comput. Algebra **47**(4), 166–169 (2013)
24. J. Jorgenson, S. Lang, On Cramér's theorem for general Euler products with functional equation. Math. Ann. **297**(3), 383–416 (1993)
25. J. Jorgenson, S. Lang, *Explicit Formulas for Regularized Products and Series*. Lecture Notes in Math., vol. 1593 (Springer, Berlin, Heidelberg 1994), pp. 1–134
26. J. Kaczorowski, *Axiomatic Theory of L-Functions: The Selberg Class*. Lecture Notes in Math., vol. 1891 (Springer, Berlin, Heidelberg, 2006), pp. 133–209
27. J. Kaczorowski, A. Perelli, On the prime number theorem for the Selberg class. Arch. Math. **80**(3), 255–263 (2003)
28. J.B. Keiper, Power series expansion of Riemann's ξ-function. Math. Comput. **58**, 765–773 (1992)
29. J.C. Lagarias, Li's coefficients for automorphic L-functions. Ann. Inst. Fourier **57**, 1689–1740 (2007)
30. X.-J. Li, The positivity of a sequence of numbers and the Riemann hypothesis. J. Number Theory **65**, 325–333 (1997)
31. X.-J. Li, Explicit formulas for Dirichlet and Hecke L-functions. Ill. J. Math **48**, 491–503 (2004)
32. X.-J. Li, An arithmetic formula for certain coefficients of the Euler product of Hecke polynomials. J. Number Theory **113**(1), 175–200 (2005)
33. X.-J. Li, An explicit formula for Hecke L-functions. Integers Electron. J. Combin. Number Theory **8**(2), #A8 (2008)
34. K. Maslanka, Li's criterion for the Riemann hypothesis - numerical approach. Opuscula Math. **24**(1), 103–114 (2004)
35. K. Mazhouda, The saddle-point method and the Li coefficients. Canad. Math. Bull. **54**(2), 316–329 (2011)
36. K. Mazhouda, On the Li coefficients for the Dirichlet L-functions. Monatsh. Math. **170**(3–4), 405–423 (2013)
37. K. Mazhouda, On the τ-Li coefficients for automorphic L-functions. Rocky Mountain J. Math. **47**(6), 1987–2011 (2017)
38. K. Mazhouda, B. Sodaïgui, The Li-Sekatskii coefficients for the Selberg class. Int. J. Math. **33**(12), Paper No. 2250075, 23 pp. (2022)
39. K. Mazhouda, L. Smajlović, On relations equivalent to the generalized Riemann hypothesis for the Selberg class. Funct. Approx. Comment. Math. **56**(1), 67–93 (2017)
40. K. Mazhouda, L. Smajlović, Evaluation of the Li coefficients on function fields and applications. Eur. J. Math. **5**, 540–550 (2019)
41. A. Odlyzko, Tables of zeros of the Riemann zeta function. http://www.dtc.umn.edu/~odlyzko//zetatables/index.html
42. A. Odžak, On the asymptotic criterion for the zero-free regions of certain L-functions. Turkish J. Math. **40**(3), 688–702 (2016)

43. A. Odžak, L. Smajlović, On Li's coefficients for the Rankin-Selberg L -functions. Ramanujan J. **21**, 303–334 (2010)
44. A. Odžak, L. Smajlović, On interpolation functions for generalized Li coefficients in the Selberg class. Int. J. Number Theory **7**, 771–792 (2011)
45. A. Odžak, L. Smajlović, On asymptotic behavior of generalized Li coefficients in the Selberg class. J. Number Theory **131**, 519–535 (2011)
46. A. Odžak, M. Zubača, On the discretized Li coefficients for a certain class of L-functions. Bull. Malays. Math. Sci. Soc. **44**, 3601–3627 (2021)
47. S. Omar, On a class of Epstein zeta functions. Tokyo J. Math. **40**(2), 339–351 (2017)
48. S. Omar, S. Bouanani, Li's criterion and the Riemann hypothesis for function fields. Finite Fields Appl. **16**, 477–485 (2010)
49. S. Omar, S. Bouanani, Li's criterion and the Riemann hypothesis for function fields II. Finite Fields Appl. **45**, 53–58 (2017)
50. S. Omar, K. Mazhouda, Le critère de Li et l'hypothèse de Riemann pour la classe de Selberg. J. Number Theory **125**(1), 50–58 (2007)
51. S. Omar, K. Mazhouda, Corrigendum et addendum à "Le critère de Li et l'hypothèse de Riemann pour la classe de Selberg" [J. Number Theory 125 (1) (2007) 50–58]. J. Number Theory **130**(4), 1109–1114 (2010)
52. S. Omar, K. Mazhouda, The Li criterion and the Riemann hypothesis for the Selberg class. II. J. Number Theory **130**(4), 1098–1108 (2010)
53. S. Omar, R. Ouni, A note on Li's coefficients for Hecke L-functions. Int. J. Number Theory **13**(7), 1747–1753 (2017)
54. S. Omar, R. Ouni, K. Mazhouda, On the Li coefficients for the Hecke L-functions. Math. Phys. Anal. Geom. **17**, 67–81 (2014)
55. N. Palojärvi, Explicit zero-free regions and a τ-Li-type criterion. Albanian J. Math. **14**(1), 47–77 (2020)
56. A. Perelli, A survey of the Selberg class of L-functions, part I. Milan J. Math. **73**, 1–28 (2004)
57. A. Perelli, A survey of the Selberg class of L-functions, part II. Riv. Mat. Univ. Parma (7) **3***, 83–118 (2004)
58. S.K. Sekatskii, Generalized Bombieri–Lagarias' theorem and generalized Li's criterion with its arithmetic interpretation. Ukrainian Math. J. **66**, 415–431 (2014)
59. A. Selberg, Old and new conjectures and results about a class of Dirichlet series, in *Proc. Amalfi Conf. Analytic Number Theory*, ed. by E. Bombieri et al. (Universitia di Salerno, 1992), pp. 367–385
60. L. Smajlović, On Li's criterion for the Riemann hypothesis for the Selberg class. J. Number Theory **130**, 828–851 (2010)
61. A. Voros, Sharpenings of Li's criterion for the Riemann hypothesis. Math. Phys. Anal. Geom. **9**, 53–63 (2006)
62. A. Voros, Discretized Keiper/Li approach to the Riemann hypothesis. Exp. Math. **29**(4), 452–469 (2020)

Real Quadratic Singular Moduli and p-Adic Families of Modular Forms

Paulina Fust, Judith Ludwig, Alice Pozzi, Mafalda Santos, and Hanneke Wiersema

1 Introduction

The goal of this chapter is to present recent developments in the theory of *singular moduli for real quadratic fields* introduced in [20] in a parallel perspective to the classical theory of complex multiplication. The theory of elliptic curves with complex multiplication has yielded some spectacular arithmetic applications. Kronecker tackled the problem of constructing abelian extensions of imaginary quadratic fields, known as the *Kronecker Jugendtraum*, via singular moduli, i.e., values of modular functions at imaginary quadratic points of the complex upper half-plane $\mathcal{H}_\infty$. While class field theory provides a description of the Galois groups classifying abelian extensions of arbitrary number fields, finding an explicit recipe for their generators is more elusive beyond the CM setting.

P. Fust
Universität Duisburg-Essen, Fakultät für Mathematik, Essen, Germany

J. Ludwig (✉)
Institute for Mathematics, Heidelberg University, Heidelberg, Germany
e-mail: judith.ludwig@iwr.uni-heidelberg.de

A. Pozzi
University of Bristol, School of Mathematics, Bristol, UK
e-mail: alice.pozzi@bristol.ac.uk

M. Santos
Department of Mathematics, Imperial College London, London, UK
e-mail: m.santos18@imperial.ac.uk

H. Wiersema
Department of Pure Mathematics and Mathematical Statistics, Centre for Mathematical Sciences, University of Cambridge, Cambridge, UK
e-mail: hw600@cam.ac.uk

R. Abdellatif et al. (eds.), *Women in Numbers Europe IV*, Association for Women in Mathematics Series 32, https://doi.org/10.1007/978-3-031-52163-8_9

Another problem for which in current approaches CM theory is an essential tool is the Birch and Swinnerton–Dyer Conjecture. Given an elliptic curve E over a number field K, the Birch and Swinnerton–Dyer Conjecture (BSD) predicts an equality between the rank of its group of K-rational points and the order of vanishing of its L-function $L(E/K, s)$ at $s = 1$. The only known approach to systematically producing global points is via the theory of Heegner points. The works of Gross–Zagier [22] and Kolyvagin [26] settle many instances of BSD in rank 1 by exploiting properties of this supply of global points.

The construction of singular moduli and Heegner points share some common features. They both arise from an appropriate evaluation process at CM points on the complex upper half-plane. The corresponding values are defined over abelian extensions of imaginary quadratic fields K. We will refer to this type of construction as *Heegner constructions*. It is natural to ask if a similar approach can be implemented to construct collections of objects lying in abelian extensions of other number fields.

Let K be a real quadratic field. From a naive perspective, the idea of evaluating modular functions at real quadratic points (or RM, by analogy with CM points) of $\mathcal{H}_\infty$ already fails because RM points lie on the real line. This suggests that an appropriate analogue of CM theory should be sought by replacing the role played by the Archimedean place with a finite place that does not split in K. This is the point of view proposed by Darmon and Vonk in their theory of singular moduli for real quadratic fields.

For a finite prime p, the Drinfeld p-adic upper half-plane is the rigid analytic space $\mathcal{H}_p$ whose $\mathbb{C}_p$-valued points are $\mathbb{P}^1(\mathbb{C}_p) \setminus \mathbb{P}^1(\mathbb{Q}_p)$. It contains RM points over real quadratic fields K in which p is ramified or inert. In analogy with CM theory, it would be natural to attempt to define singular moduli as values of suitable modular functions at RM points in $\mathcal{H}_p$, for some action of a modular group $\Gamma \subset \mathrm{GL}_2(\mathbb{Q}_p)$. Typically, one would expect this group to arise as the elements of norm one in a $\mathbb{Z}[1/p]$-order of a quaternion algebra $B/\mathbb{Q}$ that splits at p. On the one hand, if the quaternion algebra is definite, the quotient $\Gamma\backslash\mathcal{H}_p$ admits the structure of a rigid analytic space. However, in that setting, B does not contain any real quadratic subalgebra. Therefore one would not expect such quotients to appear in the framework of RM theory. On the other hand, if B is indefinite, it contains real quadratic subalgebras, but the action of Γ on $\mathcal{H}_p$ is not discrete, so that there are no suitable modular functions. Darmon and Vonk fix the indefinite quaternion algebra $B = \mathrm{M}_2(\mathbb{Q}_p)$, so that $\Gamma = \mathrm{SL}_2(\mathbb{Z}[1/p])$ is the Ihara group, and they define rigid meromorphic cocycles as classes in $H^1(\Gamma, \mathcal{M}^\times)$, where $\mathcal{M}$ is the ring of rigid meromorphic functions on $\mathcal{H}_p$. They show that these classes (or more precisely, the quasi-parabolic classes, see Sect. 3.3) can be meaningfully evaluated at RM points. They then conjecture that their RM values lie in composita of abelian extensions of real quadratic fields, making them suitable candidates for an analogue of singular moduli in the real quadratic setting. The theory of real quadratic singular moduli fits into a program launched by Darmon in [9] attempting to define conjectural analogues of Heegner constructions for real quadratic fields via p-adic methods.

Most notably, the construction of a putative analogue of Heegner points, known as Stark–Heegner points, has since seen many generalisations. While the theoretical evidence on the rationality of these points is scarce, the computational results are extensive.

In the last 50 years, new aspects of the theory of complex multiplication have emerged, in which automorphic methods are used substantially. An important theme is the study of modular generating series constructed out of CM cycles on Shimura curves, pioneered by the work of Gross and Zagier on the factorisation of differences of singular moduli [22]. The crucial idea of relating generating series to derivatives of real analytic families of Eisenstein series culminated in the proof of many cases of the Birch and Swinnerton–Dyer Conjecture for a modular elliptic curve in analytic rank 1 in [23] (see Theorem 9). Although the algebraicity properties of the analogues of the Heegner constructions in the real quadratic setting are conjectural, one can package RM cycles on $\mathcal{H}_p$ into modular generating series. It is natural to speculate that these generating series would be related to derivatives of p-adic families of modular forms. This perspective has been explored by Darmon, Pozzi and Vonk in [16] and [17].

In the Archimedean setting, only real analytic Eisenstein series admit variations in families. By contrast, the work of Hida and its generalisation via the theory of eigenvarieties introduced by Coleman–Mazur show that p-adic families of automorphic forms abound. A useful feature of the p-adic theory is that cusp forms vary in p-adic families. In addition, their corresponding Galois representations can be interpolated. This aspect has been leveraged into breakthroughs in Iwasawa theory and the theory of p-adic L-functions, for example, the proof of the Iwasawa Main Conjecture over totally real fields by Mazur–Wiles, or the proof of the Mazur–Tate–Teitelbaum Conjecture by Greenberg–Stevens.

In the context of modular generating series for RM values, the additional flexibility of the p-adic setting has been used by Darmon–Pozzi–Vonk to prove some instances of the expected algebraicity results for the analogues of elliptic units for real quadratic fields (see Theorem 11). Their approach is somewhat indirect. On the one hand, derivatives of certain p-adic families of modular forms are used to produce modular generating series for RM values. On the other hand, the derivatives of cuspidal families are related to global cohomology classes informing the arithmetic of abelian extensions of real quadratic fields. This strategy towards the algebraicity of Heegner constructions in the RM setting suggests that the richer theory of p-adic automorphic forms can sometimes make up for the lack of an a priori global description of the corresponding RM objects.

Overview of the Chapter This chapter is structured as follows. Section 2 contains a review of some aspects of the classical theory of complex multiplication, with an emphasis on singular moduli, elliptic units, and Heegner points. Section 3 focuses on the construction of their conjectural analogues for real quadratic fields. For the reader's convenience, we include some background on the Drinfeld p-adic upper half-plane as a rigid space and introduce the theory of rigid meromorphic cocycles. Sections 4 and 5 are devoted to the interplay between derivatives of families of modular forms and singular moduli, in the Archimedean and non-Archimedean setting, respectively. In particular, Sect. 4 treats Gross and Zagier's work on the factorisation of singular moduli and the BSD Conjecture in analytic rank 1 [22] and [23], while Sect. 5 focuses on the articles [16] and [17]. Sections 2 and 3 are meant to be mostly self-contained, while Sects. 4 and 5 aim to be an overview of various arithmetic applications to Heegner constructions in a unified framework.

2 Heegner Constructions for Imaginary Quadratic Fields

In this section, we will give a brief overview of results in the classical theory of elliptic curves with complex multiplication and its applications to the construction of arithmetic objects over abelian extensions of imaginary quadratic fields. At the heart of these results lies the fact that modular curves, quotients of the complex upper half-plane by congruence subgroups, admit a moduli interpretation that endows them with a rational structure. The moduli interpretation of CM points can be used to determine their images under field automorphisms, which allows one to deduce that they define global points on modular curves.

We will begin by recalling some facts about the moduli interpretation of CM points on modular curves and the Main Theorem of Complex Multiplication. We will then discuss in more detail the Heegner constructions of singular moduli, elliptic units, and Heegner points.

2.1 *Modular Curves and CM Points*

2.1.1 Elliptic Curves with Complex Multiplication

Let E be a complex elliptic curve, which we view as a Riemann surface. It admits a uniformisation $E(\mathbb{C}) \simeq \mathbb{C}/\Lambda$, where $\Lambda \subset \mathbb{C}$ is its period lattice, well-defined up to homothety; by rescaling, the latter can be chosen of the form $\Lambda = \tau\mathbb{Z}+\mathbb{Z}$, for $\tau \in \mathbb{C}$ with positive imaginary part. Endomorphisms of E are given by endomorphisms of its universal covering $\mathbb{C}$ that preserve the lattice Λ. Denote by $\mathcal{R}_E$ the set of endomorphisms of E. For a generic τ in the upper half-plane, all endomorphisms of E are given by multiplication by an integer. Otherwise, there exist $\alpha \in \mathbb{C} \setminus \mathbb{Z}$, and integers a, b, c, d satisfying

$$c\tau + d = \alpha, \qquad a\tau + b = \alpha\tau.$$

This forces $K = \mathbb{Q}(\tau)$ to be an imaginary quadratic field and α to be an algebraic integer in K. As a result $\mathcal{R}_E$ is isomorphic to an order O in K, for which the lattice

Λ is a proper fractional ideal. In this case we say that E has complex multiplication by $\mathcal{O}$.

2.1.2 The Main Theorem of Complex Multiplication

Let $\mathcal{H}_\infty$ be the Poincaré complex upper half-plane, on which the group $\mathrm{GL}_2^+(\mathbb{R})$ of invertible real matrices with positive determinant acts by Möbius transformation. For a positive integer N, consider the order

$$\mathrm{M}_0(N) = \left\{ \begin{pmatrix} a & b \\ c & d \end{pmatrix} \in \mathrm{M}_2(\mathbb{Z}) : N \mid c \right\}$$

in the ring of 2×2-matrices with integer coefficients, and let $\Gamma_0(N)$ be the group of matrices of determinant 1 in $M_0(N)$. The assignment $\tau \mapsto (\tau\mathbb{Z} + \mathbb{Z}, \tau\mathbb{Z} + 1/N\mathbb{Z})$ identifies the quotient $\Gamma_0(N)\backslash\mathcal{H}_\infty$ with the moduli space of pairs of lattices (Λ, Λ') with

$$\Lambda \subset \Lambda' \text{ such that } \Lambda'/\Lambda \simeq \mathbb{Z}/N\mathbb{Z},$$

defined up to homothety.

The moduli problem parametrising isomorphism classes of pairs of elliptic curves (E, E') with a cyclic degree N-isogeny $E \to E'$ admits a coarse moduli space $Y_0(N)$ over $\mathbb{Q}$ which we refer to as the (open) modular curve of the level $\Gamma_0(N)$. We let $X_0(N)$ be its compactification, obtained by adding cusps.

There is a bijection

$$\iota_N : \Gamma_0(N)\backslash\mathcal{H}_\infty \to Y_0(N)(\mathbb{C}), \qquad \tau \mapsto (\mathbb{C}/(\tau\mathbb{Z}+\mathbb{Z}),\ \mathbb{C}/(\tau\mathbb{Z}+1/N\mathbb{Z})),$$

which provides a uniformisation of $Y_0(N)(\mathbb{C})$ by the complex upper half-plane.

We say that $\tau \in \mathcal{H}_\infty$ is imaginary quadratic of discriminant D if $[\tau, 1]$ is a solution to a quadratic equation of the form $AX^2 + BXY + CY^2 = 0$ for some coprime integers A, B, C, and $D = B^2 - 4AC < 0$. For $\tau \in \mathcal{H}_\infty$ and $N \in \mathbb{Z}_{\geq 1}$, let

$$\mathcal{O}_\tau^{(N)} := \left\{ \begin{pmatrix} a & b \\ c & d \end{pmatrix} \in \mathrm{M}_0(N) : a\tau + b = c\tau^2 + d\tau \right\}.$$

The image of the inclusion $\mathcal{O}_\tau^{(N)} \hookrightarrow \mathbb{C}$ given by $\left(\begin{smallmatrix} a & b \\ c & d \end{smallmatrix}\right) \mapsto c\tau + d$ is either isomorphic to $\mathbb{Z}$ or to an order in the imaginary quadratic field $K = \mathbb{Q}(\tau)$. In the latter case, we say that τ is a CM point on $Y_0(N)$ and denote by $[\tau]$ its $\Gamma_0(N)$-orbit, when the level $\Gamma_0(N)$ is understood. For an order $\mathcal{O}$ in an imaginary quadratic field K of discriminant prime to N, we denote by $\mathcal{H}_\infty^{\mathcal{O}}$ the set of points $\tau \in \mathcal{H}_\infty$ with

$$\mathcal{O}_\tau^{(N)} = \mathcal{O}_\tau^{(1)} \cap \mathcal{O}_{N\tau}^{(1)} \simeq \mathcal{O}.$$

Such points are in bijection with the pairs of elliptic curves $(\mathbb{C}/\Lambda, \mathbb{C}/\Lambda')$ whose corresponding lattices (Λ, Λ') are contained in K (up to homothety) and for which the set

$$\{\alpha \in K : \alpha\Lambda \subset \Lambda,\ \alpha\Lambda' \subset \Lambda'\} \tag{1}$$

is equal to $\mathcal{O}$. The set $\mathcal{H}_\infty^{\mathcal{O}}$ is preserved by the action of $\Gamma_0(N)$ and the quotient $\Gamma_0(N)\backslash\mathcal{H}_\infty^{\mathcal{O}}$ is finite. We let $\Gamma_0(N)\backslash\mathcal{H}_\infty^{\mathrm{CM}}$ be the union of $\Gamma_0(N)$-orbits of points with complex multiplication by $\mathcal{O}$ for some imaginary quadratic order $\mathcal{O}$.

Denote by $\mathrm{Cl}(\mathcal{O})$ the class group of $\mathcal{O}$. Given a proper fractional ideal $\mathfrak{a}$ in $\mathcal{O}$ and a pair of lattices (Λ, Λ'), condition (1) is stable under multiplication by $\mathfrak{a}$. As a result

$$\mathfrak{a} * (\mathbb{C}/\Lambda, \mathbb{C}/\Lambda') := (\mathbb{C}/(\mathfrak{a} * \Lambda), \mathbb{C}/(\mathfrak{a} * \Lambda'))$$

defines an action of the class group $\mathrm{Cl}(\mathcal{O})$ on $\Gamma_0(N)\backslash\mathcal{H}_\infty^{\mathcal{O}}$. For $[\tau] \in \Gamma_0(N)\backslash\mathcal{H}_\infty^{\mathcal{O}}$, we write $\mathfrak{a} * \tau$ for $\mathfrak{a} * (\mathbb{C}/(\mathbb{Z} + \tau\mathbb{Z}), \mathbb{C}/(1/N\mathbb{Z} + \tau\mathbb{Z}))$ by abuse of notation.

The Main Theorem of Complex Multiplication ensures that CM points on modular curves are defined over suitable ring class fields and describes the Galois action on CM points. More precisely, let $\mathcal{O}$ be an order in an imaginary quadratic field K. The Artin reciprocity map yields an isomorphism

$$\mathrm{rec}\colon \mathrm{Cl}(\mathcal{O}) \to \mathrm{Gal}(H/K),$$

where H is the ring class field of $\mathcal{O}$. The Main Theorem of Complex Multiplication can be stated as follows.

Theorem 1 *Let $\mathcal{O}$ be an order of discriminant D in an imaginary quadratic field K, and let H be its corresponding ring class field. Let N be a positive integer coprime to D.*

For every point $[\tau] \in \Gamma_0(N)\backslash\mathcal{H}_\infty^{\mathcal{O}}$, we have:

(i) $\iota_N([\tau]) \in Y_0(N)(H)$.
(ii) (Shimura reciprocity law): Every $\mathfrak{a} \in \mathrm{Cl}(\mathcal{O})$ *satisfies* $\iota_N(\mathfrak{a} * [\tau]) = \mathrm{rec}(\mathfrak{a})^{-1} \iota_N([\tau])$.

For a reference, see [8, Cor. 11.37]. In Sect. 2.2, we will explain how Theorem 1 can be exploited towards the construction of global units and global points on elliptic curves over ring class fields of imaginary quadratic fields.

2.2 Heegner Constructions

Theorem 1 endows the modular curve $Y_0(N)$ with a systematic supply of global points. Given a group scheme $X/\mathbb{Q}$ together with a $\mathbb{Q}$-rational morphism

$Y_0(N) \rightarrow X$, one obtains a collection of global points on X, satisfying similar properties to those of CM points on modular curves. We refer to constructions of this flavour as *Heegner constructions.*

In the following examples, Heegner constructions are obtained from an a priori holomorphic map

$$\Phi\colon \Gamma_0(N)\backslash\mathcal{H}_\infty \rightarrow X(\mathbb{C})$$

of Riemann surfaces that is shown to arise from a $\mathbb{Q}$-rational morphism of curves (an important example being the modular parametrisation of an elliptic curve). As a consequence of Theorem 1, the following properties are satisfied:

(i) If $[\tau] \in \Gamma_0(N)\backslash\mathcal{H}_\infty^{\mathcal{O}}$, then $\Phi([\tau]) \in X(H)$.
(ii) The action of the class group $\mathrm{Cl}(\mathcal{O})$ is compatible with the Galois action on $X(H)$ under a Shimura reciprocity law.

Below we will give three examples of Heegner constructions. The following table summarises how these examples fit into this perspective.

	X	Φ
Singular moduli	Affine line $\mathbb{A}^1/\mathbb{Q}$	j-invariant
Elliptic units	Multiplicative group $\mathbb{G}_m/\mathbb{Q}$	Modular unit
Heegner points	Elliptic curve $E/\mathbb{Q}$	Modular parametrisation

The setting of singular moduli is obtained by specialising Theorem 1 to the case of level $\Gamma_0(1)$. Elliptic units are constructed from the evaluation of a modular unit at CM points. Roughly speaking, Heegner points are obtained by projecting CM cycles on the Jacobian of the modular curve onto its f-isotypic component for a newform f of weight 2.

Remark 1 The constructions presented in Sect. 3 will mimic the ones above by replacing $\mathcal{H}_\infty$ with the Drinfeld p-adic upper half-plane $\mathcal{H}_p$ and imaginary quadratic with real quadratic points. The Heegner constructions in the RM setting are obtained via techniques of rigid analytic geometry and are expected to satisfy similar properties to those of classical Heegner constructions. However, there are significant differences. The analogues of the map Φ described above will only be defined at real multiplication points. More crucially, the RM constructions are purely local, and no analogues of the modular curve are available in that setting.

We will now proceed to describe the inputs of these constructions in some detail.

2.2.1 Singular Moduli

Given a lattice Λ in $\mathbb{C}$, the theory of elliptic curves over the complex numbers yields an identification between the complex torus $\mathbb{C}/\Lambda$ and the points of an elliptic curve with equation

$$Y^2 = 4X^3 - g_2(\Lambda)X - g_3(\Lambda),$$

where the constants are defined as

$$g_2(\Lambda) = 60 \sum_{\omega \in \Lambda - \{0\}} \frac{1}{\omega^4}, \qquad g_3(\Lambda) = 140 \sum_{\omega \in \Lambda - \{0\}} \frac{1}{\omega^6}.$$

For $i = 2, 3$, and τ in the upper half-plane, the assignment $g_i(\tau) := g_i(\mathbb{Z} + \tau\mathbb{Z})$ defines a modular form of weight $2i$ and level $\mathrm{SL}_2(\mathbb{Z})$. The complex functions

$$\Delta := g_2^3 - 27g_3^2, \qquad j := 1728\frac{g_2^3}{\Delta}$$

are the *modular discriminant* and the *modular j-function*, respectively. The former is a nowhere vanishing cusp form of weight 12. The latter is a holomorphic $\mathrm{SL}_2(\mathbb{Z})$-invariant function on $\mathcal{H}_\infty$. The map

$$j \colon \mathrm{SL}_2(\mathbb{Z})\backslash\mathcal{H}_\infty \to \mathbb{A}^1(\mathbb{C}) \tag{2}$$

extends to an isomorphism of Riemann surfaces between the compactifications of $\mathrm{SL}_2(\mathbb{Z})\backslash\mathcal{H}_\infty$ and $\mathbb{A}^1(\mathbb{C})$, respectively. The crucial property of the j-invariant is that it characterises elliptic curves up to isomorphism over an algebraically closed field and captures the minimal field of definition of an elliptic curve. As a result, the coarse moduli space for elliptic curves $Y_0(1)$ can be identified with the affine line via (2). For $N \geq 1$, the image of the morphism

$$\Gamma_0(N)\backslash\mathcal{H}_\infty \to \mathbb{A}^2(\mathbb{C})\,, \qquad \tau \mapsto (j(\tau), j(N\tau))$$

is cut out by a single equation given by $\varphi_N \in \mathbb{Q}[X, Y]$, referred to as the N-th modular polynomial [18, Section 7.5]. The corresponding plane curve is birational to the modular curve $X_0(N)$.

Definition 1 *Singular moduli* are values of the j-function at points in $\mathrm{SL}_2(\mathbb{Z})\backslash\mathcal{H}_\infty^{\mathrm{CM}}$.

For $N = 1$, Theorem 1 directly translates into properties of singular moduli. The general level case easily follows from the level 1 setting, together with properties of the modular polynomials. An important application is the following corollary.

Corollary 1 *Let $D < 0$ be a fundamental discriminant, let $\tau \in \mathcal{H}_\infty$ be a point of discriminant D, and denote $K = \mathbb{Q}(\tau)$. Then the Hilbert class field of K is $K(j(\tau))$.*

This result (and its analogues for non-fundamental discriminants) can be interpreted as providing explicit generators for certain abelian extensions of imaginary quadratic fields. More generally, Hilbert's Twelfth Problem consists in finding explicit formulae for generators of all abelian extensions of number fields. The motivating example is the Kronecker–Weber Theorem. It states that all abelian extensions of $\mathbb{Q}$ can be obtained by adjoining roots of unity, which can be viewed as values of the function $z \mapsto e^{2\pi i z}$ at rational arguments. However, for an imaginary quadratic field K, singular moduli do not suffice to generate the maximal abelian extension, as one would also need to consider values of coordinates of torsion points of elliptic curves with CM by orders in K (see [31, Chp. II, Cor. 5.7]).

2.2.2 Elliptic Units

Singular moduli give generators of certain abelian extensions of imaginary quadratic fields. For arithmetic applications, it can be useful to construct generators with controlled integrality properties, by replacing the j-function with suitable units on modular curves of higher level.

A modular unit u is a nowhere vanishing function on the open modular curve $Y_0(N)$ extending meromorphically to the cusps. Examples of modular units can be constructed as products of pullbacks of the modular discriminant via projection maps from $Y_0(N)$ to the modular curve of level 1. For a fixed level N, let Σ_N be the set of divisors of N and consider a degree zero formal linear combination $\alpha = \sum_{d \in \Sigma_N} m_d[d]$. Then the holomorphic function

$$u_\alpha(z) := \prod_d \Delta(dz)^{m_d}$$

is $\Gamma_0(N)$-invariant because α has degree 0 and nowhere vanishing because Δ is. The modular unit u_α can be viewed as a morphism from the open modular curve of level $\Gamma_0(N)$ to the multiplicative group.

Definition 2 Given a divisor α as above, *elliptic units* are values of u_α at points in $\Gamma_0(N)\backslash\mathcal{H}_\infty^{\mathrm{CM}}$.

Theorem 1 implies that elliptic units are defined over ring class fields of imaginary quadratic fields. However, a careful study of the integrality of modular units (cfr. [25, Chp. 1, Lemma 1.1]) yields the following result.

Theorem 2 *Let $\mathcal{O}$ be an order in an imaginary quadratic field K, of discriminant prime to N. Let u_α be a modular unit of level $\Gamma_0(N)$ and let $\tau \in \Gamma_0(N)\backslash\mathcal{H}_\infty^{\mathcal{O}}$. Then*

$$u_\alpha(\tau) \in \mathcal{O}_H[1/N]^\times,$$

where $\mathcal{O}_H$ is the ring of integers of the ring class field H of $\mathcal{O}$.

Remark 2 In fact, the values $u_\alpha(\tau)$ for $\tau \in \Gamma_0(N)\backslash\mathcal{H}_\infty^{\mathcal{O}}$ can be used to produce genuine units in the ring class field of $\mathcal{O}$. More precisely, one can show that for every $\sigma \in \mathrm{Gal}(H/K)$, we have

$$(\sigma - 1)u_\alpha(\tau) \in \mathcal{O}_H^\times.$$

2.2.3 Heegner Points

The construction of Heegner points on an elliptic curve $E/\mathbb{Q}$ relies crucially on the *modularity* of the elliptic curve E. We briefly recall this notion.

Let $f = \sum_{n\geq 1} a_n q^n$ be a newform of weight 2 and level $\Gamma_0(N)$ for some $N \geq 1$; suppose that its Fourier coefficients are integers. The Eichler–Shimura construction attaches to f an elliptic curve, as we now recall. Let $\mathbb{T}_2(\Gamma_0(N))$ be the Hecke algebra acting on cusp forms of weight 2 and level $\Gamma_0(N)$ over $\mathbb{Z}$. Let I_f be the ideal given by the kernel of the ring morphism $\mathbb{T}_2(\Gamma_0(N)) \to \mathbb{Z}$ sending the Hecke operator T_n to a_n. Let $J_0(N)$ be the Jacobian of the modular curve $X_0(N)$ of level $\Gamma_0(N)$. It is an abelian variety of dimension equal to the genus of $X_0(N)$, endowed with an action of $\mathbb{T}_2(\Gamma_0(N))$. The quotient

$$E_f := J_0(N)/I_f J_0(N)$$

is the elliptic curve attached to f. The composition of the morphism $X_0(N) \to J_0(N)$ given by $P \mapsto (P) - (\infty)$ with the projection $J_0(N) \to E_f$ yields a surjective $\mathbb{Q}$-rational morphism

$$\pi_{E_f} \colon X_0(N) \to E_f.$$

Definition 3 Given an elliptic curve E, we say that E admits a modular parametrisation if there exists a surjective $\mathbb{Q}$-rational morphism $\pi_E \colon X_0(N) \to E$. When this is the case, we say that E is *modular*.

The work of Wiles [34], Taylor–Wiles [33] on Fermat's Last Theorem, and its generalisations by Breuil, Conrad, Diamond, and Taylor [2] culminated in a full proof of the following theorem, formerly known as the Shimura–Taniyama Conjecture.

Theorem 3 *Let $E/\mathbb{Q}$ be an elliptic curve. Then E is modular.*

Remark 3 For an alternative formulation of modularity via L-functions, see [12].

Definition 4 Let E be a modular elliptic curve, and let $\pi_E \colon X_0(N) \to E$ be its modular parametrisation. *Heegner points* on E are images of points in $\Gamma_0(N)\backslash\mathcal{H}_\infty^{\mathrm{CM}}$ under π_E.

Theorem 1 implies that Heegner points are defined over ring class fields of imaginary quadratic fields. Their importance in relation to the Birch and Swinnerton–Dyer conjecture will be discussed in Sect. 4.2.

3 Heegner Constructions for Real Quadratic Fields

Given the importance of singular moduli and more general Heegner constructions for algebraic number theory, it is desirable to have an analogue for real quadratic fields. While no direct analogue is available for real quadratic fields, a (conjectural) theory of "real multiplication" relying on p-adic methods was recently proposed by Darmon and Vonk in [20]. In this section we will present the theory of *rigid meromorphic cocycles*, which is the main new tool in this approach. A rigid meromorphic cocycle is a class in the first cohomology of the Ihara group $\mathrm{SL}_2(\mathbb{Z}[1/p])$ acting on the non-zero rigid meromorphic functions on the Drinfeld p-adic upper half-plane $\mathcal{H}_p$. These rigid meromorphic cocycles can be evaluated at real quadratic points of $\mathcal{H}_p$. Conjecturally, their values display striking similarities with the classical singular moduli. In a similar vein, the theory of analytic theta cocycles, obtained by considering cocycles for the Ihara group with values in analytic functions on $\mathcal{H}_p$ defined up to scalars, can be used to reinterpret constructions of Darmon–Dasgupta [10] and Darmon [9] of conjectural analogues of elliptic units and Heegner points for real quadratic fields.

Before introducing rigid meromorphic cocycles, we will review the relevant background on the Drinfeld p-adic upper half-plane.

3.1 *Drinfeld p-Adic Upper Half-Plane*

In search for an analogue of singular moduli for real quadratic fields, the first obstacle one runs into is the absence of real quadratic points on the complex upper half-plane. However, there is a p-adic analogue of the complex upper half-plane containing many real quadratic points, the *Drinfeld upper half-plane*, which we briefly describe. The Drinfeld upper half-plane $\mathcal{H}_p$ is a rigid analytic space,[1] whose $\mathbb{C}_p$-points are given by

$$\mathcal{H}_p = \mathbb{P}^1(\mathbb{C}_p) \setminus \mathbb{P}^1(\mathbb{Q}_p).$$

In the following, we abuse notation and write $\mathcal{H}_p$ for the rigid space as well as its $\mathbb{C}_p$-points. The space $\mathcal{H}_p$ has an admissible covering by an increasing sequence of open affinoid subdomains $\mathcal{H}_p^{\leq n}$ that are constructed by removing balls of decreasing

[1] For an introduction to rigid analytic geometry, see, e.g., [7, 4].

radius around all $\mathbb{Q}_p$-rational points. More explicitly, for each $z \in \mathbb{P}^1(\mathbb{C}_p)$, we fix homogeneous coordinates $z = [z_0 : z_1]$ such that $\max\{|z_0|, |z_1|\}=1$. Then

$$\mathcal{H}_p^{\leq n} = \{[z_0 : z_1] \in \mathbb{P}^1(\mathbb{C}_p) \mid \mathrm{ord}_p(x_0 z_1 - x_1 z_0) \leq n \ \forall [x_0 : x_1] \in \mathbb{P}^1(\mathbb{Q}_p)\}.$$

For example, when $n = 0$, the open affinoid $\mathcal{H}_p^{\leq 0}$ consists of the points $[z_0 : z_1] \in \mathbb{P}^1(\mathbb{C}_p)$ such that $\mathrm{ord}_p(x_0 z_1 - x_1 z_0) \leq 0$ for any $[x_0 : x_1] \in \mathbb{P}^1(\mathbb{Q}_p)$. Hence, $\mathcal{H}_p^{\leq 0}$ is obtained from $\mathbb{P}^1(\mathbb{C}_p)$ by removing balls of radius 1 around the $\mathbb{Q}_p$-rational points. The latter are precisely the points that are mapped to the $\mathbb{F}_p$-rational points under the projection map $\pi : \mathbb{P}^1(\mathbb{C}_p) \to \mathbb{P}^1(\overline{\mathbb{F}}_p)$, so that

$$\mathcal{H}_p^{\leq 0} = \mathbb{P}^1(\mathbb{C}_p) \setminus \pi^{-1}(\mathbb{P}^1(\mathbb{F}_p)).$$

For convenience, let us include the following ad hoc description of the rigid analytic functions on $\mathcal{H}_p$, which the reader who is unfamiliar with rigid analytic geometry can take as a definition.

Definition 5

- A *rigid analytic function* on $\mathcal{H}_p^{\leq n}$ is a limit with respect to uniform convergence of rational functions on $\mathbb{P}^1(\mathbb{C}_p)$ with poles in $\mathbb{P}^1(\mathbb{C}_p) \setminus \mathcal{H}_p^{\leq n}$.
- A *rigid analytic function* on $\mathcal{H}_p$ is a function $f : \mathcal{H}_p \to \mathbb{C}_p$, such that the restriction of f to $\mathcal{H}_p^{\leq n}$ is a rigid analytic function for every $n \geq 0$.
- A *rigid meromorphic function* on $\mathcal{H}_p$ is a ratio $f = \frac{g}{h}$ of rigid analytic functions g and h with $h \neq 0$.

We denote by $\mathcal{A}$ the set of rigid analytic functions on $\mathcal{H}_p$ and by $\mathcal{M}$ the set of rigid meromorphic functions.

The key structural properties of the rigid space $\mathcal{H}_p$ are that it is a smooth 1-dimensional Stein space and that its algebra of rigid analytic functions $\mathcal{A}$ is a reflexive Fréchet space.

The group $\mathrm{GL}_2(\mathbb{Q}_p)$ acts on $\mathcal{H}_p$ by Möbius transformations, i.e., for $\gamma = \left(\begin{smallmatrix} a & b \\ c & d \end{smallmatrix}\right) \in \mathrm{GL}_2(\mathbb{Q}_p)$ and $z \in \mathcal{H}_p$, the action of γ on z is given by

$$\gamma \cdot z = \frac{az+b}{cz+d}.$$

As the centre of $\mathrm{GL}_2(\mathbb{Q}_p)$ acts trivially, this also defines an action of $\mathrm{PGL}_2(\mathbb{Q}_p)$.

3.1.1 Bruhat–Tits Tree for $\mathrm{PGL}_2(\mathbb{Q}_p)$

A useful tool for studying the Drinfeld upper half-plane is the Bruhat–Tits tree $\mathcal{T}$ attached to the group $\mathrm{PGL}_2(\mathbb{Q}_p)$. We recall its construction.

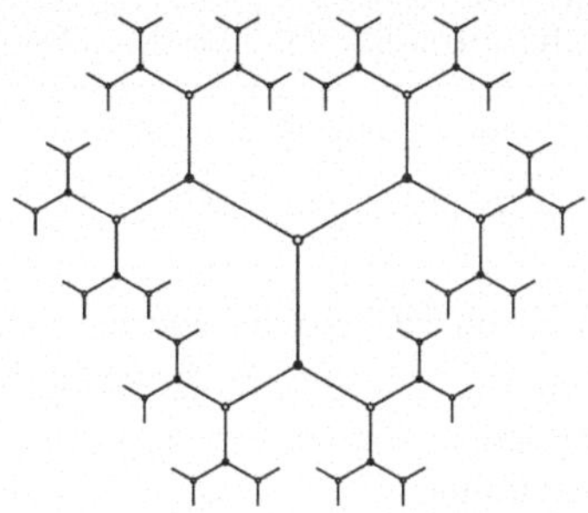

Fig. 1 The Bruhat–Tits tree of $\mathrm{PGL}_2(\mathbb{Q}_2)$

We call two $\mathbb{Z}_p$-lattices Λ and Λ' in $\mathbb{Q}_p^2$ equivalent if they are homothetic. This defines an equivalence relation on the set of all $\mathbb{Z}_p$-lattices in $\mathbb{Q}_p^2$. The set of vertices of the tree $\mathcal{T}$ is defined to be the set of equivalence classes of lattices:

$$V(\mathcal{T}) := \{\Lambda \subseteq \mathbb{Q}_p^2 \ \mathbb{Z}_p\text{-lattice}\}/\sim .$$

By definition two vertices $v, v' \in V(\mathcal{T})$ are connected by an edge in the graph $\mathcal{T}$ if there are representatives $v = [\Lambda]$ and $v' = [\Lambda']$, such that $p\Lambda \subsetneq \Lambda' \subsetneq \Lambda$. One can show that the graph defined by this is a homogeneous tree of degree $p+1$. For example, Fig. 1 shows a part of the Bruhat–Tits tree for $p = 2$.

The group $\mathrm{GL}_2(\mathbb{Q}_p)$ acts transitively on the set of all $\mathbb{Z}_p$-lattices in $\mathbb{Q}_p^2$, preserving the equivalence classes. This induces a transitive action of $\mathrm{GL}_2(\mathbb{Q}_p)$ and of $\mathrm{PGL}_2(\mathbb{Q}_p)$ on $\mathcal{T}$.

Example 1 The lattice $\mathbb{Z}_p^2$ gives rise to a vertex called the *standard vertex* $v_0 = [\mathbb{Z}_p^2]$. The adjacent vertices of v_0 are given by

$$\begin{pmatrix} 0 & -1 \\ 1 & 0 \end{pmatrix} \begin{pmatrix} 1 & 0 \\ 0 & p \end{pmatrix} v_0 \text{ and } \begin{pmatrix} 1 & 0 \\ \lambda & 1 \end{pmatrix} \begin{pmatrix} 1 & 0 \\ 0 & p \end{pmatrix} v_0 \text{ for } \lambda \in \mathbb{F}_p.$$

Moreover, if we consider the action of $\mathrm{SL}_2(\mathbb{Z}[1/p])$ on $\mathcal{T}$, the stabiliser of the standard vertex is the group $\mathrm{SL}_2(\mathbb{Z})$ and the stabiliser of the edge e_0 from v_0 to $\left(\begin{smallmatrix} 1 & 0 \\ 0 & p \end{smallmatrix}\right) v_0$ is the congruence subgroup $\Gamma_0(p)$.

There is the so-called reduction map

$$\mathrm{red} : \mathcal{H}_p \to |\mathcal{T}|$$

from the p-adic upper half-plane to the geometric realisation of $\mathcal{T}$. We refer to [19] for details regarding its construction. Via this map, the tree $\mathcal{T}$ serves as a nice combinatorial tool to study the Drinfeld upper half-plane. Let us indicate a few features of the reduction map. It is equivariant for the action of $\mathrm{PGL}_2(\mathbb{Q}_p)$.

Moreover, we can recover the affinoids $\mathcal{H}_p^{\leq n}$ in the admissible covering described above as inverse images

$$\mathcal{H}_p^{\leq n} = \mathrm{red}^{-1}(|\mathcal{T}^{\leq n}|),$$

where $\mathcal{T}^{\leq n}$ is the subtree of $\mathcal{T}$ consisting of all vertices and edges that have distance $\leq n$ to the standard vertex $v_0 = [\mathbb{Z}_p^2]$. Finally, the reduction map allows us to view $\mathcal{H}_p$ as a tubular neighbourhood of the Bruhat–Tits tree. For example, the inverse image of the edge e_0 is the annulus

$$\mathrm{red}^{-1}(e_0) = \{z \in \mathbb{C}_p | 1 < |z| < p\}.$$

3.1.2 RM Points on the Drinfeld p-Adic Upper Half-Plane

We introduce some notation for real quadratic points on $\mathcal{H}_p$ that will later be used to define the evaluation of rigid meromorphic cocycles.

Let $\tau \in \mathbb{C}_p$. We say that τ is a real quadratic point of discriminant D if $[\tau : 1]$ satisfies the homogeneous quadratic equation $AX^2 + BXY + CY^2 = 0$ with $A, B, C \in \mathbb{Z}$ coprime integers satisfying $D = B^2 - 4AC > 0$, with D not a square.

Then $K = \mathbb{Q}(\tau)$ is a real quadratic extension of $\mathbb{Q}$, and the point $[\tau : 1] \in \mathbb{P}^1(\mathbb{C}_p)$ lies in $\mathcal{H}_p$ if and only if the prime p does not split in K. We refer to points arising in this way as *real multiplication* points or RM points of $\mathcal{H}_p$ and denote the set of these points as $\mathcal{H}_p^{\mathrm{RM}}$. For $\tau \in \mathcal{H}_p^{\mathrm{RM}}$, define

$$\mathcal{O}_\tau := \left\{ \begin{pmatrix} a & b \\ c & d \end{pmatrix} \in \mathrm{M}_2(\mathbb{Z}[1/p]) : a\tau + b = c\tau^2 + d\tau \right\}.$$

Then $\mathcal{O}_\tau$ is isomorphic to a $\mathbb{Z}[1/p]$ order in the real quadratic field $K = \mathbb{Q}(\tau)$ via the inclusion

$$\mathcal{O}_\tau \hookrightarrow K, \quad \gamma = \begin{pmatrix} a & b \\ c & d \end{pmatrix} \mapsto c\tau + d.$$

Let $\Gamma := \mathrm{SL}_2(\mathbb{Z}[1/p])$ denote the Ihara group. Let $\mathcal{O}_{\tau,1}^\times$ denote the elements in $\mathcal{O}_\tau^\times$ of reduced norm one. The Dirichlet S-unit theorem implies that the stabiliser

$$\mathrm{Stab}_\Gamma(\tau) \cong \mathcal{O}_{\tau,1}^\times \cong \{\pm 1\} \times \langle \gamma_\tau \rangle$$

is up to torsion a free group of rank 1. It is generated by a fundamental unit γ_τ of norm 1, called the *automorph* of τ. Moreover, the points of $\mathcal{H}_p^{\mathrm{RM}}$ are precisely those for which the stabiliser in Γ has rank 1. This feature is crucial to the evaluation of the rigid meromorphic cocycles defined in the subsequent sections. We note that the action of Γ on $\mathcal{H}_p^{\mathrm{RM}}$ need not preserve the discriminant, but it does preserve its prime-to-p part.

3.2 Group Actions, Curves, and p-Adic Theta Functions

We review the theory of p-adic theta functions arising in the context of p-adic uniformisation of Shimura curves. While the construction of rigid cocycles does not fit in this setting, the cocycles arising in Sect. 3.3 are obtained via similar methods, after an appropriate degree shift in cohomology.

Let $\Gamma_0 \subset \mathrm{SL}_2(\mathbb{Q}_p)$ be any subgroup. We say that Γ_0 acts discretely on $\mathcal{H}_p$ if the orbit $\Gamma_0\tau$ of each element $\tau \in \mathcal{H}_p$ intersects each of the affinoids $\mathcal{H}_p^{\leq n}$ only in finitely many points.

Note that the groups $\mathrm{SL}_2(\mathbb{Z})$, $\mathrm{SL}_2(\mathbb{Z}[1/p])$, and $\mathrm{SL}_2(\mathbb{Z}_p)$ all act non-discretely on $\mathcal{H}_p$. To see this, recall that $\mathrm{SL}_2(\mathbb{Z})$ is contained in the stabiliser of the standard vertex v_0 of the Bruhat–Tits tree and that any element $\tau \in \mathcal{H}_p^{\leq 0}$ is mapped to v_0 via the reduction map. So the whole orbit $\mathrm{SL}_2(\mathbb{Z})\tau$ is contained in $\mathcal{H}_p^{\leq 0} = \mathrm{red}^{-1}(\mathcal{T}^{\leq 0})$, and the set $\mathrm{SL}_2(\mathbb{Z})\tau \cap \mathcal{H}_p^{\leq 0}$ contains infinitely many elements.

The upper half-plane $\mathcal{H}_p$ is of major significance for arithmetic geometry as it can be used to construct families of algebraic curves (the so-called Mumford curves) that uniformise certain Shimura curves p-adically. These curves arise as quotients of $\mathcal{H}_p$ by discrete group actions. We refer to [19, §2.4] and references therein for details. Here we only mention the following class of examples.

Example 2 Let B be the definite quaternion algebra $\mathbb{Q}[i, j] \subset \mathbb{H}$ inside the Hamilton quaternions and let $R = \mathbb{Z}[i, j, \frac{i+j+k+1}{2}]$ be a maximal order. Choose an odd prime p at which B splits and consider the group $\Gamma_B = (R[1/p])_1^\times$ of units $\gamma \in R[1/p]^\times$ of reduced norm 1. After choosing an isomorphism $(B \otimes_{\mathbb{Q}} \mathbb{Q}_p)^\times \cong \mathrm{GL}_2(\mathbb{Q}_p)$, the group Γ_B embeds into $\mathrm{SL}_2(\mathbb{Q}_p)$ and acts discretely on $\mathcal{H}_p$. Let $\Gamma' \subset \Gamma_B$ be a subgroup of finite index that has no fixed points on $\mathcal{H}_p$. Then the quotient space $X_{\Gamma'} = \Gamma' \backslash \mathcal{H}_p$ is a rigid analytic space that admits a model as a Shimura curve.

Note that a definite quaternion algebra does not admit an embedding of a real quadratic field. Essentially due to this fact the quotients $X_{\Gamma'}$ are not so relevant for a theory of real multiplication. As we will see below, a better setup for these purposes is to consider the indefinite quaternion algebra $B = \mathrm{M}_2(\mathbb{Q})$ and the (non-discrete) actions of groups like $\mathrm{SL}_2(\mathbb{Z})$ and $\mathrm{SL}_2(\mathbb{Z}[1/p])$ on $\mathcal{H}_p$.

Nevertheless, we want to briefly discuss one aspect from the theory of Mumford curves, the construction of p-adic theta functions. These are important tools in the study of the Jacobian of the compactification of a Mumford curve. Key ideas of their construction can be modified to construct interesting functions also in situations where a non-discrete group action is involved. Assume that $\Gamma_0 \subset \mathrm{SL}_2(\mathbb{Q}_p)$ is a subgroup that acts discretely and without fixed points on $\mathcal{H}_p$. The p-adic theta functions for Γ_0 are meromorphic functions on $\mathcal{H}_p$, which are invariant under Γ_0 up to multiplicative scalars and which can be constructed as follows.

For an element $w \in \mathbb{P}^1(\mathbb{C}_p)$, define the rational functions t_w on $\mathbb{P}^1(\mathbb{C}_p)$ by

$$t_w(z) = \begin{cases} z - w, & \text{if } |w| \le 1, \\ z/w - 1, & \text{if } |w| > 1, \\ 1, & \text{if } w = \infty. \end{cases}$$

Note that for two distinct elements w^+, w^- in $\mathcal{H}_p$ and $\gamma \in \Gamma_0$ the quotient function

$$\frac{t_{w^+}(\gamma z)}{t_{w^-}(\gamma z)}$$

has a simple zero at $\gamma^{-1}w^+$ and a simple pole at $\gamma^{-1}w^-$, as does the function $\frac{t_{\gamma^{-1}w^+}(z)}{t_{\gamma^{-1}w^-}(z)}$. Therefore, they are equal up to a constant.

Now consider the degree zero divisor $\Delta = w^+ - w^-$. Then the product

$$f_\Delta(z) = \prod_{\gamma \in \Gamma_0} \frac{t_{\gamma w^+}(z)}{t_{\gamma w^-}(z)}$$

converges to a rigid meromorphic function on $\mathcal{H}_p$. (For more details, see Section 2.2 in [20].) By the observation above, the function f_Δ is Γ_0-invariant up to multiplication by a constant in $\mathbb{C}_p$. Hence, it defines an element of $(\mathcal{M}^\times/\mathbb{C}_p^\times)^{\Gamma_0} = H^0(\Gamma_0, \mathcal{M}^\times/\mathbb{C}_p^\times)$. The function f_Δ is called the *p-adic theta function* associated to the degree zero divisor Δ.

3.3 *Rigid Meromorphic Cocycles*

One way of looking at the j-function is as an $\mathrm{SL}_2(\mathbb{Z})$-invariant function on the complex upper half-plane $\mathcal{H}_\infty$, i.e., an element of $H^0(\mathrm{SL}_2(\mathbb{Z}), \mathrm{Hol}(\mathcal{H}_\infty))$, where $\mathrm{Hol}(\mathcal{H}_\infty)$ denotes the holomorphic functions on $\mathcal{H}_\infty$. Now consider $\Gamma := \mathrm{SL}_2(\mathbb{Z}[1/p])$ acting on $\mathcal{H}_p$. In search for direct analogues of j in the p-adic setting, one might first naturally want to consider the groups $H^0(\Gamma, \mathcal{A})$ or $H^0(\Gamma, \mathcal{M})$. However, it follows essentially from the Weierstrass preparation theorem that any Γ-invariant rigid analytic function must be constant (cf. [20, Lemma 1.9]); hence, $H^0(\Gamma, \mathcal{A}) \cong \mathbb{C}_p \cong H^0(\Gamma, \mathcal{M})$, and so these groups do not contain interesting functions. Considering the corresponding cohomology groups in degree one on the other hand turns out to be a successful strategy, even more so if one furthermore switches to the multiplicative theory. This leads to the following definition:

Definition 6

- A *rigid meromorphic* (resp., *analytic*) *cocycle* is an element in $H^1(\Gamma, \mathcal{M}^\times)$ (resp., in $H^1(\Gamma, \mathcal{A}^\times)$).
- A *rigid meromorphic* (resp., *analytic*) *theta cocycle* is an element in $H^1(\Gamma, \mathcal{M}^\times/\mathbb{C}_p^\times)$ (resp., in $H^1(\Gamma, \mathcal{A}^\times/\mathbb{C}_p^\times)$).

Let $\Gamma_\infty \subset \Gamma$ be the subgroup of upper triangular matrices. A rigid meromorphic cocycle is called *parabolic* if its restriction to Γ_∞ is trivial. It is called *quasi-parabolic* if its restriction to Γ_∞ lies in the image of $H^1(\Gamma_\infty, \mathbb{C}_p^\times)$. The groups of such cohomology classes are denoted by $H^1_{\mathrm{par}}(\Gamma, \mathcal{M}^\times)$ and $H^1_f(\Gamma, \mathcal{M}^\times)$, respectively. One can show that any cohomology class in $H^1_f(\Gamma, \mathcal{M}^\times)$ has a unique representative, whose values on Γ_∞ consist of constant functions.

Remark 4 One reason for considering the multiplicative group $\mathcal{M}^\times$ as coefficients rather than the additive group $\mathcal{M}$ is the connection to the theory of modular symbols and the so-called rigid meromorphic period functions. More precisely, the subgroup $H^1_{\mathrm{par}}(\Gamma, \mathcal{M}^\times)$ can be identified with the group of Γ-invariant modular symbols $\mathcal{MS}^\Gamma(\mathcal{M}^\times)$ [20, Section 1]. This connection to modular symbols is useful for establishing structural results. It also allows one to produce many interesting elements in $H^1(\Gamma, \mathcal{M}^\times)$. Contrary to this, the group $H^1(\Gamma, \mathcal{M})$ contains no *parabolic* elements [20, Remark 1.8].

In practice, it is easier to write down examples of theta cocycles (we will give a few of these below). Note that there is a natural map $H^1(\Gamma, \mathcal{M}^\times) \to H^1(\Gamma, \mathcal{M}^\times/\mathbb{C}_p^\times)$, and one would like to lift theta cocycles to rigid meromorphic cocycles. Unfortunately, this is in general not possible as there is an obstruction to lifting. What is possible though is to lift the restriction of a theta cocycle to the group $\mathrm{SL}_2(\mathbb{Z})$ to an element in $H^1(\mathrm{SL}_2(\mathbb{Z}), \mathcal{M}^\times)$, as $H^2(\mathrm{SL}_2(\mathbb{Z}), \mathbb{C}_p^\times) = 0$. In some situations, this is good enough.

3.3.1 Examples

Let us discuss some examples.

Example 3 (The Trivial Cocycle) Fix a base point $\xi \in \mathbb{P}^1(\mathbb{Q}_p)$. For any $\gamma \in \Gamma$, define the rational function

$$J_{\mathrm{triv}}(\gamma)(z) := \frac{z - \gamma\xi}{z - \xi}.$$

Up to a scalar, this function is uniquely determined as the rational function with divisor $(\gamma\xi) - (\xi)$. Then for two elements $\gamma_1, \gamma_2 \in \Gamma$, the function $J_{\mathrm{triv}}(\gamma_1\gamma_2)$ is the function with divisor $(\gamma_1\gamma_2\xi) - (\xi) = (\gamma_1\gamma_2\xi) - (\gamma_1\xi) + (\gamma_1\xi) - (\xi)$. Hence up to a scalar, $J_{\mathrm{triv}}(\gamma_1\gamma_2)$ is the same as $J_{\mathrm{triv}}(\gamma_1)\gamma_1 J_{\mathrm{triv}}(\gamma_2)$. So we obtain a rigid analytic theta cocycle $J_{\mathrm{triv}} \in H^1(\Gamma, \mathcal{A}^\times/\mathbb{C}_p^\times)$.

Example 4 (Theta Cocycles Attached to RM Points) This example is inspired by a similar construction in the theory of rational modular cocycles (see [20, Section 1.4]). As we have observed above, Γ does not act discretely on $\mathcal{H}_p$. However, it does act discretely on the product $\mathcal{H}_p \times \mathcal{H}_\infty$ of the Drinfeld upper half-plane and the complex upper half-plane $\mathcal{H}_\infty$. We can use this to construct examples of theta cocycles as follows.

For a tuple $(r, s) \in \mathbb{P}^1(\mathbb{Q})^2$ and w real quadratic, let w' be its algebraic conjugate. The geodesic from r to s in $\mathcal{H}_\infty$ cuts $\overline{\mathcal{H}_\infty} = \mathcal{H}_\infty \cup \mathbb{R} \cup \{\infty\}$ into two connected components, say "on the right hand side" of the geodesic is the $-$-component $\mathcal{H}_\infty^-$ and on the left-hand side is the $+$-component $\mathcal{H}_\infty^+$ (see Fig. 2). This allows us to define the following symbol:

$$(r, s)(w, w') := \begin{cases} 0, \text{ if } w, w' \text{ are in the same connected component} \\ 1, \text{ if } w \in \mathcal{H}_\infty^+, \;\; w' \in \mathcal{H}_\infty^- \\ -1, \text{ if } w \in \mathcal{H}_\infty^-, \;\; w' \in \mathcal{H}_\infty^+. \end{cases} \tag{3}$$

We refer to it as the signed intersection number of the geodesics from r to s and from w to w'. For example, the intersection number of the geodesics in Fig. 2 is -1.

For an RM point $\tau \in \mathcal{H}_p^{RM}$, define the function

$$\begin{aligned} J_\tau : \Gamma &\to \mathcal{M}^\times \\ \gamma &\mapsto \prod_{w \in \Gamma\tau} t_w(z)^{(\gamma\infty,\infty)(w,w')}, \end{aligned}$$

where $t_w(z)$ is as in Sect. 3.2. As it turns out, the set of elements $w \in \Gamma\tau$ such that $(\gamma\infty, \infty)(w, w') \neq 0$ is a discrete subset of $\mathcal{H}_p$. One can show that the product converges, that indeed $J_\tau(\gamma)$ defines a non-zero rigid meromorphic function, and that furthermore J_τ defines a class in $H^1(\Gamma, \mathcal{M}^\times/\mathbb{C}_p^\times)$.

Example 5 (The Winding Cocycle) Let us also mention the so-called *winding cocycle* from [16, Section 2.3], which is an example of a rigid analytic theta cocycle. To define it, choose a base point $\xi = (\xi_p, \xi_\infty) \in \mathcal{H}_p \times \mathcal{H}_\infty$, such that $\xi_p \in \mathcal{H}_p^{\leq 0}$ and ξ_∞ does not lie in any geodesic in the Γ-orbit of $[0, \infty]$. Denote by Σ the Γ-orbit of the pair $(0, \infty)$ in $\mathbb{P}^1(\mathbb{Q})^2$, and define for every $\gamma \in \Gamma$ and $z \in \mathcal{H}_p$

$$J^{\xi}_{(0,\infty)}(\gamma)(z) := \prod_{(r,s)\in\Sigma} \left(\frac{\xi_p - r}{\xi_p - s} \cdot \frac{z - s}{z - r} \right)^{(r,s)(\xi_\infty, \gamma\xi_\infty)},$$

using the usual conventions when some of the points are ∞ and where $(r, s)(\xi_\infty, \gamma\xi_\infty)$ denotes the signed intersection number as in (3). The function $J^{\xi}_{(0,\infty)}(\gamma)$ converges to a rigid analytic function on $\mathcal{H}_p$, and $\gamma \mapsto J^{\xi}_{(0,\infty)}(\gamma)$ satisfies the cocycle condition up to scalars in $\mathbb{C}_p^\times$ (cf. [16, Proposition 2.5]).

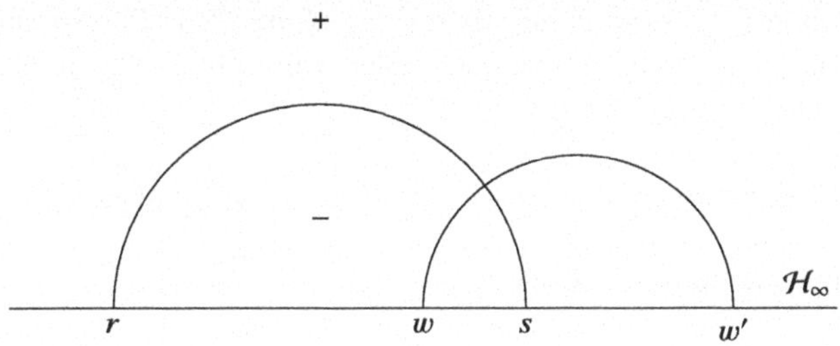

Fig. 2 Intersection of geodesics

The corresponding class in $H^1(\Gamma, \mathcal{A}^\times/\mathbb{C}_p^\times)$ is independent of the choice of base point [16, Prop. 2.6] and is called the winding cocycle, and we denote it by $J_{(0,\infty)}$.

3.3.2 Hecke Module Structure of Rigid Meromorphic Cocycles

On the space $H^1(\Gamma, \mathcal{M}^\times)$ of rigid meromorphic cocycles and on the space $H^1(\Gamma, \mathcal{M}^\times/\mathbb{C}_p^\times)$ of rigid meromorphic theta cocycles, one can define Hecke operators T_n for any $n \geq 1$. This is done by decomposing a relevant union of double cosets into cosets and following a recipe of Shimura. We refer to [16, §2.4] for details.

Definition 7 Given a quasi-parabolic rigid meromorphic cocycle J, its *rigid meromorphic period function* is the value at $S := \left(\begin{smallmatrix} 0 & -1 \\ 1 & 0 \end{smallmatrix}\right) \in \Gamma$ of the quasi-parabolic representative of J.

Let us define the abstract Hecke algebra as $\mathbb{T} := \mathbb{Z}[T_n \ , \ n \geq 1]$. One can then prove the following structural result.

Theorem 4 ([20, Theorem 1 and Remark 2]) *The group $H^1(\Gamma, \mathcal{M}^\times)$ is of infinite rank. Any finite rank $\mathbb{T}$-stable submodule of $H^1_f(\Gamma, \mathcal{M}^\times)$ is contained in $H^1_f(\Gamma, \mathcal{A}^\times)$.*

Analytic theta cocycles behave differently. As Hecke modules, they are closely related to modular forms. In [21], Darmon and Vonk prove a classification result using ideas of Stevens and Schneider–Teitelbaum. We briefly review the key points (see also [16, Section 3.1]).

Consider the annulus $U := \{z \in \mathbb{C}_p | \ 1 < |z| < p\}$. Its stabiliser in Γ is $\Gamma_0(p)$. Denote $\operatorname{dlog} f := \frac{f'(z)}{f(z)} dz$. The group homomorphism

$$\mathcal{A}^\times \to \mathbb{C}_p, \qquad f \mapsto \operatorname{res}_U(\operatorname{dlog} f),$$

where res_U denotes the p-adic annular residue along U, is $\Gamma_0(p)$-equivariant and trivial on constants. It can be shown that it takes values in $\mathbb{Z}$. This induces a map on cohomology

$$\delta_U := \mathrm{res}_U \circ \mathrm{dlog}\colon H^1(\Gamma, \mathcal{A}^\times/\mathbb{C}_p^\times) \to H^1(\Gamma_0(p), \mathbb{Z}).$$

Theorem 5 ([21, Section 3]) *The map*

$$\delta_U \otimes 1 : H^1(\Gamma, \mathcal{A}^\times/\mathbb{C}_p^\times) \otimes \mathbb{Q} \to H^1(\Gamma_0(p), \mathbb{Z}) \otimes \mathbb{Q}$$

is surjective and has a 1*-dimensional kernel, generated by the analytic theta cocycle* J_{triv}.

The map δ_U *admits a Hecke-equivariant section*

$$\mathrm{ST}^\times : H^1(\Gamma_0(p), \mathbb{Z}) \to H^1(\Gamma, \mathcal{A}^\times/\mathbb{C}_p^\times).$$

The section $\mathrm{ST}^\times$ above is called the multiplicative *Schneider–Teitelbaum lift*, and one can use it to classify the analytic theta cocycles as follows.

Theorem 6 ([21, Section 3]) *The space* $H^1(\Gamma, \mathcal{A}^\times/\mathbb{C}_p^\times)\otimes\mathbb{Q}$ *is of dimension* $2g+2$, *where* g *is the genus of the modular curve* $X_0(p)$. *The Hecke action factors through the faithful Hecke algebra of modular forms* $M_2(\Gamma_0(p))$.

3.4 Heegner Constructions

The crucial motivation for the study of rigid meromorphic cocycles and analytic theta cocycles is the construction of conjectural analogues of Heegner objects for real quadratic fields. In analogy with the Heegner constructions, they can be viewed as images of maps

$$\Phi\colon \Gamma\backslash\mathcal{H}_p^{\mathrm{RM}} \to X(\mathbb{C}_p)$$

for the group schemes $X/\mathbb{Q}$ appearing in Sect. 2.2.

The corresponding images are expected to be global. Note that however in the RM settings these maps will be defined exclusively at RM points in $\mathcal{H}_p$ via the evaluation of rigid meromorphic or theta cocycles, which we now describe.

3.4.1 Evaluation at RM Points

We now define the evaluation of quasi-parabolic rigid meromorphic cocycles at RM points. Recall that for $\tau \in \mathcal{H}_p^{\mathrm{RM}}$, $\gamma_\tau \in \Gamma$ denotes the automorph of τ as defined in Sect. 3.1.2.

Definition 8 Let $[J] \in H^1_f(\Gamma, \mathcal{M}^\times)$ be a quasi-parabolic rigid meromorphic cocycle. Consider a representative $J \in [J]$ given by a map $J : \Gamma \to \mathcal{M}^\times$ such that $J(\Gamma_\infty) \subset \mathbb{C}_p^\times$ and let $\tau \in \mathcal{H}_p^{\mathrm{RM}}$. Define the *evaluation of J at τ* as

$$J[\tau] := J(\gamma_\tau)(\tau) \in \mathbb{C}_p \cup \{\infty\}.$$

Remark 5 One can check that this is constant on Γ-orbits.

Remark 6 Defining the evaluation of a theta cocycle at an RM point is also possible but requires some extra work involving the restriction to (a conjugate of) $\mathrm{SL}_2(\mathbb{Z})$. We refer to Section 2.3 of [21] for the definition. When the theta cocycle lifts to a rigid meromorphic cocycle, the values essentially agree. When the cocycle does not lift, the values are expected to be transcendental in general but are still of arithmetic interest as they are related to Gross–Stark units (see Conjecture 2) and Stark–Heegner points (see Conjecture 3) below.

3.4.2 Singular Moduli for Real Quadratic Fields

Let $\tau \in \mathcal{H}_p^{\mathrm{RM}}$ and write H_τ for the narrow ring class field corresponding to $\mathcal{O}_\tau$. For a quasi-parabolic rigid meromorphic cocycle J, we let j be its associated period function (see Definition 7). Then we define a field

$$H_j = H_J := \mathrm{Compositum}_{j(\tau)=\infty}(H_\tau),$$

which is a finite extension of $\mathbb{Q}$.

Definition 9 Let $\tau \in \Gamma \backslash \mathcal{H}_p^{\mathrm{RM}}$. For a quasi-parabolic meromorphic cocycle J, we say that $J[\tau]$ is a *real quadratic singular modulus.*

The following conjecture, due to Darmon and Vonk, justifies the analogy with the values of the j-function at CM points.

Conjecture 1 ([20]) If $J \in H^1_f(\Gamma, \mathcal{M}^\times)$, then $J[\tau]$ is algebraic and lies in the compositum of H_J and H_τ.

Note that the analogy with classical singular moduli is imperfect because the choice of cocycle J impacts the field of definition of its value. A perhaps more appropriate analogy is that these values should be thought of as multiplicative analogues of differences of singular moduli, lying in composita of the corresponding ring class fields.

3.4.3 Elliptic Units for Real Quadratic Fields

Consider the unique (normalised) Eisenstein eigenform of weight 2 and level $\Gamma_0(p)$ with Fourier expansion

$$E_2^{(p)} = \frac{p-1}{12} + \sum_{n\geq 1}\left(\sum_{d|n,\, p\nmid d} d\right) q^n.$$

For a choice of base point $z_0 \in \mathcal{H}_\infty$, the function computing path integrals of the differential attached to the Eisenstein series

$$\gamma \mapsto \int_{z_0}^{\gamma z_0} \frac{1}{2\pi i} E_2^{(p)}(z)dz$$

for $\gamma \in \Gamma_0(p)$ takes integer values and transforms as a (non-parabolic) cocycle $\varphi_{\mathrm{DR}} \in H^1(\Gamma_0(p), \mathbb{Z})$. Its values can be described explicitly in terms of Dedekind sums (see [10, §2.5]).

The Schneider–Teitelbaum lift

$$J_{\mathrm{DR}} := \mathrm{ST}^\times(\varphi_{DR})$$

of φ_{DR} to $H^1(\Gamma, \mathcal{A}^\times/\mathbb{C}_p^\times)$ is called the *Dedekind–Rademacher theta cocycle.*

Definition 10 Let $\tau \in \Gamma\backslash\mathcal{H}_p^{\mathrm{RM}}$ be a point of discriminant D prime to p. The value $J_{\mathrm{DR}}[\tau]$ is called the *real quadratic elliptic unit* attached to τ.

These values should be thought of as analogues of the elliptic units defined in Sect. 2, as the following conjecture suggests.

Conjecture 2 ([10, Conj. 2.14]) Let $\tau \in \mathcal{H}_p$ be a real quadratic point of discriminant D prime to p and let H_τ be the narrow ring class field of the order defined by τ. Then the real quadratic elliptic unit attached to τ satisfies

$$J_{\mathrm{DR}}[\tau] \in \mathcal{O}_{H_\tau}[1/p]^\times.$$

Note that unlike in the CM setting, these values are expected to be p-units rather than genuine units in the Hilbert class field. The analogy with the construction of elliptic units is perhaps not obvious on the surface. However, the Eisenstein series $E_2^{(p)}$ whose periods appear in the definition above can be viewed as the logarithmic derivative of $\Delta(pz)/\Delta(z)$, a typical example of the modular unit featuring in Sect. 2.2.2.

3.4.4 Stark–Heegner Points

Let f be a newform of weight 2 and level $\Gamma_0(p)$, with Fourier coefficients valued in $\mathbb{Z}$. Consider the differential form $\omega_f = 2\pi i f(z)dz$, and let

$$\omega_f^+ = \omega_f + \overline{\omega}_f, \text{ and } \omega_f^- = \omega_f - \overline{\omega}_f$$

be its real and imaginary parts, respectively. Given $z_0 \in \mathcal{H}_\infty$, the maps $\tilde{\phi}_f \colon \Gamma_0(p) \to \mathbb{R}$ defined by

$$\gamma \mapsto \int_{z_0}^{\gamma z_0} \omega_f^{\pm}$$

can be rescaled by suitable periods $\Omega_f^{\pm}$ in such a way that the pair of cocycles $\varphi_f^{\pm} := 1/\Omega_f^{\pm}\tilde{\varphi}_f^{\pm}$ take values in $\mathbb{Z}$. The Schneider–Teitelbaum lifts of the cocycles defined above

$$J_f^{\pm} := \mathrm{ST}^{\times}(\varphi_f^{\pm})$$

are called the *elliptic theta cocycles* attached to f.

By the Eichler–Shimura construction reviewed in Sect. 2.2.3, a newform $f \in S_2(\Gamma_0(p))$ as above corresponds to an elliptic curve $E := E_f$ with multiplicative reduction at p. For such an elliptic curve, Tate showed that its $\mathbb{C}_p$-valued points can be uniformised as

$$\Psi_{\mathrm{Tate}} \colon \mathbb{C}_p^{\times}/q_E^{\mathbb{Z}} \xrightarrow{\sim} E(\mathbb{C}_p)$$

for some parameter $q_E \in \mathbb{C}_p^{\times}$ satisfying $|q_E| < 1$.

Definition 11 Let $\tau \in \Gamma\backslash\mathcal{H}_p^{\mathrm{RM}}$ be a point of discriminant D prime to p. A *Stark–Heegner point* for E attached to τ is

$$P_\tau^{\pm} := \Psi_{\mathrm{Tate}}(J_f^{\pm}[\tau]) \in E(\mathbb{C}_p).$$

The following conjecture originally formulated by Darmon in [9] motivates the conjectural analogy between Stark–Heegner points and the classical Heegner points in CM theory.

Conjecture 3 ([9],[21, Conjecture 3.18]) Let $\tau \in \Gamma\backslash\mathcal{H}_p^{\mathrm{RM}}$ be a point of discriminant D prime to p. Denote H_τ the narrow ring class field of the order $\mathcal{O}_\tau$ attached to τ. Then

$$P_\tau^{\pm} \in E(H_\tau).$$

In addition, the point $P_\tau^{\pm}$ is expected to lie in the (± 1)-part of $E(H_\tau)$ for the action of the complex conjugation in the narrow Hilbert class field of τ.

Remark 7 The above conjecture (and suitable generalisations) has been verified extensively, see, for example, [15]. The theoretical evidence is much more fragmentary, and no general approach is known at this stage. However, an important case was showed in the article [3] for Stark–Heegner points defined over genus fields of real quadratic fields by relating them to Heegner points in this special setting.

4 Families of Real Analytic Eisenstein Series and CM Theory

In this section, we introduce the second main theme of this chapter, which is how modular generating series constructed from Heegner objects appear in relation to derivatives of families of modular forms. We will start by reviewing some Archimedean examples; we will then move to their p-adic counterparts in Sect. 5.

In the Archimedean setting, derivatives of real analytic families of Eisenstein series fit prominently in the Kudla program. Roughly speaking, this vast program seeks to show that certain formal q-expansions encoding special cycles on orthogonal Shimura varieties are modular. The most basic instances of these statements involve CM cycles on modular curves. The works [22], [23] of Gross and Zagier were indispensable to the emergence of this program. In Sect. 4.1, we will give an account of their work on the differences of singular moduli, which will be close in spirit to the p-adic analogues discussed in Sect. 5. We will then give a brief overview of the work of Gross and Zagier on the derivative of the L-function attached to an elliptic curve [23]. Finally in Sect. 4.3, we explain how the two articles can be interpreted under a unified perspective, which will be translated in the p-adic setting in Sect. 5.

4.1 Gross–Zagier's Factorisation Formula

We review the work of Gross and Zagier on the factorisation of differences of singular moduli, the CM values of the j-function introduced in Sect. 2.2.1. Their landmark work unveiled new phenomena in the theory of complex multiplication, initiating the study of the intersection theory of CM cycles on modular curves via automorphic methods.

Let us introduce the main result. Let D_1 and D_2 be two negative relatively prime fundamental discriminants. Let $D = D_1 D_2$ and let w_i be the number of roots of unity in the quadratic orders of discriminant D_i. For $\tau \in \mathcal{H}_\infty$, denote by $[\tau]$ the equivalence class of τ under the action of $\mathrm{SL}_2(\mathbb{Z})$. The main theorem of complex multiplication implies that the product

$$J(D_1, D_2) = \prod_{\substack{[\tau_1],\ \mathrm{disc}\tau_1 = D_1 \\ [\tau_2],\ \mathrm{disc}\tau_2 = D_2}} (j(\tau_1) - j(\tau_2))^{\frac{4}{w_1 w_2}}$$

(or more precisely, its square for $D_i \geq -4$) is an integer. Gross and Zagier are motivated by the deceptively simple question of determining its prime factorisation.

For primes ℓ with $\left(\frac{D_1 D_2}{\ell}\right) \neq -1$, define

$$\epsilon(\ell) = \begin{cases} \left(\frac{D_1}{\ell}\right) & \text{if } (\ell, D_1) = 1, \\ \left(\frac{D_2}{\ell}\right) & \text{if } (\ell, D_2) = 1. \end{cases}$$

Now if $n = \prod_i \ell_i^{a_i}$ is such that $\left(\frac{D_1 D_2}{\ell_i}\right) \neq -1$ for all i, then we define $\epsilon(n) = \prod_i \epsilon(\ell_i)^{a_i}$. Gross and Zagier prove the following theorem.

Theorem 7 ([22, Thm 1.3])

$$J(D_1, D_2)^2 = \pm \prod_{\substack{x,n,n' \in \mathbb{Z} \\ n,n' > 0, \\ x^2 + 4nn' = D}} n^{\epsilon(n')}. \tag{4}$$

In *loc. cit*, the authors provide two alternative proofs of this result. The first one is essentially algebraic and relies on the moduli interpretation of CM points on the modular curve. The second is purely analytic and makes use of automorphic techniques to obtain a formula for $J(D_1, D_2)^2$. While one can easily deduce the algebraicity of $J(D_1, D_2)$ from the theory of complex multiplication, no use of this fact is made in the analytic proof, where instead it can be seen as a byproduct of the proof itself. In view of translating the Gross–Zagier approach to the RM setting, the analytic strategy is of course more relevant, as a moduli interpretation of RM points of the Drinfeld p-adic upper half-plane is lacking.

We will briefly outline some ideas behind the algebraic proof before focusing on the analytic one. The general idea is rather natural: If a prime p divides $J(D_1, D_2)^2$, there is a pair of elliptic curves with CM by two orders $\mathcal{O}_{K_1}$ and $\mathcal{O}_{K_2}$ with isomorphic reductions over $\overline{\mathbb{F}}_p$. This forces the endomorphism ring of the reduction to be a maximal order in the quaternion algebra ramified at p and ∞.

The existence of embeddings of the orders $\mathcal{O}_{K_i}$ into an order in a quaternion algebra of discriminant p is rather restrictive on the prime. Working out explicit conditions, one determines the primes contributing to the factorisation of $J(D_1, D_2)^2$. A delicate calculation using Deuring's theory of endomorphisms of elliptic curves in finite characteristic is necessary to determine the precise exponent in the factorisation formula.

The first step of the analytic proof consists in producing a reformulation of (4) more suitable to analytic manipulation. Let $F = \mathbb{Q}(\sqrt{D})$ and consider the diagram of quadratic extensions:

$$\begin{array}{ccccc} & & \mathbb{Q}(\sqrt{D_1}, \sqrt{D_2}) & & \\ & \diagup & | & \diagdown & \\ \mathbb{Q}(\sqrt{D_1}) & & F & & \mathbb{Q}(\sqrt{D_2}) \\ & \diagdown & | & \diagup & \\ & & \mathbb{Q} & & \end{array}$$

where $\mathbb{Q}(\sqrt{D}_1, \sqrt{D}_2)$ is an unramified extension of the real quadratic field F. By Artin reciprocity, the genus character cutting out this extension can be viewed as a quadratic character of the narrow class group $\mathrm{Cl}^+(\mathcal{O}_F)$ of F, which we denote by χ. An elementary calculation shows that the explicit factorisation formula for $J(D_1, D_2)$ is equivalent to

$$-\log |J(D_1, D_2)|^2 = \sum_{\nu \in \mathfrak{d}_+^{-1}, \mathrm{Tr}\nu = 1} \sum_{\mathfrak{n} | (\nu \mathfrak{d})} \chi(\mathfrak{n}) \log \mathrm{Nm}(\mathfrak{n}), \tag{5}$$

where $\mathfrak{d}_+^{-1}$ denotes the totally positive elements in the inverse different $\mathfrak{d}^{-1}$ of $\mathcal{O}_F$. The right-hand side of (5) evokes formulae appearing in the work of Siegel on special values of Dedekind ζ-functions for totally real fields [30, §2]. For $\nu \in F$, denote (ν_1, ν_2) its real embeddings. For $\mathbf{z} = (z_1, z_2) \in \mathcal{H}_\infty^2$ and (ν, μ) in F^2, we let

$$\mathrm{Nm}(\mu \mathbf{z} + \nu) := (\mu_1 z_1 + \nu_1) \cdot (\mu_2 z_2 + \nu_2).$$

The formulae appearing in *loc. cit.* arise as restrictions of Hilbert Eisenstein series of weight (k, k), where $k \geq 2$ is an even integer and level $\mathrm{SL}_2(\mathcal{O}_F)$, which are given by

$$E_{F,k}(\mathbf{z}) = \sum_{\mathfrak{a} \in \mathrm{Cl}(\mathcal{O}_F)} \mathrm{Nm}(\mathfrak{a})^k \sum_{(m,n) \in (\mathfrak{a}^2)' / \mathcal{O}_F^\times} \mathrm{Nm}(m\mathbf{z}+n)^{-k}, \qquad \text{for } \mathbf{z} \in \mathcal{H}_\infty^2, \tag{6}$$

along the diagonal embedding $\Delta \colon \mathcal{H}_\infty \to \mathcal{H}_\infty^2$. The diagonal restriction inherits modular properties from those of $E_{F,k}$. It is an elliptic modular form of weight $2k$ and level 1. Siegel's formulae are obtained by expressing the resulting form, whose constant term in the Fourier expansion encodes the Dedekind special value, in terms of a basis of the space of elliptic modular forms for small values of k.

The strategy of Gross and Zagier consists of mimicking Siegel's approach in the degenerate case in which $k = 1$. The series in (6) would not converge for $k = 1$. The problem can be obviated by considering the family of real analytic Hilbert Eisenstein series of parallel weight one

$$\mathcal{E}_{F,1,s}(\mathbf{z}) = \sum_{\mathfrak{a} \in \mathrm{Cl}(\mathcal{O}_F)} \chi(\mathfrak{a})\,\mathrm{Nm}(\mathfrak{a})^{1+2s} \sum_{(m,n)\in(\mathfrak{a}^2)'/\mathcal{O}^\times} \mathrm{Nm}(m\mathbf{z}+n)^{-k}\,|\mathrm{Nm}(m\mathbf{z}+n)|^{-s}\,\mathbf{y}^s \tag{7}$$

for a complex variable s with $\mathrm{Re}(s) > 0$. For fixed values $\mathbf{z} \in \mathcal{H}_\infty^2$, this defines a holomorphic function of s, and it can be extended meromorphically to all of $\mathbb{C}$. The strategy for constructing holomorphic weight 1 Eisenstein series, also known as *Hecke's trick*, consists in taking the limit of $\mathcal{E}_{F,1,s}$ as s tends to 0. In this setting, Hecke's trick fails at producing an interesting weight one Hilbert Eisenstein series: An unexpected cancellation occurs when computing the corresponding Fourier expansion. Gross and Zagier are drawn to consider instead the derivative $\mathcal{E}'_{F,1,s}$ at $s = 0$. This is a real analytic weight 1 Hilbert modular form; it can be restricted along the diagonal embedding Δ to obtain an (non-holomorphic) elliptic modular form of weight 2, as in [30]. From this real analytic form $\Delta^*\mathcal{E}'_{F,1}$, one can extract a holomorphic one by applying a projector onto the holomorphic subspace as in [32]. The resulting modular form, denoted by $(\Delta^*\mathcal{E}'_{F,1})_{\mathrm{hol}}$, is the object of the following theorem, summarising the calculations in [22, §7].

Theorem 8 *The first Fourier coefficient a_1 of the holomorphic modular form $(\Delta^*\mathcal{E}'_{F,1})_{\mathrm{hol}}$ of weight 2 and level $\mathrm{SL}_2(\mathbb{Z})$ satisfies the equation*

$$a_1 \cdot \lambda = \log|J(D_1, D_2)|^2 + \sum_{\nu \in \mathfrak{d}_+^{-1},\, \mathrm{Tr}\nu = 1} \sum_{\mathfrak{n}|(\nu\mathfrak{d})} \chi(\mathfrak{n}) \log \mathrm{Nm}(\mathfrak{n}),$$

for some $\lambda \in \mathbb{C}^\times$.

Now as $(\Delta^*\mathcal{E}'_{F,1})_{\mathrm{hol}}$ is of weight 2 and level $\mathrm{SL}_2(\mathbb{Z})$, one abstractly knows that it is the zero function; hence, $a_1 = 0$, and the equality (5) follows.

4.2 *Gross–Zagier's Work on BSD in Analytic Rank 1*

Beyond the factorisation of differences of singular moduli, the circle of ideas of [22] yielded some striking developments towards the Birch and Swinnerton–Dyer Conjecture.

Let $E/\mathbb{Q}$ be an elliptic curve. The L-function $L(E, s)$ is defined as a convergent infinite product for $\mathrm{Re}(s) > 3/2$. If E is modular, that is, $L(f, s) = L(E, s)$ for a suitable modular form f, the L-function $L(E, s)$ admits analytic continuation to all of $\mathbb{C}$ and functional equation with centre $s = 1$. The Birch and Swinnerton–Dyer Conjecture states that the algebraic rank, that is the rank of the group $E(\mathbb{Q})$, is equal to the analytic rank, the order of vanishing of $L(E, s)$ at $s = 1$.

The main result in [23] is the following.

Theorem 9 ([23, Thm. 7.3]) *Suppose that $L(E, 1) = 0$. There exists a point $P \in E(\mathbb{Q})$ such that*

$$L'(E, 1) \doteq \Omega_E \langle P, P \rangle$$

where Ω_E is the real period of a regular differential on E, the pairing $\langle \cdot, \cdot \rangle$ denotes the Néron–Tate height pairing on $E(\mathbb{Q}) \otimes \mathbb{R}$, and the equality denoted by $\doteq$ is defined up to scalars in $\mathbb{Q}^\times$.

In particular, if the analytic rank is one, the rank of the Mordell–Weil group $E(\mathbb{Q})$ is positive. This result was later complemented by Kolyvagin's work bounding the algebraic rank in analytic rank 0 and 1. The point P arises as a Heegner construction (see below). Combining the results of Gross–Zagier and Kolyvagin, one can conclude that this Heegner construction produces non-torsion points in $E(\mathbb{Q})$ precisely when the algebraic (or analytic) rank is 1.

Let us discuss some ingredients of the proof. Let N be the conductor of E. The assumption that E is modular implies that up to isogeny E is a direct factor of the Jacobian $J_0(N)$ of the modular curve of level $X_0(N)$. Let $D < 0$ be a fundamental discriminant coprime to N, and let $K = \mathbb{Q}(\sqrt{D})$. Let $\tau \in \mathcal{H}_\infty$ be an element with discriminant D defining a CM point of the modular curve of level $\Gamma_0(N)$. Then the divisor $c_\tau := (\tau) - (\infty)$ defines an H-rational point of $J_0(N)$, where H is the Hilbert class field of K. By Shimura reciprocity, the divisor

$$c_D = \sum_{[\tau]:\ \mathrm{disc}(\tau)=D} c_\tau,$$

where the sum runs over a set of representatives of the $\Gamma_0(N)$-orbits of points of discriminant D in $\mathcal{H}_\infty$ is K-rational. The point P appearing in the statement of Theorem 9 is constructed as a suitable trace of the divisor c_D.

The vector space $J_0(N)(H) \otimes \mathbb{R}$ is endowed with a canonical height pairing, which can be described as a sum of local heights and is equivariant for the Hecke action.

The formal q-expansion

$$G_D(q) = \sum_{n \geq 1} \langle c_D, T_n c_D \rangle q^n \tag{8}$$

is the q-expansion of a modular form of weight 2 and level $\Gamma_0(N)$. This formal series $G_D(q)$ is a prototypical example of a *modular generating series*. It packages pairings of Hecke translates of Heegner objects into a formal q-expansion. Its modularity follows from the fact that the Hecke action on the Jacobian of the modular curve factors through the Hecke algebra of modular forms of weight 2 and level $\Gamma_0(N)$.

The proof of Theorem 9 hinges on providing an alternative construction of the f-isotypic component of the series G_D that can be related to L-values. For this, the key input is the Rankin–Selberg method. It allows to reinterpret the derivative $L'(f/K, s)$ at $s = 1$ as the f-isotypic component of the holomorphic projection of the product of:

- A weight 1 theta series θ_1 attached to K and
- The derivative at $s = 0$ of a family of real analytic Eisenstein series $\mathcal{E}_{1,s}$ of constant weight 1, parametrised by a complex variable s

The comparison between the modular generating series G_D and the holomorphic projection of $\theta_1 \cdot \mathcal{E}'$ involves matching local contributions for both. In particular, it requires computing local height pairings for the divisor c_D at non-Archimedean places via intersection theory, and at Archimedean places via Green functions.

4.3 *Biquadratic Extensions and Modular Generating Series*

While the arithmetic applications of [22] and [23] are on the surface fairly different, the strategies of proof fit into a general framework. The diagram 4.1 can be viewed as a special case of a diagram of quadratic field extensions (or, more generally, of étale algebras of degree 2) of the form

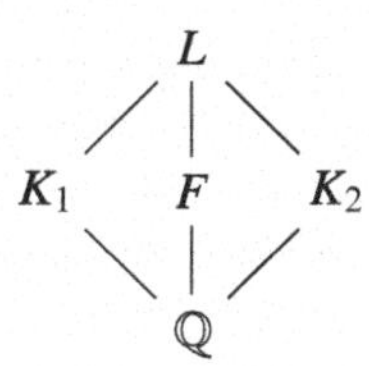

where F is a real quadratic field (or a split quadratic algebra over $\mathbb{Q}$, that is $F \simeq \mathbb{Q} \times \mathbb{Q}$) and $K_i = \mathbb{Q}(\sqrt{D_i})$ are imaginary, so that L is forced to be a CM extension of F.

For the imaginary quadratic fields K_1 and K_2, one can produce Heegner divisors c_{D_i} attached to points of discriminant D_i in the Jacobian of modular curves of suitable level. As in (8), the global height pairings on the Hecke translates of c_{D_i} can be parlayed into a modular generating series $G_{D_1,D_2}(q) = \sum_{n\geq 1}\langle c_{D_1}, T_n c_{D_2}\rangle q^n$. The setting of Gross–Zagier's work on BSD in analytic rank 1 corresponds to the degenerate case in which the imaginary quadratic fields are equal, F is the split quadratic algebra over $\mathbb{Q}$, and $L = K_1 \times K_1$.

Obtaining an explicit characterisation of the Fourier coefficients of G_{D_1,D_2} is an essential step towards the desired arithmetic applications:

- In [22], the quantity $J(D_1, D_2)$ can be read off the Archimedean contribution of the height pairing $\langle c_{D_1}, c_{D_2}\rangle$ for the modular curve of level 1.
- In [23], for a modular elliptic curve E corresponding to a newform f, the f-isotypic component of G_{D_1,D_2} encodes the height of the Heegner point appearing in the statement of 9.

The crucial idea in both [22] and [23] is to express the modular generating series G_{D_1,D_2} in an alternative form by exploiting the derivative of a suitable family $\mathcal{G}_s$ of real analytic Hilbert modular forms of constant parallel weight 1 for the quadratic

$\mathbb{Q}$-algebra F, parametrised by the variable $s \in \mathbb{C}$. In [22], the role of $\mathcal{G}_s$ is the Eisenstein series $E_{F,1,s}$. In the degenerate setting of [23], the role of $\mathcal{G}_s$ is played by the pair of elliptic modular forms $(\mathcal{E}_{1,s}, \theta_1)$ viewed as a pair of functions on two copies of $\mathcal{H}_\infty$-invariant under the action of a congruence subgroup of $\mathrm{GL}_2(\mathbb{Q} \times \mathbb{Q})$.

The modular generating series G_{D_1,D_2} is then shown to be equal to the formal q-expansion of $(\Delta^*\mathcal{G}')_{\mathrm{hol}}$, the elliptic (holomorphic) modular form obtained by:

(i) Computing the *derivative* $\mathcal{G}'$ of $\mathcal{G}_s$ at $s = 0$. This yields a real analytic Hilbert modular form of parallel weight 1.
(ii) Pulling back $\mathcal{G}'$ under the diagonal embedding $\Delta\colon \mathcal{H}_\infty \to \mathcal{H}_\infty^2$. The resulting *diagonal restriction* $\Delta^*\mathcal{G}'$ is a real analytic elliptic weight 2 modular form.
(iii) Applying a *holomorphic projection* to $\Delta^*\mathcal{G}'$.

The resulting holomorphic projection $(\Delta^*\mathcal{G}')_{\mathrm{hol}}$ is finally shown to equal the modular generating series G_{D_1,D_2}.

In Sect. 5 the structure of these proofs will be mimicked in the p-adic setting, by replacing CM cycles with their real quadratic counterparts and real analytic families with p-adic families of modular forms.

Remark 8 In [22], the modular generating series G_{D_1,D_2} does not appear explicitly. Because the calculation is carried out for the modular curve of level 1, which has genus 0, such a generating series is identically zero. The calculation in the analytic proof of *loc. cit.* amounts to determining the degree 1 coefficient of the modular generating series as a sum of local contributions, thereby leading to the desired factorisation formula.

Remark 9 Note that in [23], the theta series θ_1 appearing as the second component of the family $\mathcal{G}_s$ should be thought of as constant in the variable s. This reflects the fact that in the Archimedean setting, only real analytic Eisenstein series admit variations in families. This imposes some restrictions on the Heegner cycles for which the Gross–Zagier strategy can be carried out. In general, one would expect the relevant Hilbert family $\mathcal{G}_s$ to deform a parallel weight 1 Hilbert theta series attached to the CM extension L/F. However, these theta series are usually cuspidal, and as such they do not admit Archimedean deformations. By contrast, in the p-adic setting, cusp forms can often be p-adically deformed. This will allow additional flexibility in the settings arising in Sect. 5.

5 p-Adic Families of Modular Forms and RM Theory

As we have seen in the last section, real analytic families of Eisenstein series play a vital role in the works of Gross and Zagier on CM theory. In the p-adic setting there is a direct analogue, the p-adic family of Eisenstein series, and below we will review how it is used in RM theory. Before we go into details we provide a bit of context for the much richer theory of *p-adic* families of modular forms.

5.1 A Brief Overview of p-Adic Families of Modular Forms

The theory of p-adic families of modular forms (or more generally of automorphic forms) has its origins in the 70s when Serre observed that the Hecke eigenvalues of Eisenstein series can be p-adically interpolated and hence that Eisenstein series can be viewed as a p-adic family parametrised by the weight. This simple yet striking observation combined with a general interest in establishing congruences of modular forms motivated the search for a theory of p-adic variations of modular forms.

A p-adic modular form (of level $\mathrm{SL}_2(\mathbb{Z})$) as defined by Serre in [29] is a power series $f = \sum_{n\geq 0} a_n q^n \in \mathbb{Q}_p[[q]]$ such that there is a sequence of classical modular forms $f_i = \sum_{n\geq 0} a_{i,n} q^n$ that converges to f (uniformly in the coefficients). There is an alternative geometric definition due to Katz [24], which involves the (rigid analytic versions of) modular curves and generalises the classical definition of modular forms as sections of certain line bundles.

In a family of p-adic modular forms, the weight is one of the crucial p-adically varying parameters. In fact this parameter can be viewed to vary in the so-called *weight space*, which is a rigid analytic space $\mathcal{W}$ whose $\mathbb{C}_p$-points are given by

$$\mathcal{W}(\mathbb{C}_p) = \mathrm{Hom}_{\mathrm{cts}}(\mathbb{Z}_p^\times, \mathbb{C}_p^\times) ,$$

the group of continuous $\mathbb{C}_p^\times$-valued characters of $\mathbb{Z}_p^\times$ and which identifies with a finite union of open unit discs. The weight $k \in \mathbb{Z}$ of a classical modular form is viewed as a $\mathbb{Q}_p$-point of $\mathcal{W}$ by identifying it with the character $z \mapsto z^{k-1}$, and a general p-adic modular form then comes with a weight $\kappa \in \mathcal{W}(\mathbb{C}_p)$.

For building a good geometric theory of families, the space of p-adic modular forms turns out to be too big, a problem that both Hida and Coleman managed to overcome using the Hecke operator U_p that acts on the space of p-adic modular forms. In a nutshell, Hida studies the *ordinary subspace*, i.e., the subspace of the space of p-adic modular forms on which U_p acts invertibly. Coleman on the other hand works with certain subspaces of the space of p-adic forms, the so-called *overconvergent* p-adic modular forms. Overconvergence was already studied by Katz and is an extra condition that decreases the size of these spaces (although the result is still an infinite-dimensional space). Now, on these spaces of overconvergent forms, the operator U_p acts as a compact operator, and the complement of the kernel of U_p is well behaved. Coleman managed to use this fact to build p-adic families of modular forms [6].

From a geometric point of view, the theory for p-adic families of modular forms then culminates in the construction of the so-called eigencurve by Coleman and Mazur [5]. The eigencurve is a rigid analytic curve C that lives over $\mathcal{W}$ and whose points correspond to overconvergent p-adic modular Hecke eigenforms that are not in the kernel of the U_p operator. The *classical points*, i.e., the points corresponding to classical modular forms, form a Zariski-dense subset of C. The global geometry of the Coleman–Mazur eigencurve remains a fascinating and challenging subject in current research (see [27] for very recent progress on some of the original questions of Coleman and Mazur).

Let us highlight one important feature of the theory of eigencurves, namely the connection to the theory of Galois representations. To a classical modular eigenform f, one can associate a 2-dimensional Galois representation ρ_f, which is characterised by identifying the Hecke eigenvalues with traces of Frobenius elements [13], [18, Chapter 9]. As it turns out, one can interpolate these Galois representations (or rather, their traces and determinants) on the eigencurve to a family of pseudorepresentations. Moreover, the infinitesimal neighbourhood of a classical point on the eigencurve has an interpretation in terms of a (suitably refined) Galois deformation space. This allows the usage of tools like Galois cohomology in the study of the local geometry of the eigencurve.

Since the works of Serre, Katz, Coleman and Mazur, p-adic forms have been introduced for other reductive groups G over a number field F by different techniques. In particular the definition of overconvergent modular forms has been generalised to the setting of more general Shimura varieties. Furthermore, we have constructions of the so-called eigenvarieties generalising the Coleman–Mazur eigencurve. Let us emphasise (as this is relevant below) that we have a good theory of p-adic families of Hilbert modular forms at hand.

5.2 A Modular Generating Series for RM Values

While a priori singular moduli for real quadratic fields do not enjoy the same algebraicity properties as their imaginary quadratic counterparts, they can be packaged into modular generating series, which one can hope to relate to derivatives of modular forms in the p-adic setting. In [16], Darmon, Pozzi and Vonk imitate the strategy of [22] outlined in Sect. 4.1 by replacing the real analytic family of Hilbert Eisenstein series appearing in (7) with a p-adic family parametrised by the weight, and relate it to a modular generating series for RM values of theta cocycles. We will describe their result and compare it with the formalism presented in Sect. 4.3.

Let F be a real quadratic field of discriminant D, let χ be an odd character of $\mathrm{Cl}^+(\mathcal{O}_F)$, and let p be a prime unramified in F. Given an odd integer $k \in \mathbb{Z}_{\geq 1}$, consider the parallel weight k Hilbert Eisenstein series in the variable $\mathbf{z} = (z_1, z_2)$ in $\mathcal{H}_\infty^2$ with Fourier expansion

$$E^{(p)}_{F,\chi,k}(\mathbf{z}) = L^{(p)}(F, \chi, 1-k) + 4 \sum_{\nu \in \mathfrak{d}_+^{-1}} \sum_{\mathfrak{n} | (\nu\mathfrak{d}),\ p \nmid \mathrm{Nm}(\mathfrak{n})} \chi(\mathfrak{n})\, \mathrm{Nm}(\mathfrak{n})^{k-1} e^{2\pi i \mathrm{Tr}(\nu \cdot \mathbf{z})}, \tag{9}$$

where $\mathrm{Tr}(\nu \cdot \mathbf{z}) := \nu_1 z_1 + \nu_2 z_2$ for the real embeddings ν_1, ν_2 of $\nu \in F$ and

$$L^{(p)}(F,\chi,s) = \sum_{\mathfrak{n},\ p\nmid \mathrm{Nm}(\mathfrak{n})} \chi(\mathfrak{n})\mathrm{Nm}(\mathfrak{n})^{s}.$$

These modular forms are obtained as p-stabilisations of Hilbert modular forms of level $\mathrm{SL}_2(\mathcal{O}_F)$ defined similarly to (6) at primes above p in F. The p-stabilisation procedure allows their Fourier coefficients to interpolate into continuous functions of the variable $k \in \mathcal{W}$, at the cost of introducing p into the level. This can be verified directly for the Hecke coefficients attached to $\nu \in \mathfrak{d}_+^{-1}$ and can be deduced for the constant term following Serre's approach to the analyticity of the Kubota–Leopoldt p-adic ζ-function. We denote the family as $\mathcal{E}_{F,\chi,k}$, where the weight k is thought of as a variable $k \in \mathcal{W}$, as explained in Sect. 5.1.

Darmon, Pozzi and Vonk translate the steps outlined in Sect. 4.3 to the current setting. In analogy to Gross–Zagier, who consider a family of constant weight 1, it is natural to analyse the specialisation of the family $\mathcal{E}_{F,\chi,k}$ at $k = 1$, in view of applying a derivative. While the Fourier expansion of (9) may not vanish at $k = 1$, there is no harm in first considering the pullback of the family $\mathcal{E}_{F,\chi,k}$ along the diagonal embedding $\Delta\colon \mathcal{H}_\infty \hookrightarrow \mathcal{H}_\infty^2$. For every classical weight k, the diagonal restriction of the weight k-specialisation $E^{(p)}_{F,\chi,k}$ gives a classical elliptic modular form of weight $2k$, and its Fourier coefficients interpolate p-adically as functions of $k \in \mathcal{W}$. This yields a p-adic family of elliptic modular forms $\Delta^*\mathcal{E}_{F,\chi,k}$. Its behaviour at $k = 1$ depends on the splitting of p in F (see [16, Thm. A]):

- When p splits in F, the specialisation of $\Delta^*\mathcal{E}_{F,\chi,k}$ need not vanish. The coefficients of its Fourier expansion can be expressed in terms of intersection pairings of the geodesic attached to real quadratic points of discriminant D on the modular curve of level $\Gamma_0(p)$ and the path between $(0,\infty)$ on the modular curve of level $\Gamma_0(p)$.
- When p is inert in F, the specialisation at $k = 1$ vanishes identically.

The split setting is rather classical in flavour. In the inert setting, the vanishing happens for similar reasons as in [22], as the weight 1 specialisation of $\Delta^*\mathcal{E}_{F,\chi,k}$ can itself be viewed as the p-stabilisation of the diagonal restriction of an Eisenstein series of weight 1 and trivial level, which must be identically 0. In the latter setting, it is tempting to consider the derivative $\Delta^*\mathcal{E}'_{F,\chi,k}$ of the above family at $k = 1$. Exploiting the vanishing of $\Delta^*\mathcal{E}_{F,\chi,k}$ at $k = 1$, one can show that this derivative is a p-adic (and, in fact, even overconvergent) modular form of weight 2 and trivial tame level [16, Thm. 2.1]. Following the Gross–Zagier strategy, one wishes to tweak the resulting p-adic modular form and produce a classical one. This can be achieved by means of Hida's ordinary projector, constructed by taking limits of suitable iterates of the U_p operator. We denote the resulting modular form by $(\Delta^*\mathcal{E}'_{F,\chi,k})_{\mathrm{ord}}$. It turns out to be a modular generating series for the RM values of the winding theta cocycle $J_{(0,\infty)}$ at real quadratic points of discriminant D.

Theorem 10 ([16, Thm. B]) *Suppose that p is inert in F. The ordinary projection of the diagonal restriction of $\mathcal{E}'_{F,\chi,k}$ is a classical modular form of weight 2 and level $\Gamma_0(p)$ with q-expansion*

$$(\Delta^* \mathcal{E}'_{F,\chi,k})_{\mathrm{ord}} = L'_p(F,\chi,0) + \sum_{n\geq 1} q^n \sum_{[\tau],\ \mathrm{disc}\,\tau = D} \chi(\mathfrak{c}_\tau) \log_p \left(\mathrm{Nm}_{\mathbb{Q}_p}^{\mathbb{Q}_{p^2}} (T_n J_{(0,\infty)})[\tau] \right),$$

where the sum runs over $\mathrm{SL}_2(\mathbb{Z})$*-classes of points of discriminant* D, $\mathfrak{c}_\tau$ *denotes the fractional ideal* $\mathbb{Z} + \tau\mathbb{Z}$ *if* $\tau - \tau' > 0$, *and* $\sqrt{D}(\mathbb{Z} + \tau\mathbb{Z})$ *otherwise, and* $\mathbb{Q}_{p^2}$ *is the unique unramified quadratic extension of* $\mathbb{Q}_p$.

Note that the points τ in the above sum belong to $\mathcal{H}_p$ precisely because p is inert in F. Unlike in Theorem 8, the constructed modular form has no a priori reason to vanish since it has level $\Gamma_0(p)$.

Remark 10 The modular form in Theorem 10 is in general non-zero. Its constant coefficient is the leading term of a p-adic L-function that can be expressed as the p-adic logarithm of certain global units, which will be the subject of Theorem 11.

Remark 11 The values $J_{(0,\infty)}[\tau]$ of the winding theta cocycle at RM points τ in $\mathcal{H}_p$ should be thought of as multiplicative analogues of the differences of singular moduli $j(\tau_1) - j(\tau_2)$ appearing in Gross–Zagier [22], where the pair $(0, \infty)$ corresponds to the split quadratic form $Q_{(0,\infty)}(X, Y) = XY$. However, this setting is degenerate, because the solutions to $Q_{(0,\infty)}$ lie at the boundary of the p-adic upper half-plane. In particular, the corresponding theta cocycle is not expected to have algebraic values at real quadratic points in general. In the general framework described in Sect. 4.3, this setting should correspond to the diagram of biquadratic $\mathbb{Q}$-algebras of the form

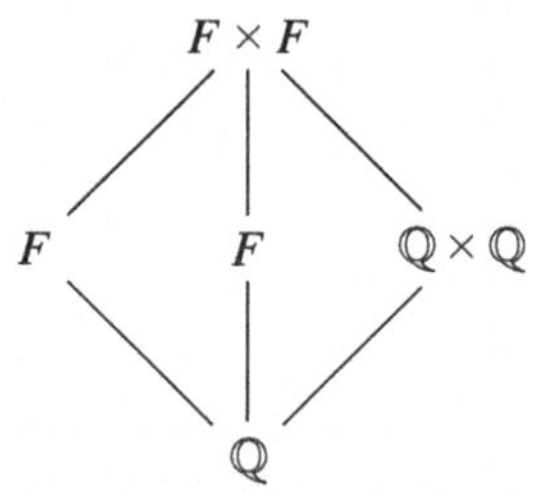

where the leftmost and rightmost algebras in the middle row correspond to quadratic forms of discriminant D and $Q_{(0,\infty)}$, respectively. The analogy with the settings of Gross–Zagier [22] and [23] is imperfect: In their work the Archimedean place does not split in the quadratic fields K_1 and K_2. By contrast, in the setting of [16] the prime p splits in the algebra $\mathbb{Q} \times \mathbb{Q}$. This is related to the fact that the Eisenstein series appearing in [16] is attached to an arbitrary totally odd character of the class group $\mathrm{Cl}^+(\mathcal{O}_F)$, while in [22] only genus characters are considered.

5.3 The Values of the Dedekind–Rademacher Theta Cocycle

In the Archimedean setting, we discussed how modular generating series for Heegner cycles on the Jacobian of modular curves can be leveraged into arithmetic applications for the Heegner constructions discussed in Sect. 2. Particularly, the main result in [22] towards the Birch and Swinnerton–Dyer Conjecture is obtained by projecting a suitable modular generating series into the f-isotypic component for a newform f corresponding to an elliptic curve via the Eichler–Shimura construction.

Similarly, we will now discuss an arithmetic application of the ideas appearing in Sect. 5.2 to the algebraicity of the RM values of the Dedekind–Rademacher theta cocycle J_{DR} defined in Sect. 3.4. The values of the Dedekind–Rademacher cocycle at real multiplication points should be thought of as real quadratic analogues of the elliptic units, in light of Conjecture 2. The cocycle J_{DR} is the Eisenstein class in the module of analytic theta cocycles $H^1(\Gamma, \mathcal{A}^\times/\mathbb{C}_p^\times)$, as its definition involves the periods of the weight 2 Eisenstein series of level $\Gamma_0(p)$. Accordingly, the main result of [17] is obtained by projecting a suitable modular generating series similar to the one appearing in Theorem 10 onto its Eisenstein subspace.

However, a significant new element occurs in this setting that does not have an Archimedean counterpart. The modular generating series in question is obtained by considering the derivative of a p-adic family of *cusp forms* deforming over an appropriate weight space for Hilbert modular forms for a real quadratic field F. The coefficients of its derivative are richer than those of Eisenstein series and encode information about the arithmetic of the Hilbert class field of F.

The main result [17] can be formulated as follows.

Theorem 11 *Let $D > 0$ be a fundamental discriminant, and let $p \nmid D$ be a prime number. Let $\tau \in \mathcal{H}_p$ be a point of discriminant D. Let $F = \mathbb{Q}(\tau)$ and let H be the narrow Hilbert class field of F. The values of the Dedekind–Rademacher cocycle at τ satisfy*

$$J_{DR}[\tau] = u_\tau^{12} \text{ modulo } \left(p^{\mathbb{Z}} \times \text{ roots of unity in } \mathbb{Q}_{p^2}^\times\right)$$

for an element $u_\tau \in \mathcal{O}_H[1/p]^\times \otimes \mathbb{Q}$.

Prior to this work, the equality of local norms

$$\mathrm{Nm}_{\mathbb{Q}_p}^{\mathbb{Q}_p^2}(J_{\mathrm{DR}}(\tau)) \doteq \mathrm{Nm}_{\mathbb{Q}_p}^{\mathbb{Q}_p^2}\left(u_\tau^{12}\right) \tag{10}$$

up to p-powers and roots of unity in $\mathbb{Q}_{p^2}$ was already known, as both sides of (10) had been related to the derivative of a p-adic L-function (for a definition, see [10, §4]). More precisely, Thm. 4.1 in *loc. cit.* relates the derivative of this p-adic L-function to the norm of the RM values of the Dedekind–Rademacher cocycle. On the other hand, the p-unit u_τ is an example of the Gross–Stark units, appearing in the conjectures of Stark and Gross about derivatives of Artin L-functions and their p-adic analogues. In particular, the formula for the derivative of the p-adic zeta function above and the norm of the p-unit u_τ were proved by Darmon, Dasgupta and Pollack in their work on the Gross–Stark Conjecture.

The main contribution of the refinement of (10) in Theorem 11 is proving the *algebraicity* of the RM values of the Dedekind–Rademacher cocycle itself, for which the equality up to local norms would not suffice. This establishes cases of Conjecture 2, providing theoretical evidence in favour of pursuing a theory of real multiplication via p-adic methods.

Remark 12 The unit u_τ in the statement of Theorem 11 can be characterised more precisely. It is trivial if F has a unit of negative norm, or equivalently, if the narrow class field of H is totally real. Otherwise, it is a non-trivial element in the subspace of $\mathcal{O}_H[1/p]^\times \otimes \mathbb{Q}$ on which the complex conjugation in H acts as -1. The recent breakthroughs of Dasgupta and Kakde on the Brumer–Stark Conjecture imply that this unit in fact lies in $\mathcal{O}_H[1/p]^\times \otimes \mathbb{Z}[1/2]$.

The crucial intuition behind the strategy in [17] is that removing the norm in (10) could be achieved by removing the norms in the modular generating series in Theorem 10 obtained from the derivative of the p-adic family of Eisenstein series. This approach requires finding a suitable family of p-adic modular forms and exploiting its derivatives to produce the desired modular generating series. Fortunately, the setting allows a lot of flexibility due to a lucky coincidence. The weight 1 Eisenstein series $E^{(p)}_{F,\chi,1}$ considered in (9) happens to be cuspidal when viewed as an overconvergent p-adic Hilbert modular form. In fact, it is an example of a critical p-stabilisation of a classical Eisenstein series (considered, for example, by Bellaiche and Chenevier for elliptic modular forms [1]). As such, it defines a point on the eigenvariety parametrising p-adic families of Hilbert cusp forms for F. Unlike Eisenstein series, which only vary in parallel weight, Hilbert families of cusp forms range over a larger weight space parametrising pairs of weights (k_1, k_2).

Cuspidal variations of Hilbert modular forms are, in general, entirely inexplicit. However, their attached Galois representations provide a useful tool to tackle a local description. In particular, they can be exploited to entirely determine the Fourier expansion of the derivative of a family $\mathcal{F}_{k,2-k}$ specialising to $E^{(p)}_{F,\chi,1}$ at $k = 1$, in the *antiparallel* direction of weights (k_1, k_2) satisfying $k_1 + k_2 = 2$. The diagonal restriction of the derivative $\mathcal{F}_{k,2-k}$ with respect to k, denoted by $\Delta^*\mathcal{F}'$, is again an overconvergent p-adic modular form. Its ordinary projection $(\Delta^*\mathcal{F}')_{\mathrm{ord}}$ (up to an explicit correction term obtained from the diagonal restriction of explicit Eisenstein series) yields the following theorem, which is the key stepping towards Theorem 11.

Theorem 12 ([17, Thm. C]) *There is a classical modular form of weight 2 and level $\Gamma_0(p)$ with q-expansion*

$$\log_p(u_\tau) + \sum_{n\geq 1} \log_p(T_n J_{(0,\infty)}[\tau])q^n,$$

for a unit $u_\tau \in \mathcal{O}_H[1/p]^\times \otimes \mathbb{Q}$.

From this result, the relation between the unit u_τ and the RM values of the Dedekind–Rademacher cocycle can easily be deduced by decomposing the winding cocycle $J_{(0,\infty)}$ with respect to an eigenbasis of the space of analytic theta cocycles, and in particular determining its projection onto the Eisenstein eigenspace spanned by J_{DR}.

Remark 13 The relevance of cuspidal deformations of the Eisenstein series $E^{(p)}_{F,\chi,1}$ towards Theorem 11 should not be surprising. A cuspidal variation of this Eisenstein series in the parallel weight direction played a crucial role in the proof of the Gross–Stark Conjecture in [11] implying the equality (10). This fits into a general theme relating congruences between cusp forms and Eisenstein series to L-functions, initiated by Ribet in [28] and often referred to as *Ribet's method*. Theorem 11 is a special case of a far more general result of Dasgupta and Kakde [14], providing explicit p-adic formulae for units in abelian extensions of totally real fields. Their result can be seen as the culmination of a program of Dasgupta and collaborators settling Hilbert's Twelfth problem for totally real fields via p-adic methods. Their approach makes use of generalisations of Ribet's method. However, unlike in [17], the strategy is applied to variations of Hilbert modular forms over finite group rings, rather than over a p-adic weight space.

Acknowledgments We would like to thank the organisers of the WIN-Europe 4 conference for inviting us to contribute to these proceedings. We would like to thank Henri Darmon and Jan Vonk for enlightening explanations of their work. We would also like to thank Sandra Rozensztajn, Ana Caraiani, and Catherine Hsu for helpful conversations during the preparation of this manuscript. We would like to thank Oğuz Gezmiş and the anonymous referees for many helpful comments and suggestions.

P.F. was funded by the DFG Graduiertenkolleg 2553. J.L. acknowledges support from the Deutsche Forschungsgemeinschaft (DFG, German Research Foundation) through TRR 326 *Geometry and Arithmetic of Uniformized Structures*, project number 444845124. A.P. was funded by the Leverhulme Trust (via the Leverhulme Prize of Ana Caraiani). M.S. was funded by the Engineering and Physical Sciences Research Council [EP/L015234/1], the EPSRC Centre for Doctoral Training in Geometry and Number Theory (The London School of Geometry and Number Theory), Imperial College London, and by the Max-Planck-Institut für Mathematik. H.W. acknowledges support from the Herchel Smith Postdoctoral Fellowship Fund and the Engineering and Physical Sciences Research Council (EPSRC) grant EP/W001683/1.

References

1. J. Bellaïche, G. Chenevier, Lissité de la courbe de Hecke de GL_2 aux points Eisenstein critiques. J. Inst. Math. Jussieu **5**(2), 333–349 (2006)
2. C. Breuil, B. Conrad, F. Diamond, R. Taylor, On the modularity of elliptic curves over **Q**: wild 3-adic exercises. J. Am. Math. Soc. **14**(4), 843–939 (2001)
3. M. Bertolini, H. Darmon, The rationality of Stark-Heegner points over genus fields of real quadratic fields. Ann. Math. (2) **170**(1), 343–369 (2009)

4. S. Bosch. *Lectures on Formal and Rigid Geometry*, volume 2105 of *Lect. Notes Math.* (Springer, Cham, 2014)
5. R. Coleman, B. Mazur, The eigencurve, in *Galois Representations in Arithmetic Algebraic Ggeometry. Proceedings of the Symposium*, Durham, UK, July 9–18, 1996 (Cambridge University Press, Cambridge, 1998), pp. 1–113
6. R.F. Coleman, p-adic Banach spaces and families of modular forms. Invent. Math. **127**(3), 417–479 (1997)
7. B. Conrad, Several approaches to non-Archimedean geometry, in *p-Adic Geometry. Lectures from the 2007 10th Arizona Winter School*, Tucson, AZ, USA, March 10–14, 2007 (American Mathematical Society (AMS), Providence, RI, 2008), pp. 9–63
8. D.A. Cox, *Primes of the Form* $x^2 + ny^2$*. Fermat, Class Field Theory and Complex Multiplication* (Wiley, New York, etc., 1989)
9. H. Darmon, Integration on $\mathcal{H}_p \times \mathcal{H}$ and arithmetic applications. Ann. Math. (2) **154**(3), 589–639 (2001)
10. H. Darmon, S. Dasgupta, Elliptic units for real quadratic fields. Ann. Math. (2) **163**(1), 301–346 (2006)
11. S. Dasgupta, H. Darmon, R. Pollack, Hilbert modular forms and the Gross-Stark conjecture. Ann. Math. (2) **174**(1), 439–484 (2011)
12. H. Darmon, F. Diamond, R. Taylor, Fermat's last theorem, in *Elliptic Curves, Modular Forms & Fermat's Last Theorem (Hong Kong, 1993)* (Int. Press, Cambridge, MA, 1997), pp. 2–140
13. P. Deligne, Modular forms and ℓ-adic representations. Sémin. Bourbaki 1968/69, No. 355. Lect. Notes Math. **179**, 139–172 (1971)
14. S. Dasgupta, M. Kakde, Brumer–Stark units and Hilbert's 12th problem (2021). arxiv.2103.02516
15. H. Darmon, R. Pollack, Efficient calculation of Stark-Heegner points via overconvergent modular symbols. Isr. J. Math. **153**, 319–354 (2006)
16. H. Darmon, A. Pozzi, J. Vonk, Diagonal restrictions of p-adic Eisenstein families. Math. Ann. **379**(1–2), 503–548 (2021)
17. H. Darmon, A. Pozzi, J. Vonk, The values of the Dedekind–Rademacher cocycle at real multiplication points. J. Eur. Math. Soc. (2023)
18. F. Diamond, J. Shurman, *A First Course in Modular Forms*, volume 228 of *Grad. Texts Math.* (Springer, Berlin, 2005)
19. S. Dasgupta, J. Teitelbaum, The p-adic upper half-plane, in *p-adic Geometry*, volume 45 of *Univ. Lecture Ser.* (Amer. Math. Soc., Providence, RI, 2008), pp. 65–121
20. H. Darmon, J. Vonk, Singular moduli for real quadratic fields: a rigid analytic approach. Duke Math. J. **170**(1), 23–93 (2021)
21. H. Darmon, J. Vonk, Real quadratic Borcherds products. Pure Appl. Math. Q. **18**(5), 1803–1865 (2022)
22. B.H. Gross, D.B. Zagier, On singular moduli. J. Reine Angew. Math. **355**, 191–220 (1985)
23. B. Gross, D. Zagier, Heegner points and derivatives of l-series. Invent. Math. **84**, 225–320 (1986)
24. N.M. Katz, p-adic properties of modular schemes and modular forms, in *Modular Functions of one Variable III, Proc. internat. Summer School*, Univ. Antwerp 1972, Lect. Notes Math., vol. 350, pp. 69–190 (1973)
25. D.S. Kubert, S. Lang, *Modular Units*. Grundlehren der mathematischen Wissenschaften. (Springer, 1981)
26. V.A. Kolyvagin, Finiteness of $E(\mathbb{Q})$ and $(E, \mathbb{Q})$ for a subclass of Weil curves. Math. USSR, Izv. **32**(3), 523–541 (1989)
27. R. Liu, N.X. Truong, L.X. Xiao, B. Zhao, Slopes of modular forms and geometry of eigencurves (2023). arxiv.2302.07697
28. K.A. Ribet, A modular construction of unramified p-extensions of $Q(\mu_p)$. Invent. Math. **34**(3), 151–162 (1976)
29. J.-P. Serre, Formes modulaires et fonctions zêta p-adiques, in *Modular Functions of One Variable III*, ed. by W. Kuijk, J.-P. Serre (Springer, Berlin, Heidelberg, 1973), pp. 191–268

30. C.L. Siegel, Berechnung von Zetafunktionen an ganzzahligen Stellen. Nachr. Akad. Wiss. Gött., II. Math.-Phys. Kl. **1969**, 87–102 (1969)
31. J.H. Silverman, *Advanced Topics in the Arithmetic of Elliptic Curves*, volume 151 of *Graduate Texts in Mathematics* (Springer, New York, 1994)
32. J. Sturm, Projections of C^{∞}-automorphic forms. Bull. Am. Math. Soc., New Ser. **2**, 435–439 (1980)
33. R. Taylor, A. Wiles, Ring-theoretic properties of certain Hecke algebras. Ann. Math. (2) **141**(3), 553–572 (1995)
34. A. Wiles, Modular elliptic curves and Fermat's Last Theorem. Ann. Math. (2) **141**(3), 443–551 (1995)

Elliptic Fibrations and Involutions on K3 Surfaces

Alice Garbagnati and Cecília Salgado

1 Introduction

K3 surfaces occupy a prominent role in the classification of algebraic surfaces, lying in the middle of Enriques' classification. Their rich geometry has earned the attention of mathematicians in different fields, ranging from number theory, algebraic, differential, and symplectic geometry to differential equations and dynamical systems. One of the implications of this rich geometry is that K3 surfaces are the only class in which surfaces might admit more than one elliptic fibration that is not of product type and has a section. Sterk showed in [35] that on a K3 surface there are finitely many elliptic fibrations modulo the automorphism group. It is therefore natural to search for a classification of such fibrations when they exist.

This survey aims to review our recent contributions to this problem.

1.1 Historical Background

Oguiso was one of the first to tackle this classification problem. In [30], he provides a classification, up to automorphisms, of all elliptic fibrations on Kummer surfaces X of the product of two non-isogenous elliptic curves. In particular, he shows that, under generality conditions, such surfaces admit 11 elliptic fibrations which he describes by providing their Mordell–Weil groups and types of singular fibers

A. Garbagnati (✉)
Dipartimento di Matematica, Univ. Statale di Milano, Milan, Italy
e-mail: alice.garbagnati@unimi.it

C. Salgado
Bernoulli Institute, University of Groningen, Groningen, The Netherlands
e-mail: c.salgado@rug.nl

R. Abdellatif et al. (eds.), *Women in Numbers Europe IV*, Association for Women in Mathematics Series 32, https://doi.org/10.1007/978-3-031-52163-8_10

(i.e., describing the frame lattice). Moreover, he shows that there are 59 distinct fibrations up to automorphisms. Oguiso's method is geometric and relies on the presence of a non-symplectic involution on X which he uses as a starting point for his classification. By definition, non-symplectic automorphisms on a K3 surface do not preserve the symplectic form on it, and hence, the quotient of a K3 surface by a non-symplectic involution has no symplectic structure. More precisely, if the fixed locus of a non-symplectic involution is non-empty, the quotient surface is rational. In particular, the non-symplectic involution on the K3 surface studied by Oguiso exhibits the surface as a specific double cover of a rational surface with a prescribed branch locus. Generalizations of Oguiso's method are based on the construction of K3 surfaces as similar double covers.

A few years later, Nishiyama takes an algebraic approach, in [29], and provides a classification of all possible frame lattices on K3 surfaces that are cyclic quotients of products of elliptic curves. These are Kummer or singular K3 surfaces. His approach is lattice theoretic and can be applied in a much larger generality to other K3 surfaces.

Oguiso's techniques were extended by Kloosterman in [23] in the particular case of K3 surfaces that are double covers of the projective plane ramified over six lines in general position. These have Picard number 16, and moreover, the cover involution acts trivially on the Néron–Severi group. The presence of this involution allows him to apply Oguiso's method.

Later, Kumar also tackled the problem by providing a classification, up to automorphisms, of all 25 elliptic fibrations on a generic Jacobian Kummer surface, in [24]. Kumar's method is algebraic. More precisely, he exhibits candidates for divisor classes of an elliptic fibration, up to the automorphism group of the K3, in its Néron–Severi group. This is feasible thanks to the knowledge of the full automorphism group of the surface.

Comparin and Garbagnati generalized Kloosterman's results, in [13], expanding the classification for K3 surfaces that are double cover of the plane branched over the union of rational curves. The cover involution still acts trivially on the Néron–Severi group. As a consequence of their classification, they obtained a classification of van Geemen–Sarti involutions on such surfaces.

Nishiyama's method was used by several authors to obtain a list of all possible frame lattices for elliptic fibrations on K3 surfaces. Examples include the K3 associated to the modular group $\Gamma_1(8)$ by Bertin and Lecacheux [5] and the singular K3 with transcendental lattice $\langle 2\rangle \oplus \langle 6\rangle$, which also admits an interpretation as a modular surface, as described in [7].

From a lattice theoretic point of view, the class of the fiber of an elliptic fibration on a K3 surface X is an isotropic vector (i.e., a vector with self-intersection 0) in its Néron–Severi group. These vectors can be found in [3] if X admits a non-symplectic involution that acts trivially on the Néron–Severi group and whose fixed locus on X contains a curve of genus $g \geq 2$. The cases $g = 0, 1$ are considered in [1] and [2].

Shimada adopted a different point of view on the problem providing a classification of all frame lattices of elliptic fibrations on K3 surfaces with finite Mordell–Weil groups in [32]. In particular, he does not consider a fixed the K3 surface but imposes conditions on the elliptic fibrations. Similarly to Nishiyama's treatment, his method is purely lattice theoretic.

As Oguiso's results discussed above show, the classification of possible frames of elliptic fibrations is usually coarser than the one up to automorphisms of the surfaces. The different types of classifications of elliptic fibrations on K3 surfaces are described in [8]. There, the authors also discuss when certain classifications coincide. Recently, in [16], Festi and Veniani provide a method to count elliptic fibrations on K3 surfaces up to automorphisms given the knowledge of the frame lattices.

1.2 Our Contributions

The study of K3 surfaces that can be realized as double covers of rational elliptic surfaces and the interplay between elliptic fibrations on the K3 surface and linear systems on the rational surface is the starting point of our series of papers, [19] and [18], and the *Women in Numbers Europe 2 and 3* research projects [6] and [9].

Our first contribution, inspired by Oguiso's work (see [30] and Remark 5), distinguishes the elliptic fibrations in three classes, according to the action of the cover involution on them. These yield three types of linear systems on the rational quotient. In [19], we describe such linear systems, which are not necessarily complete, and provide explicit examples of each of them on K3 surfaces, namely, double covers of a rational elliptic surface with reducible fibers I_3+I_6 and Mordell–Weil group $\mathbb{Z} \times \mathbb{Z}/3\mathbb{Z}$. In particular, we show that, under certain hypotheses, the elliptic fibrations on a K3 surface induce fibrations in curves of genus 0 or 1 on the rational elliptic surface (cf. [19, Thm. 4.2]). We also describe the relationship between the singular fibers of the aforementioned fibrations on the rational surfaces and of the elliptic fibrations on the K3 surfaces (cf. [19, Thms. 5.3 and 5.5]).

The paper [18] puts our classification program to work by giving a complete treatment of general K3 surfaces with non-symplectic involutions that fix a curve of positive genus, since the case of a fixed locus formed only by rational curves was already analyzed in [13, 23, 30]. We first focus on the next case, namely of curves of genus 1, which is naturally related to rational elliptic surfaces. Indeed, the presence of a genus 1 curve in the fixed locus of a non-symplectic involution assures that the elliptic fibration determined by it on the K3 is carried on to the rational quotient. One of the features of our method is that it oftentimes naturally yields a Weierstrass equation for the elliptic fibrations. This is usually not evident, and, as far as we know, most papers around the classification of elliptic fibrations do not provide a clear path. If the fixed locus of the chosen non-symplectic involution contains a curve of genus at least 2, then the relation between fibrations on the K3 and on the rational quotient is less evident since the quotient surfaces are not naturally endowed with elliptic or rational fibrations. For this reason, if the non-symplectic involution fixes a curve of genus at least 2, we use Nishiyama's method to give a list of all possible frame lattices of elliptic fibrations on the K3 surface.

In [6], we develop our classification program introduced in [19] further by classifying elliptic fibrations on certain K3 surfaces that are not general among the ones with a non-symplectic involution (so they do not have a 2-elementary Néron–Severi lattice). More precisely, we give a list of all frame lattices of elliptic fibrations on the K3 surfaces that are covers of the modular elliptic surface of level 5. We also describe an algorithm to determine the Weierstrass equations for many of the elliptic fibrations considered (see [6, Secs. 5.2 and 6.1.2]).

Our work takes an arithmetic turn in [9] when we use our classification to determine fields of definition of elliptic fibrations and their Mordell–Weil groups for certain K3 surfaces. More precisely, we consider K3 surfaces that are quadratic base changes of extremal rational elliptic surfaces either with one reducible fiber or with configurations (III^*, I_2) or (III^*, III), branched over smooth fibers. We show that the degree of the field of definition of the fibration and the Mordell–Weil group is related to the type of elliptic fibration according to the classification introduced in [19].

1.3 Advantages, Limitations, and Related Open Problems

Our work translates the problem of determining elliptic fibrations on a K3 surface X with a non-symplectic involution ι to that of describing linear systems on the quotient surface X/ι, which is rational and hence has an easier geometry. In particular, the realization of the quotient as a blow-up of the projective plane allows us to give an explicit description of the linear systems that induce elliptic fibrations on the K3. One of the main advantages of this approach is that it often yields Weierstrass equations for the elliptic fibrations that are preserved by the given non-symplectic involution ι. Another advantage of the aforementioned explicit description is that it allows one to determine an upper bound for the degree of the fields of definitions of the elliptic fibrations on the K3. This was explored to a minor extent in [9].

One of the apparent restrictions in our previous contributions is the requirement of the presence of a non-symplectic involution with a non-empty fixed locus. It turns out that this is not really a restriction, since all the K3 surfaces admitting an elliptic fibration admit also a non-symplectic involution with non-empty fixed locus, namely the so-called elliptic involution (discussed in Sect. 2.2). Nevertheless, in many cases, there is more than one non-symplectic involution with a non-empty fixed locus on the K3 surface, and in order to give a geometric interpretation to all the elliptic fibrations on the K3 surface, it could be more convenient to consider a different involution, perhaps one which is not the elliptic involution of any elliptic fibration on the K3 surface. For example, the K3 surfaces with a 2-elementary Néron–Severi lattice, double covers of rational elliptic surfaces and Kummer surfaces are naturally endowed with an involution whose geometry is well-understood and that helps in the classification of all the elliptic fibrations on the surface.

Therefore, it is natural to assume as a starting point the choice of a K3 surface with an involution, and then determine the elliptic fibrations. Each elliptic fibration induces an involution on the surface (the elliptic involution), which could be the original one or not. In the first part of this survey, we impose no condition to the involution, i.e., we do not require that it is non-symplectic nor that its fixed locus is non-empty, putting our previous result in a more natural and non-restrictive setting.

One of the main limitations of our method that remains in the more general framework of this survey is the difficulty in describing elliptic fibrations of one of types we introduced, more specifically the elliptic fibrations that are not preserved by the involution. In particular, when the non-symplectic involution ι acts non-trivially on $NS(X)$, then one is bounded to using another method, such as Nishiyama's, to check that the list of elliptic fibrations is complete. This could be circumvented to some extent by exhibiting another non-symplectic involution for which the fibration is of a different type. However, when the action of ι is trivial on $NS(X)$, i.e., when $NS(X)$ is 2-elementary, there are no fibrations of type 3, and our method determines all elliptic fibrations and moreover Weierstrass equations if the genus of the curves in the fixed locus is at most 1 (cf. Theorem 3).

The discussion above yields natural open problems. First, one might ask whether there are elliptic fibrations that can be realized as all three types by three different involutions. We expect an affirmative answer. By Garbagnati and Salgado [19, Table 1], certain fiber types are excluded for they are never realized by fibrations of type 1. In Sect. 4.3, we give examples of fibrations that can be realized as two different types. For fibrations of type 3, i.e., whose class of fiber is not preserved by the chosen involution, a bound for the intersection multiplicity of the two elliptic fibrations that are exchanged by the involution would be handy in describing the associated linear systems more explicitly.

Another interesting problem is finding an algorithm that yields a Weierstrass equation of the elliptic fibrations classified by us in the case the fixed locus of the involution contains a curve of genus at least 2. Finally, a very natural generalization of the problem we considered concerns the classification of elliptic fibrations on K3 surfaces endowed with a non-symplectic automorphism of higher order. This is work in progress by F. Zingali Meira, see [36].

1.4 Applications and Related Work

Our work has been applied in different contexts. In [14], the knowledge of two distinct elliptic fibrations on a K3 provided in [19] is used to show that certain classes of K3 surfaces have the Hilbert Property. In [25], the list of elliptic fibrations given in [18] is used to construct a fourfold birational to a Borcea–Voisin orbifold endowed with a flat elliptic fibration morphism. The relation of our work with F-theory and string theory becomes apparent in [12]. In [10], the authors study certain K3 surfaces with a 2-elementary Néron–Severi lattice. For these, they provide equations for projective models using the elliptic fibrations described in

[18] and show that the classification of frame lattices of elliptic fibrations given in [18] is actually a classification of all elliptic fibrations up to automorphisms applying the results of [16]. In [11], the authors reconsider the classification of the elliptic fibrations given in [13] and [18], providing also the classification up to automorphisms for general members of the considered families of K3 surfaces.

1.5 Outline of the Paper

In Sect. 2, we cover the necessary background on fibrations on rational and K3 surfaces.

Section 3 focuses on K3 surfaces with involutions, without any extra assumption on the involution, adopting a more general approach to the problem of classifying elliptic fibrations. We outline a classification of elliptic fibrations depending on the action of the given involution in this general setting that restricts to the classification introduced in [19] for non-symplectic involutions with non-empty fixed locus. Moreover, we determine the corresponding linear systems on the quotient surface, which in this general setting can be K3, Enriques, or rational, depending on the type of involution and on its fixed locus. We also discuss the interplay between singular fibers on the elliptic fibration on the K3 and its type with respect to the given involution. We give a more detailed account of elliptic fibrations that are preserved by the involution, dealing with those for which the involution acts on the basis in Sect. 3.2 and those for which the involution acts only on the fibers in Sect. 3.1. In Sect. 3.3, we restrict to K3 surfaces with a 2-elementary Néron–Severi lattice, recalling the results of [18].

In Sect. 4, we restrict the results from Sect. 3 to K3 surfaces that are double covers of rational elliptic surfaces, as in our previous contributions [19], [18], [6], and [9]. In this setting, if the involution acts only on the fibers, then quotient fibration is a conic bundle (see Sect. 4.1), and if the involution acts on the base of the fibration, then the quotient fibration has genus 1 (see Sect. 4.2). In Sect. 4.3, we show that a given elliptic fibration can be of different types with respect to different involutions. In fact, we provide examples of elliptic fibrations that realize simultaneously any pair of types.

Section 5 deals with extremal elliptic surfaces. More precisely, we list the extremal elliptic K3 surfaces that are quadratic base changes of extremal rational elliptic surfaces, whose complete classification is given in [27].

In Sect. 6, we revisit work from [6, Sec. 5.2]. More precisely, we present an algorithm to determine Weierstrass equations for certain elliptic fibrations on K3 surfaces that are double covers of rational elliptic surfaces. This is done by providing an alternative description of such K3 surfaces as double covers of minimal rational surfaces with special branch loci.

2 Preliminaries

Throughout this chapter, we work over an algebraically closed field of characteristic zero. In what follows, we introduce the definitions that remain in force for the rest of the chapter. The reader will find a detailed account of the theory of elliptic and K3 surfaces in [26], [31], and [22].

Definition 1 Let S be a smooth projective surface and C a smooth projective curve. We say that $\mathcal{E} : S \to C$ is a genus 1 fibration if it is a surjective map with connected fibers, such that the generic fiber is a smooth genus 1 curve.

We say that a genus 1 fibration $\mathcal{E}$ is an elliptic fibration if it has a distinguished section $\sigma : C \to S$.

We say that an elliptic fibration is relatively minimal if there are no (-1)-curves contained in the fibers.

We say that a surface is an elliptic surface if it admits an elliptic fibration.

In what follows, we always assume the base curve to be rational, i.e., $C \simeq \mathbb{P}^1$. The reason of this assumption is that in all the cases considered in this chapter, it is satisfied, and indeed, the following very well-known result holds.

Lemma 1 *If $\mathcal{E} : S \to C$ is an elliptic fibration and $H^1(S, \mathcal{O}_S) = 0$, then $C \simeq \mathbb{P}^1$. In particular, if S is a K3 or a rational surface, $C \simeq \mathbb{P}^1$.*

Proof By Miranda [26, (III.4.1)], the genus g of the base curve is related with the irregularity of the surface S, $q := \dim(H^1(S, \mathcal{O}_S))$, by the formula $q = g$ if the fibration is not trivial, and $q = g + 1$ otherwise. By hypothesis $H^1(S, \mathcal{O}_S)$ is trivial; hence, $\mathcal{E}$ is non-trivial and C is rational.

The hypothesis that $\mathcal{E}$ is relatively minimal implies that S is minimal unless S is rational. In that case, the fibration $\mathcal{E}$ is the unique elliptic fibration on S, and the negative curves of S are either (-1) or (-2)-curves. The former are sections, the latter components of reducible fibers. In what follows, relatively minimal rational elliptic surfaces are called RES for short. Since the canonical divisor of a RES is the opposite of the class of the fiber of the fibration, the elliptic fibration on an RES is uniquely determined. In particular, we might refer to S as a rational elliptic surface, keeping the morphisms $\mathcal{E}$ and σ implicit. If S is a K3 surface, then it might admit several relatively minimal elliptic fibrations. In that case, we make explicit which fibration we are referring to.

The reader can find more details, including examples, basic constructions, and a table of all possible singular fibers in [26].

Remark 1 Given two plane cubics F, G without a common component, assume that F is smooth. The blow-up of $\mathbb{P}^2$ along the possibly non-reduced scheme $F \cap G$ is a rational elliptic surface. Conversely, over an algebraically closed field, any RES is isomorphic to the blow-up of the projective plane in the 9, not necessarily distinct, base points of a pencil of cubics. See [26, Lem. IV.1.2]. □

2.1 Weierstrass Equations

The presence of a distinguished section on an elliptic fibration $\mathcal{E} : S \to C$ allows us to identify a specific point on each fiber of the fibration and therefore to define a group law on it. Hence, each smooth fiber is not just a genus 1 curve but an elliptic curve (with a fixed group structure). In particular, smooth fibers admit a Weierstrass equation. This allows us to define an equation for a (possibly) singular model of S, which is called the Weierstrass model. If $C \simeq \mathbb{P}^1$ and $\mathcal{E}$ is not a trivial fibration, this equation, called the Weierstrass equation of the fibration, has the following form (see [26, Chapter II]):

$$y^2 = x^3 + A(t:s)x + B(t:s), \quad A, B \in \mathbb{C}_{\text{hom}}[t:s], \ \deg(A) = 4m, \ \deg(B) = 6m, \tag{1}$$

for a certain $m \in \mathbb{N}$.

We notice that $m = 1$ if, and only if, S is rational and that $m = 2$ if, and only if, S is a K3 surface (see [26, III.1 and III.4.6] or [31, 4.10]). For each point $(\bar{t} : \bar{s})$, the Weierstrass equation exhibits the corresponding fiber as a double cover of $\mathbb{P}^1_{(x:z)}$ branched on the zero locus of $\left(x^3 + A(\bar{t} : \bar{s})xz^2 + B(\bar{t} : \bar{s})\right) z$. A section is a map $(\bar{t} : \bar{s}) \mapsto (x(\bar{t} : \bar{s}), z(\bar{t} : \bar{s}), y(\bar{t} : \bar{s}), (\bar{t} : \bar{s}))$, and the distinguished section σ is $(\bar{t} : \bar{s}) \mapsto (0, 0, 1, (\bar{t} : \bar{s}))$.

2.2 Automorphisms on Surfaces with an Elliptic Fibration

The presence of a group law on all but finitely many fibers of the elliptic fibration $\mathcal{E} : S \to C$ also allows one to define an involution on the surface S, which restricts to the elliptic involution on each fiber, i.e., to the involution mapping each point to its opposite in the group law. On the variables appearing in the Weierstrass equation (1), this involution acts as follows $\alpha : (x, y, (t : s)) \to (x, -y, (t : s))$. On each fiber, this is precisely the cover involution of the double cover map from the fiber to $\mathbb{P}^1_x$ (see [26, III.2]). In the case S is a K3 surface, α is non-symplectic, and the quotient S/α is a rational surface.

The set of sections of an elliptic fibration $\mathcal{E} : S \to C$ is naturally endowed with a group law (induced by one of the fibers), and it is called Mordell–Weil group of the fibration. The choice of name is justified by the $1-1$ correspondence between sections defined over a given field k and $k(C)$-points in the generic fiber of $\mathcal{E}$ [26, VII.2]. If the elliptic fibration $\mathcal{E} : S \to C$ admits a section $\lambda : C \to S$, different from the distinguished section σ, one defines the automorphism t_λ, called translation by λ, which restricts to each fiber to the translation by the intersection of the fiber and the section λ. The automorphism t_λ has the same order as the order of λ in the Mordell–Weil group. If S is a K3 surface, then the automorphism t_λ is symplectic.

In the following, we consider K3 surfaces with an involution and an elliptic fibration structure; α or t_λ are examples of automorphisms on elliptic K3 surfaces, but the ones appearing in the following are not necessarily of these types.

Notation Throughout this chapter, the letter R is reserved for rational elliptic surfaces and X for K3 surfaces.

Despite not allowing for different relatively minimal elliptic fibrations, rational surfaces are, clearly, the only class of surfaces on which elliptic fibrations and fibrations by rational curves might coexist. In what follows, we introduce a special class of rational fibrations, namely conic bundles, in a way that fits our purposes.

Definition 2 Let R be a rational surface. A conic bundle on R is a dominant morphism $f : R \to \mathbb{P}^1$ such that its generic fiber is a smooth curve of genus 0. □

If R is a RES and D is a fiber of a conic bundle, then $D \cdot (-K_R) = 2$ by the adjunction formula. Hence, alternatively, by Riemann–Roch, a conic bundle is determined by an effective divisor class D such that $D^2 = 0$ and $D \cdot (-K_R) = 2$. We call the class of such D in $\mathrm{NS}(R)$ a conic class and use D for the divisor or its class interchangeably.

Example 1 **[Conic Bundles on RES]** The theories of elliptic surfaces and conic bundles meet in the guise of rational elliptic surfaces. Let $(R, \mathcal{E}, \sigma)$ be a rational elliptic surface. If $\mathcal{E}$ has positive Mordell–Weil rank, then any two sections C_1, C_2 that intersect transversally at a point yield a conic class $D = C_1 + C_2$. Indeed,

$$D^2 = C_1^2 + 2C_1 \cdot C_2 + C_2^2 = -1 + 2 - 1 = 0 \text{ and } -K_R \cdot D = -K_R \cdot (C_1 + C_2) = 1 + 1 = 2.$$

For elliptic surfaces with trivial Mordell–Weil rank, conic classes are given by sums of fiber components with sections. Alternatively, taking as starting point the description in Remark (1), any pencil of lines through one of the blown-up points yields a conic bundle in R.

In [19, Figure 1] and [15] one finds the description of conic classes that are sums of negative curves. □

3 K3 Surfaces with Involutions

In what follows, we consider K3 surfaces with an involution. We distinguish the elliptic fibrations on the K3 according to the action of the involution on their fibers and study the corresponding linear systems on the quotient surface.

Let X be a K3 surface and $\iota \in \mathrm{Aut}(X)$ an involution. Let $q : X \to X/\iota$ be the quotient map. Then one of the following holds:

(i) X/ι is birational to a K3 surface, if ι is symplectic.
(ii) X/ι is an Enriques surface, if ι is non-symplectic and $\mathrm{Fix}(\iota) = \emptyset$.
(iii) X/ι is a rational surface, if ι is non-symplectic and $\mathrm{Fix}(\iota) \neq \emptyset$.

The involution ι is symplectic if, and only if, X/ι is singular, cf. case (i) above. In this case, we denote by Y its minimal model. For simplicity, we use the letter Y to denote X/ι also when this quotient is smooth. As we already observed, the elliptic involution α mentioned in Sect. 2.2 is as in (iii), and the involutions that are translation t_λ (again as in Sect. 2.2) for sections of order two λ are as in (i).

Let $F \subset X$ be a smooth genus 1 curve. Then, by the Riemann–Roch theorem, F spans a base-point-free 1-dimensional linear system $|F|$ on X. In other words, we have a genus 1 fibration $\varphi_{|F|} : X \to \mathbb{P}^1$. We denote by F_t the fiber of $\varphi_{|F|}$ over a point $t \in \mathbb{P}^1$.

Let F be as above. Then one of the following holds:

(a) $\iota(F) = F$.
(b) $\iota(F) = G$, for some genus 1 curve $G \neq F$ with $F \cdot G = m \geq 0$. □

Let $D = q(F)$; then D is a curve of genus 0 or 1. The following proposition gives a characterization of the case $g(D) = 0$.

Proposition 1 *Let X be a K3 surface, $\iota \in$ Aut(X) an involution, F, D, and q as above. Then the following are equivalent:*

(1) $g(D) = 0$.
(2) $F \not\subset Fix(\iota)$, $\iota(F) = F$, and $\iota_{|F}$ has 4 fixed points.
(3) Either ι is non-symplectic and $\iota(E) = E$ for each $E \in |F|$ or ι is symplectic, and there exist $E, E' \in |F|$ such that $\iota(E) = E'$ and $E \neq E'$.

Proof We observe that if $\iota(F) = G$ where $F \neq G$, then $q(F) \simeq q(G) \simeq F$; hence, $g(D) = 1$. So (1) implies that ι acts on F, i.e., $\iota(F) = F$, and $D = q(F)$ is $F/\iota_{|F}$. In particular, $g(D) = 0$ if and only if $\iota_{|F}$ is not the identity and is not fixed-point-free. Since ι is an involution, this is equivalent to $\iota_{|F}$ fixes 4 points. If (2) holds, the Riemann–Hurwitz formula allows to compute the genus of D proving (1). Indeed, since $\iota_{|F}$ is an involution of F, $q_{|F} : F \to D$ is the quotient map $F \to F/\iota_{|F}$ and so $D = q(F) = F/\iota_{|F}$.

We show that (1) implies (3). If $g(D) = 0$, then $\iota(F) = F$, so the linear system $|F|$ is preserved by ι. Therefore either each member of the linear system is preserved or it is mapped to another member of the same linear system. In the first case, there exists $E \in |F|$ such that $\iota_{|E} \neq id_{|E}$ (otherwise ι would be the identity on X) and ι cannot be a translation on the fibers, since $g(D) = 0$. So ι is the elliptic involution on each fiber (possibly composed with a translation). This implies that $\iota_{|E}$ does not preserve the period of E, and hence, ι does not preserve the period of X, i.e., it is non-symplectic. Otherwise, there exist $E, E' \in |F|$ such that $\iota(E) = E'$, so ι is not the identity on the basis of the fibration and it preserves exactly two fibers: F is one of them. We already observed that $\iota_{|F}$ is the elliptic involution (possibly composed with a translation), so ι is the composition of two non-symplectic involutions, one acting on the basis, the other on the fibers. This implies that ι is symplectic.

Vice versa (3) implies (1). Indeed, if ι is non-symplectic and it is the identity on $|F|$, it is the elliptic involution (possibly composed with a translation) on each fiber, so $D \simeq F/\iota_{|F}$ is a rational curve. If ι is symplectic and acts as a non-trivial involution on the basis, then ι is the composition of an involution on the basis and a non-period preserving involution on the fiber. This implies that $\iota_{|F}$ is the elliptic involution and $D \simeq F/\iota_F$ is rational. □

Remark 2 We observe that in the case ι is symplectic and $g(D) = 0$, the surface X/ι is singular in 8 points, and D passes through 4 of them. We consider the divisor $B = \beta^*(D)$ where $\beta : Y \to X/\iota$ is the minimal resolution. If $E \in |F|$ is a curve not preserved by ι, then $E \sim F$, so $\beta^*(D) = \beta^*(q(F)) \sim \beta^*(q(E))$, and since $q(E)$ does not pass through the singular points of X/ι, $\beta^*(D)^2 = \beta^*(q(E))^2 = q(E)^2 = F^2 = 0$. In particular, $|B|$ defines a fibration on Y. □

Remark 3 If $F \subset \mathrm{Fix}(\iota)$, i.e., $\iota|_F = id|_F$, then ι is non-symplectic, since it fixes a curve and $D \cong F$. □

If $\iota(F) = G$ and $F \cdot G = m > 0$, as in (b) above, then there exists an invariant curve M that is multisection of both fibrations $\varphi_{|F|}$ and $\varphi_{|G|}$. Indeed, let us consider two curves $F_t \in |F|$ and $G_t \in |G|$ such that $\iota(F_t) = G_t$. Denote $P_1^{(t)}, \cdots, P_m^{(t)}$ the m intersection points in $F_t \cap G_t$. This set is invariant for ι. Let $s \neq t$; then $F_s \cap F_t = \emptyset$. Hence, the sets $\{P_1^{(t)}, \cdots, P_m^{(t)}\}$ and $\{P_1^{(s)}, \cdots, P_m^{(s)}\}$ are disjoint. In particular, varying $t \in \mathbb{P}^1$, one obtains a curve M, which is invariant under ι and intersects the fibers of $\varphi_{|F|}$ and of $\varphi_{|G|}$ in m points.

Definition 3 Let ι be an involution on a K3 surface X, $\varphi_{|F|} : X \to \mathbb{P}^1$ a genus 1 fibration, F_t the fiber over a point $t \in \mathbb{P}^1$. We say that $\varphi_{|F|}$ is of:

Type 1) with respect to ι, if $\forall t \in \mathbb{P}^1$, $\iota(F_t) = F_t$, i.e., ι acts as an involution on F_t

Type 2) with respect to ι, if $\exists t \in \mathbb{P}^1$ such that $\iota(F_t) = F_s$, with $s \neq t$, ($F_s := \varphi_{|F|}^{-1}(s)$)

Type 3) with respect to ι, if $\exists t \in \mathbb{P}^1$ such that $\iota(F_t) = G$, for G a genus 1 curve that is not a fiber of $\varphi_{|F|}$ □

We observe that for a fibration of Type 3), ι^* maps the class F to the class of the curve $G \neq F$, in particular $\iota^*(F) \neq F$, and hence, there are no curves in the linear system of F mapped to curves in the same linear system. We conclude that if a fibration is of type 3, then $\forall t \in \mathbb{P}^1$, $\iota(F_t) \notin |F|$, excluding the possibility that a fibration is simultaneously of two different types with respect to the same involution.

Remark 4 If $\varphi_{|F|}$ is of type 1 or 2 with respect to ι, then:

(i) ι preserves the class of a fiber of $\varphi_{|F|}$. If it is of type 1, then ι acts trivially on the base of $\varphi_{|F|}$. If it is of type 2, then it acts as an involution of the base of $\varphi_{|F|}$, and in particular, ι preserves two fibers of $\varphi_{|F|}$.

(ii) If X/ι is smooth (i.e., ι is non-symplectic), and $D = q(F)$, then $\varphi_{|D|} : Y \to \mathbb{P}^1$ is a fibration. Indeed, $D^2 = 0$ since the class of the divisor q^*D is a multiple of the class of F.

(iii) If X/ι is singular (i.e., ι is symplectic), the linear system B as in Remark 2 defines a fibration on the desingularization Y. □

Remark 5 The classification in Definition 3 was inspired by Oguiso's work [30]. More precisely, Oguiso considered K3 surfaces with a non-symplectic involution acting trivially on the Néron–Severi group and distinguished the elliptic fibrations according to the presence of a section in the fixed locus of the involution. The presence of a section in the fixed locus implies that the involution acts trivially on the basis of the elliptic fibration. These are what we called fibrations of type 1 in Definition 3. For the remaining fibrations in [30], the involution gives an involution on the basis. These are called type 2 in Definition 3. Finally, we observe that there is no analog of the fibration of type 3 in [30] thanks to the trivial action of the involution on the Néron–Severi group. □

The curve $D = q(F)$ defines a linear system on Y, which endows Y with a fibration structure if $\varphi_{|F|}$ is of types 1 OR 2, by Remark 4. The following theorem describes more explicitly the properties of the linear system $|D|$ and, if $\varphi_{|F|}$ is of types 1 and 2, the interactions between the fibrations $\varphi_{|D|}$ and $\varphi_{|F|}$.

Theorem 2 *Let ι be an involution on X, $q : X \to X/\iota$ the quotient map, and $\beta : Y \to X/\iota$ the minimal resolution. Let F be a genus 1 curve on X, $\varphi_{|F|} : X \to \mathbb{P}^1$ the induced fibration, $D := q(F)$, and $B := \beta^*(D)$. Then one of the following holds:*

1. *If $\varphi_{|F|}$ is of type 1, then $\varphi_{|B|} : Y \to \mathbb{P}^1$ is a fibration and either:*
   (a) Y is a rational surface and $\varphi_{|B|}$ is a conic bundle.
   (b) Y is a K3 surface, and $\varphi_{|B|}$ is a genus 1 fibration.
2. *If $\varphi_{|F|}$ is of type 2, then $\varphi_{|B|} : Y \to \mathbb{P}^1$ is a genus 1 fibration.*
3. *If $\varphi_{|F|}$ is of type 3, then* $\dim(|D|) > 1$.

Proof First we observe that if Y is rational, ι is non-symplectic, and hence, X/ι is smooth, so β is not necessary, and we can assume it is the identity. In particular, B and D coincide:

(1) If Y is rational and $\varphi_{|F|}$ is of type 1, by Remark 4, $\varphi_{|D|} = \varphi_{|B|}$ is a fibration on Y and the genus of its generic fiber is given in Proposition 1. If Y is a K3 surface and $\varphi_{|F|}$ is of type 1, then $\varphi_{|B|}$ is a fibration on a K3 surface, by Remark 2. In particular, it is a genus 1 fibration, since all the fibrations on K3 surfaces are genus 1 fibration. So in case $\varphi_{|F|}$ is of type 1, it remains only to show that Y is not an Enriques surface, i.e., that ι has a non-empty fixed locus. By (i) in Remark 4, ι acts as the identity on the base of $\varphi_{|F|}$. Let us consider the restriction of ι to the smooth fibers, i.e., $\iota|_{F_t}$ for almost all $t \in \mathbb{P}^1$. If $\iota|_{F_t}$ acts as a translation on F_t for all the smooth fibers F_t, then it preserves their period and so it is symplectic and cannot be an Enriques involution. Otherwise, ι_{F_t} has some fixed points on F_t. In other words, either ι is symplectic or it is non-symplectic with $\mathrm{Fix}(\iota) \neq \emptyset$.

(2) If $\varphi_{|D|}$ is of type 2, by (i) in Remark 4, ι is an involution on the basis and preserves two fibers. If $\iota(F) \neq F$, then $F \simeq \iota(F) \simeq D$, so $g(D) = g(F)$, and since F is mapped to another fiber, it does not contain points fixed by ι. So $g(B) = g(D) = g(F) = 1$. The fixed locus of ι is contained in the two preserved fibers. In particular, if ι is non-symplectic and F is a preserved

fiber, then either F is a fixed curve or $\iota_{|F}$ is fixed-point-free. In both the cases, $g(F) = g(D)$. It remains to discuss the case where ι is symplectic and F is a preserved fiber. In Remark 2, it is proved that the linear system $|B|$ defines a fibration on Y that is in our case a K3 surface. The fibrations on K3 surfaces are necessarily genus 1 fibrations, by adjunction.

(3) If $\varphi_{|F|}$ is of type 3, then $F_t \cdot G_t = m > 0$, and we observe that $q^*(D) = F + G$ and $2D^2 = (q^*(F + G))^2 = 2m > 0$. Hence, $D^2 = m > 0$, and $|D|$ does not induce a fibration on Y.

Constructions of elliptic fibrations of type 1 with respect to a symplectic or to a non-symplectic involution are provided in Sects. 3.1 and 4.1 and are basically obtained by considering automorphisms that act trivially on the basis and act either as translations or as the elliptic involution on the fibers, respectively.

Constructions of elliptic fibrations of type 2 with respect to an involution are provided in Sects. 3.2 and 4.2 and are essentially obtained by considering automorphisms that act non-trivially on the basis, possibly composed with the elliptic involution on the fiber (in particular, this is necessary if one requires that the automorphism is symplectic). An interesting example of elliptic fibrations of type 2 with respect to a fixed-point-free involution is due to Hulek and Schuett, see [21, Section 3.4], and is obtained by considering an involution that is the composition of an involution on the basis and a translation by a special section (not necessarily of order 2) on the fibers.

Genus 1 fibrations of type 3 with respect to an involution ι have in general worse geometric descriptions, since the involution does not act on the fibration (but maps it to another fibration). Nevertheless, some interesting examples are known, and one is presented in [17, Section 3.5], where the authors find two genus 1 fibrations permuted both by a symplectic involution and by a non-symplectic one. The construction is the following: Let X be a K3 surface that is a double cover of $\mathbb{P}^1 \times \mathbb{P}^1$ branched on a curve of bidegree $(4, 4)$. If the branch curve is invariant under the automorphism γ of $\mathbb{P}^1 \times \mathbb{P}^1$ that switches the two rulings in $\mathbb{P}^1 \times \mathbb{P}^1$, then γ lifts to two involutions on X, one is symplectic and the other is not (indeed their composition is the cover involution of the map $X \to \mathbb{P}^1 \times \mathbb{P}^1$, which is a non-symplectic involution). By construction, both these involutions switch the genus 1 fibrations induced on X by the two projections $\mathbb{P}^1 \times \mathbb{P}^1 \to \mathbb{P}^1$. Hence, these fibrations are of type 3 with respect to both involutions.

Other examples of type 3 fibration are presented in [19]. By the proof of (3) in Theorem 2, we observe that if F is the class of the fiber of a fibration of type 3 with respect to ι, then ι preserves the linear system $|F + G|$. Hence, ι acts on each curve in this linear system either mapping it to another curve of the same linear system or as an involution of the curve. We underline that to compute the genus of the general member of the pushforward of the linear system $|F + G|$, in [(4.1)19] we wrongly assume that ι acts always as in the latter case, but this is not necessarily true.

3.1 Fibrations of Type 1

Given an involution $\iota \in \mathrm{Aut}(X)$, we discuss the properties of fibrations of type 1 with respect to ι. We first deal with symplectic involutions and later with non-symplectic ones.

Proposition 2 *Let X be a K3 surface and $\iota \in \mathrm{Aut}(X)$ a symplectic involution. Let $\varphi_{|F|}$ be an elliptic fibration on X of type 1 with respect to ι, and $\varphi_{|D|}$ the respective elliptic fibration on Y. Then both $\varphi_{|F|}$ and $\varphi_{|D|}$ admit a 2-torsion section. In particular, both fibrations cannot have fibers of types II^* or IV^*.* □

Proof Since $\varphi_{|F|}$ is of type 1 with respect to ι, this involution acts as the translation by a 2-torsion point in each fiber of $\varphi_{|F|}$ (see proof of (1) in Theorem 2). Hence, it acts as a translation by a 2-torsion point on the generic fiber, which clearly implies the presence of a 2-torsion section for $\varphi_{|F|}$. In particular, we can write a (generalized) Weierstrass equation for $\varphi_{|F|}$ as follows:

$$y^2 = x(x^2 + a(t)x + b(t)).$$

Taking the quotient by the 2-torsion section $(x, y, t) = (0, 0, t)$ yields a (generalized) Weierstrass equation for $\varphi_{|D|}$ (see [33, Proposition 3.7]):

$$y^2 = x(x^2 - 2a(t)x + (a(t)^2 - 4b(t)),$$

which clearly admits a 2-torsion section of the same shape. For the statement on the fiber types, recall that fibers of type II^* and IV^* are incompatible with the presence of a 2-torsion section [26, Table VII.2.6.]. □

By Proposition 2, ι is a translation by a 2-torsion section, which we denote by λ. The fiber types of $\varphi_{|D|}$ can be readily determined by the Weierstrass equation given in the proof. For completion, we present them below. We adopt the standard notation for the fiber components, namely, Θ_i denote the components and Θ_0, the neutral component, i.e., the one met by the zero section. In a fiber of type I_n^*, we denote by Θ_{near} the component connected to Θ_0 by a chain of three (-2)-curves (counting both Θ_0 and Θ_{near}) and by Θ_{far} the ones connected to Θ_0 by a chain of $n+3$ (-2)-curves.

Corollary 1 *Let $\varphi_{|F|}$ be an elliptic fibration on X, ι the translation by a 2-torsion section of $\varphi_{|F|}$, and $\varphi_{|D|}$ the induced fibration on the quotient. Then the fiber types of $\varphi_{|F|}$ and $\varphi_{|D|}$ are related as follows:*

F_t	III^*	I_{2n+1}	$I_{2n}, \lambda \cdot \Theta_0 = 1$	$I_{2n}, \lambda \cdot \Theta_n = 1$	I^*_{2n+1}	$I^*_{2n}, \lambda \cdot \Theta_{near} = 1$	$I^*_{2n}, \lambda \cdot \Theta_{far} = 1$
$D_{(t)}$	III^*	I_{4n+2}	I_{4n}	I_n	I^*_{4n+2}	I^*_{4n}	I^*_n

.

□

We move to study fibrations of type 1 with respect to a given non-symplectic involution. We first obtain results on the structure of the Mordell–Weil group and inadmissible singular fibers. Finally, we provide the reader with references for an account of the admissible types of singular fibers.

Proposition 3 *Let X be a K3 surface, $\iota \in \mathrm{Aut}(X)$ a non-symplectic involution, and $\varphi_{|F|}$ a fibration of type 1 with respect to ι:*

(i) If ι^* acts trivially on $\mathrm{NS}(X)$, then $\mathrm{MW}(\varphi_{|F|}) \subset (\mathbb{Z}/2\mathbb{Z})^2$.
(ii) If ι fixes two smooth genus 1 curves, then $\varphi_{|F|}$ has at most 6 irreducible singular fibers, and it admits no fibers of type $II^*, IV^*, II, III, IV, I_n$, for n odd and $n > 7$, and I_m^*, for m odd and $m > 5$ or m even and $m > 12$. □

Proof For part (i), see [18, Prop. 2.5] and [30, Lemma 2.4]. Part (ii) follows from [19, Remarks 5.6 and 5.7] and depends by the comparison between the singular fibers of the elliptic fibration on X and the singular fibers of fibration induced on the quotient, whose fibers are rational, by point (a) in Theorem 2. □

Remark 6 The reader will find a table for all admissible singular fibers of $\varphi_{|F|}$ in [19, Table 1]. This is also under the hypothesis as in (ii) Proposition 3. □

Remark 7 Comparing Propositions 2 and 3, the reader might notice the presence of extra assumptions on $\mathrm{Fix}(\iota)$ and on the action of ι^* on $\mathrm{NS}(X)$ in the statements for the non-symplectic case. These were not present in the symplectic case as there the fixed locus is always given by 8 points and ι^* never acts trivially on $\mathrm{NS}(X)$. □

3.2 *Fibrations of Type 2*

By Proposition 2, part (2), fibrations of type 2 with respect to ι induce genus 1 fibrations on Y. We have a closer look at such fibrations in what follows. Let $q : X \to Y$ be the quotient map, possibly composed with a birational map. By an abuse of notation, we denote $|D|$ both the linear system $|D|$ when X/ι is smooth and the linear system $|\beta^* D| = |B|$ when $\beta : Y \to X/\iota$ is the desingularization of Y. We have the following commutative diagram:

$$\begin{array}{ccc} X & \overset{q}{\underset{2:1}{\dashrightarrow}} & Y \\ {\scriptstyle \varphi_{|F|}}\downarrow & & \downarrow{\scriptstyle \varphi_{|D|}} \\ \mathbb{P}^1 & \overset{\psi}{\underset{2:1}{\longrightarrow}} & \mathbb{P}^1 \end{array}$$

where ψ is the map induced by q on the base of $\varphi_{|F|}$. In particular, ψ is branched above two points $t_0, t_1 \in \mathbb{P}^1$. Let D_{t_0} and D_{t_1} be the fibers of $\varphi_{|D|}$ above them. Then the restriction $q|_{F_{t_i}} : F_{t_i} \to D_{t_i}$ is $1:1$ for $i = 0, 1$, if, and only if, D_{t_i} is in the branch locus of q, or equivalently, if, and only if, D_{t_i} is not a multiple fiber of $\varphi_{|D|}$.

Proposition 4 *With the notation as above, let $\varphi_{|F|}$ be a fibration of type 2 with respect to ι. Then the following holds:*

(i) If Y is a K3 surface, then $\varphi_{|D|}$ has no multiple fibers, and at least one among D_{t_0} and D_{t_1} is reducible, with a non-reduced component.
(ii) If Y is an Enriques surface, then both D_{t_0} and D_{t_1} are multiple fibers.
(iii) If Y is rational, then at most one among D_{t_0} and D_{t_1} is a multiple fiber. □

Proof First observe that $\mathrm{Fix}(\iota) \subset F_{t_0} \cup F_{t_1}$. If $\mathrm{Fix}(\iota) = \emptyset$, then ι is an Enriques involution, and the branch locus of q is empty. Hence, D_{t_0} and D_{t_1} are multiple fibers of $\varphi_{|D|}$. This is case (ii). Assume that $\mathrm{Fix}(\iota) \neq \emptyset$. If ι is symplectic, then $\iota|_{F_{t_i}}$ fixes 4 points, for $i = 0, 1$. Hence, $q(F_{t_i})$ is a singular curve of multiplicity 2. After resolving its singularities, we obtain a non-multiple fiber D_{t_i} with at least one non-reduced component. If ι is non-symplectic, then at least one among F_{t_0} and F_{t_1}, say F_{t_0}, has some components contained in $\mathrm{Fix}(\iota)$. Hence, D_{t_0} shares a common component with the branch locus of q and, in particular, it cannot be a multiple fiber. □

Under the assumption that $\varphi_{|D|}$ has a section, then the statement of Proposition 4 becomes simpler.

Corollary 2 *With the notation as above, let $\varphi_{|F|}$ be a fibration of type 2 with respect to ι. Suppose that $\varphi_{|D|}$ has a section. Then:*

(i) Y is not an Enriques surface.
(ii) D_{t_0} and D_{t_1} are both branch curves for q.
(iii) $F_{t_i} \in \{I_{2n}, IV, I_0^*, IV^*\}$, for $i = 0, 1$.
(iv) If Y is rational, then D_{t_0} and D_{t_1} are reduced fibers of $\varphi_{|D|}$. □

Proof The presence of a section prevents the existence of multiple fibers; hence, case (ii) of Proposition 4 cannot happen and D_{t_i} are both branch fibers. This implies (i) and that F_{t_i} are both obtained by a quadratic base change on possibly singular fibers of an elliptic fibration, which proves (iii). Moreover, the presence of a section of $\varphi_{|D|}$ also assures that $\varphi_{|F|}$ admits a section. If Y is rational, and one of D_{t_i} is non-reduced, then an Euler number computation implies that the Euler characteristic $e(X) = 12$ or 0, and hence, X would not be a K3, proving (iv). □

Remark 8 Corollary 2 allows us to identify candidates for fibrations of type 2 with respect to a given involution among the list of all elliptic fibrations on a given K3 surface. Indeed, any fiber type that is not of the type listed in (iii) of Corollary 2 occurs in pairs, and there are at most two fiber types that occur alone. □

3.3 *Non-symplectic Involutions Acting Trivially on* NS(X)

The assumption that the action of ι^* on $\mathrm{NS}(X)$ is trivial simplifies the study of elliptic fibrations on X. Indeed, not only do we have the previous results at hand but, moreover, there are no fibrations of type 3 with respect to ι. If ι acts trivially on $NS(X)$, the lattice theoretic invariants that identify $NS(X)$ also identify the topology of the fixed locus and allow one to give an explicit description of X as double cover of a rational surface with a prescribed branch locus.

We summarize our results on the elliptic fibrations on such special K3 surfaces in what follows.

Theorem 3 *Let X be a K3 surface. Then there is a non-symplectic involution* $\iota \in \mathrm{Aut}(X)$ *such that* ι^* *acts trivially on* $NS(X)$ *if, and only if,* $NS(X)$ *is 2-elementary. In that case,* ι *is non-symplectic; X does not admit elliptic fibrations of type 3 with respect to* ι*; all elliptic fibrations on X are classified.* □

Proof The first assertion and the fact that ι is non-symplectic are classic and proved by Nikulin in [28] (see also [4, Thm. 0.1]). The nonexistence of fibrations of type 3 follows directly from its definition. Finally, the classification of elliptic fibrations on such surfaces is given in a series of papers by different authors. More precisely, the 2-elementary lattices that can be realized as the Néron–Severi lattice of a K3 are determined by 3 invariants, namely, r, its rank, a, its length, and δ, which encodes information on the discriminant form. If $(r, a, \delta) = (18, 4, 0)$, the K3 surface is the Kummer surface of the product of two non-isogenous elliptic curves and the classification was given by Oguiso in [30]. If $(r, a, \delta) = (16, 6, 1)$, the K3 surface is the double cover of the plane branched along 6 lines in general position, and the classification was given by Kloosterman in [23]. More generally, if $r+a = 22$, then the fixed locus of the involution consists only of rational curves and the classification was given in [13]. The remaining cases are contained in [18]. We observe that the fixed locus contains a curve of genus $g = \frac{22-r-a}{2}$. In particular, if $r+a = 22$, or 20, then the K3 can be realized as a double cover of a RES, branched along two fibers, which might be reducible or not, depending on $r + a$. □

Proposition 5 *Let X be a K3 surface. Assume that* $NS(X)$ *is 2-elementary. Then there is an involution* $\iota \in \mathrm{Aut}(X)$ *such that there are no elliptic fibrations on X of type 3 with respect to* ι*, and every elliptic fibration of type 1 has a finite Mordell–Weil group. In particular, if there is an elliptic fibration with infinite Mordell–Weil group, then it is of type 2 with respect to* ι*.* □

Proof Under the hypothesis, there exists an involution ι such that ι^* acts trivially on $NS(X)$ (as in Theorem 3). If $\mathcal{E}$ is an elliptic fibration of type 1 with a section C of infinite order, then C is not preserved by ι since it is mapped by ι to its opposite in the Mordell–Weil group. In particular, $\iota^*(C) \neq C$, and this contradicts the triviality of ι^*. □

4 K3s That Are Base Change of RES

In the previous sections, our input data have been that of a K3 surface with an involution. In our series of papers [19], [18], [6], and [9], we adopted a different viewpoint by considering a pair (X, R), with R a RES and X a K3 surface obtained as a double cover of R. This is equivalent to fixing a K3 surface with a non-symplectic involution whose fixed locus is composed of curves of genus at most 1.

In what follows, we give a brief account of our contributions and place them in the light of the previous sections.

Let $\mathcal{E}_R : R \to B_0 = \mathbb{P}^1$ be a relatively minimal RES and $\psi : B = \mathbb{P}^1 \to B_0$, a degree 2 morphism. We have the following commutative diagram:

$$\begin{array}{ccccc} X & \longrightarrow & R \times_{\psi} \mathbb{P}^1 & \xrightarrow[2:1]{q} & R \\ & \searrow^{\mathcal{E}_X} & \downarrow & & \downarrow \mathcal{E}_R \\ & & \mathbb{P}^1 & \xrightarrow[2:1]{\psi} & \mathbb{P}^1 \end{array}$$

Diagram 1

where X is a minimal model of $R \times_{\psi} \mathbb{P}^1$. If q is branched over reduced fibers of $\mathcal{E}_R$, then X is a K3 surface, and in that case, it is the minimal model of $R \times_{\psi} \mathbb{P}^1$. We assume that this is the case from now on.

The elliptic fibration $\mathcal{E}_R$ naturally induces an elliptic fibration on X, which is denoted by $\mathcal{E}_X$ above, and in what follows, in particular, we observe that the distinguished section of $\mathcal{E}_R$ determines a distinguished section of $\mathcal{E}_X$[26, Lemma 1.3.2].

The degree 2 map $\psi : B \to B_0$ induces an involution on B that lifts to the deck involution of the cover $X \to R$. It clearly preserves the class of a fiber of $\mathcal{E}_X$ in $\mathrm{NS}(X)$.

Context Let X be a K3 surface constructed as above and ι be the deck involution. In what follows, we denote by $\nu : X \to \mathbb{P}^1$ an elliptic fibration on X, possibly different from $\mathcal{E}_X$, and we analyze ν with respect to ι. By construction X/ι is rational, or equivalently, ι is non-symplectic with a non-empty fixed locus.

We start by observing that $\mathrm{Fix}(\iota)$ is contained in the union of the two fibers over the ramification points of ψ. Hence, it consists of a union of curves of genus at most 1. In particular, a genus 1 curve contained in it must be a fiber of $\mathcal{E}_X$.

Remark 9 If the branch fibers of q are smooth, then $R \times_{\psi} \mathbb{P}^1$ is smooth and X coincides with it. □

Corollary 3 *The elliptic fibration $\mathcal{E}_X$ is of type 2 with respect to ι. Moreover, if* $\mathrm{Fix}(\iota)$ *contains at least one smooth fiber of $\mathcal{E}_X$, then $\mathcal{E}_X$ is the unique fibration of type 2 with respect to ι.* □

Proof By construction, the involution ι on X is induced by the deck involution of the cover $\psi : \mathbb{P}^1 \to \mathbb{P}^1$. Hence, ι acts as an involution on the base of the fibration $\mathcal{E}_X$. Therefore $\mathcal{E}_X$ is of type 2 with respect to ι. Let F be the smooth fiber of $\mathcal{E}_X$ contained in the fixed locus. If $\nu : X \to \mathbb{P}^1$ is a fibration of type 2 with respect to ι, the fixed locus of ι is contained in fibers of ν, and if it contains a genus 1 curve, then it is a fiber of the fibration. This implies that $\mathcal{E}_X$ and ν coincide, since they share a fiber. □

If $\mathrm{Fix}(\iota)$ contains only rational curves, then there could be other elliptic fibrations of type 2 with respect to ι. This gives other realizations of X as a base change of an RES. Indeed, in that case, $R \times_{\psi} \mathbb{P}^1$ is singular and X is a blow-up of $R \times_{\psi} \mathbb{P}^1$. Hence, there is a commutative diagram

$$\begin{array}{ccc} X & \xrightarrow{\quad f \quad} & R \times_{\psi} \mathbb{P}^1 \\ {\scriptstyle q}\big\downarrow & & \big\downarrow{\scriptstyle \tilde{q}} \\ Y = X/\iota & \xrightarrow{\quad \tilde{f} \quad} & R \end{array}$$

Diagram 2

where q is the quotient map, f is a birational morphism, and Y is a rational surface with $K_Y^2 < 0$, and $\tilde{f}$ and $\tilde{q}$ are such that $\tilde{q} \circ f = \tilde{f} \circ q$. In particular, being a blow-up of R, the surface Y might have more than one non-relatively minimal elliptic fibration. In that case, there is more than one way to realize X as a double cover of an RES, and by construction, each non-relatively minimal elliptic fibration on Y yields a fibration of type 2 on X.

Remark 10 Let X be a K3 with 2-elementary $\mathrm{NS}(X)$. Assume that there is an elliptic fibration $\mathcal{E}_X : X \to \mathbb{P}^1$ with infinite Mordell–Weil group. Proposition 5 implies that there exists a RES $\mathcal{E}_R : R \to \mathbb{P}^1$ such that $\mathcal{E}_X$ is a quadratic base change of $\mathcal{E}_R$. □

4.1 *Fibrations of Type 1*

There are always fibrations of type 1 on K3 surfaces that arise as a double cover of an RES. Indeed, as discussed in Example 1, RESs admit conic bundles, and their pull-back yields elliptic fibrations on the K3 surface. The following is a reciprocal of (1) in Theorem 2 restricted to our context of K3 surfaces that are base change of RES.

Proposition 6 *Let Y be the quotient X/ι as in Diagram* (2) *and $|D|$ a conic bundle on Y. Then $|q^{-1}(D)|$ is a genus 1 fibration on X. Moreover, it is of type 1.* □

Proof It is enough to notice that $q^*(D)^2 = D^2 = 0$. By adjunction $q^*(D)$ has genus 1, and its class spans a genus 1 fibration on X. It is of type 1 by Theorem 2, point (2) and (3). □

If $D \subset Y$ is a genus zero curve such that $D^2 = 0$ and $D \cdot (-K_Y) = 2$, then D is a conic class in Y, and hence, it endows Y with a conic bundle structure. The linear system $|\tilde{f}(D)|$ in R has base points if, and only if, D intersects at least one of the exceptional curves of the blow-up map $\tilde{f}$. In this case, $|\tilde{f}(D)|$ is clearly not a conic bundle on R. If the latter holds, we called the divisor class $\tilde{f}(D)$ in R a *generalized conic bundle* in [19], to stress the fact that D is a conic bundle only on Y (i.e., on a blow-up of R). We reformulate Proposition 6 and its reciprocal in that language.

Theorem 4 ([19, Proposition 3.8 and Theorem 4.2]) *Let X, R, and ι be as in Diagram (2). Let $\nu : X \to \mathbb{P}^1$ be an elliptic fibration. Then ν is of type 1 with respect to ι if, and only if, $|\tilde{f} \circ q(\nu^{-1}(t))|$ is a conic bundle or a generalized conic bundle on R.*

4.2 Fibrations of Type 2

While R admits only one elliptic fibration, $\mathcal{E}_R$, which is by construction a relatively minimal elliptic fibration, the quotient surface Y, if different from R, might admit several genus 1 fibrations, both with and without sections, that are not relatively minimal. Given such a genus 1 fibration $\nu_Y : Y \to \mathbb{P}^1$ different from $\mathcal{E}_R$, consider R' the blow-down of Y to a surface on which ν_Y is relatively minimal. If ν_Y is induced on Y by an elliptic fibration $\nu : X \to \mathbb{P}^1$ on X that is of type (2) w.r.t. ι, then this produces a realization of X as a double cover of a different RES, R'. Similarly to what happens for generalized conic bundles, the curves $\tilde{f}(\nu_Y^{-1}(s))$ span a linear system that has base points on R and in particular is not a fibration. In [19] and [18], we called such linear systems *splitting genus 1 fibrations* on R.

We can now reformulate the point (2) of Theorem 2 and its reciprocal in this language.

Proposition 7 *Let X, R, and ι be Diagram (2). Let $\nu : X \to \mathbb{P}^1$ be an elliptic fibration. Then ν is of type 2 with respect to ι if, and only if, the linear system $|\tilde{f} \circ q(\nu^{-1}(s))|$ is a (splitting) genus 1 fibration on R.* □

4.3 Examples of Fibrations That Are of Different Types with Respect to Different Involutions

Given a K3 surface X as in Diagram (1) and its deck involution ι, we observed that any elliptic fibration is of types 1, 2, or 3 with respect to ι. The type of a fibration is, of course, relative to the involution we are considering. Hence, a priori, a given elliptic fibration on X can be of different types with respect to different involutions. A natural question is whether two different types can be simultaneously realized for the same elliptic fibration on a given K3 surface. We provide some examples, gathered from our previous papers [9] and [6], that give a positive answer to this question.

A Fibration Realized as Types 2 and 3 Let us consider the K3 surface X whose transcendental lattice is $U \oplus U(2)$. The frame lattices of the elliptic fibrations on X are listed in [18, Table 1, case $k = 8$, $r = 8$, $a = 2$, $\delta = 0$], and [11] shows that each frame corresponds to one elliptic fibration on X up to automorphisms, with the exception of the case with reducible fibers $2I_4^*$. In [9], the surface X is obtained as a quadratic base change of two different elliptic fibrations: R_3, whose singular fibers are $III^* + I_2 + I_1$ and R_4, whose singular fibers are $I_4^* + 2I_1$. In both cases, the branch fibers are smooth, and the deck involution of $X \to R_i$ will be denoted by ι_i, for $i = 3, 4$.

In [9, Tables 5 and 6], the types of each elliptic fibration on X with respect to ι_3 and ι_4 are determined. In particular, by Cantoral-Farfán et al. [9, Tables 5 and 6], one observes that X admits an elliptic fibration whose singular fibers are $2III^* + 2I_2 + 2I_1$. It is of type 2 with respect to ι_3 and of type 3 with respect to ι_4.

A Fibration Realized as Types 1 and 3 Consider the same surface and involutions as above. Then it admits another fibration whose singular fibers are I_{12}^*. It is of type 3 with respect to ι_3 and of type 1 with respect to ι_4.

A Fibration Realized as Types 1 and 2 In [6, Table 6.1], the frames of the elliptic fibrations on a surface denoted by $S_{5,5}$ are listed, and [11] shows that each frame corresponds to one elliptic fibration on X up to automorphisms, with the exception of the case with reducible fibers $III^*+I_6^*+I_2$. The surface $S_{5,5}$ is obtained by a base change on a RES whose singular fibers are $2I_5 + 2I_1$ and the branch fibers are the I_5-fibers. Let us denote by $\iota_{5,5}$ the deck involution. In [6, Table 6.1], we observe that on $S_{5,5}$ there is a fibration with singular fibers $2I_4^* + 2I_2$ and that it is of type 1 with respect to $\iota_{5,5}$. The surface $S_{5,5}$ is the unique K3 surface with transcendental lattice $\begin{bmatrix} 2 & 0 \\ 0 & 2 \end{bmatrix}$ and coincides with the surface that one finds as a quadratic base change of a RES with reducible fibers $I_4^* + 2I_1$ by considering the I_1-fibers as branch fibers (see also Table (1)). If we denote by $\iota_{1,1}$ the deck transformation, we observe the fibration with fibers $2I_4^* + 2I_2$ is of type 2 with respect to $\iota_{1,1}$.

5 Extremal Elliptic Fibrations

We consider special K3 surfaces, namely extremal elliptic K3 surfaces obtained as a quadratic base change of extremal rational elliptic surfaces, studying a problem analogous to the one treated in [27, Section 7].

Definition 4 An elliptic fibration $\mathcal{E}_Z : Z \to \mathbb{P}^1$ on a surface Z is called extremal if $rk(MW(\mathcal{E}_Z)) = 0$, and the Picard number is the largest possible, i.e., $rk(Pic(Z)) = h^{1,1}(Z)$. □

In particular, we say that a RES is extremal if the rank of the Mordell–Weil group of the unique elliptic fibration defined on it is 0. If a K3 surface admits an extremal elliptic fibration, then its Picard number is 20.

Extremal elliptic fibrations on rational elliptic surfaces are classified (see [27]) and fit in a list of 16 cases.

Extremal elliptic fibrations on K3 surfaces are classified (see [34]) and fit in a list of 325 cases.

Remark 11 An extremal RES contains a finite number of negative curves. This allows one to classify all the conic bundles on it, see [19]. Since the conic bundles on RES induce elliptic fibrations on the K3 surface obtained by a quadratic base change, it seems of particular interest to consider quadratic base changes on extremal RES.

K3 surfaces admitting an extremal elliptic fibration could contain infinitely many negative curves. □

In [27, Section 7], the authors determine what extremal RESs can be obtained as a base change of another extremal RES. Here we reconsider the analogous problem, namely, which extremal elliptic fibrations on K3 surfaces can be obtained as a quadratic base change of an extremal RES?

Theorem 6 *A quadratic base change* $\psi : \mathbb{P}^1 \to \mathbb{P}^1$ *of a RES* $\mathcal{E}_R : R \to \mathbb{P}^1$ *as in Diagram (1) produces an extremal K3 surface if, and only if, the following conditions are satisfied:*

(a) $\mathcal{E}_R : R \to \mathbb{P}^1$ *is an extremal RES.*
(b) ψ *is branched along singular reduced fibers of* $\mathcal{E}_R$.

There are 14 K3 surfaces admitting at least one extremal elliptic fibration obtained by a base change on an extremal RES. Some of these K3 surfaces admit more than one elliptic fibration, and one finds 25 different frames for extremal elliptic fibrations on K3 surfaces obtained as a quadratic base change by an extremal RES, see Table (1).

Proof Let X be an extremal elliptic K3 that is a quadratic base change of a RES R as in Diagram (1). Since every section of $\mathcal{E}_R$ induces a section of $\mathcal{E}_X$, and $\text{rank}(MW(\mathcal{E}_X)) = 0$, we must have $\text{rank}(MW(\mathcal{E}_R)) = 0$, and hence, $\mathcal{E}_R$ has to be extremal.

If the base change was branched on two smooth fibers, then $\mathcal{E}_X$ has twice as many reducible fibers as R. The lattice generated by the non-trivial components of the reducible fibers of $\mathcal{E}_X$ has rank 8 (since $\mathcal{E}_R$ is extremal and $\rho(R) = 10$), so the one generated by the non-trivial components of the reducible fibers of $\mathcal{E}_X$ has rank 16. Hence, $\rho(X) = 2 + 16 = 18$, which contradicts the fact that $\mathcal{E}_X$ is extremal. Similarly one excludes the case in which only one of the branch fibers is singular. We recall that the branch fibers are necessarily reduced; otherwise, X would not be a K3 surface.

Vice versa, by a case-by-case analysis, one can directly check that if $\mathcal{E}_R$ is an extremal RES and one considers a base change branched over two singular reduced fibers, one obtains an elliptic K3 surface $\mathcal{E}_X : X \to \mathbb{P}^1$ such that the trivial lattice has rank 20. This implies that $\rho(X) = 20$ and $\text{rank}(MW(\mathcal{E}_X)) = 0$.

This case-by-case analysis produces the following table, which allows one also to determine all the extremal fibration on K3 surfaces that can be obtained as a quadratic base change of a RES.

In the first column, there are all the extremal rational elliptic surfaces that have at least two singular reduced fibers, in the second the admissible choice for the branch fibers, in the third the singular fibers of X, which completely determines the transcendental lattice of X, given in the fourth column and obtained by comparing the fibrations with the one listed in [34]. The transcendental lattice is represented by a matrix in the form $\begin{bmatrix} a & b \\ b & c \end{bmatrix}$, and we express it as (a, b, c). The singular fibers of $\mathcal{E}_X$ are obtained as follows: Each singular fiber of $\mathcal{E}_R$, which is not a branch fiber, corresponds to two fibers of the same type on $\mathcal{E}_X$; a branch fiber of type I_n in $\mathcal{E}_R$ corresponds to a fiber of type I_{2n} on $\mathcal{E}_X$, by Miranda [26, (VI.4.1)].

Table 1 Extremal elliptic K3s X as double covers of extremal RES R

#	singular fibers $\mathcal{E}_R$	branch fibers	singular fibers $\mathcal{E}_X$	T_X
1	$II^* + 2I_1$	$2I_1$	$2II^* + 2I_2$	(2, 0, 2)
2	$I_4^* + 2I_1$	$2I_1$	$2I_4^* + 2I_2$	(2, 0, 2)
3	$I_9 + 3I_1$	$2I_1$	$2I_9 + 2I_2 + 2I_1$	(4, 2, 10)
4	$I_9 + 3I_1$	$I_9 + I_1$	$I_{18} + I_2 + 4I_1$	(2, 0, 2)
5	$III^* + I_2 + I_1$	$I_2 + I_1$	$2III^* + I_4 + I_2$	(2, 0, 4)
6	$I_8 + I_2 + 2I_1$	$2I_1$	$2I_8 + 4I_2$	(4, 0, 4)
7	$I_8 + I_2 + 2I_1$	$I_2 + I_1$	$2I_8 + I_4 + I_2 + 2I_1$	(2, 0, 4)
8	$I_8 + I_2 + 2I_1$	$I_8 + I_1$	$I_{16} + 3I_2 + 2I_1$	(2, 0, 4)
9	$I_8 + I_2 + 2I_1$	$I_8 + I_2$	$I_{16} + I_4 + 4I_1$	(2, 0, 2)
10	$IV^* + I_3 + I_1$	$I_3 + I_1$	$2IV^* + I_6 + I_2$	(2, 0, 6)
11	$I_2^* + 2I_2$	$2I_2$	$2I_2^* + 2I_4$	(4, 0, 4)
12	$I_1^* + I_4 + I_1$	$I_4 + I_1$	$2I_1^* + I_8 + I_2$	(2, 0, 8)
13	$I_6 + I_3 + I_2 + I_1$	$I_2 + I_1$	$2I_6 + I_4 + 2I_3 + I_2$	(6, 0, 12)
14	$I_6 + I_3 + I_2 + I_1$	$I_3 + I_1$	$3I_6 + 3I_2$	(2, 0, 6)
15	$I_6 + I_3 + I_2 + I_1$	$I_6 + I_1$	$I_{12} + 2I_3 + 3I_2$	(2, 0, 12)
16	$I_6 + I_3 + I_2 + I_1$	$I_3 + I_2$	$3I_6 + I_4 + 2I_1$	(4, 0, 6)
17	$I_6 + I_3 + I_2 + I_1$	$I_6 + I_2$	$I_{12} + I_4 + 2I_3 + 2I_1$	(4, 2, 4)
18	$I_6 + I_3 + I_2 + I_1$	$I_6 + I_3$	$I_{12} + I_6 + 2I_2 + 2I_1$	(2, 0, 4)
19	$2I_5 + 2I_1$	$2I_1$	$4I_5 + 2I_2$	(10, 0, 10)
20	$2I_5 + 2I_1$	$I_5 + I_1$	$I_{10} + 2I_5 + I_2 + 2I_1$	(2, 0, 10)
21	$2I_5 + 2I_1$	$2I_5$	$2I_{10} + 4I_1$	(2, 0, 2)
22	$2I_4 + 2I_2$	$2I_2$	$6I_4$	(4, 0, 4)
23	$2I_4 + 2I_2$	$I_4 + I_2$	$I_8 + 3I_4 + 2I_2$	(4, 0, 8)
24	$2I_4 + 2I_2$	$2I_4$	$2I_8 + 4I_2$	(4, 0, 4)
25	$4I_3$	$2I_3$	$2I_6 + 4I_3$	(6, 0, 6)

One observes that cases 1, 2, 4, 9, 21 correspond to the same K3 surface since the transcendental lattice identifies the K3 surface uniquely. The same holds true for cases 5, 7, 8, 18, for cases 6, 11, 22, 24, and for cases 10 and 14. Hence, we obtained 14 K3 surfaces, some of which admit more than one extremal elliptic fibration. □

Remark 12 The fact that on the same K3 surface, there are inequivalent elliptic fibrations (in particular extremal elliptic fibrations), which are obtained as a base change by a RES, implies that there are fibrations of type 2 with respect to the involution related with one of these base changes. More explicitly, we can interpret the K3 surface X as a base change of a specific RES. In particular, X is endowed with the cover involution ι. All the other inequivalent elliptic fibrations on X obtained as base changes from other rational elliptic surfaces are of type 2 with respect to ι. □

We observe that in [6] we considered the base change in case 21 of the previous table, denoted by $S_{5,5}$ in [6]. The cases 1, 2, 4, 9, 21 correspond to a K3 surface with a 2-elementary Néron–Severi lattice. The elliptic fibrations for the latter are classified (see, e.g., [13]).

Remark 13 In [27, Theorem 7.3], the authors proved that the extremal RES with singular fibers $I_8+I_2+2I_1$ and $2I_4+2I_2$ can be obtained by a quadratic base change on another extremal RES. Composing the quadratic base change described in [27, Theorem 7.3] with the one considered in Table 1, one obtains a $4:1$ base change from an extremal RES that produces an extremal elliptic K3 surface. The cover group can be either $(\mathbb{Z}/2\mathbb{Z})^2$ or $\mathbb{Z}/4\mathbb{Z}$, and both these cases appear (for example, the K3 surface in line 6 can be obtained as $(\mathbb{Z}/2\mathbb{Z})^2$ cover of the RES with singular fibers $I_4^* + 2I_1$, the one in line 9 as $\mathbb{Z}/4\mathbb{Z}$ cover of the same RES). □

6 K3s as Double Cover of Minimal Rational Surfaces

As in the previous two sections, we consider elliptic K3 surfaces obtained as a quadratic base change of a RES $\mathcal{E}_R : R \to \mathbb{P}^1$. In what follows, we present a different point of view on the same surfaces, namely, such K3 surfaces can be naturally realized as a double cover of a minimal rational surface whose branch locus satisfies certain properties. We first describe the relationship between these different points of view and then show that their interactions allow one to write explicit equations for the elliptic fibrations on X, following [6].

Proposition 8 *Let $\mathcal{E}_R : R \to \mathbb{P}^1$ be a relatively minimal RES. Let $\beta : R \to R^o$ be a composition of contractions of (-1)-curves to a minimal surface R^o; then R^o is either $\mathbb{P}^2$ or $\mathbb{P}^1 \times \mathbb{P}^1$ or $\mathbb{F}_2$. In the latter case $\mathcal{E}_R : R \to \mathbb{P}^1$ admits at least one reducible fiber.* □

Proof Since there are no curves on R with self-intersection less than -2, also on R^o, there cannot be curves with self-intersection less than -2. This excludes all the Hirzebruch surfaces $\mathbb{F}_n$ with $n \geq 3$ as possible minimal models of R. For the same reason, if $R^o \simeq \mathbb{F}_2$, then R contains a (-2)-curve, and hence, it admits a reducible fiber. The surface $\mathbb{F}_0$ is $\mathbb{P}^1 \times \mathbb{P}^1$, and the surface $\mathbb{F}_1$ is not minimal. □

Proposition 9 *Let X be obtained as a quadratic base change by a RES $\mathcal{E}_R : R \to \mathbb{P}^1$ as in Diagram (1). Let $\beta : R \to R^o$ be the birational morphism described in Proposition 8. Then X is the minimal resolution of a double cover of R^o branched on a reducible curve with at least two irreducible components.* □

Proof By Diagram (1), we obtain a generically $2 : 1$ map $q_X : X \to R$ branched along two fibers of R. The birational morphism $\beta : R \to R^o$ is the blow-up of R^o in the base points of a pencil $\mathcal{P}$ of genus 1 curves and so β maps two fibers of $\mathcal{E}_R$ to two curves in this pencil. Hence, $\beta \circ q_X : X \to R^o$ is a generically $2 : 1$ map branched on two curves in the pencil $\mathcal{P}$. □

We consider a reciprocal of Proposition 8 that allows one to interpret certain double covers of minimal rational surfaces as base change of rational elliptic surfaces. To state the result, we denote: by H the class of a line in $\mathbb{P}^2$; by h_1, h_2 the generators of $\mathrm{Pic}(\mathbb{P}^1 \times \mathbb{P}^1)$ such that $h_1^2 = h_2^2 = 0$ and $h_1 \cdot h_2 = 1$; by Γ, Φ the generators of $\mathrm{Pic}(\mathbb{F}_n)$ such that $\Phi^2 = 0$, $\Gamma^2 = -n$, $\Phi \cdot \Gamma = 1$.

Corollary 4 *Let X be a K3 surface and $X \to W$ be a double cover of a minimal rational surface W branched on the union of two (possibly reducible) curves C_1 and C_2 without common components. If $W = \mathbb{P}^2$ (resp., $\mathbb{P}^1 \times \mathbb{P}^1$, $\mathbb{F}_2$) and $C_1 \sim C_2 \in |3H|$ (resp., $|2h_1+2h_2|$, $|2\Gamma+4\Phi|$), then X is obtained by a quadratic base change on the RES obtained blowing up W in the points $C_1 \cap C_2$.*

Proof By Hartshorne [20, Chapter V Corollary 2.18], the linear system $|2\Gamma + 4\Phi|$ contains smooth curves. We observe that all linear systems in the statement are the anticanonical linear system in the respective surface, and hence, the genus of a general member is 1. Hence, it suffices to consider two smooth curves in such linear systems to find the required pencil on the minimal surface W. □

Remark 14 We observe that:

(i) By choosing C_1 and C_2 general in their linear system in $\mathbb{P}^2$ or $\mathbb{P}^1 \times \mathbb{P}^1$ (resp., $\mathbb{F}_2$), one can recover a general RES with no reducible fibers (resp., with exactly one reducible fiber). The other configurations of singular fibers can then be obtained by specializing C_1 and C_2.
(ii) Explicit examples of contractions of the RES to different minimal models are provided in [9, Figures 1 and 3].

□

The following example deals with a well-known case and shows the relevance of Corollary 4.

Example 2 Let X be the double cover of $\mathbb{P}^2$ branched on six lines in general position $\ell_1, \ldots, \ell_6$. The elliptic fibrations on X are classified by Kloosterman, [23]. In order to show the existence of the elliptic fibrations, he considered pencils of curves in $\mathbb{P}^2$ whose pull-back to X give pencils of genus 1 curves. A slightly different point of view is the following: Let us consider the curves $C_1 = \ell_1 \cup \ell_2 \cup \ell_3$ and $C_2 = \ell_4 \cup \ell_5 \cup \ell_6$. The pencil generated by C_1 and C_2 is a pencil of genus 1 curves, and the blow-up of $\mathbb{P}^2$ in its base point yields a RES $\mathcal{E}_R : R \to \mathbb{P}^1$ with two fibers of type I_3 (corresponding to the curves C_1 and C_2). By Corollary 4, X is the base change of $\mathcal{E}_R : R \to \mathbb{P}^1$ branched on the two I_3-fibers, so X is naturally endowed with an elliptic fibration with two fibers I_6 (the ones obtained as a base change of R branched on the two I_3-fibers). The other elliptic fibrations on X are either induced by a generalized conic bundle (classified in [23, Section 9]) or by splitting genus 1 pencil (classified in [23, Section 8]).

The described point of view, i.e., considering K3 surfaces that are a double cover of minimal rational surfaces and elliptic fibrations induced by linear systems of plane curves, is the one adopted in [23] and [13]. In both these papers, the surface R^o is isomorphic to $\mathbb{P}^2$. In [9], we also considered elliptic fibrations related with a model of X as double cover of $R^o \simeq \mathbb{P}^1 \times \mathbb{P}^1$ and $R^o \simeq \mathbb{F}_2$ (see, e.g., [9, Figures 1 and 3]). In particular, in [9], one considers elliptic fibrations whose field of definition is not necessarily algebraically closed; in this case, it is possible that R cannot be contracted over the field of definition to $\mathbb{P}^2$, but it can be contracted to some other minimal models of rational surfaces.

6.1 An Algorithm to Determine Weierstrass Equations

In [6], the authors consider certain K3 surfaces obtained as a quadratic base change of a RES, classify conic bundles $|B|$, and construct generalized conic bundles on the RES. They consider the push-forward of these conic bundles to R^o obtaining pencils of curves in $\mathbb{P}^2$, hence intertwining the two points of view. We observe that the same can be reproduced for $R^o = \mathbb{P}^1 \times \mathbb{P}^1$ or $\mathbb{F}_2$. The advantage of this point of view is that the classification of the conic bundles is quite easy, in particular when one has a strong control on the negative curves on R, as in the case of extremal rational elliptic fibrations. On the other hand, one is able to explicitly write equations for pencils of plane curves (or of curves in $\mathbb{P}^1 \times \mathbb{P}^1$) with some prescribed properties of the base points. This allows the authors to apply an algorithm that gives the Weierstrass equation of some of the classified elliptic fibrations on X.

By Corollary 4, we know that X is a double cover of $\mathbb{P}^2$ branched on the union of two cubics, so its equation is of the form

$$w^2 = f_3(x_0 : x_1 : x_2)g_3(x_0 : x_1 : x_2). \tag{1}$$

Let B be a conic bundle on R, i.e., a base-point-free linear system of rational curves giving $R \to \mathbb{P}^1_\tau$. Pushing forward to $\mathbb{P}^2$, B is given by a pencil of plane rational curves with equation $h(x_0 : x_1 : x_2, \tau) = 0$. The polynomial $h(x_0 : x_1 : x_2, \tau)$ is homogeneous in x_0, x_1, x_2, say of degree $e \geq 1$ and linear in τ.

For general τ, we must find an isomorphism of the curve $h(x_0 : x_1 : x_2, \tau) = 0$ with $\mathbb{P}^1$ and extract the images of the four intersection points with $f_3g_3 = 0$.

When $e \leq 3$, an isomorphism with $\mathbb{P}^1$ is provided by projection from a point of order $e - 1$ on the curve (e.g., any point in $\mathbb{P}^2$ if $e = 1$, a point on the conic if $e = 2$, and a double point of the cubic if $e = 3$). Such a point necessarily exists (in the case $e = 3$, the singularity must be a base point of the pencil) and is also necessarily a base point of the original pencil of cubics giving $\mathcal{E}_R$. Up to acting by $\mathrm{PGL}_3(\mathbb{C})$, we may assume that this point is $(0 : 1 : 0)$.

Algorithm when $e \leq 3$**:**

1. Compute the resultant of the polynomials $f_3(x_0 : x_1 : x_2)g_3(x_0 : x_1 : x_2)$ and $h(x_0 : x_1 : x_2, \tau)$ with respect to the variable x_1. The result is a polynomial $r(x_0 : x_2, \tau)$ that is homogeneous in x_0 and x_2, corresponding to the images of *all* the intersection points $\{f_3g_3 = 0\} \cap \{h_\tau = 0\}$ after projection from $(0 : 1 : 0)$.
2. Since B is a conic bundle, $r(x_0 : x_2, \tau)$ will be of the form $a(x_0 : x_2, \tau)^2 b(x_0 : x_2, \tau)c(\tau)$, where a and b are homogeneous in x_0 and x_2, the degree of a depends upon e, and the degree of b in x_0 and x_2 is 4.
3. The equation of X is now given by $w^2 = r(x_0 : x_2, \tau)$, which is birationally equivalent to
$$w^2 = c(\tau)b(x_0 : x_2, \tau), \tag{2}$$
by the change of coordinates $w \mapsto wa(x_0 : x_2, \tau)$. Since for almost every τ, Eq. (2) is the equation of a $2 : 1$ cover of $\mathbb{P}^1_{(x_0:x_2)}$ branched in 4 points, (2) is the equation of the genus 1 fibration on the K3 surface X induced by the conic bundle $|B|$.
4. If there is a section of fibration (2), then it is possible to obtain the Weierstrass form by standard transformations.

Remark 15 The algorithm is not restricted to conic bundles. It can be applied exactly in the same way to the generalized conic bundles. □

When $e \geq 4$, the projection from a point may suffice, for example, if all curves have a base point of degree $e - 1$. However, there are several conic bundles whose general members cannot be parametrized by lines, and in this case, one has to consider conics. In [6, Section 5.2], an algorithm is presented also in this case. Moreover, also in the case of elliptic fibrations induced by splitting genus 1 pencils, similar algorithms exist, see [6, Section 6.1.2].

An Example To illustrate how the algorithm works, we consider the RES R given by the blow-up of $\mathbb{P}^2$ in the nine base points of the pencil of cubics $\mathcal{P}$ generated by the cubics C_0 and C_∞ that are, respectively, the zero locus of the polynomials xyz and $x^2(ay+bz)+x(cy^2+dyz+ez^2)+(fy^3+gy^2z+hyz^2+lz^3)$. We observe that C_0 is the union of three lines and that, for a generic choice of the coefficients, C_∞ is a smooth cubic passing through $(1 : 0 : 0)$. The pencil $\mathcal{P}$ induces an elliptic fibration $\mathcal{E}_R$ on R whose fibers are the strict transforms of the cubics in $\mathcal{P}$. In particular, there are two fibers that are the strict transforms of C_0 and C_∞. The quadratic base change branched on these fibers produces a K3 surface X naturally endowed with an elliptic fibration, and as above an equation for a birational model of such a K3 surface is

$$w^2 = xyz\left(x^2(ay+bz)+x(cy^2+dyz+ez^2)+(fy^3+gy^2z+hyz^2+lz^3)\right).$$

The deck involution of the cover $X \to R$ induces on this model of X the involution ι acting as -1 on w. It is non-symplectic, and its fixed locus consists of the smooth cubic C_∞, the three lines components of C_0, and another rational curve (due to the blow-ups of $(1:0:0)$, which are needed to obtain X as a smooth model of the double cover of $\mathbb{P}^2$). One can show that the involution ι acts trivially on $NS(X)$, so one deduces the invariants of $NS(X)$ by the knowledge of the fixed locus. In particular, $NS(X)$ is a 2-elementary lattice of rank 14 and length 6 (see e.g., [4, Figure 1]), and the elliptic fibrations on X are classified [18, Table 1 case $r = 14$, $a = 6$, $\delta = 1$].

A conic bundle on R is the pull-back of a line through a base point of the pencil $\mathcal{P}$. Let us consider, e.g., the lines through $(1:0:0)$, whose equations are of the form $y = \lambda z$. We plug this in the equation of the K3 surface to obtain

$$\begin{aligned} w^2 = \lambda x z^2 &\left(x^2(a\lambda z+bz)+x(c\lambda^2z^2+d\lambda z^2+ez^2)\right.\\ &\left.+(f\lambda^3z^3+g\lambda^2z^3+h\lambda z^3+lz^3)\right). \end{aligned}$$

Taking $z = 1$, which corresponds to fixing the section at infinity, we obtain

$$w^2 = \lambda x\left(x^2(a\lambda+b)+x(c\lambda^2+d\lambda+e)+(f\lambda^3+g\lambda^2+h\lambda+l)\right).$$

The change of coordinates $w \mapsto w\lambda^2(a\lambda+b)^2$, $x \mapsto \lambda(a\lambda+b)$ gives, after dividing all the equations by $\lambda^4(a\lambda+b)^4$,

$$w^2 = x\left(x^2+x\frac{(c\lambda^2+d\lambda+ez^2)}{\lambda(a\lambda+b)^2}+\frac{(f\lambda^3+g\lambda^2+h\lambda+l)}{\lambda^2(a\lambda+b)^3}\right).$$

This is an extended Weierstrass form for an elliptic fibration on the K3 surface X, which in particular shows that the fibration has not only the distinguished section, but also a 2-torsion one (given by $(x, w, \lambda) = (0, 0, \lambda)$).

If one prefers to write the coefficients of the extended Weierstrass equation as polynomials (instead of the quotient of polynomials), one can consider another change of coordinates, namely, $w \mapsto \frac{w}{\lambda^3(a\lambda+b)^3}$, $x \mapsto \frac{x}{\lambda^2(a\lambda+b)^2}$, and then multiply all the equations by $\lambda^6(a\lambda+b)^6$.

This gives

$$w^2 = x\left(x^2+x\lambda(c\lambda^2+d\lambda+e)+\lambda^2(a\lambda+b)(f\lambda^3+g\lambda^2+h\lambda+l)\right).$$

For $\lambda = 0$ and $\lambda = \infty$, one finds reducible fibers that correspond to the reducible fibers of the conic bundle. Indeed, $\lambda = 0$ and $\lambda = \infty$ correspond to the lines

$y = 0$ and $z = 0$ in $\mathbb{P}^2$, i.e., to fibers of the conic bundle that are also components of the fibers of the elliptic fibration $\mathcal{E}_R$. By computing the discriminant of the Weierstrass equation, one finds all the reducible fibers, which, for a general choice of coefficients, are $2I_0^* + 4I_2 + 4I_1$. In particular, the elliptic fibration induced by the conic bundle from lines through $(1 : 0 : 0)$ corresponds to case (4.6) in [18, Table 1].

Acknowledgments Our collaboration started in the first edition of *Women in Numbers Europe*. We are happy to thank the organizers of that and all the other *Women in Numbers Europe* editions. It has been a great experience to participate in these collaborative workshops. We thank the referees for their careful reading and suggestions that improved the manuscript.

References

1. V. Alexeev, Reflective hyperbolic 2-elementary lattices, K3 surfaces and hyperkahler varieties. Preprint (2022). arXiv:2209.09110
2. V. Alexeev, P. Engel, Compactifications of moduli spaces of K3 surfaces with a nonsymplectic involution. Preprint (2022). arxiv:2208.10383
3. V. Alexeev, V.V. Nikulin, Classification of log del Pezzo surfaces of index ≤ 2 and applications. MSJ Memoirs, 104–116 (2006)
4. M. Artebani, A. Sarti, S. Taki, K3 surfaces with non-symplectic automorphisms of prime order. Math. Zeitschrift **268**, 507–533 (2011)
5. M.J. Bertin, O. Lecacheux, *Elliptic Fibrations on the Modular Surface Aassociated to* $\Gamma_1(8)$, in *Arithmetic and geometry of K3 surfaces and Calabi–Yau threefolds*. Fields Inst. Commun., vol. 67 (Springer, New York, 2013), pp. 153–199
6. F. Balestrieri, J. Desjardins, A. Garbagnati, C. Maistret, C. Salgado, I. Vogt, *Elliptic Fibrations on Covers of the Elliptic Modular Surface of Level 5*. Women in numbers Europe II, Assoc. Women Math. Ser., vol. 11 (Springer, Cham, 2018), pp. 159–197
7. M.J. Bertin, A. Garbagnati, R. Hortsch, O. Lecacheux, M. Mase, C. Salgado, U. Whitcher, *Classifications of Elliptic Fibrations of a Singular K3 Surface*. Women in numbers Europe, Assoc. Women Math. Ser., vol. 2 (Springer, Cham, 2015), pp. 17–49
8. A.P. Braun, Y. Kimura, T. Watari, On the classification of elliptic fibrations modulo isomorphism on K3 surfaces with large Picard number. Preprint (2013). arXiv:1312.4421v1
9. V. Cantoral-Farfán, A. Garbagnati, C. Salgado, A. Trbović, R. Winter, *Fields of Definition of Elliptic Fibrations on Covers of Certain Extremal Rational Elliptic Surfaces*. Women in numbers Europe III, Research directions in number theory. Assoc. Women Math. Ser., vol. 24 (Springer, Cham, 2021), pp. 171–205
10. A. Clingher, A. Malmendier, On K3 surfaces of Picard rank 14. Preprint (2020). arXiv:2009.09635
11. A. Clingher, A. Malmendier, On Neron-Severi lattices of Jacobian elliptic K3 surfaces. Preprint (2021). arXiv:2109.01929
12. A. Clingher, A. Malmendier, On the duality of F-theory and the CHL string in seven dimensions. Commun. Math. Phys. **393**, 631–667 (2022)
13. P. Comparin, A. Garbagnati, van Geemen–Sarti involutions and elliptic fibrations on K3 surfaces double cover of $\mathbb{P}^2$. J. Math. Soc. Jpn. **66**, 479–522 (2014)
14. J.L. Demeio, Elliptic fibrations and the Hilbert property. IMRN **2021**(13), 10260–10277 (2021)
15. R. Dias Costa, Classification of conic bundles on a rational elliptic surface in any characteristic. Preprint (2022). arXiv:2206.03549

16. D. Festi, D.C. Veniani, Counting elliptic fibrations on K3 surfaces. J. Math. Soc. Jpn. **75**(4), 1195-1225 (2023)
17. B. van Geemen, A. Sarti, Nikulin involutions on K3 surfaces. Math. Z. **255**, 731–753 (2007)
18. A. Garbagnati, C. Salgado, Elliptic fibrations on K3 surfaces with a non-symplectic involution fixing rational curves and a curve of positive genus. Rev. Mat. Iberoam. **36**, 1167–1206 (2020)
19. A. Garbagnati, C. Salgado, Linear systems on rational elliptic surfaces and elliptic fibrations on K3 surfaces. J. Pure Appl. Algebra **223**, 277–300 (2019)
20. R. Hartshorne, *Algebraic Geometry*. Graduate Texts in Math (Springer, 1977)
21. K. Hulek, M. Schuett, Enriques Surfaces and Jacobian elliptic K3 surfaces. Math. Z. **268**, 1025–1056 (2011)
22. D. Huybrechts, *Lectures on K3 Surfaces*. Cambridge Studies in Advanced Mathematics (Cambridge University Press, Cambridge, 2016)
23. R. Kloosterman, Classification of all Jacobian elliptic fibrations on certain K3 surfaces. J. Math. Soc. Jpn. **58**, 665–680 (2006)
24. A. Kumar, Elliptic fibrations on a generic Jacobian Kummer surface. J. Algebraic Geom. **23**, 599–667 (2014)
25. K. Kanno, T. Watari, W = 0 complex structure moduli stabilization on CM-type K3 x K3 orbifolds. Commun. Math. Phys. **398**(6), 703–756 (2023)
26. R. Miranda, *The Basic Theory of Elliptic Surfaces*. Dottorato di Ricerca in Matematica, ETS Editrice Pisa (1989)
27. R. Miranda, U. Persson, On extremal rational elliptic surfaces. Math. Z. **193**, 537–558 (1986)
28. V.V. Nikulin, *Quotient-Groups of Ggroups of Automorphisms of Hyperbolic Forms by Subgroups Generated by 2-Reflections. Algebro-Geometric Applications*. Current Problems in Mathematics, vol. 18. Akad. Nauk SSSR, Vsesoyuz (Inst. Nauchn. i Tekhn. Informatsii, Moscow, 1981), pp. 3–114
29. K. Nishiyama, The Jacobian fibrations on some K3 surfaces and their Mordell–Weil groups. Jpn. J. Math. (N.S.) **22**, 293–347 (1996)
30. K. Oguiso, On Jacobian fibrations on the Kummer surfaces of the product of non-isogenous elliptic curves. J. Math. Soc. Jpn. **41**, 651–680 (1989)
31. M. Schuett, T. Shioda, Elliptic surfaces. Adv. Stud. Pure Math. **2010**, 51–160 (2010)
32. I. Shimada, On elliptic K3 surfaces. Michigan Math. J. **47**(3), 423–446 (2000)
33. J. Silverman, J.T. Tate, *Rational Points on Elliptic Curves*, 2nd edn. Undergraduate Texts in Mathematics (Springer, Cham, 2015)
34. I. Shimada, D.Q. Zhang, Classification of extremal elliptic K3 surfaces and fundamental groups of open K3 surfaces. Nagoya Math. J. **161**, 23–54 (2001)
35. H. Sterk, Finiteness results for algebraic K3 surfaces. Math. Zeitschrift **189**(4), 507–513 (1985)
36. F. Zingali Meira, Jacobian elliptic fibrations on K3s with a non-symplectic automorphism of order 3. Preprint (2024). arXiv:2403.10712v1

Mirror Constructions for K3 Surfaces from Bimodal Singularities

Makiko Mase and Ursula Whitcher

1 Introduction

Given a K3 surface realized as a hypersurface in a weighted projective space or a Gorenstein Fano toric variety, one may construct a mirror K3 surface in various ways. Depending on the precise model, available descriptions of mirror symmetry include the Greene–Plesser mirror, the Berglund–Hübsch transpose construction for invertible polynomials, Dolgachev–Nikulin–Pinkham's lattice-polarized K3 surfaces, and Batyrev's reflexive polytope construction. The multitude of descriptions raises the question of whether mirror constructions are consistent. Comparing different mirror constructions often entails making choices—one might need to specify a family containing a K3 surface or a lattice polarization, for example—and thus it is important to establish systematic methods for making these choices.

Ebeling and Ploog studied invertible polynomials in three variables obtained from Arnold's classification of bimodal singularities in [9]; we list these polynomials in Table 1.

The second author would like to thank the Isaac Newton Institute for Mathematical Sciences, Cambridge, for support and hospitality during the K-theory, algebraic cycles, and motivic homotopy theory programme, where work on this paper was undertaken. This work was supported by EPSRC grant no EP/R014604/1. Both authors are indebted to the anonymous referees for helpful comments that improved the readability of this chapter.

M. Mase (✉)
Tokyo Metropolitan University, Tokyo, Japan
e-mail: mtmase@arion.ocn.ne.jp

U. Whitcher
Mathematical Reviews (American Mathematical Society), Ann Arbor, MI, USA
e-mail: uaw@umich.edu

R. Abdellatif et al. (eds.), *Women in Numbers Europe IV*, Association for Women in Mathematics Series 32, https://doi.org/10.1007/978-3-031-52163-8_11

Table 1 Strange duality of the bimodal singularities

Name	f	f^T	Dual
$J_{3,0}$	$x^6+xy^3+z^2$	$x^6y+y^3+z^2$	Z_{13}
$Z_{1,0}$	$x^5y+xy^3+z^2$	$x^5y+xy^3+z^2$	$Z_{1,0}$
$Q_{2,0}$	$x^4z+xy^3+z^2$	$x^4y+y^3+xz^2$	Z_{17}
$W_{1,0}$	$x^6+y^2z+z^2$	$x^6+y^2+yz^2$	$W_{1,0}$
$S_{1,0}$	$x^5y+y^2z+z^2$	$x^5+xy^2+yz^2$	W_{17}
$U_{1,0}$	$x^3y+y^2z+z^3$	$x^3+xy^2+yz^3$	$U_{1,0}$
E_{18}	$x^5+y^3+xz^2$	$x^5z+y^3+z^2$	Q_{12}
E_{19}	$x^7+xy^3+z^2$	$x^7y+y^3+z^2$	$Z_{1,0}$
E_{20}	$x^{11}+y^3+z^2$	$x^{11}+y^3+z^2$	E_{20}
Z_{18}	$x^6y+xy^3+z^2$	$x^6y+xy^3+z^2$	Z_{18}
Z_{19}	$x^9y+y^3+z^2$	$x^9+xy^3+z^2$	E_{25}
Q_{16}	$x^4z+y^3+xz^2$	$x^4z+y^3+xz^2$	Q_{16}
Q_{17}	$x^5z+xy^3+z^2$	$x^5y+y^3+xz^2$	$Z_{2,0}$
Q_{18}	$x^8z+y^3+z^2$	$x^8+y^3+xz^2$	E_{30}
W_{17}	$x^5+xz^2+y^2z$	$x^5z+yz^2+y^2$	$S_{1,0}$
W_{18}	$x^7+y^2z+z^2$	$x^7+y^2+yz^2$	W_{18}
S_{16}	$x^4y+xz^2+y^2z$	$x^4y+xz^2+y^2z$	S_{16}
S_{17}	$x^6y+y^2z+z^2$	$x^6+xy^2+yz^2$	$X_{2,0}$
U_{16}	$x^5+y^2z+yz^2$	$x^5+y^2z+yz^2$	U_{16}

In a series of papers, the first author and her collaborators have compared mirror constructions for K3 surfaces obtained by extending the bimodal singularity polynomials to an invertible polynomial in four variables. In [18], they observed there is an extension to an invertible polynomial defining a K3 surface in weighted projective space for all but 4 examples, showed Berglund–Hübsch duality for these K3 surface invertible polynomials can be viewed as a special case of Batyrev mirror symmetry, and commented on potential relationships with homological mirror symmetry. These observations are consistent with broader efforts to unify Berglund–Hübsch and Batyrev mirror symmetry, such as the construction of Clarke in [5]; the relationship between Berglund–Hübsch, Batyrev, and homological mirror symmetry is treated in depth in [10] and [8].

In [15–17], the first author studied Dolgachev–Nikulin–Pinkham's mirror symmetry construction for K3 surfaces obtained from bimodal singularities using an invertible polynomial in four variables. Let $\Lambda_{K3} \cong U \oplus U \oplus U \oplus E_8 \oplus E_8$ be the K3 lattice, the unique (up to isomorphism) unimodular lattice of signature $(3, 19)$; here U is the unimodular lattice of signature $(1, 1)$, and we take E_8 to be negative definite. According to [7], if two K3 surfaces X and $\check{X}$ are mirror, there should exist lattice polarizations $M \hookrightarrow H^2(X, \mathbb{Z})$ and $\check{M} \hookrightarrow H^2(\check{X}, \mathbb{Z})$ such that $M^\perp \cong \check{M} \oplus nU$, where n is a positive integer and the perpendicular complement is computed using the isomorphism $H^2(X, \mathbb{Z}) \cong \Lambda_{K3}$. In particular, rank M + rank $\check{M}$ must be equal to 20. Given a reflexive polytope Δ, we may obtain a K3 surface X as a hypersurface in the toric variety $\mathbb{P}_\Delta$ obtained by an appropriate resolution of singularities of the normal fan of Δ. In this case, the obvious lattice polarization to choose is Pic_Δ, the lattice generated by intersecting X with the divisors of $\mathbb{P}_\Delta$. The main results of [15–17] identify polar dual pairs of reflexive polytopes Δ and Δ° and associated K3 hypersurfaces in Gorenstein Fano toric varieties that satisfy Dolgachev–Nikulin–Pinkham-style mirror symmetry using the lattices Pic_Δ and $\mathrm{Pic}_{\Delta^\circ}$ for all but five of the bimodal singularity pairs studied in [18].

Our first main theorem shows that the remaining five pairs of bimodal singularity polynomials cannot be extended to invertible polynomials whose Newton polytopes are polar dual reflexive polytopes Δ and Δ° such that Pic_Δ and $\mathrm{Pic}_{\Delta^\circ}$ yield mirror lattice polarizations in the sense of Dolgachev–Nikulin–Pinkham.

Theorem A *For each of the bimodal singularity mirror pairs* (B', B) *being*

$$(Z_{13}, J_{3,0}), (X_{2,0}, S_{17}), (W_{18}, W_{18}), (W_{17}, S_{1,0}), (U_{16}, U_{16}),$$

let f *be the defining equation of* B*. Then, there does not exist an invertible deformation* F *of* f *such that the Newton polytope of* F *is a reflexive polytope* Δ *and* $\operatorname{rank}\mathrm{Pic}_\Delta + \operatorname{rank}\mathrm{Pic}_{\Delta^\circ} = 20$. □

Remark 1 Let Δ be a reflexive polytope obtained as the Newton polytope of a four-variable invertible deformation of one of the bimodal singularities listed in Theorem A, let $\mathbb{P}_\Delta$ be the smooth toric variety determined by a maximal projective subdivision of the normal fan of Δ, and let X be a regular K3 hypersurface in $\mathbb{P}_\Delta$. The map

$$r : H^{1,1}(\mathbb{P}_\Delta) \to H^{1,1}(X)$$

is a natural restriction of the Hodge (1, 1)-components.

The rank of the cokernel of r is known to be bounded by a *toric correction term* representing divisors of $\mathbb{P}_\Delta$ whose intersection with X has multiple components:

$$\operatorname{coker}(r) \geq \sum_{\Gamma} \ell^*(\Gamma)\ell^*(\Gamma^\circ),$$

where the sum is over all edges in Δ, $\ell^*(\Gamma)$ represents the number of lattice points in the relative interior of an edge, and Γ° is the polar dual of Γ. By a direct computation, in the course of the proof of the main theorem, we will show that the toric correction term is nonzero for all Δ associated to bimodal singularities from the main theorem. Therefore the main theorem can be rephrased as the statement that we cannot choose a reflexive Newton polytope for the pairs

$$(Z_{13}, J_{3,0}), (X_{2,0}, S_{17}), (W_{18}, W_{18}), (W_{17}, S_{1,0}), (U_{16}, U_{16}),$$

such that the natural restriction map r of the Hodge (1, 1)-components is surjective.

We can relax the requirements of Theorem A in two ways. First, it might happen that the Newton polytope associated to an invertible polynomial is simply not reflexive. In such cases, one may include the Newton polytope in a larger reflexive polytope; the first author has pursued this strategy in past work, including [18]. A reflexive Newton polytope may also admit inclusion in a larger reflexive polytope, yielding a different polarizing lattice.

Alternatively, we may fix our choice of reflexive polytope but choose the polarizing lattice more carefully. This strategy has previously been pursued in order to resolve the apparent contradiction between the Dolgachev–Nikulin–Pinkhamand Batyrev mirror symmetry constructions for K3 hypersurfaces in toric varieties. A proposal introduced by Rohsiepe in the preprint [23] and reviewed in [25] describes an appropriate choice of polarizing lattices for K3 surfaces realized as hypersurfaces in Gorenstein Fano toric varieties obtained from reflexive polytopes. The idea is to consider a sublattice $(\mathrm{Pic}_\Delta)_{\mathrm{tor}}$ of Pic_Δ given by the so-called *toric divisors*; these are the divisors given by the pullback of the divisors of the ambient toric variety. The *toric correction term* measures the difference between the rank of Pic_Δ and the rank of $(\mathrm{Pic}_\Delta)_{\mathrm{tor}}$. As we will discuss in Sect. 2, the toric correction term is symmetric, so it also measures the difference between the rank of $\mathrm{Pic}_{\Delta^\circ}$ and the rank of $(\mathrm{Pic}_{\Delta^\circ})_{\mathrm{tor}}$.

Theorem 1 ([23]) *Let Δ and Δ° be polar dual three-dimensional reflexive polytopes. Then* $(\mathrm{Pic}_\Delta)^\perp \cong (\mathrm{Pic}_{\Delta^\circ})_{\mathrm{tor}} \oplus U$ *and* $((\mathrm{Pic}_\Delta)_{\mathrm{tor}})^\perp \cong \mathrm{Pic}_{\Delta^\circ} \oplus U$.

Note that because the rank of the unimodular lattice U is 2 and the rank of the K3 lattice is 22, we have $\mathrm{rank}(\mathrm{Pic}_\Delta) + \mathrm{rank}\,(\mathrm{Pic}_{\Delta^\circ}) \geq 20$, with equality if and only if the toric correction term vanishes.

It follows from Theorem 1 that any of the bimodal singularities studied in [18] may be extended to a polynomial defining a K3 hypersurface in a Gorenstein Fano toric variety in such a way that Batyrev mirror symmetry for the hypersurface induces lattice mirror symmetry in the sense of Dolgachev–Nikulin–Pinkham. Our second main theorem shows that we may make these choices in a way compatible with Berglund–Hübsch duality.

Theorem B *Let (B', B) be one of the bimodal singularity mirror pairs*

$$(Z_{13}, J_{3,0}),\ (X_{2,0}, S_{17}),\ (W_{18}, W_{18}),\ (W_{17}, S_{1,0}),\ (U_{16}, U_{16}),$$

and let f be the defining equation of B. Then there exist a reflexive polytope Δ and an invertible deformation F of f such that the Newton polytope of F is a subpolytope of Δ, the Newton polytope of F^T is a subpolytope of Δ°, and $(\mathrm{Pic}_\Delta)^\perp \cong (\mathrm{Pic}_{\Delta^\circ})_{\mathrm{tor}} \oplus U$. □

The plan of the paper is as follows. In Sect. 2 we review the various mirror constructions, remark on the connection between Rohsiepe's mirror proposal and Morrison's monomial-divisor mirror map, and establish notation. In Sect. 3 we prove Theorem A by a case-by-case analysis. In Sect. 4 we prove Theorem B and describe an explicit choice of polytopes and lattices in each case. Some of our analysis uses computations in the computer algebra systems Mathematica and SageMath. See the ancillary files attached to the preprint version of this paper, [19], for illustrative code.

2 Mirror Constructions

In this section, we review the mirror constructions we are comparing and establish notation.

2.1 Berglund–Hübsch Duality

We briefly review the Berglund–Hübsch duality construction in the form we will use it. For more details, see, for example, [1]. Let $A = (a_{ij})$ be a matrix of nonnegative integers, and consider the corresponding polynomial F_A given by the sum of $n + 1$ monomials in $n + 1$ variables:

$$F_A = \sum_{i=0}^{n} \prod_{j=0}^{n} x_j^{a_{ij}}.$$

Definition 1 We say the polynomial F_A is *invertible* if the matrix A is invertible, there exist positive integers q_j such that $\sum_{j=0}^{n} q_j a_{ij}$ is the same constant for all i, and the polynomial F_A has exactly one critical point, at the origin.

Invertible polynomials were classified in [12]. They may be written as sums of polynomials in disjoint variables of three *atomic types*: Fermat polynomials, of the form $x_1^{a_1}$, loops, of the form $x_1^{a_1}x_2 + x_2^{a_2}x_3 + \cdots + x_{m-1}^{a_{m-1}}x_m + x_m^{a_m}x_1$, and chains, of the form $x_1^{a_1}x_2 + x_2^{a_2}x_3 + \cdots + x_{m-1}^{a_{m-1}}x_m + x_m^{a_m}$.

Definition 2 We say an invertible polynomial F_A satisfies *the Calabi–Yau condition* if $\sum_{j=0}^{n} q_j a_{ij} = \sum_{j=0}^{n} q_j$.

If F_A is an invertible polynomial satisfying the Calabi–Yau condition, then the weights $(q_0, \ldots, q_n)$ determine a weighted projective space $\mathbb{WP}(q_0, \ldots, q_n)$ and F_A determines a Calabi–Yau hypersurface X_A in this weighted projective space. When $n = 3$, X_A is a K3 surface.

In general, Berglund–Hübsch duality is a duality of orbifolds. Given an appropriately chosen group of discrete symmetries G acting on a Calabi–Yau hypersurface X_A, we obtain a mirror pair X_A/G and X_{A^T}/G^T, where X_{A^T} is the Calabi–Yau hypersurface determined by the transpose matrix A^T and G^T can be computed from the data of G and A. We will focus on the case where G is trivial. In this case, G^T

consists of automorphisms of X_{A^T} that are induced by the multiplicative action of the torus $(\mathbb{C}^*)^{n+1}$ on the ambient weighted projective space and act *symplectically* on X_{A^T}, preserving the holomorphic form.

2.2 Batyrev's Duality and the Monomial-Divisor Mirror Map

We review the mirror symmetry construction for Calabi–Yau hypersurfaces in toric varieties described in [3], examine the differences that arise in the case of K3 surfaces, and establish notation. For a more detailed exposition, see [6] for Calabi–Yau varieties or [25] for the K3 surface case. Let $N \cong \mathbb{Z}^k$ be an integral lattice, and let $M = \text{Hom}(N, \mathbb{Z})$ be the dual lattice. Let $N_{\mathbb{R}} = N \otimes \mathbb{R}$ and $M_{\mathbb{R}} = M \otimes \mathbb{R}$ be the corresponding vector spaces. The duality between N and M induces a real-valued pairing $\langle v, w \rangle$ between elements of $N_{\mathbb{R}}$ and $M_{\mathbb{R}}$.

A *lattice polytope* in $N_{\mathbb{R}}$ or $M_{\mathbb{R}}$ is the finite hull of a convex set of vectors in the lattice. Given a lattice polytope Δ in $M_{\mathbb{R}}$ containing the origin, we define its *polar dual* Δ° in $N_{\mathbb{R}}$ as

$$\{v \in N \mid \langle v, w \rangle \geq -1 \text{ for all } w \in \Delta\}.$$

If Δ° is also a lattice polytope, we say Δ and Δ° are *reflexive*. Note that $(\Delta^\circ)^\circ = \Delta$. Reflexive polytopes in three and four dimensions are classified in [13]: Up to lattice isomorphism, there are 4,319 reflexive polytopes in three dimensions and 473,800,776 in four dimensions. The authors of [13] also specify a normal form that selects a unique representative of each equivalence class.

Now, assume Δ is a reflexive polytope in $M_{\mathbb{R}}$. We may obtain a fan R in $N_{\mathbb{R}}$ by taking the normal fan to Δ, or, equivalently, by taking the fan over the faces of Δ°. A fan obtained in this way determines a Gorenstein Fano toric variety. A simplicial refinement Σ of the fan R such that the one-dimensional cones are the lattice points of Δ° is called a *maximal projective subdivision* of the fan. In three dimensions, a maximal projective subdivision Σ determines a smooth toric variety, which we call $\mathbb{P}_\Delta$. In four dimensions, the resulting toric variety $\mathbb{P}_\Delta$ may have orbifold singularities. However, in either case, a general anticanonical hypersurface in such a toric variety will be smooth.

The hypersurfaces obtained using three-dimensional reflexive polytopes are smooth K3 surfaces, while four-dimensional reflexive polytopes determine Calabi–Yau threefolds. Such a hypersurface will be a semiample divisor in the ambient toric variety $\mathbb{P}_\Delta$, but need no longer be ample. Given a maximal projective subdivision Σ of the normal fan to Δ and a hypersurface X in $\mathbb{P}_\Delta$, we say X is Σ-*regular*, or, if the context is clear, *regular*, if the intersection of X with the torus $\mathbf{T}_\sigma$ is either empty or a smooth subvariety of codimension 1 for every cone σ in Σ.

Given a face Γ of a lattice polytope, we write $\ell(\Gamma)$ for the number of lattice points in Γ, and $\ell^*(\Gamma)$ for the number of lattice points in the relative interior of Γ. In [3], for $k \geq 4$, Batyrev derived the following formulas for the Hodge numbers of a regular Calabi–Yau $k-1$-fold X obtained from an n-dimensional reflexive polytope $\Delta \subset M_{\mathbb{R}}$ by the above procedure:

$$h^{1,1}(X) = \ell(\Delta^\circ) - k - 1 - \sum_{\text{codim }\Gamma^\circ=1} \ell^*(\Gamma^\circ) + \sum_{\text{codim }\Gamma^\circ=2} \ell^*(\Gamma^\circ)\ell^*(\Gamma) \quad (1)$$

$$h^{k-2,1}(X) = \ell(\Delta) - k - 1 - \sum_{\text{codim }\Gamma=1} \ell^*(\Gamma) + \sum_{\text{codim }\Gamma^\circ=2} \ell^*(\Gamma^\circ)\ell^*(\Gamma). \quad (2)$$

By reversing the roles of Δ and Δ°, we obtain formulas for the Hodge numbers of the mirror $k-1$-fold X°; we see $h^{1,1}(X) = h^{k-2,1}(X^\circ)$ and $h^{k-2,1}(X) = h^{1,1}(X^\circ)$.

Batyrev's computation highlights the role of two important subspaces. The so-called *toric* subspace $H^{1,1}_{\text{tor}}(X)$ of $H^{1,1}(X)$ is given by the pullback of $H^{1,1}(\mathbb{P}_\Delta)$ along the natural inclusion map and has dimension $\ell(\Delta^\circ) - k - 1 - \sum_{\text{codim }\Gamma^\circ=1} \ell^*(\Gamma^\circ)$. The *polynomial deformation space* $H^{k-2,1}_{\text{poly}}(X)$ is a subspace of $H^{k-2,1}(X)$ isomorphic to the first-order polynomial deformations of X and has dimension $\ell(\Delta) - k - 1 - \sum_{\text{codim }\Gamma=1} \ell^*(\Gamma)$. We call $\sum_{\text{codim }\Gamma^\circ=2} \ell^*(\Gamma^\circ)\ell^*(\Gamma)$, which measures the difference between $H^{1,1}_{\text{tor}}(X)$ and $H^{1,1}(X)$ or $H^{k-2,1}_{\text{poly}}(X)$ and $H^{k-2,1}(X)$, the *toric correction term*. Aspinwall, Greene, and Morrison showed in [2] that there is a natural isomorphism between $H^{1,1}_{\text{tor}}(X)$ and $H^{k-2,1}_{\text{poly}}(X^\circ)$, induced by a correspondence between divisors and monomials that they termed the *monomial-divisor mirror map*.

In the case of a K3 hypersurface in a three-dimensional toric variety, we have $k-2=1$. Since all K3 surfaces have the same Hodge diamond, the equality of Hodge numbers is trivial. As Dolgachev observed in [7], and as we will discuss in more detail in § 2.3, one may instead study the structure of the Picard group of a K3 surface. For a regular K3 surface X obtained from a three-dimensional reflexive polytope $\Delta \subset M_{\mathbb{R}}$, we have the following inequality (see [23] for a detailed discussion):

$$\operatorname{rank} \operatorname{Pic}(X) \geq \ell(\Delta^\circ) - 4 - \sum_{\text{codim }\Gamma^\circ=1} \ell^*(\Gamma^\circ) + \sum_{\text{codim }\Gamma^\circ=2} \ell^*(\Gamma^\circ)\ell^*(\Gamma). \quad (3)$$

Here, $\ell(\Delta^\circ) - 4 - \sum_{\text{codim }\Gamma^\circ=1} \ell^*(\Gamma^\circ)$ measures the rank of the subgroup of the so-called *toric divisors* $(\operatorname{Pic}_\Delta)_{\text{tor}}$, generated by the pullback of the divisors in the ambient space. The toric correction term measures the rank of a sublattice $L_0(\Delta)$ corresponding to divisors of the ambient space whose intersection with X splits into multiple components. Together, $\operatorname{Pic}_{\text{tor}}(X)$ and $L_0(\Delta)$ generate the lattice that we shall call $\operatorname{Pic}_\Delta$. We denote the rank of this lattice by ρ_Δ. We have the formula

$$\rho_\Delta = \ell(\Delta^\circ) - 4 - \sum_{\text{codim }\Gamma^\circ=1} \ell^*(\Gamma^\circ) + \operatorname{rank} L_0(\Delta). \quad (4)$$

The inequality rank Pic$(X) \geq \rho_\Delta$ is inevitable because non-isotrivial families of K3 surfaces do not have constant Picard rank: Oguiso showed in [22] that any analytic neighborhood in the base of a one-parameter, non-isotrivial family of K3 surfaces has a dense subset where the Picard ranks of the corresponding surfaces are strictly greater than the minimum Picard rank of that family. On the other hand, Bruzzo and Grassi show in [4] that when the K3 hypersurfaces are both smooth and ample, $\text{Pic}(X) = (\text{Pic}_\Delta)_{\text{tor}}$ for very general X. In this case, the toric correction term is 0, so $(\text{Pic}_\Delta)_{\text{tor}} = \text{Pic}_\Delta$. (Recall that a very general property holds outside a countable union of proper closed subvarieties.)

In our comparison of mirror constructions, we often begin with a polynomial f and then explore different ways that this polynomial could be used to represent a hypersurface in a toric variety. Combinatorially, the polynomial corresponds to a choice of lattice points determined by its monomials, and we are choosing different ways to include these points in a lattice polytope (typically a reflexive polytope). Hypersurfaces obtained in this way need not be isomorphic. However, by [8, Theorem 1.1], as long as f is not divisible by one of its coordinates x_i, they are birational.

2.3 Reconciling Batyrev and Dolgachev–Nikulin–Pinkham Mirror Symmetry

Dolgachev proposed in [7] that one should view mirror symmetry for K3 surfaces as a relationship between moduli spaces of *lattice-polarized* K3 surfaces. Recall that a *lattice* is a finitely generated free $\mathbb{Z}$-module equipped with a non-degenerate, symmetric integer-valued form. Let E_8 be the unimodular, negative-definite ADE lattice of rank 8, and let U be the indefinite unimodular *hyperbolic lattice*, with intersection form given by $\left(\begin{smallmatrix} 0 & 1 \\ 1 & 0 \end{smallmatrix}\right)$. Then for any K3 surface X, there exists an isomorphism $\phi : H^2(X, Z) \to \Lambda_{K3}$, where $\Lambda_{K3} := U \oplus U \oplus U \oplus E_8 \oplus E_8$ is the *K3 lattice*. A choice of such a ϕ is called a *marking*.

Suppose we are given an even, non-degenerate lattice M of signature $(1, t)$ and a primitive embedding $M \hookrightarrow \Lambda_{K3}$. (Recall that a lattice embedding is primitive if the quotient of the ambient lattice by the image of the embedding is a free abelian group.) We say X is *M-polarized* if there exists a primitive embedding $M \hookrightarrow \text{Pic}(X)$, and we say X is *marked M-polarized* if there exists a marking ϕ such that $\phi^{-1}(M) \subseteq \text{Pic}(X)$. In [7], Dolgachev constructed moduli spaces of marked M-polarized K3 surfaces satisfying an appropriate pseudo-ampleness condition.

We say M is *m-admissible* if $M^\perp = J \oplus \check{M}$ for some lattice J isomorphic to mU. In this situation, we call $\check{M}$ the *mirror* of M. Note that rank$M$ + rank$\check{M} = 20$. The simplest case, and the one we will be concerned with in what follows, is when M is 1-admissible. When this happens, $(\check{M})^\perp = J \oplus M$, so not only is $\check{M}$ the mirror of M, but M is the mirror of $\check{M}$. Furthermore, one may view the moduli spaces of M- and $\check{M}$-polarized K3 surfaces as mirror families.

Now, suppose we are given a family of K3 surfaces realized as hypersurfaces in a toric variety obtained from a reflexive polytope. To study Dolgachev–Nikulin–Pinkham's mirror symmetry construction in this context, one must choose a polarizing lattice. The choice is simplest when the toric correction term is 0. When we calculate the toric correction term for each of the 4319 isomorphism classes of three-dimensional reflexive polytopes, we find that 1863 yield a toric correction term of 0, of which 53 correspond to self-dual reflexive polytopes; thus, the 0 toric correction term case is common, but not the majority. Given a reflexive polytope Δ in one of these 1863 isomorphism classes, we let M be the space of divisors induced by the ambient toric variety, $\text{Pic}_\Delta = (\text{Pic}_\Delta)_{\text{tor}}$. Rohsiepe showed in [23] that in this case $M^\perp$ is isomorphic to $U \oplus \text{Pic}_{\Delta^\circ}$, so $\check{M}$ is $\text{Pic}_{\Delta^\circ} = (\text{Pic}_{\Delta^\circ})_{\text{tor}}$. This choice of polarizing lattice M is clearly symmetric: If we swap the roles of Δ and Δ°, we also swap the roles of M and $\check{M}$. We may view this computation as a confirmation that for toric K3 hypersurfaces, the monomial-divisor mirror map is not merely an identification of groups but extends to the lattice structure.

For the remaining 2456 isomorphism classes, we cannot set M to $(\text{Pic}_\Delta)_{\text{tor}}$ and $\check{M}$ to $(\text{Pic}_{\Delta^\circ})_{\text{tor}}$ because the ranks of the lattices are too small, and we cannot set M to Pic_Δ and $\check{M}$ to $\text{Pic}_{\Delta^\circ}$ because the ranks of those lattices are too big. However, a computation described in [23] shows that if we choose M to be $(\text{Pic}_\Delta)_{\text{tor}}$, then $M^\perp$ is always isomorphic to $U \oplus \text{Pic}_{\Delta^\circ}$. Thus, we may reconcile the Batyrev and Dolgachev–Nikulin–Pinkham mirror symmetry constructions if we polarize by the pullback of the divisors of the ambient space on one side, and the full lattice generated by intersecting our K3 surface with the ambient space divisors on the other side.

2.4 *Picard Lattices and Intersection Numbers*

In Sect. 4, we illustrate the computation of Pic_Δ and $(\text{Pic}_\Delta)_{\text{tor}}$ for specific cases of interest. We will use the following facts about lattices:

Corollary 1 (Corollary 1.6.2 [20]) *Let lattices S and T be primitively embedded into the $K3$ lattice. Then S and T are orthogonal to each other in the $K3$ lattice if and only if $q_S \simeq -q_T$, where q_S (resp., q_T) is the discriminant form of S (resp., T).*

Corollary 2 (Corollary 1.12.3 [20]) *Let S be an even lattice of signature (t_+, t_-) and Λ be an even unimodular lattice of signature (l_+, l_-). There exists a primitive embedding of S into Λ if and only if the following three conditions are simultaneously satisfied:*

(1) $l_+ - l_- \equiv 0 \mod 8$.
(2) $l_- - t_- \geq 0$ *and* $l_+ - t_+ \geq 0$.
(3) $\text{rank}\,\Lambda - \text{rank}\,S > l(A_S)$.

Here A_S denotes the discriminant group of S, which is finitely generated abelian, and $l(A_S)$ is the minimal number of generators of A_S.

We will use the following fundamental lemma.

Lemma 1 *Let L be a sublattice of the K3 lattice $\Lambda_{K3} := U^{\oplus 3} \oplus E_8^{\oplus 2}$. If the discriminant number* $\det(L)$ *of L is prime, then L has no proper overlattices.*

Proof Let $\det(L) = p$ be a prime number. If it were an overlattice $L \subset N \subsetneq \Lambda_{K3}$ (the N is an overlattice), then we would have $[N : L]^2 \det(N) = \det(L) = 1 \cdot p$. Thus, one has $[N : L] = 1$. Therefore, there does not exist such a lattice N. In particular, the discriminant group A_L of L is isomorphic to $\mathbb{Z}/p\mathbb{Z}$.

We will use the following formulas to compute the Picard lattices from toric data.

Let $\mathbf{M}$ be a three-dimensional lattice. Let $\{\mathbf{e}_1, \mathbf{e}_2, \mathbf{e}_3\}$ be the standard basis of $\mathbb{R}^3 \simeq \mathbf{M} \otimes_{\mathbb{Z}} \mathbb{R} =: \mathbf{M}_{\mathbb{R}}$. Let Δ be a reflexive polytope in $\mathbf{M}_{\mathbb{R}}$. Denote by $\mathrm{Div}_{\mathbb{T}}(\widetilde{\mathbb{P}}_\Delta)$ the set of all toric divisors $\widetilde{D}_i$ in the MPCP toric resolution $\widetilde{\mathbb{P}}_\Delta$ of the projective toric variety $\mathbb{P}_\Delta$ [3]. The lattice $\mathbf{M}$ is explicitly given in Section 3, using the description given in [18]. It is defined by the weight system $\boldsymbol{w} = (w_0, w_1, w_2, w_3)$ such that the deformation F of the defining polynomial f of a bimodal singularity is an anticanonical member of the weighted projective space of weight system $\boldsymbol{w}$.

The toric divisors are related by a linear system with three equations:

$$\sum_{i=1}^{d+3} (v_i, \mathbf{e}_j)\widetilde{D}_i = 0 \qquad j = 1, 2, 3, \tag{5}$$

where (,) is the standard inner product in $\mathbb{R}^3$.

If one restricts a toric divisor $\widetilde{D}_i \in \mathrm{Div}_{\mathbb{T}}(\widetilde{\mathbb{P}}_\Delta)$ to the minimal model $\widetilde{X}$ of a generic section X by the MPCP toric resolution of the ambient space, one has a divisor on $\widetilde{X}$, which we denote by D_i.

As to the intersection numbers,

$$D_i^2 = \begin{cases} 2\ell^*(\psi_i) - 2 & \text{if } v_i \text{ is a vertex.} \\ -2 & \text{if } v_i \text{ is in the interior of an edge.} \end{cases} \tag{6}$$

Here ψ_i is the face in Δ^* that is dual to v_i, and $\ell^*(\psi_i)$ is the number of lattice points in the interior of the face ψ_i. Moreover,

$$D_i.D_j = \begin{cases} 1 & \text{if } v_i \text{ and } v_j \text{ are next to each other on an edge.} \\ \ell^*(m_{ij}^*) + 1 & \text{if } v_i \text{ and } v_j \text{ are vertices that are connected} \\ & \text{by an edge } m_{ij} \text{ whose dual is } m_{ij}^*. \\ 0 & \text{otherwise.} \end{cases} \tag{7}$$

For the above formulas (2), (3), see [11, Chapter 5]; for further illustrations of their use in the case of K3 hypersurfaces, see [14] or [23].

3 Invertible Deformations

In the following subsections, we prove our first Main Theorem by a case-by-case analysis:

Theorem A *For each of the bimodal singularity mirror pairs* (B', B) *being*

$$(Z_{13}, J_{3,0}), (X_{2,0}, S_{17}), (W_{18}, W_{18}), (W_{17}, S_{1,0}), (U_{16}, U_{16}),$$

let f *be the defining equation of* B*. Then, there does not exist an invertible deformation* F *of* f *such that the Newton polytope of* F *is a reflexive polytope* Δ *and* $\operatorname{rank}\operatorname{Pic}_\Delta + \operatorname{rank}\operatorname{Pic}_{\Delta^\circ} = 20$. □

3.1 Z_{13} and $J_{3,0}$

The singularity $J_{3,0}$ is defined by $f = x^6 + xy^3 + z^2$.

First, we choose a basis

$$e_1 = (-1, 5, -1, -1),\ e_2 = (-1, 0, 2, -1),\ e_3 = (-1, -1, -1, 1)$$

of the group

$$\mathbf{M} := \left\{(i, j, k, l) \in \mathbb{Z}^4 \mid i + 3j + 5k + 9l = 0 \text{ and } k + l \equiv 0 \mod 2\right\}.$$

Then, the monomials X^6, XY^3, and Z^2 correspond to the lattice points $(1, 0, 0)$, $(0, 1, 0)$, and $(0, 0, 1)$, respectively.

On the other hand, for the singularity of type Z_{13} defined by $f^T = x^6y + y^3 + z^2$, choose a basis

$$e_1' = (-2, 1, 0, 0),\ e_2' = (-6, 0, 1, 0),\ e_3' = (-9, 0, 0, 1)$$

of the group

$$\mathbf{M}' := \left\{(i, j, k, l) \in \mathbb{Z}^4 \mid i + 2j + 6k + 9l = 0\right\}.$$

Then the monomials X^6Y, Y^3, and Z^2 correspond to the lattice points $(5, 0, -1)$, $(-1, 2, -1)$, $(-1, -1, 1)$, respectively.

The possible choices of F for $J_{3,0}$ are given by adding the monomial W^9Z or W^{18} to the polynomial f, since F should be an invertible polynomial.

Case (i): In the case that the deformation is $F = X^6 + XY^3 + Z^2 + W^9Z$, the Newton polytope of F is not reflexive. Indeed, the Newton polytope of F contains a facet

$$\Phi = \mathrm{Conv}\{(1,0,0),\ (0,1,0),\ (-1,-3,-4)\}.$$

However, the polar dual of the facet Φ is easily computed to be the non-integral vertex $(-1,\ -1,\ 5/4) \notin \mathbb{Z}^3$.

Case (ii): In the case that the deformation is $F = X^6 + XY^3 + Z^2 + W^{18}$, the Newton polytope of F is reflexive. Indeed, the Newton polytope of F is the convex hull

$$\Delta = \mathrm{Conv}\left\{(1,0,0),\ (0,1,0),\ (0,0,1),\ (-2,-6,-9)\right\}.$$

Then, the polar dual polytope is given by

$$\Delta^\circ = \mathrm{Conv}\left\{(8,-1,-1),\ (-1,2,-1),\ (-1,-1,1),\ (-1,-1,-1)\right\}.$$

Since Δ° is an integral polytope, the polytope Δ is reflexive. In the [24] database of reflexive polytopes, Δ and Δ° are indices 745 and 4282, respectively. The first author and a collaborator showed in [18] that Δ° contains the Newton polytope of F^T as a subpolytope.

Next, we compute the toric contribution rank $L_0(\Delta)$.

Let Γ be the edge of Δ given by

$$\Gamma = \mathrm{Conv}\left\{(0,0,1),\ (-2,-6,-9)\right\}.$$

One has its polar dual

$$\Gamma^\circ = \mathrm{Conv}\left\{(-1,2,-1),\ (8,-1,-1)\right\}.$$

No other edge contributes to rank $L_0(\Delta)$. Thus,

$$\mathrm{rank}\, L_0(\Delta) = 2.$$

3.2 $X_{2,0}$ and S_{17}

The singularity S_{17} is defined by $f = x^6y + y^2z + z^2$.

First, we choose a basis

$$e_1 = (1, 1, -1, 0),\ e_2 = (-6, 0, 1, 1),\ e_3 = (-12, 0, 0, 3)$$

of the group

$$\mathbf{M} := \left\{(i, j, k, l) \in \mathbb{Z}^4 \mid i + j + 2k + 4l = 0 \text{ and } j - i \equiv 0 \mod 3\right\}.$$

Then, the monomials X^6Y, Y^2Z, and Z^2 correspond to the lattice points $(5, 5, -2)$, $(-1, 0, 0)$, and $(-1, -2, 1)$, respectively.

On the other hand, for the singularity of type $X_{2,0}$ defined by $f^T = x^6 + xy^2 + yz^2$, choose a basis

$$e_1' = (0, 5, -1, -1),\ e_2' = (-1, 5, 0, -2),\ e_3' = (0, -2, 0, 1)$$

of the group

$$\mathbf{M}' := \left\{(i, j, k, l) \in \mathbb{Z}^4 \mid i + j + 3k + 2l = 0\right\}.$$

Then, the monomials XY^2, YZ^2, and WX^6 correspond to the lattice points $(-1, 1, 0)$, $(0, 1, 3)$, and $(1, 0, 0)$, respectively.

The only possible choice of F is given by adding the monomial W^7X to the polynomial f, since F should be an invertible polynomial. Note that this monomial corresponds to the lattice point $(0, -1, 0)$. The Newton polytope of F is not reflexive. Indeed, the Newton polytope of F contains a facet

$$\Phi = \text{Conv}\{(5, 5, -2),\ (-1, 0, 0),\ (0, -1, 0)\}.$$

However, the polar dual of the facet Φ is easily computed to be the non-integral vertex $(1,\ 1,\ 11/2) \notin \mathbb{Z}^3$.

Similarly, the Newton polytope of F^T is not reflexive. Indeed, the Newton polytope of F^T contains a facet

$$\Phi' = \text{Conv}\{(-1, 1, 0),\ (0, 1, 3),\ (1, 0, 0)\}.$$

However, the polar dual of the facet Φ' is easily computed to be the non-integral vertex $(-1,\ -2,\ 1/3) \notin \mathbb{Z}^3$.

3.3 W_{18}

The singularity W_{18} is defined by $f = x^7 + y^2z + z^2$. This singularity is dual to $f^T = x^7 + y^2 + yz^2$, another realization of W_{18}.

First, we choose a basis

$$e_1 = (0, 0, 2, -1),\ e_2 = (1, 1, 1, -1),\ e_3 = (6, -1, 0, -1)$$

of the group

$$\mathbf{M} := \left\{(i, j, k, l) \in \mathbb{Z}^4 \mid 3i + 4j + 7k + 14l = 0\right\}.$$

Then the monomials X^7, Y^2Z, and Z^2 correspond to the lattice points $(-3, 5, -1)$, $(1, -1, 0)$, and $(0, -1, 0)$, respectively.

On the other hand, choose a basis

$$e'_1 = (-1, -3, 1, 0),\ e'_2 = (1, 5, -1, -1),\ e'_3 = (1, -1, 0, 0)$$

of the group

$$\mathbf{M}' := \left\{(i, j, k, l) \in \mathbb{Z}^4 \mid i + j + 4k + 2l = 0\right\}.$$

In this basis, the monomials WX^7, Y^2, and YZ^2 correspond to the lattice points $(0, 1, -1)$, $(2, 1, 0)$, and $(-1, -1, -1)$, respectively.

The possible choices of F are given by adding the monomial W^7Y or W^8X to the polynomial f, since F should be an invertible polynomial.

Case (i): In the case that the deformation is $F = X^7 + Y^2Z + Z^2 + W^7Y$, the Newton polytope of F is not reflexive. Indeed, the Newton polytope of F contains a facet

$$\Phi = \mathrm{Conv}\{(-3, 5, -1),\ (1, -1, 0),\ (0, 0, 1)\}.$$

However, the polar dual of the face Φ is easily computed to be the non-integral vertex $(-7/2,\ -5/2,\ -1) \notin \mathbb{Z}^3$.

Case (ii): In the case that the deformation is $F = X^7 + Y^2Z + Z^2 + W^8X$, the Newton polytope of F is reflexive. Indeed, the Newton polytope of F is the convex hull

$$\Delta = \mathrm{Conv}\{(-3, 5, -1),\ (1, -1, 0),\ (-1, 1, 1),\ (0, -1, 0)\}.$$

Then, the polar dual polytope is given by

$$\Delta^\circ = \mathrm{Conv}\{(0, 1, -2),\ (0, 1, 6),\ (2, 1, 0),\ (-4, -3, -2)\}.$$

Since Δ° is an integral polytope, the polytope Δ is reflexive. In the [24] database, Δ and Δ° are polytopes 9 and 4312, respectively.

Next, we compute the toric contribution rank $L_0(\Delta)$. Let Γ be the edge of Δ given by

$$\Gamma = \text{Conv}\{(-1, 1, 1), (-3, 5, -1)\}.$$

The polar dual of Γ is

$$\Gamma^{\circ} = \text{Conv}\{(2, 1, 0), (-4, -3, -2)\},$$

and

$$\ell^*(\Gamma) = \ell^*(\Gamma^{\circ}) = 1.$$

No other edge contributes to rank $L_0(\Delta)$. Thus,

$$\text{rank}\, L_0(\Delta) = 1.$$

In this case, the polytope Δ^T corresponding to F^T has vertices $(0, 1, -1)$, $(2, 1, 0)$, $(-1, -1, -1)$, and $(0, 1, 6)$. This polytope is not reflexive. Furthermore, one may use the normal form for three-dimensional lattice polytopes described in [12] and implemented in [24] to check that Δ^T is not isomorphic to any lattice subpolytope of Δ°. Thus, for this choice of Δ, Batyrev's duality does not appear consistent with Berglund–Hübsch duality. In [18], the first author and her collaborator showed how to choose a reflexive polytope containing Δ that resolves this discrepancy. We will study that polytope in more detail in Sect. 4.

3.4 W_{17} and $S_{1,0}$

The singularity $S_{1,0}$ is defined by $f = x^5y + y^2z + z^2$.

First, we choose a basis

$$e_1 = (0, 5, -1, -1),\ e_2 = (-1, 4, 0, -1),\ e_3 = (-1, -1, -1, 1)$$

of the group

$$\mathbf{M} := \left\{(i, j, k, l) \in \mathbb{Z}^4 \mid 2i + 3j + 5k + 10l = 0\right\}.$$

Then, the monomials X^5Y, Y^2Z, and Z^2 correspond to the lattice points $(0, 1, 0)$, $(-1, 1, 0)$, and $(0, 0, 1)$, respectively.

On the other hand, choose a basis

$$e_1' = (4, 0, -1, 0),\ e_2' = (-6, 1, 1, 0),\ e_3' = (-3, 0, 0, 1)$$

of the group

$$\mathbf{M}' := \left\{ (i, j, k, l) \in \mathbb{Z}^4 \mid i + 2j + 4k + 3l = 0 \right\}.$$

Then the monomials X^5, XY^2, and YZ^2 correspond to the lattice points $(5, 4, -1)$, $(-1, 0, -1)$, and $(-1, -1, 1)$, respectively.

The only possible choice of F is given by adding the monomial W^{10} to the polynomial f, since F should be an invertible polynomial. Note that on this side the monomial W^{10} corresponds to the lattice point $(4, -6, -3)$. However, with this choice, the Newton polytope of F is not reflexive. Indeed, the Newton polytope of F contains the facet

$$\Phi = \mathrm{Conv}\,\{(-1, 1, 0),\ (0, 1, 0),\ (4, -6, -3)\}.$$

However, the polar dual of the face Φ is easily computed to be the non-integral vertex $(0,\ -1,\ 7/3) \notin \mathbb{Z}^3$.

We consider whether we can obtain a reflexive Newton polytope by working on the other side, using the polynomial $f^T = X^5 + XY^2 + YZ^2$ defining the singularity W_{17}. There are two possibilities for a projectivization G of f^T, namely, $G = f^T + W^{10}$ or $G = f^T + W^7 Z$. We claim that in both cases, the Newton polytope of G is not reflexive. Note that the monomials W^{10} and $W^7 Z$ are respectively corresponding to lattice points $(0, -1, -1)$ and $(0, -1, 0)$.

If a projectivization is $G = f^T + W^{10}$, the dual of the facet

$$\mathrm{Conv}\{(-1, 0, -1), (-1, -1, 1), (0, -1, -1)\} = \mathrm{Conv}\{XY^2, YZ^2, W^{10}\}$$

in the Newton polytope of G is the vertex $1/3(2, 2, 1) \notin \mathbb{Z}^3$. Thus, the Newton polytope of G is not reflexive.

If a projectivization is $G = f^T + W^7 Z$, the dual of the facet

$$\mathrm{Conv}\{(-1, 0, -1), (5, 4, -1), (0, -1, 0)\} = \mathrm{Conv}\{XY^2, X^5, W^7 Z\}$$

in the Newton polytope of G is the vertex $1/3(-2, 3, 5) \notin \mathbb{Z}^3$. Thus, the Newton polytope of G is not reflexive.

3.5 U_{16}

The singularity U_{16} is defined by $f = x^5 + y^2 z + yz^2$. This singularity is self-dual. First, we choose a basis

$$e_1 = (4, -1, -1, 0),\ e_2 = (4, -1, 0, -1),\ e_3 = (5, 0, -1, -1)$$

of the group

$$\mathbf{M} := \left\{(i, j, k, l) \in \mathbb{Z}^4 \mid 2i + 3j + 5k + 5l = 0\right\}.$$

Then the monomials X^5, Y^2Z, and YZ^2 correspond to the lattice points $(-2, -2, 3)$, $(0, 1, -1)$, and $(1, 0, -1)$, respectively.

On the other hand, choose a basis

$$e_1' = (0, -2, 0, 1),\ e_2' = (0, -2, 1, 0),\ e_3' = (1, 3, -1, -1)$$

of the group

$$\mathbf{M}' := \left\{(i, j, k, l) \in \mathbb{Z}^4 \mid i + j + 2k + 2l = 0\right\}.$$

Then the monomials WX^5, Y^2Z, and YZ^2 correspond to the lattice points $(-1, -1, 0)$, $(-1, 0, -1)$, and $(0, -1, -1)$, respectively.

The only possible choice of F is given by adding the monomial W^6X to the polynomial f since F should be an invertible polynomial; thus, we have $F = X^5Y + Y^2Z + Z^2 + W^6X$. Note that the monomial W^6X corresponds to the lattice point $(0, 0, 1)$.

The Newton polytope of F is the polytope Δ given by

$$\Delta = \mathrm{Conv}\,\{(1, 0, -1),\ (0, 0, 1),\ (-2, -2, 3),\ (0, 1, -1)\}.$$

The polar dual polytope of Δ is given by

$$\Delta^\circ = \mathrm{Conv}\,\{(-2, -2, -1),\ (-2, 1, -1),\ (4, 4, 5),\ (1, -2, -1)\}.$$

Since the polytope Δ° is an integral polytope, the polytope Δ is reflexive. The index of Δ in the [24] database is 1, and the index of Δ° is 4281.

Next, we compute the toric contribution rank $L_0(\Delta)$.

Let Γ be the edge in Δ given by

$$\Gamma = \mathrm{Conv}\,\{(0, 0, 1),\ (-2, -2, 3)\}.$$

The polar dual edge is

$$\Gamma^\circ = \mathrm{Conv}\,\{(-2, 1, -1),\ (1, -2, -1)\},$$

and

$$l^*(\Gamma) = 1, \qquad l^*(\Gamma^\circ) = 2.$$

No other edge contributes to rank $L_0(\Delta)$. Thus,

$$\text{rank } L_0(\Delta) = l^*(\Gamma)l^*(\Gamma^\circ) = 2.$$

4 Mirror Polytopes and Mirror Lattices

In this section, we examine Theorem B:

Theorem B *Let (B', B) be one of the bimodal singularity mirror pairs*

$$(Z_{13}, J_{3,0}), (X_{2,0}, S_{17}), (W_{18}, W_{18}), (W_{17}, S_{1,0}), (U_{16}, U_{16}),$$

and let f be the defining equation of B. Then there exists a reflexive polytope Δ and an invertible deformation F of f such that the Newton polytope of F is a subpolytope of Δ, the Newton polytope of F^T is a subpolytope of Δ°, and $(\mathrm{Pic}_\Delta)^\perp \cong (\mathrm{Pic}_{\Delta^\circ})_{\mathrm{tor}} \oplus U$. □

In practice, several different polytopes Δ may be available. In each case, we describe a specific choice of Δ and verify that the lattices $(\mathrm{Pic}_{\Delta^\circ})_{\mathrm{tor}}$ and $U \oplus \mathrm{Pic}_\Delta$ are orthogonal to each other in the $K3$ lattice, using the formulas in Sect. 2.4.

4.1 Z_{13} and $J_{3,0}$

The $J_{3,0}$ singularity is defined by $f = x^6+xy^3+z^2$, and Z_{13} by $f^T = x^6y+y^3+z^2$. Recall from Sect. 3.1 that the singularity $J_{3,0}$ is defined by $f = x^6 + xy^3 + z^2$ and may be extended to an invertible deformation $F = X^6 + XY^3 + Z^2 + W^{18}$ with a reflexive Newton polytope Δ given by

$$\Delta = \mathrm{Conv}\,\{(1, 0, 0), (0, 1, 0), (0, 0, 1), (-2, -6, -9)\}.$$

We label the lattice points on the edges of Δ as

$$\begin{aligned} &m_1 = (1, 0, 0), \quad m_2 = (0, 1, 0), \quad m_3 = (0, 0, 1),\\ &m_4 = (-2, -6, -9), m_5 = (0, -2, -3), m_6 = (-1, -4, -6),.\\ &m_7 = (-1, -3, -4).\end{aligned}$$

The polar dual polytope of Δ is given by

$$\Delta^\circ = \mathrm{Conv}\,\{(8, -1, -1), (-1, 2, -1), (-1, -1, 1), (-1, -1, -1)\}.$$

Label the lattice points on the edges of Δ° by

$$\begin{array}{lll} v_1 = (-1,-1,1), & v_2 = (-1,2,-1), & v_3 = (-1,-1,-1), \\ v_4 = (8,-1,-1), & v_5 = (-1,-1,0), & v_6 = (-1,0,-1), \\ v_7 = (-1,1,-1), & v_8 = (0,-1,-1), & v_9 = (1,-1,-1), \\ v_{10} = (2,-1,-1), & v_{11} = (3,-1,-1), & v_{12} = (4,-1,-1), \\ v_{13} = (5,-1,-1), & v_{14} = (6,-1,-1), & v_{15} = (7,-1,-1), \\ v_{16} = (5,0,-1), & v_{17} = (2,1,-1). & \end{array}.$$

The Newton polytope of the deformation $F^T = X^6Y + Y^3 + Z^2 + W^{18}$ of $f^T = x^6y + y^3 + z^2$ is a subpolytope of the polar dual Δ° of Δ.

As we have seen in the previous section, the toric contribution is rank $L_0(\Delta) = 2$. Moreover, one has

$$\rho_\Delta = 4 + 13 + 2 - 3 = 16, \quad \rho_{\Delta^\circ} = 4 + 3 + 2 - 3 = 6.$$

We now compute $(\mathrm{Pic}_{\Delta^\circ})_{\mathrm{tor}}$. The collection $\{m_5, m_6, m_7\}$ consists of linearly independent vectors. Let L be a lattice that is generated by divisors D_1, D_2, D_3 and D_4. By taking the new generators $\{D_1, -D_2 + D_3, -2D_1 + 3D_2 - 2D_3 + D_4, D_1 + D_4\}$, we have

$$L \simeq U \oplus A_2.$$

By a direct computation, we have $\det L = -3$ and $\mathrm{sgn}\, L = (1, 3)$.

Note that the intersection matrix of L with respect to the generators $\{D_1, D_2, D_3, D_4\}$ is given by

$$\begin{pmatrix} 0 & 2 & 3 & 0 \\ 2 & 6 & 9 & 1 \\ 3 & 9 & 12 & 0 \\ 0 & 1 & 0 & -2 \end{pmatrix}.$$

The lattice $L \simeq U \oplus A_2$ is a primitive sublattice of the $K3$ lattice $\Lambda_{K3} = U^{\oplus 3} \oplus E_8^{\oplus 2}$. By definition, the lattice $(\mathrm{Pic}_{\Delta^\circ})_{\mathrm{tor}}$ is actually equal to the lattice L. Moreover, since the order of the discriminant group $A_L = L^*/L$ coincides with $|\det L| = 3$, and the group A_L is a finitely generated Abelian group, we have $A_L \simeq \mathbb{Z}/3\mathbb{Z}$.

We compute Pic_Δ. The vectors $\{v_1, v_5, v_8\}$ are linearly independent. Let L' be the lattice that is generated by the divisors

$$\mathcal{B}' = \{V_2, V_3, V_4, V_6, V_7, V_9, \ldots, V_{15}, V_{16}^{(1)}, V_{16}^{(2)}, V_{17}^{(1)}, V_{17}^{(2)}\}.$$

Here, $V_i = V_i^{(1)} + V_i^{(2)}$ is a restricted toric divisor for $i = 17, 18$. By a direct computation, we have $\det L' = -3$ and $\mathrm{sgn}\, L' = (1, 15)$ with the aid of Mathematica.

Note that the intersection matrix of L' with respect to the basis $\mathcal{B}'$ is given by

$$\begin{pmatrix}
0 & 0 & 0 & 0 & 1 & 0 & 0 & 0 & 0 & 0 & 0 & 0 & 0 & 0 & 1 & 1 \\
0 & -2 & 0 & 1 & 0 & 0 & 0 & 0 & 0 & 0 & 0 & 0 & 0 & 0 & 0 & 0 \\
0 & 0 & -2 & 0 & 0 & 0 & 0 & 0 & 0 & 0 & 0 & 1 & 1 & 1 & 0 & 0 \\
0 & 1 & 0 & -2 & 1 & 0 & 0 & 0 & 0 & 0 & 0 & 0 & 0 & 0 & 0 & 0 \\
1 & 0 & 0 & 1 & -2 & 0 & 0 & 0 & 0 & 0 & 0 & 0 & 0 & 0 & 0 & 0 \\
0 & 0 & 0 & 0 & 0 & -2 & 1 & 0 & 0 & 0 & 0 & 0 & 0 & 0 & 0 & 0 \\
0 & 0 & 0 & 0 & 0 & 1 & -2 & 1 & 0 & 0 & 0 & 0 & 0 & 0 & 0 & 0 \\
0 & 0 & 0 & 0 & 0 & 0 & 1 & -2 & 1 & 0 & 0 & 0 & 0 & 0 & 0 & 0 \\
0 & 0 & 0 & 0 & 0 & 0 & 0 & 1 & -2 & 1 & 0 & 0 & 0 & 0 & 0 & 0 \\
0 & 0 & 0 & 0 & 0 & 0 & 0 & 0 & 1 & -2 & 1 & 0 & 0 & 0 & 0 & 0 \\
0 & 0 & 0 & 0 & 0 & 0 & 0 & 0 & 0 & 1 & -2 & 1 & 0 & 0 & 0 & 0 \\
0 & 0 & 1 & 0 & 0 & 0 & 0 & 0 & 0 & 0 & 1 & -2 & 0 & 0 & 0 & 0 \\
0 & 0 & 1 & 0 & 0 & 0 & 0 & 0 & 0 & 0 & 0 & 0 & -2 & 0 & 1 & 0 \\
0 & 0 & 1 & 0 & 0 & 0 & 0 & 0 & 0 & 0 & 0 & 0 & 0 & -2 & 0 & 1 \\
1 & 0 & 0 & 0 & 0 & 0 & 0 & 0 & 0 & 0 & 0 & 0 & 1 & 0 & -2 & 0 \\
1 & 0 & 0 & 0 & 0 & 0 & 0 & 0 & 0 & 0 & 0 & 0 & 0 & 1 & 0 & -2
\end{pmatrix}.$$

By Lemma 1, the lattice L' is a primitive sublattice of the $K3$ lattice $\Lambda_{K3} = U^{\oplus 3} \oplus E_8^{\oplus 2}$. By definition, the lattice Pic_Δ is actually equal to the lattice L'.

Since $\mathrm{rank}\, L' > 12$, there exists an even negative-definite lattice L'' of rank 14 and $\det L'' = 3$ such that $L' = U \oplus L''$ holds. Moreover, since the order of the discriminant group $A_{L'} = L'^*/L'$ coincides with $|\det L'| = 3$, and the group $A_{L'}$ is a finitely generated Abelian group, we have $A_{L'} \simeq \mathbb{Z}/3\mathbb{Z}$.

We have seen that:

(i) $\det L = -3 = -\det(U \oplus L')$.
(ii) $A_L \simeq \mathbb{Z}/3\mathbb{Z} \simeq A_{U \oplus L'}$.

By Corollary 1, we conclude that the lattice $L = (\mathrm{Pic}_{\Delta^\circ})_{\mathrm{tor}}$ and the lattice $U \oplus L' = U \oplus \mathrm{Pic}_\Delta$ are orthogonal to each other in the $K3$ lattice Λ_{K3}.

In fact, one finds $L'' \simeq E_6 \oplus E_8$.

4.2 $X_{2,0}$ and S_{17}

The S_{17} singularity is defined by $f = x^6 y + y^2 z + z^2$, and $X_{2,0}$ by $f^T = x^6 + xy^2 + yz^2$. Recall that the only possible invertible deformation of S_{17} is given by $F = X^6 Y + Y^2 Z + Z^2 + W^7 X$, with monomials corresponding to the lattice points $(5, 5, -2)$, $(-1, 0, 0)$, $(-1, -2, 1)$, and $(0, -1, 0)$. We showed in Sect. 3.2 that the Newton polytope of F is not reflexive. We take the larger polytope Δ given by

$$\Delta = \mathrm{Conv}\,\{(-1, -2, 1),\ (0, -1, 0),\ (6, 5, -2),\ (5, 5, -2),\ (-1, 2, -1)\}.$$

This choice is motivated by the consideration that one should take a reflexive polytope that contains two faces

$$\Phi_1 = \mathrm{Conv}\,\{(-1,-2,1),\ (-1,2,-1),\ (5,5,-2)\}\,,$$

and

$$\Phi_2 = \mathrm{Conv}\,\{(-1,-2,1),\ (-1,2,-1),\ (0,-1,0)\}\,.$$

By construction, the Newton polytope of F is a subpolytope of Δ. We label the lattice points on the edges of Δ as

$$\begin{aligned} m_1 &= (-1,-2,1),\ \ m_2 = (0,-1,0),\ \ m_3 = (6,5,-2),\\ m_4 &= (5,5,-2),\ \ m_5 = (-1,2,-1),\ m_6 = (3,2,-1),.\\ m_7 &= (-1,0,0) \end{aligned}$$

The polar dual polytope of Δ is given by

$$\Delta^\circ = \mathrm{Conv}\,\{(0,1,3),\ (1,-6,-12),\ (1,1,2),\ (-1,1,0),\ (0,-3,-7)\}\,.$$

Since the polytope Δ° is an integral polytope, the polytope Δ is reflexive. In the [24] database of reflexive polytopes, Δ has index 760 and Δ° has index 3046. We label the lattice points on the edges of Δ° as

$$\begin{aligned} v_1 &= (0,1,3), & v_2 &= (1,1,2), & v_3 &= (-1,1,0),\\ v_4 &= (0,-3,-7), & v_5 &= (1,-6,-12), & v_6 &= (0,1,1),\\ v_7 &= (0,-1,-2), & v_8 &= (1,0,0), & v_9 &= (1,-1,-2),\\ v_{10} &= (1,-2,-4), & v_{11} &= (1,-3,-6), & v_{12} &= (1,-4,-8),\\ v_{13} &= (1,-5,-10). \end{aligned}$$

The Newton polytope of the deformation $F^T = WX^6 + XY^2 + YZ^2 + W^7$ of $f^T = x^6 + xy^2 + yz^2$ is a subpolytope of the polar dual Δ° of Δ.

Next, we compute the toric contribution rank $L_0(\Delta)$.

Let Γ be the edge in Δ given by

$$\Gamma = \mathrm{Conv}\,\{(-1,-2,1),\ (-1,2,-1)\}\,.$$

One has its polar dual

$$\Gamma^\circ = \mathrm{Conv}\,\{(1,1,2),\ (1,-6,-12)\}\,,$$

and

$$\ell^*(\Gamma) = 1, \qquad l^*(\Gamma^\circ) = 6.$$

In fact, no other edge that contributes to rank $L_0(\Delta)$. Thus,

$$\operatorname{rank} L_0(\Delta) = 6.$$

Moreover, one has

$$\rho(\Delta) = 5 + 8 + 6 - 3 = 16, \quad \rho(\Delta^\circ) = 5 + 2 + 6 - 3 = 10.$$

We now compute $(\mathrm{Pic}_{\Delta^\circ})_{\mathrm{tor}}$. Let L be the lattice that is generated by the divisors D_2, D_5, D_6, and D_7, where the divisor D_i corresponds to the lattice point m_i. By taking the new generators $\{D_2, D_2 + D_5, -D_2 - D_5 + D_6 - D_7, D_7 - D_2\}$, we have

$$L \simeq U \oplus \begin{pmatrix} -4 & 1 \\ 1 & -2 \end{pmatrix}.$$

By a direct computation, we have $\det L = -7$ and $\operatorname{sgn} L = (1, 3)$.

By Lemma 1, the lattice L is a primitive sublattice of the $K3$ lattice $\Lambda_{K3} = U^{\oplus 3} \oplus E_8^{\oplus 2}$. By definition, the lattice $(\mathrm{Pic}_{\Delta^\circ})_{\mathrm{tor}}$ is actually equal to the lattice L. Moreover, since the order of the discriminant group $A_L = L^*/L$ coincides with $|\det L| = 7$, and the group A_L is a finitely generated Abelian group, we have $A_L \simeq \mathbb{Z}/7\mathbb{Z}$.

We compute Pic_Δ. Let L' be a lattice that is generated by divisors

$$\mathscr{B}' = \{V_1, V_3, V_4, V_5, V_8^{(1)}, \ldots, V_{13}^{(1)}, V_8^{(2)}, \ldots, V_{13}^{(2)}\}.$$

Here, $V_i = V_i^{(1)} + V_i^{(2)}$ is a restricted toric divisor for $i = 8, \ldots, 13$. By a direct computation, we have $\det L' = -7$ and $\operatorname{sgn} L' = (1, 15)$ with the aid of Mathematica.

By Lemma 1, the lattice L' is a primitive sublattice of the $K3$ lattice Λ_{K3}. By definition, the lattice Pic_Δ is actually equal to the lattice L'.

Since $\operatorname{rank} L' > 12$, there exists an even negative-definite lattice L'' of rank 14 and $\det L'' = 7$ such that $L' = U \oplus L''$ holds. Moreover, since the order of the discriminant group $A_{L'} = L'^*/L'$ coincides with $|\det L'| = 7$, and the group $A_{L'}$ is a finitely generated Abelian group, we have $A_{L'} \simeq \mathbb{Z}/7\mathbb{Z}$.

We have seen that:

(i) $\det L = -7 = -\det(U \oplus L')$.
(ii) $A_L \simeq \mathbb{Z}/7\mathbb{Z} \simeq A_{U \oplus L'}$.

By Corollary 1, we conclude that the lattice $L = (\mathrm{Pic}_{\Delta^\circ})_{\mathrm{tor}}$ and the lattice $U \oplus L' = U \oplus \mathrm{Pic}_\Delta$ are orthogonal to each other in the $K3$ lattice Λ_{K3}.

In fact, using a list of [21], one finds

$$L' = U \oplus A_6 \oplus E_8.$$

4.3 W_{18}

The singularity W_{18} is defined by $f = x^7 + y^2z + z^2$. There are two possible deformations, $F_1 = X^7 + Y^2Z + Z^2 + W^7Y$ and $F_2 = X^7 + Y^2Z + Z^2 + W^8X$. In Sect. 3.3, we showed that the Newton polytope of F_1 is not reflexive, and although the Newton polytope of F_2 is reflexive, its polar dual does not contain the Newton polytope of F_2^T. Consider the larger lattice polytope Δ given by

$$\Delta = \text{Conv}\,\{(-3, 5, -1),\ (2, -1, 0),\ (-1, 1, 1),\ (0, -1, 0),\ (0, 0, 1)\}\,.$$

The polytope Δ is reflexive, and the Newton polytopes of F_1 and of F_2 are both subpolytopes of Δ. The Newton polytope of the deformation $F_2^T = WX^7 + Y^2 + YZ^2 + W^8$ of $f^T = x^7 + y^2z + z^2$ is a subpolytope of the polar dual Δ° of Δ. In the [24] database of reflexive polytopes, Δ has index 760 and Δ° has index 3046. These are precisely the same indices as the pair of polytopes we identified for the $X_{2,0}$ and S_{17} singularity in Sect. 4.2. Thus, the reflexive polytopes are isomorphic, and as before we have lattices given by

$$(\text{Pic}_{\Delta^\circ})_{\text{tor}} \simeq U \oplus \begin{pmatrix} -4 & 1 \\ 1 & -2 \end{pmatrix}$$

and

$$\text{Pic}\,\Delta = U \oplus A_6 \oplus E_8.$$

4.4 W_{17} and $S_{1,0}$

The $S_{1,0}$ singularity is defined by $f = x^5y + y^2z + z^2$, and W_{17} by $f^T = x^5 + xy^2 + yz^2$. Recall that the only possible invertible deformation of the singularity $S_{1,0}$ is given by $F = X^5Y + Y^2Z + Z^2 + W^{10}$, with monomials corresponding to the lattice points $(0, 1, 0)$, $(-1, 1, 0)$, $(0, 0, 1)$, and $(4, -6, -3)$, respectively. As we have seen in the previous section, the Newton polytope Δ_F of F is not reflexive. So, we take the larger reflexive polytopes that have Δ_F as a subpolytope. In fact, there are two such possibilities as we shall see below. In the following, for a lattice L, we denote by $L_{\mathcal{B}}$ the intersection matrix of L with respect to the set of generators $\mathcal{B}$. Case (i) Let the polytope Δ be given by

$$\Delta = \text{Conv}\,\{(0, 1, 0),\ (0, 0, 1),\ (2, -2, -1),\ (-2, 2, -1),\ (4, -6, -3)\}\,.$$

By construction, the Newton polytope of F is a subpolytope of Δ. We label the lattice points on the edges of Δ as

$$\begin{array}{lll} m_1 = (0,1,0), & m_2 = (0,0,1), & m_3 = (2,-2,-1), \\ m_4 = (-2,2,-1), & m_5 = (4,-6,-3), & m_6 = (3,-4,-2), \\ m_7 = (2,-3,-1), & m_8 = (1,-1,0), & m_9 = (1,-2,-2), \\ m_{10} = (-1,1,0) & & \end{array}.$$

The polar dual polytope of Δ is given by

$$\Delta^\circ = \mathrm{Conv}\left\{(-1,-1,1),\ (-2,-1,-1),\ (0,-1,-1),\ (5,4,-1),\ (-1,0,-1)\right\}.$$

We label the lattice points on the edges of Δ° as

$$\begin{array}{lll} v_1 = (-1,-1,1), & v_2 = (5,4,-1), & v_3 = (-1,0,-1), \\ v_4 = (-2,-1,-1), & v_5 = (0,-1,-1), & v_6 = (2,2,-1), \\ v_7 = (-1,-1,-1), & v_8 = (4,3,-1), & v_9 = (3,2,-1), \\ v_{10} = (2,1,-1), & v_{11} = (1,0,-1) & \end{array}.$$

Since the polytope Δ° is an integral polytope, the polytope Δ is reflexive. The index of Δ in the SageMath database is 1959, while the index of Δ° is 1960. The Newton polytope of the deformation $F^T = X^5 + XY^2 + YZ^2 + W^{10}$ of $f^T = x^5 + xy^2 + yz^2$ is a subpolytope of the polar dual Δ° of Δ.

Next, we compute the toric contribution rank $L_0(\Delta)$.

Let Γ_1 be an edge in Δ given by

$$\Gamma_1 = \mathrm{Conv}\left\{(0,0,1),\ (-2,2,-1)\right\}.$$

One has its polar dual

$$\Gamma_1^\circ = \mathrm{Conv}\left\{(0,-1,-1),\ (5,4,-1)\right\},$$

and

$$l^*(\Gamma_1) = 1, \qquad l^*(\Gamma_1^\circ) = 4.$$

Let Γ_2 be an edge in Δ given by

$$\Gamma_2 = \mathrm{Conv}\left\{(0,0,1),\ (4,-6,-3)\right\}.$$

One has its polar dual

$$\Gamma_2^\circ = \mathrm{Conv}\left\{(-1,0,-1),\ (5,4,-1)\right\},$$

and

$$l^*(\Gamma_2) = 1, \qquad l^*(\Gamma_2^\circ) = 1.$$

In fact, no other edge contributes to rank $L_0(\Delta)$. Thus,

$$\operatorname{rank} L_0(\Delta) = l^*(\Gamma_1)l^*(\Gamma_1^\circ) + l^*(\Gamma_2)l^*(\Gamma_{12}^\circ) = 4 + 1 = 5.$$

Moreover, one has

$$\rho(\Delta) = 11 + 5 - 3 = 13, \quad \rho(\Delta^\circ) = 10 + 5 - 3 = 12.$$

We now compute $(\mathrm{Pic}_{\Delta^\circ})_{\mathrm{tor}}$. Let L be the lattice that is generated by the divisors $D_1, D_2, D_4, D_5, D_6, D_9, D_{10}$, where the divisor D_i is on a generic hypersurface corresponding to the lattice point m_i. By a direct computation, we have $\det L = 20$ and $\operatorname{sgn} L = (1, 6)$. Note that the lattice L contains primitively the hyperbolic lattice $U \simeq \langle D_1+D_6-D_9,\ D_2-D_9\rangle_{\mathbb{Z}}$. By definition, a lattice generated by the restrictions of toric divisors is primitive in the $K3$ lattice Λ_{K3}. We have that the discriminant group A_L of L is given by $A_L \simeq \mathbb{Z}/20\mathbb{Z}$ since the element associated to the divisor $-D_2 + 2D_9 + D_{10}$ in A_L is of order 20. Indeed, let $\phi : \mathbb{Z}^7 \to \mathbb{Q}^7$ be a map for $\boldsymbol{x} = (x_1, \dots, x_7) \in \mathbb{Z}^7$,

$$\phi(\boldsymbol{x}) := \boldsymbol{x}.L_{\mathcal{B}}^{-1}.$$

We can show directly that

$$20 \cdot \phi(0, 0, 1, 0, 0, 0, 1) \in \mathbb{Z}^7,$$
$$k \cdot \phi(0, 0, 1, 0, 0, 0, 1) \in \mathbb{Q}^7 \backslash \mathbb{Z}^7,\ k = 2, 4, 5, 10.$$

We calculate Pic_{Δ}. Let L' be a lattice that is generated by divisors on a generic hypersurface

$$\mathcal{B}' = \{V_1,\ V_2,\ V_4,\ V_6^{(1)},\ V_6^{(2)},\ V_8^{(1)},\ V_9^{(1)},\ V_{10}^{(1)},\ V_{11}^{(1)},\ V_8^{(2)},\ V_9^{(2)},\ V_{10}^{(2)},\ V_{11}^{(2)}\}.$$

Here, $V_i = V_i^{(1)} + V_i^{(2)}$ is a restricted toric divisor for $i = 6, 8, 9, 10, 11$. By a direct computation, we have $\det L' = 20$ and $\operatorname{sgn} L' = (1, 12)$. We have that the discriminant group $A_{L'}$ of L' is given by $A_{L'} \simeq \mathbb{Z}/20\mathbb{Z}$ since the element associated to the divisor $V_6^{(1)} + V_8^{(1)}$ in $A_{L'}$ is of order 20. Indeed, let $\psi : \mathbb{Z}^{13} \to \mathbb{Q}^{13}$ be a map for $\boldsymbol{x} = (x_1, \dots, x_{13}) \in \mathbb{Z}^{13}$,

$$\psi(\boldsymbol{x}) := \boldsymbol{x}.L'^{-1}_{\mathcal{B}'}.$$

We can show directly that

$$20 \cdot \phi(0, 0, 0, 1, 0, 1, 0, \dots, 0) \in \mathbb{Z}^{13},$$
$$k \cdot \phi(0, 0, 0, 1, 0, 1, 0, \dots, 0) \in \mathbb{Q}^{13} \backslash \mathbb{Z}^{13},\ k = 2, 4, 5, 10.$$

We claim that the lattice L' is a primitive lattice of the $K3$ lattice $\Lambda_{K3} = U^{\oplus 3} \oplus E_8^{\oplus 2}$, which is of signature $(l_+, l_-) = (3, 19)$. We have $l_+ - t_+ = 3 - 1 = 2 \geq 0$, $l_- - t_- = 19 - 12 = 7 \geq 0$, and $\operatorname{rank} \Lambda_{K3} - \operatorname{rank} L' = 22 - 13 = 9 > 1 = l(A_{L'})$. Thus by Corollary 2, the lattice L' is indeed a primitive sublattice of the $K3$ lattice.

We have seen that:

(i) $\det L = 20 = -\det (U \oplus L')$.
(ii) $A_L \simeq \mathbb{Z}/20\mathbb{Z} \simeq A_{L'}$.

By Corollary 1, we conclude that the lattices $L = (\mathrm{Pic}_{\Delta^*})_{\mathrm{tor}}$ and $U \oplus L' = U \oplus \mathrm{Pic}_\Delta$ are orthogonal to each other in the $K3$ lattice Λ_{K3}.
Case (ii) Let the polytope Δ be given by

$$\Delta = \Delta_{(2,3,5,10)} = \mathrm{Conv}\left\{(1,0,0),\ (0,1,0),\ (0,0,1),\ (-2,2,-1),\ (4,-6,-3)\right\}.$$

By construction, the Newton polytope of F is a subpolytope of Δ. We label the lattice points on the edges of Δ as

$$\begin{gathered} m_1 = (0,1,0), \qquad m_2 = (0,0,1), \qquad m_3 = (1,0,0), \\ m_4 = (-2,2,-1),\ m_5 = (4,-6,-3),\ m_6 = (3,-4,-2), \\ m_7 = (2,-2,-1),\ m_8 = (2,-3,-1),\ m_9 = (1,-2,-2), \\ m_{10} = (-1,1,0) \end{gathered}.$$

The polar dual polytope of Δ is given by

$$\Delta^\circ = \mathrm{Conv}\left\{(-1,-1,1),\ (-1,-1,-1),\ (0,-1,-1),\ (5,4,-1),\ (-1,0,-1)\right\}.$$

We label the lattice points on the edges of Δ° as

$$\begin{gathered} v_1 = (-1,-1,1), \quad v_2 = (5,4,-1), \quad v_3 = (-1,0,-1), \\ v_4 = (-1,-1,-1),\ v_5 = (0,-1,-1),\ v_6 = (2,2,-1), \\ v_7 = (-1,-1,0), \quad v_8 = (4,3,-1), \quad v_9 = (3,2,-1), \\ v_{10} = (2,1,-1), \quad v_{11} = (1,0,-1) \end{gathered}.$$

Since the polytope Δ° is an integral polytope, the polytope Δ is reflexive. The Newton polytope of the deformation $F^T = X^5 + XY^2 + YZ^2 + W^{10}$ of $f^T = x^5 + xy^2 + yz^2$ is a subpolytope of the polar dual Δ° of Δ.

The edges Γ_1 and Γ_2 defined in Case (i) are also edges of Δ, and there is no other contribution to $\operatorname{rank} L_0(\Delta)$. Therefore, we have $\operatorname{rank} L_0(\Delta) = 5$.

Moreover, one has

$$\rho(\Delta) = 11 + 5 - 3 = 13, \quad \rho(\Delta^\circ) = 10 + 5 - 3 = 12.$$

We now compute $(\mathrm{Pic}_{\Delta^\circ})_{\mathrm{tor}}$. Let L be the lattice that is generated by the divisors $D_1, D_2, D_3, D_7, D_6, D_5, D_9$, where the divisor D_i is on a generic hypersurface corresponding to the lattice point m_i. By a direct computation, we have $\det L = 20$ and $\mathrm{sgn}\, L = (1,\ 6)$. Note that the lattice L contains primitively the hyperbolic lattice $U \simeq \langle D_1 + D_6 - D_7,\ D_1 + D_2 - 2D_3 \rangle_{\mathbb{Z}}$.

By definition, a lattice generated by the restrictions of toric divisors is primitive in the $K3$ lattice Λ_{K3}. We have that the discriminant group A_L of L is given by $A_L \simeq \mathbb{Z}/20\mathbb{Z}$ since the element associated to the divisor $D_6 + D_7$ in A_L is of order 20. Indeed, let $\phi : \mathbb{Z}^7 \to \mathbb{Q}^7$ be a map for $\boldsymbol{x} = (x_1, \ldots, x_7) \in \mathbb{Z}^7$,

$$\phi(\boldsymbol{x}) := \boldsymbol{x}.L_{\mathcal{B}}^{-1}.$$

We can show directly that

$$20 \cdot \phi(0,\ 0,\ 0,\ 0,\ 1,\ 0,\ 0) \in \mathbb{Z}^7,$$
$$k \cdot \phi(0,\ 0,\ 0,\ 0,\ 1,\ 0,\ 0) \in \mathbb{Q}^7 \backslash \mathbb{Z}^7,\ k = 2, 4, 5, 10.$$

Note that the lattices L in Cases (i) and (ii) are isometric.

We calculate Pic_Δ. Let L' be a lattice that is generated by divisors on a generic hypersurface

$$\mathcal{B}' = \{V_1,\ V_2,\ V_4,\ V_6^{(1)},\ V_6^{(2)},\ V_8^{(1)},\ V_9^{(1)},\ V_{10}^{(1)},\ V_{11}^{(1)},\ V_8^{(2)},\ V_9^{(2)},\ V_{10}^{(2)},\ V_{11}^{(2)}\}.$$

Here, $V_i = V_i^{(1)} + V_i^{(2)}$ is a restricted toric divisor for $i = 6, 8, 9, 10, 11$.

By a direct computation, we have $\det L' = 20$ and $\mathrm{sgn}\, L' = (1,\ 12)$. We have that the discriminant group $A_{L'}$ of L' is given by $A_{L'} \simeq \mathbb{Z}/20\mathbb{Z}$ since the element associated to the divisor $V_6^{(1)} + V_8^{(1)}$ in $A_{L'}$ is of order 20. Indeed, let $\psi : \mathbb{Z}^{13} \to \mathbb{Q}^{13}$ be a map for $\boldsymbol{x} = (x_1, \ldots, x_{13}) \in \mathbb{Z}^{13}$,

$$\psi(\boldsymbol{x}) := \boldsymbol{x}.L'^{-1}_{\mathcal{B}'}.$$

We can show directly that

$$20 \cdot \phi(0,\ 0,\ 0,\ 1,\ 0,\ 1,\ 0, \ldots,\ 0) \in \mathbb{Z}^{13},$$
$$k \cdot \phi(0,\ 0,\ 0,\ 1,\ 0,\ 1,\ 0, \ldots,\ 0) \in \mathbb{Q}^{13} \backslash \mathbb{Z}^{13},\ k = 2, 4, 5, 10.$$

We claim that the lattice L' is a primitive lattice of the $K3$ lattice $\Lambda_{K3} = U^{\oplus 3} \oplus E_8^{\oplus 2}$, which is of signature $(l_+,\ l_-) = (3,\ 19)$. We have $l_+ - t_+ = 3 - 1 = 2 \geq 0$, $l_- - t_- = 19 - 12 = 7 \geq 0$, and $\mathrm{rank}\, \Lambda_{K3} - \mathrm{rank}\, L = 22 - 13 = 9 > 1 = l(A_L)$. By Corollary 2, the lattice L' is indeed a primitive sublattice of the $K3$ lattice.

We have seen that:

(i) $\det L = 20 = -\det(U \oplus L')$.
(ii) $A_L \simeq \mathbb{Z}/20\mathbb{Z} \simeq A_{L'}$.

By Corollary 1, we conclude that the lattice $L = (\mathrm{Pic}_{\Delta^\circ})_{\mathrm{tor}}$ and the lattice $U \oplus L' = U \oplus \mathrm{Pic}_\Delta$ are orthogonal to each other in the $K3$ lattice Λ_{K3}.

We can easily observe that in cases (i) and (ii), although the polytopes are not isomorphic, but the lattices $(\mathrm{Pic}_{\Delta^\circ})_{\mathrm{tor}}$ and Pic Δ are respectively isometric.

4.5 U_{16}

The singularity U_{16} is defined by $f = x^5 + y^2z + yz^2$, and consider a deformation $F = X^5 + Y^2Z + YZ^2 + W^6X$, and let the polytope Δ be given by

$$\Delta := \Delta_F = \mathrm{Conv}\,\{(1, 0, -1),\ (0, 0, 1),\ (-2, -2, 3),\ (0, 1, -1)\}\,.$$

By construction, the Newton polytope of F is a subpolytope of Δ. We label the lattice points on the edges of Δ as

$$m_1 = (1, 0, -1), \quad m_2 = (0, 0, 1), \quad m_3 = (-2, -2, 3),$$
$$m_4 = (0, 1, -1),\ m_5 = (-1, -1, 2).$$

The polar dual polytope of Δ is given by

$$\Delta^\circ = \mathrm{Conv}\,\{(-2, -2, -1),\ (-2, 1, -1),\ (4, 4, 5),\ (1, -2, -1)\}\,.$$

We label the lattice points on the edges of Δ° as

$$\begin{array}{llll} v_1 = (-2, -2, -1), & v_2 = (-2, 1, -1), & v_3 = (4, 4, 5), & v_4 = (1, -2, -1), \\ v_5 = (-2, -1, -1), & v_6 = (-2, 0, -1), & v_7 = (0, 2, 1), & v_8 = (2, 3, 3), \\ v_9 = (-1, -2, -1), & v_{10} = (0, -2, -1), & v_{11} = (3, 2, 3), & v_{12} = (2, 0, 1), \\ v_{13} = (-1, -1, 0), & v_{14} = (0, 0, 1), & v_{15} = (1, 1, 2), & v_{16} = (2, 2, 3), \\ v_{17} = (3, 3, 4), & v_{18} = (-1, 0, -1), & v_{19} = (0, -1, -1) & \end{array}.$$

Since the polytope Δ° is an integral polytope, the polytope Δ is reflexive. The Newton polytope of the deformation $F^T = WX^5 + Y^2Z + YZ^2 + W^6$ of $f^T = x^5 + y^2z + yz^2$ is a subpolytope of the polar dual Δ° of Δ. As we have seen in the previous section, the toric contribution is rank $L_0(\Delta) = 2$.

Moreover, one has

$$\rho(\Delta) = 19 + 2 - 3 = 18, \ \ \rho(\Delta^\circ) = 5 + 2 - 3 = 4.$$

We compute $(\mathrm{Pic}_{\Delta^\circ})_{\mathrm{tor}}$. A collection of $\{m_1,\ m_2,\ m_5\}$ are linearly independent vectors. Let L be a lattice that is generated by divisors D_3 and D_4. By taking the new generators $\{D_3,\ -D_3 + D_4\}$, we have $L \simeq U(3)$. It is known that the discriminant group of $U(3)$ is isomorphic to $\mathbb{Z}/9\mathbb{Z}$. By definition, the lattice $(\mathrm{Pic}_{\Delta^\circ})_{\mathrm{tor}}$ is actually equal to the lattice L.

We compute Pic_Δ. A collection of $\{v_1, v_5, v_9\}$ are linearly independent vectors. Let L' be a lattice that is generated by divisors

$$\mathcal{B}' = \{V_2, V_3, V_4, V_6, V_7, V_8, V_{10}, \ldots, V_{17}, V_{18}^{(1)}, V_{18}^{(2)}, V_{19}^{(1)}, V_{19}^{(2)}\}.$$

Here, $V_i = V_i^{(1)} + V_i^{(2)}$ is a restricted toric divisor for $i = 18, 19$.

By a direct computation, we have $\det L' = -9$ and $\mathrm{sgn}\, L' = (1, 17)$ with the aid of Mathematica.

Since $\mathrm{rank}\, L' > 12$, there exists an even negative-definite lattice L'' of rank 16 and $\det L'' = 9$ such that $L' = U \oplus L''$ holds. Moreover, set $L'_{\mathbb{B}'}$ to be the intersection matrix of L' with respect to the basis $\mathcal{B}'$, and by defining a map $\phi : \mathbb{Z}^{18} \to \mathbb{Q}^{18}$

$$\phi(x_1, \ldots, x_{18}; s) := s \cdot (x_1 \cdots x_{18}).{L'}_{\mathbb{B}'}^{-1},$$

and one can see that

$$\phi(x_1, \ldots, x_{18}; 9) \in \mathbb{Z}.$$

This means that there exists an element of order 9 in the discriminant group of L'. Therefore, $L'^*/L' \simeq \mathbb{Z}/9\mathbb{Z}$, so $L'^*/L' \simeq (L \oplus U)^*/(L \oplus U)$. Thus, the lattice L' is also a primitive sublattice of Λ_{K3}, and $L' \simeq (L \oplus U)^\perp$ holds.

We have seen that:

(i) $\det L = -9 = -\det(U \oplus L')$.
(ii) $A_L \simeq \mathbb{Z}/9\mathbb{Z} \simeq A_{U \oplus L'}$.

By Corollary 1, we conclude that the lattices $L = (\mathrm{Pic}_{\Delta^\circ})_{\mathrm{tor}}$ and $U \oplus L' = U \oplus \mathrm{Pic}_\Delta$ are orthogonal to each other in the $K3$ lattice Λ_{K3}. □

References

1. M. Artebani, S. Boissière, A. Sarti, The Berglund-Hübsch-Chiodo-Ruan mirror symmetry for K3 surfaces. J. Math. Pures Appl. (9) **102**(4), 758–781 (2014)
2. P.S. Aspinwall, B.R. Greene, D.R. Morrison, The monomial-divisor mirror map. Int. Math. Res. Notices **12**, 319–337 (1993)
3. V.V. Batyrev, Dual polyhedra and mirror symmetry for Calabi-Yau hypersurfaces in toric varieties. J. Alg. Geom. **3**, 493–535 (1994)
4. U. Bruzzo, A. Grassi, Picard group of hypersurfaces in toric 3-folds. Int. J. Math. **23**(2), 1–14 (2012)
5. P. Clarke, Duality for toric Landau-Ginzburg models. Adv. Theor. Math. Phys. **21**(1), 243–287 (2017)

6. D.A. Cox, S. Katz, *Mirror Symmetry and Algebraic Geometry* (American Mathematical Society, Providence, RI, 1999)
7. I.V. Dolgachev, Mirror symmetry for lattice polarized $K3$ surfaces. J. Math. Sci. **81**(3), 2599–2630 (1996)
8. C.F. Doran, D. Favero, T.L. Kelly, Equivalences of families of stacky toric Calabi-Yau hypersurfaces. Proc. Am. Math. Soc. **146**(11), 4633–4647 (2018)
9. W. Ebeling, D. Ploog, A geometric construction of Coxeter-Dynkin diagrams of bimodal singularities. Manuscripta Math. **140**(1–2), 195–212 (2013)
10. D. Favero, T.L. Kelly, Fractional Calabi-Yau categories from Landau-Ginzburg models. Algebr. Geom. **5**(5), 596–649 (2018)
11. W. Fulton, *Introduction to Toric Varieties*, 2nd edn. Ann. Math. Stu., vol. 131 (Princeton University Press, Princeton, 1997)
12. M. Kreuzer, H. Skarke, On the classification of quasihomogeneous functions. Commun. Math. Phys. **150**(1), 137–147 (1992)
13. M. Kreuzer, H. Skarke, On the classification of reflexive polyhedra. Commun. Math. Phys. **185**(2), 495–508 (1997)
14. M. Mase, Correspondence among families of Gorenstein $K3$ surfaces in certain $\mathbf{Q}$-Fano 3-folds. Dissertation, Tokyo Metropolitan Univ., 2012. Available at https://tokyo-metro-u.repo.nii.ac.jp/?action=pages_view_main&active_action=repository_view_main_item_detail&item_id=2117&item_no=1&page_id=30&block_id=164
15. M. Mase, A mirror duality for families of K3 surfaces associated to bimodular singularities, Manuscripta Math. **149**(3–4), 389–404 (2016)
16. M. Mase, Polytope duality for families of K3 surfaces associated to transpose duality. Comment. Math. Univ. St. Pauli **65**(2), 131–139 (2016)
17. M. Mase, Lattice duality for families of K3 surfaces associated to transpose duality. Manuscripta Math. https://doi.org/10.1007/s00229-017-0936-5 (online 10 April 2017)
18. M. Mase, K. Ueda, A note on bimodal singularities and mirror symmetry. Manuscripta Math. **146**(1–2), 153–177 (2015)
19. M. Mase, U. Whitcher, Mirror constructions for K3 surfaces from bimodal singularities. Preprint (2022). arXiv:2206.14367 [math.AG]
20. V.V. Nikulin, Integral symmetric bilinear forms and some of their applications. Math. USSR-Izv **14**, 103–167 (1980)
21. K. Nishiyama, The Jacobian fibrations on some $K3$ surfaces and their Mordell-Weil groups. Jpn. J. Math. **22**(2), 293–347 (1996)
22. K. Oguiso, Picard numbers in a family of hyperkähler manifolds - A supplement to the article of R. Borcherds, L. Katzarkov, T. Pantev, N. I. Shepherd-Barron. Preprint (2000). arXiv.org:math/0011258
23. F. Rohsiepe, Lattice polarized toric K3 surfaces. Preprint (2004). arXiv:hep-th/0409290 v1
24. Sage, *SageMath, the Sage Mathematics Software System (Version 8.2)* (The Sage Developers, 2018). http://www.sagemath.org
25. U. Whitcher, Reflexive polytopes and lattice-polarized K3 surfaces. Calabi-Yau varieties: arithmetic, geometry and physics. Fields Inst. Monogr. **34**, 65–79 (2015)

A Note on the Non-vanishing of Poincaré Series

Sonja Žunar

1 Introduction

In the theory of modular forms, the most famous examples of Poincaré series are the classical holomorphic Eisenstein series E_k of even weight $k \geq 4$ on the upper half-plane $\mathcal{H} = \mathbb{C}_{\Im(z)>0}$ given by

$$E_k(z) = \sum_{\left(\begin{smallmatrix} a & b \\ c & d \end{smallmatrix}\right) \in \mathrm{SL}_2(\mathbb{Z})_\infty \backslash \mathrm{SL}_2(\mathbb{Z})} (cz+d)^{-k}, \quad z \in \mathcal{H},$$

where $\mathrm{SL}_2(\mathbb{Z})_\infty$ denotes the subgroup of upper-triangular matrices in $\mathrm{SL}_2(\mathbb{Z})$. It is well-known that the series E_k converge absolutely and uniformly on compact subsets of $\mathcal{H}$ and define modular forms of weight k for $\mathrm{SL}_2(\mathbb{Z})$. These modular forms do not vanish at the cusps, so in particular they do not vanish identically. But for cusp forms defined by Poincaré series, such as the classical Poincaré series $\Psi_{n,k} : \mathcal{H} \to \mathbb{C}$,

$$\Psi_{n,k}(z) = \sum_{\left(\begin{smallmatrix} a & b \\ c & d \end{smallmatrix}\right) \in \mathrm{SL}_2(\mathbb{Z})_\infty \backslash \mathrm{SL}_2(\mathbb{Z})} (cz+d)^{-k}\, e^{2\pi i n \frac{az+b}{cz+d}}, \tag{1}$$

where $n \in \mathbb{Z}_{>0}$, the question of non-vanishing is highly non-trivial and is yet to be solved in complete generality.

To give an example of why this question is interesting, recall that the cusp forms $\Psi_{n,k} \in S_k(\mathrm{SL}_2(\mathbb{Z}))$ are known for their following property: Denoting by

---

S. Žunar (✉)
Faculty of Geodesy, University of Zagreb, Zagreb, Croatia
e-mail: szunar@geof.hr

R. Abdellatif et al. (eds.), *Women in Numbers Europe IV*, Association for Women in Mathematics Series 32, https://doi.org/10.1007/978-3-031-52163-8_12

$\langle\cdot,\cdot\rangle_{S_k(\mathrm{SL}_2(\mathbb{Z}))}$ the Petersson inner product on $S_k(\mathrm{SL}_2(\mathbb{Z}))$ (see (5)), for every cusp form $f(z)=\sum_{n=1}^{\infty}a_n(f)e^{2\pi inz}$ in $S_k(\mathrm{SL}_2(\mathbb{Z}))$, we have

$$a_n(f)=\frac{(4\pi n)^{k-1}}{(k-2)!}\left\langle f,\Psi_{n,k}\right\rangle_{S_k(\mathrm{SL}_2(\mathbb{Z}))},\qquad n\in\mathbb{Z}_{>0}.$$

Thus, for fixed k and n, we have the following equivalence:

$$\Psi_{n,k}\equiv 0\qquad\Leftrightarrow\qquad a_n(f)=0\text{ for all }f\in S_k(\mathrm{SL}_2(\mathbb{Z})).\tag{2}$$

In the case when $k=12$, the space $S_k(\mathrm{SL}_2(\mathbb{Z}))$ is one-dimensional and spanned by the modular discriminant

$$\Delta(z)=(2\pi)^{12}q\prod_{r=1}^{\infty}\left(1-q^r\right)^{24}=(2\pi)^{12}\sum_{n=1}^{\infty}\tau(n)q^n,\quad z\in\mathcal{H},\ q=e^{2\pi iz},$$

where τ is the Ramanujan tau function. Thus, by (2), the following two claims are equivalent:

(A1) $\Psi_{n,12}\not\equiv 0$ for all $n\in\mathbb{Z}_{>0}$.
(A2) $\tau(n)\neq 0$ for all $n\in\mathbb{Z}_{>0}$.

The claim (A2) is known as Lehmer's conjecture (cf. [22]). Another take on this conjecture, described in [31, §2] (see also [3]), relates zeroes of τ to the algebraicity of certain Fourier coefficients of a harmonic weak Maass form $g\in H_{-10}(\mathrm{SL}_2(\mathbb{Z}))$ that is "dual" to Δ with respect to the differential operator $\xi_{2-k}=2iy^{2-k}\overline{\frac{\partial}{\partial\bar{z}}}$: $H_{2-k}(\Gamma_0(N))\to S_k(\Gamma_0(N))$, where $k=12$ and $N=1$.

Let us also mention that the non-vanishing of the classical Poincaré series $\Psi_{n,k}$ of even weight $k\in\{24\}\cup\mathbb{Z}_{\geq 28}$ is implied by the generalized Maeda's conjecture, which states that for each $n\in\mathbb{Z}_{>0}$, the characteristic polynomial $T_n(x)$ of the nth Hecke operator T_n on $S_k(\mathrm{SL}_2(\mathbb{Z}))$ is irreducible over $\mathbb{Q}$ with Galois group the symmetric group $\mathfrak{S}_d$ of order $d=\dim_{\mathbb{C}}S_k(\mathrm{SL}_2(\mathbb{Z}))$ (see, e.g., [1]). Indeed, suppose that the generalized Maeda's conjecture is true and that there exist an integer $n>0$ and an even integer $k\in\{24\}\cup\mathbb{Z}_{\geq 28}$ such that $\Psi_{n,k}\equiv 0$. Then, by (2), we have that $a_n(f_j)=0$ for all $j=1,\ldots,d$, where $f_1,\ldots,f_d\in S_k(\mathrm{SL}_2(\mathbb{Z}))$ are the normalized Hecke eigenforms. But, as is well-known, the numbers $a_n(f_j)$ comprise the spectrum of the operator T_n. It follows that $T_n(x)=x^d$, but this polynomial is not irreducible over $\mathbb{Q}$ since $d>1$.

In this chapter, we give a brief overview of a few existing methods for proving that a cusp form given by a Poincaré series does not vanish identically, where by the term "Poincaré series" we mean any series of the form

$$\left(P_{k,\Lambda\backslash\Gamma}f\right)(z)=\sum_{\left(\begin{smallmatrix}a&b\\c&d\end{smallmatrix}\right)\in\Lambda\backslash\Gamma}(cz+d)^{-k}f\left(\frac{az+b}{cz+d}\right),\qquad z\in\mathcal{H},$$

where $k \in \mathbb{Z}_{>0}$, Γ is a discrete subgroup of finite covolume in $\mathrm{SL}_2(\mathbb{R})$, Λ is a subgroup of Γ, and f is a suitable function $\mathcal{H} \to \mathbb{C}$. Our focus is on methods that have been used for studying the non-vanishing of the classical Poincaré series with respect to a group Γ with a cusp at ∞, such as:

(1) Finding an explicit formula for a Fourier coefficient of the cusp form in question and bounding it away from zero, as in [41]
(2) Studying arithmetic properties of Fourier coefficients, e.g., finding arithmetic relations between them or studying their congruence properties
(3) Applying an integral non-vanishing criterion, as in [29].

In Sect. 3, we briefly describe a few methods that have been used for studying the non-vanishing of the classical Poincaré series and give examples of the corresponding non-vanishing results. In Sects. 4 and 5, we turn our attention to two less investigated types of Poincaré series: the Poincaré series of elliptic type introduced by Petersson and the Poincaré series related to Kohnen's kernel function for $S_k(\Gamma)$. In the end, we briefly reflect on a few more types of Poincaré series whose non-vanishing has been partially investigated, as well as on the more general problem of determining all linear relations among the Poincaré series of a fixed type, and include some references on both for the interested reader.

The account we give here is by no means complete and is meant only as an instructive contemporary glimpse into the extensive ongoing search for a general answer to the question of non-vanishing of the cuspidal Poincaré series.

2 Basics

The group $\mathrm{SL}_2(\mathbb{R})$ acts on the upper half-plane $\mathcal{H} = \mathbb{C}_{\Im(z)>0}$ and on the Riemann sphere $\mathbb{C} \cup \{\infty\}$ by Möbius transformations:

$$g.z = \frac{az+b}{cz+d}, \qquad g = \begin{pmatrix} a & b \\ c & d \end{pmatrix} \in \mathrm{SL}_2(\mathbb{R}),\ z \in \mathbb{C} \cup \{\infty\}.$$

For every weight $k \in \mathbb{Z}_{>0}$, we also have the following right action of $\mathrm{SL}_2(\mathbb{R})$ on the space $\mathbb{C}^{\mathcal{H}}$ of functions $f : \mathcal{H} \to \mathbb{C}$:

$$\left(f|_k g\right)(z) = j(g,z)^{-k} f(g.z), \qquad f \in \mathbb{C}^{\mathcal{H}},\ g \in \mathrm{SL}_2(\mathbb{R}),\ z \in \mathcal{H}, \tag{3}$$

where $j : \mathrm{SL}_2(\mathbb{R}) \times \mathcal{H} \to \mathbb{C}$ is the standard automorphic factor given by

$$j(g,z) = cz + d, \qquad g = \begin{pmatrix} a & b \\ c & d \end{pmatrix} \in \mathrm{SL}_2(\mathbb{R}),\ z \in \mathcal{H}.$$

Let $\Gamma \subseteq \mathrm{SL}_2(\mathbb{R})$ be a Fuchsian group of the first kind, i.e., a discrete subgroup of finite covolume in $\mathrm{SL}_2(\mathbb{R})$. The space $S_k(\Gamma)$ of cusp forms of weight $k \in \mathbb{Z}_{>0}$ for Γ is the space of holomorphic functions $f : \mathcal{H} \to \mathbb{C}$ such that $\sup_{z \in \mathcal{H}} |f(z)| \, \Im(z)^{\frac{k}{2}} < \infty$ and

$$f\big|_k \gamma = f, \qquad \gamma \in \Gamma. \tag{4}$$

Let v be the standard $\mathrm{SL}_2(\mathbb{R})$-invariant Radon measure on $\mathcal{H}$ defined by $dv(x+iy) = \frac{dx\,dy}{y^2}$. Equipped with the Petersson inner product

$$\langle f, g\rangle_{S_k(\Gamma)} = \int_{\Gamma\backslash\mathcal{H}} f(z)\,\overline{g(z)}\,\Im(z)^k\,dv(z), \qquad f, g \in S_k(\Gamma), \tag{5}$$

$S_k(\Gamma)$ is a finite-dimensional complex Hilbert space.

Let Λ be a subgroup of Γ. For any $f : \mathcal{H} \to \mathbb{C}$ such that

$$f\big|_k \lambda = f, \qquad \lambda \in \Lambda,$$

we can form a Poincaré series

$$P_{k,\Lambda\backslash\Gamma} f = \sum_{\gamma \in \Lambda\backslash\Gamma} f\big|_k \gamma.$$

Note that if pointwise absolutely convergent, this series satisfies the condition (4). This is why Poincaré series are a useful tool for constructing (not necessarily cuspidal) modular forms.

3 Classical Poincaré Series

Let Γ be a Fuchsian group of the first kind with a cusp at ∞, and let $\Gamma_\infty = \{\gamma \in \Gamma : \gamma.\infty = \infty\}$. There exists a unique $h_\Gamma \in \mathbb{R}_{>0}$ such that

$$\{\pm I_2\}\Gamma_\infty = \left\{\pm \begin{pmatrix} 1 & nh_\Gamma \\ & 1 \end{pmatrix} : n \in \mathbb{Z}\right\}.$$

[25, Thm. 1.5.4(2)]. Fix $k \in \mathbb{Z}_{\geq 3}$, and suppose that $\Gamma_\infty = \left\{\begin{pmatrix} 1 & nh_\Gamma \\ & 1 \end{pmatrix} : n \in \mathbb{Z}\right\}$ if k is odd. Then, every $f \in S_k(\Gamma)$ is h_Γ-periodic and has a Fourier expansion

$$f(z) = \sum_{n=1}^{\infty} a_n(f)\, e^{2\pi i n \frac{z}{h_\Gamma}}, \qquad z \in \mathcal{H},$$

for some $(a_n(f))_{n\in\mathbb{Z}_{>0}} \subseteq \mathbb{C}$.

For any $r \in \mathbb{R}_{>0}$, let $\psi_r : \mathcal{H} \to \mathbb{C}$, $\psi_r(z) = e^{2\pi i r z}$. It is well-known that the classical Poincaré series

$$\Psi_{n,k,\Gamma}(z) = \left(P_{k,\Gamma_\infty\backslash\Gamma}\,\psi_{n/h_\Gamma}\right)(z) = \sum_{\gamma\in\Gamma_\infty\backslash\Gamma} j(\gamma,z)^{-k} e^{2\pi i n \frac{\gamma.z}{h_\Gamma}}, \qquad z \in \mathcal{H},$$

where $n \in \mathbb{Z}_{>0}$, converge absolutely and uniformly on compact subsets of $\mathcal{H}$ and span the space $S_k(\Gamma)$. We will write $\Psi_{n,k} = \Psi_{n,k,\mathrm{SL}_2(\mathbb{Z})}$ and $\Psi_{n,k,N} = \Psi_{n,k,\Gamma_0(N)}$ for $N \in \mathbb{Z}_{>0}$. The cusp forms $\Psi_{n,k,\Gamma}$ are known for their following property:

$$\left\langle f, \Psi_{n,k,\Gamma}\right\rangle_{S_k(\Gamma)} = \frac{h_\Gamma^k\,(k-2)!}{(4\pi n)^{k-1}}\, a_n(f), \qquad f \in S_k(\Gamma),\ n \in \mathbb{Z}_{>0}. \tag{6}$$

See, e.g., [25, Thm. 2.6.10].

Next, let $N \in \mathbb{Z}_{>0}$, and let us take a look at the Fourier expansion

$$\Psi_{n,k,N}(z) = \sum_{m=1}^{\infty} c_{k,N}(m,n) e^{2\pi i m z}, \qquad z \in \mathcal{H}, \tag{7}$$

of the classical Poincaré series for $\Gamma_0(N)$ of even weight $k \geq 4$. Applying (6) with $\Gamma = \Gamma_0(N)$ and $f = \Psi_{n,k,N}$, we conclude that

$$\Psi_{n,k,N} \equiv 0 \quad \Leftrightarrow \quad c_{k,N}(n,n) = 0. \tag{8}$$

In [34] (see also [17, §3.2]), Petersson proved the formula

$$c_{k,N}(m,n) = \left(\frac{m}{n}\right)^{\frac{k-1}{2}} \left(\delta_{m,n} + 2\pi i^{-k} \sum_{r=1}^{\infty} \frac{S(m,n;Nr)}{Nr} J_{k-1}\left(\frac{4\pi\sqrt{mn}}{Nr}\right)\right), \tag{9}$$

where J_{k-1} is the Bessel function of the first kind of order $k-1$ and $S(m,n;Nr)$ is a Kloosterman sum given by

$$S(m,n;s) = \sum_{\substack{h=1\\(h,s)=1}}^{s} e^{\frac{2\pi i}{s}(mh+nh')}, \qquad \text{where } hh' \equiv 1 \pmod{s}.$$

By applying (6) with $\Gamma = \Gamma_0(N)$ and $f = \Psi_{m,k,N}$, by (9), we obtain that for all $m, n, N \in \mathbb{Z}_{>0}$ and all even $k \in \mathbb{Z}_{\geq 4}$, we have

$$\begin{aligned} &\left\langle \Psi_{m,k,N}, \Psi_{n,k,N} \right\rangle_{S_k(\Gamma_0(N))} \\ &\quad = \frac{(k-2)!}{(4\pi\sqrt{mn})^{k-1}} \left(\delta_{m,n} + 2\pi i^{-k} \sum_{r=1}^{\infty} \frac{S(m,n;Nr)}{Nr} J_{k-1}\left(\frac{4\pi\sqrt{mn}}{Nr}\right)\right). \end{aligned} \tag{10}$$

Finally, we note that by (9)

$$c_{k,N}(m,n) = \left(\frac{m}{n}\right)^{k-1} c_{k,N}(n,m), \qquad m,n \in \mathbb{Z}_{>0}. \tag{11}$$

Using the above preliminaries, in the following subsections, we describe, in chronological order, several ideas for studying the non-vanishing of the classical Poincaré series, as well as a few explicit non-vanishing results to which these ideas have led.

3.1 Basic Non-vanishing Results

Of course, if $S_k(\Gamma) = 0$, then $\Psi_{n,k,\Gamma} \equiv 0$ for all $n \in \mathbb{Z}_{>0}$. This happens, for instance, if $\Gamma = \mathrm{SL}_2(\mathbb{Z})$ and $k \in \{4, 6, 8, 10, 14\}$.

On the other hand, if the genus of Γ is zero, then for every even integer $k \geq 4$, the cusp forms $\Psi_{1,k,\Gamma}, \Psi_{2,k,\Gamma}, \ldots, \Psi_{\dim_{\mathbb{C}} S_k(\Gamma),k,\Gamma}$ constitute a basis for $S_k(\Gamma)$; hence, in particular, they do not vanish identically. In the $\mathrm{SL}_2(\mathbb{Z})$ case, this was proved by Petersson [35]; for the general case, see [45].

3.2 Properties of Fourier Coefficients

Using (9) and sharp estimates of Kloosterman sums, in 1980 Rankin bounded the coefficient $c_{k,1}(n,n)$ away from zero for $k \gg 0$ and $n \ll k^{2-\varepsilon}$. In this way, he proved the $\mathrm{SL}_2(\mathbb{Z})$ case [41, Thm. 1] of the following theorem.

Theorem 1 *There exist $k_0 \in \mathbb{Z}_{\geq 4}$ and $B \in \mathbb{R}_{>\ln 16}$ such that for all subgroups Γ of finite index in $\mathrm{SL}_2(\mathbb{Z})$, all even integers $k \geq k_0$, and all $n \in \mathbb{Z}_{>0}$,*

$$n \leq k^2 e^{-B\frac{\ln k}{\ln\ln k}} \quad \Rightarrow \quad \Psi_{nh_\Gamma,k,\Gamma} \not\equiv 0.$$

Proof It was first noted in [15] that the general case follows directly from [41, Thm. 1] since

$$\sum_{\delta\in\Gamma\backslash\mathrm{SL}_2(\mathbb{Z})} \Psi_{nh_\Gamma,k,\Gamma}\big|_k\delta = \sum_{\gamma\in\Gamma_\infty\backslash\mathrm{SL}_2(\mathbb{Z})} \psi_n\big|_k\gamma$$

$$= \sum_{\delta\in\mathrm{SL}_2(\mathbb{Z})_\infty\backslash\mathrm{SL}_2(\mathbb{Z})} \left(\sum_{\gamma\in\Gamma_\infty\backslash\mathrm{SL}_2(\mathbb{Z})_\infty} \psi_n\big|_k\gamma \right)\Big|_k\delta$$

$$= |\Gamma_\infty\backslash\mathrm{SL}_2(\mathbb{Z})_\infty|\ \Psi_{n,k},$$

where the last equality holds since $\psi_n\big|_k\gamma = \psi_n$ for all $\gamma\in\mathrm{SL}_2(\mathbb{Z})_\infty$. □

In the following years, by generalizing Rankin's method of estimating the Fourier coefficient $c_{k,1}(n,n)$, several mathematicians proved various results on the non-vanishing of classical Poincaré series $\Psi_{n,k,\Gamma}$ with $\Gamma\neq\mathrm{SL}_2(\mathbb{Z})$. For instance, results proved by Lehner in [23] imply the following result, weaker than Rankin's, for a more general group Γ.

Theorem 2 *Let $\Gamma\subseteq\mathrm{SL}_2(\mathbb{R})$ be a Fuchsian group of the first kind with a cusp at ∞. Then, there exist $k_\Gamma, C_\Gamma\in\mathbb{R}_{>0}$ such that for all even integers $k\geq k_\Gamma$ and all $n\in\mathbb{Z}_{>0}$,*

$$n\leq C_\Gamma k^{\frac{4}{3}} \quad\Rightarrow\quad \Psi_{n,k,\Gamma}\not\equiv 0.$$

Let $d(n)$ denote the number of positive divisors of an integer $n\in\mathbb{Z}_{>0}$. In his 1989 paper [26], by adapting Rankin's method, Mozzochi proved new results on the non-vanishing of classical Poincaré series with respect to the Hecke congruence subgroups $\Gamma_0(N)$. In particular, he proved the following result that does not depend on the weight k.

Theorem 3 ([26, Thm. 1]) *Let $N,n\in\mathbb{Z}_{>0}$. If*

$$n\,(n,N)\left(\sum_{m|n}\frac{d(m)}{\sqrt{m}}\right)^2 \leq \frac{1}{2^{15}\pi^3}\left(\frac{N}{d(N)\ln(2N)}\right)^2,$$

then $\Psi_{n,k,N}\not\equiv 0$ for all even integers $k\geq 4$.

In 2012, Das and Ganguly proved the following improvement of Theorem 1 for congruence subgroups $\Gamma_0(N)$.

Theorem 4 ([9, Thm. 5.2]) *There exist $k_0\in\mathbb{Z}_{\geq 4}$ and $B\in\mathbb{R}_{>\ln 16}$ such that for all $N\in\mathbb{Z}_{>0}$, all even integers $k\geq k_0$, and all $n\in\mathbb{Z}_{>0}$,*

$$n\leq Nk^2e^{-B\frac{\ln k}{\ln\ln k}} \quad\Rightarrow\quad \Psi_{n,k,N}\not\equiv 0.$$

In 2019, Choi and Im proved a similar result in [4] for Poincaré series $\Psi_{n,k,\Gamma_0^+(p)}$, where p is a prime number and $\Gamma_0^+(p)$ is the group generated by $\Gamma_0(p)$ and the Fricke involution $W_p = \begin{pmatrix} & -p^{-\frac{1}{2}} \\ p^{\frac{1}{2}} & \end{pmatrix}$.

Another Rankin's insight about Fourier coefficients of the classical Poincaré series is that certain additive properties of Kloosterman sums imply that for all $n, r \in \mathbb{Z}_{>0}$ and for every prime number $p \nmid n$,

$$c_{k,1}\left(np^r, np^r\right) = c_{k,1}(n,n) + p^{k-1} c_{k,1}\left(np^{r-1}, np^{r+1}\right).$$

By (6) with $\Gamma = \mathrm{SL}_2(\mathbb{Z})$ and $f = \Psi_{n,k}$ and by (11), this implies the following theorem.

Theorem 5 ([41, Thm. 2]) *Let $k \in \mathbb{Z}_{\geq 4}$ be an even integer, and let $n \in \mathbb{Z}_{>0}$. Then:*

(i) If $\Psi_{n,k} \equiv 0$, then $\Psi_{nm,k} \equiv 0$ for all $m \in \mathbb{Z}_{>0}$ that are relatively prime to n.
(ii) If $\Psi_{n,k} \not\equiv 0$, then for every prime number $p \nmid n$ and $r \in \mathbb{Z}_{>0}$, the following implication holds:

$$\Psi_{np^r,k} \equiv 0 \quad \Rightarrow \quad \Psi_{np^{r-1},k} \not\equiv 0 \text{ and } \Psi_{np^{r+1},k} \not\equiv 0.$$

3.3 Integral Non-vanishing Criteria

In 2009, Muić proved an integral non-vanishing criterion [27, Thm. 4.1] for Poincaré series on unimodular locally compact Hausdorff groups. As a corollary, using the standard lift $f \mapsto F_f$ that takes any function $f : \mathcal{H} \to \mathbb{C}$ that is Γ-invariant under the action (3) to a function $F_f : \Gamma\backslash\mathrm{SL}_2(\mathbb{R}) \to \mathbb{C}$ given by

$$F_f(g) = \left(f\big|_k g\right)(i), \qquad g \in \mathrm{SL}_2(\mathbb{R}), \tag{12}$$

in [29] he obtained a non-vanishing criterion for Poincaré series on $\mathcal{H}$ given in the second part of the following theorem.

Theorem 6 *Let:*

(N1) $\Gamma \subseteq \mathrm{SL}_2(\mathbb{R})$ *be a Fuchsian group of the first kind with a cusp at* ∞.
(N2) $k \in \mathbb{Z}_{\geq 3}$, *where k is even if* $\Gamma \cap \left\{\begin{pmatrix} -1 & x \\ & -1 \end{pmatrix} : x \in \mathbb{R}\right\} \neq \emptyset$.

Let $f : \mathcal{H} \to \mathbb{C}$ be a holomorphic, h_Γ-periodic function such that

$$\iint_{[0,h_\Gamma]\times\mathbb{R}_{>0}} |f(z)|\,\Im(z)^{\frac{k}{2}}\,dv(z) < \infty.$$

Then:

(i) *The Poincaré series $P_{k,\Gamma_\infty\backslash\Gamma} f$ converges absolutely and uniformly on compact subsets of $\mathcal{H}$ and defines an element of $S_k(\Gamma)$.*
(ii) *The cusp form $P_{k,\Gamma_\infty\backslash\Gamma} f$ does not vanish identically if there exists a compact set $C \subseteq [0, h_\Gamma] \times \mathbb{R}_{>0}$ with the following properties:*

(C1) $\gamma.C \cap C = \emptyset$ *for all* $\gamma \in \Gamma \setminus \Gamma_\infty$.
(C2) $\iint_C |f(z)|\,\Im(z)^{\frac{k}{2}}dv(z) > \frac{1}{2}\iint_{[0,h_\Gamma]\times\mathbb{R}_{>0}} |f(z)|\,\Im(z)^{\frac{k}{2}}\,dv(z)$.

Proof This follows from [30, Lem. 2.3] and [29, Lem. 3.1]. □

For Γ as in (N1) of Theorem 6, let us define the level N_Γ of Γ by

$$N_\Gamma = \inf\left\{|c| : \begin{pmatrix} a & b \\ c & d \end{pmatrix} \in \Gamma \setminus \Gamma_\infty\right\}.$$

By [25, Lem. 1.7.3], $N_\Gamma \geq \frac{1}{h_\Gamma}$. By applying Theorem 6(ii) to the Poincaré series $\Psi_{n,k,\Gamma}$ with the set C of the form $[0, h_\Gamma] \times [\varepsilon, \delta]$, where $\varepsilon \searrow \frac{1}{N_\Gamma}$ and $\delta \nearrow \infty$, Muić proved the following theorem.

Theorem 7 *Let Γ and k be as in (N1)–(N2) of Theorem 6. For all $n \in \mathbb{Z}_{>0}$,*

$$n \leq \frac{h_\Gamma N_\Gamma}{4\pi}(k-4) \quad \Rightarrow \quad \Psi_{n,k,\Gamma} \not\equiv 0.$$

Proof This follows from [29, Thm. 1.1(iii) and Thm. 1.2(ii)]. □

Note that in the case of Poincaré series $\Psi_{n,k,N}$, the explicit sufficient non-vanishing condition in Theorem 7 is more restrictive than the one in Theorem 4, but Theorem 7 applies to a wider range of groups Γ. In fact, the results of [29] are even more general than stated here and include the case of Poincaré series with non-trivial nebentypus.

3.4 Asymptotic Orthogonality of Fourier Coefficients

In [20], Kowalski et al. proved that for even integers $k \geq 4$ and $m, n, N \in \mathbb{Z}_{>0}$,

$$\lim_{k\to\infty} c_{k,1}(m,n) = \delta_{m,n} \qquad \text{and} \qquad \lim_{N\to\infty} c_{k,N}(m,n) = \delta_{m,n},$$

where $\delta_{m,n}$ is the Kronecker delta symbol. In particular,

$$\lim_{k\to\infty} c_{k,1}(n,n) = 1 \qquad \text{and} \qquad \lim_{N\to\infty} c_{k,N}(n,n) = 1,$$

which by (8) implies that for fixed n,

$$\Psi_{n,k} \not\equiv 0 \text{ for } k \gg 0$$

(note that this follows from Theorem 1), and for fixed n and k,

$$\Psi_{n,k,N} \not\equiv 0 \text{ for } N \gg 0$$

(note that this follows from Theorem 7).

3.5 Non-vanishing on Average

In [10], Das and Ganguly showed that a positive proportion of Poincaré series $\Psi_{n,k,N}$ does not vanish identically when either the index n or the weight k varies over a sufficiently long interval, while the other parameters are kept fixed. More precisely, applying a result of Kowalski et al. [19, Cor. 1] obtained by studying $\mathfrak{B}$-free numbers in short intervals, they proved the following theorem.

Theorem 8 ([10, Thm. 4.1]) *There exists an absolute, effective constant $C \in \mathbb{R}_{>0}$ such that for all $N \in \mathbb{Z}_{>0}$ and even $k \in \mathbb{Z}_{\geq 4}$ such that* $\max\{k, N\} > C$, *the following holds: Given $\varepsilon \in \mathbb{R}_{>0}$, there exist constants $C_{k,n,\varepsilon}, x_{k,N,\varepsilon} \in \mathbb{R}_{>0}$ such that for all $x \geq x_{k,N,\varepsilon}$ and $y \geq x^{\frac{7}{17}+\varepsilon}$,*

$$\frac{\#\left\{n \in [x, x+y] \cap \mathbb{Z} : \Psi_{n,k,N} \not\equiv 0\right\}}{y} \geq C_{k,N,\varepsilon}.$$

Moreover, using (10) and Fourier analysis, they proved the following theorem.

Theorem 9 ([10, Thm. 4.2]) *There exist constants $k_0, C \in \mathbb{R}_{>0}$ such that for every choice of $n, N \in \mathbb{Z}_{>0}$ such that $(n, N) = 1$, and for every real number $K \geq k_0\,(n/N)^{\frac{2}{3}}\, d(N)$,*

$$\frac{\#\left\{k \in [K, 2K] \cap 2\mathbb{Z} : \Psi_{n,k,N} \not\equiv 0\right\}}{K} \geq \frac{C}{d(n)^2}.$$

3.6 Lehmer's Conjecture on the Ramanujan Tau Function

For cusp forms $\Psi_{n,12}$, whose non-vanishing is equivalent to Lehmer's conjecture on the non-vanishing of the Ramanujan tau function τ (see Sect. 1), stronger results are known: In [13], by computing Galois representations associated with τ modulo

$\ell \in \{11, 13, 17, 19, 29, 31, 41\}$ and using the well-known congruences for τ modulo powers of 2, 3, 5, 7, 23, and 691, the non-vanishing of $\tau(n)$ and $\Psi_{n,12}$ was proved for

$$n < 816212624008487344127999 \approx 8 \cdot 10^{23}.$$

4 Poincaré Series of Elliptic Type

Let $\Gamma \subseteq \mathrm{SL}_2(\mathbb{R})$ be a Fuchsian group of the first kind. For any $\xi \in \mathcal{H}$, $k \in \mathbb{Z}_{\geq 3}$, and $n \in \mathbb{Z}_{\geq 0}$, in [36], Petersson introduced a cusp form $\Phi_{n,k,\Gamma,\xi} \in S_k(\Gamma)$ given by

$$\Phi_{n,k,\Gamma,\xi}(z) = \left(P_{k,\Gamma}\, \phi_{n,k,\xi}\right)(z) = \sum_{\gamma \in \Gamma} j(\gamma, z)^{-k} \frac{(\gamma.z - \xi)^n}{\left(\gamma.z - \overline{\xi}\right)^{n+k}}, \qquad z \in \mathcal{H},$$

where $\phi_{n,k,\xi} : \mathcal{H} \to \mathbb{C}$, $\phi_{n,k,\xi}(z) = \frac{(z-\xi)^n}{(z-\overline{\xi})^{n+k}}$. Petersson proved that

$$\left\langle f, \Phi_{n,k,\Gamma,\xi} \right\rangle_{S_k(\Gamma)} = C_{n,k,\Gamma,\xi}\, b_n(\xi, f), \qquad f \in S_k(\Gamma),$$

where $C_{n,k,\Gamma,\xi}$ is a constant independent of f, and $b_n(\xi, f)$ is a coefficient in the expansion

$$\left(f\big|_k \sigma^{-1}\right)(w) = \sum_{m=0}^{\infty} b_m(\xi, f) w^m, \qquad \sigma = \begin{pmatrix} 1 & -\xi \\ 1 & -\overline{\xi} \end{pmatrix}, \quad w \in \mathbb{C}_{|z|<1}.$$

Using his aforementioned integral non-vanishing criterion [27, Thm. 4.1], Muić proved that for fixed n and k, the cusp form $\Phi_{n,k,\Gamma,i}$ does not vanish identically provided that $\Gamma \subseteq \Gamma(N)$ for some $N \gg 0$ [28, Prop. 6–23].

O'Sullivan and Risager found a recursive formula for coefficients $b_m(\xi, \Delta)$ for certain choices of ξ, which enabled them to prove the following theorem.

Theorem 10 ([32, Thm. 2.3 and Rem. 1]) *Let $n \in \mathbb{Z}_{\geq 0}$. Let*

$$D \in \{-3, -4, -7, -8, -11, -15, -19, -20, -24\} \quad \text{and}$$

$$\xi_D = \begin{cases} \frac{\sqrt{D}}{2}, & \text{if } D \text{ is even,} \\ \frac{1+\sqrt{D}}{2}, & \text{if } D \text{ is odd.} \end{cases}$$

Then,

$$\Phi_{n,12,\mathrm{SL}_2(\mathbb{Z}),\xi_D} \begin{cases} \equiv 0, & \text{if } D = -3 \text{ and } 3 \nmid n, \\ \equiv 0, & \text{if } D = -4 \text{ and } 2 \nmid n, \\ \not\equiv 0, & \text{otherwise.} \end{cases} \tag{13}$$

The first two cases in (13) are trivial and follow from the fact that ξ_{-3} (resp., ξ_{-4}) is a fixed point of order 3 (resp., 2) for $\mathrm{PSL}_2(\mathbb{Z})$. Non-trivial examples of vanishing of $\Phi_{n,12,\mathrm{SL}_2(\mathbb{Z}),\xi}$ are also known: For example, by [32, Thm. 1.6], there exists $y \in \left]1, \sqrt{2}\right[$ such that $\Phi_{2,12,\mathrm{SL}_2(\mathbb{Z}),iy}$ vanishes identically.

5 Kohnen's Kernel Function

Given an even integer $k \geq 4$ and a complex number $s \in \mathbb{C}_{1<\Re(z)<k-1}$, in [18] W. Kohnen studied the non-vanishing of Poincaré series $R_{k,s} \in S_k(\mathrm{SL}_2(\mathbb{Z}))$ given by

$$R_{k,s}(z) = \frac{1}{2} e^{\frac{\pi i s}{2}} \Gamma(s)\Gamma(k-s) \sum_{\gamma \in \mathrm{SL}_2(\mathbb{Z})} j(\gamma, z)^{-k} (\gamma.z)^{-s}, \quad z \in \mathcal{H}.$$

The cusp forms $R_{k,s}$ have the following property:

$$\left\langle f, R_{k,\bar{s}} \right\rangle_{S_k(\mathrm{SL}_2(\mathbb{Z}))} = \frac{(-1)^{\frac{k}{2}} \pi (k-2)!}{2^{k-2}} L^*(f, s), \qquad f \in S_k(\mathrm{SL}_2(\mathbb{Z})),$$

where $L^*(f, \cdot) : \mathbb{C} \to \mathbb{C}$ is an entire function uniquely determined by the formula

$$L^*(f, s) = (2\pi)^{-s} \Gamma(s) \sum_{n=1}^{\infty} \frac{a_n(f)}{n^s}, \qquad \Re(s) > \frac{k}{2} + \frac{1}{2}.$$

Kohnen found an explicit Fourier expansion of the functions $R_{k,s}$ in terms of hypergeometric functions and showed that the first coefficient in said expansion is non-zero for certain choices of k and s. In this way, he proved the following theorem.

Theorem 11 *For every even integer $k \geq 4$, let $f_{k,1}, \ldots, f_{k,d_k}$ be a basis of Hecke eigenforms for $S_k(\mathrm{SL}_2(\mathbb{Z}))$ such that $a_1(f_{k,r}) = 1$ for all r. Let $t_0 \in \mathbb{R}$ and $\varepsilon \in \mathbb{R}_{>0}$. Then, there exists $C_{t_0,\varepsilon} \in \mathbb{R}_{\geq 4}$ such that for all even integers $k \geq C_{t_0,\varepsilon}$, we have*

$$\sum_{r=1}^{d_k} \frac{1}{\left\langle f_{k,r}, f_{k,r} \right\rangle_{S_k(\mathrm{SL}_2(\mathbb{Z}))}} L^*(f_{k,r}, s) \neq 0, \qquad \Im(s) = t_0, \ \varepsilon < \left| \Re(s) - \frac{k}{2} \right| < \frac{1}{2}.$$

This result was generalized to the case of non-trivial level and primitive nebentypus [39], as well as to the half-integral weight case [40] and to the case of the L-function twisted by a primitive Dirichlet character [44]. Moreover, Kohnen's idea was used to prove results on cuspidal Hilbert modular forms [16], Siegel cusp forms [11], and Jacobi forms [33]. Moreover, in [30], Muić studied the Poincaré series $R_{k,s}$ using his integral non-vanishing criterion and proved new results on the non-vanishing of L-

functions of cusp forms of integral weight for a general Fuchsian group Γ of the first kind with a regular cusp at ∞. Analogous results were proved in the half-integral weight case in [46].

6 Non-vanishing Results for Other Types of Poincaré Series

Rankin's and Mozzochi's methods were used to prove results on the non-vanishing of Jacobi Poincaré series in [43] and [8], and the latter were used in [12] to prove several non-vanishing results for Siegel Poincaré series of exponential type of degree 2. In order to state an example of such a result, we recall that the group

$$\mathrm{Sp}_4(\mathbb{R}) = \left\{ g \in \mathrm{GL}_4(\mathbb{R}) : g^\top \begin{pmatrix} & I_2 \\ -I_2 & \end{pmatrix} g = \begin{pmatrix} & I_2 \\ -I_2 & \end{pmatrix} \right\}$$

acts on the Siegel upper half-space

$$\mathcal{H}_2 = \left\{ z = x + iy \in M_2(\mathbb{C}) : z^\top = z,\, y > 0 \right\}$$

by

$$g.z = (Az + B)(Cz + D)^{-1}, \qquad g \in \begin{pmatrix} A & B \\ C & D \end{pmatrix} \in \mathrm{Sp}_4(\mathbb{R}),\ z \in \mathcal{H}_2.$$

Let $\mathbf{S}_2$ denote the set of half-integral, positive definite symmetric matrices in $M_2(\mathbb{Q})$. For every matrix $T \in \mathbf{S}_2$ and every integral weight $k \geq 6$, the Siegel Poincaré series

$$P_{k,T}(z) = \sum_{\gamma = \left(\begin{smallmatrix} A & B \\ C & D \end{smallmatrix}\right) \in U \backslash \mathrm{Sp}_4(\mathbb{Z})} \det(Cz + D)^{-k} e^{2\pi i \mathrm{Tr}(T\gamma.z)}, \qquad z \in \mathcal{H}_2,$$

where $U = \left\{ \left(\begin{smallmatrix} I_2 & S \\ & I_2 \end{smallmatrix}\right) : S \in M_2(\mathbb{Z}),\ S^\top = S \right\}$, converges absolutely and uniformly on compact subsets of $\mathcal{H}_2$. For a fixed k, the functions $P_{k,T}$ span the finite-dimensional space $S_k(\mathrm{Sp}_4(\mathbb{Z}))$ of weight k Siegel cusp forms for $\mathrm{Sp}_4(\mathbb{Z})$.

Theorem 12 ([12, Thm. 5.2]) *There exist $k_0 \in \mathbb{R}_{\geq 6}$ and $B \in \mathbb{R}_{>\log 16}$ such that $P_{k,T} \not\equiv 0$ for all even integers $k \geq k_0$ and all matrices $T = \left(\begin{smallmatrix} n & r/2 \\ r/2 & m \end{smallmatrix}\right) \in \mathbf{S}_2$ such that* $\gcd(n, r, m) = 1$ *and* $\det(2T) \leq k^2 \exp(-B \log k / \log\log 2k)$.

The non-vanishing of Siegel Poincaré series of exponential type was also studied in [7]. Moreover, in [48], integral non-vanishing criteria were applied to certain Siegel Poincaré series of elliptic type and in [47] to vector-valued Poincaré series of exponential and elliptic type, providing non-vanishing results similar in flavor to Theorem 7.

The problem of non-vanishing of Poincaré series stems from Poincaré's 1882 memoir [38, p. 249], in which he studied Poincaré series of rational functions on the unit disk $\mathcal{D} = \mathbb{C}_{|z|<1}$. These Poincaré series and their non-vanishing were thoroughly studied by Ljan, Hejhal, Metzger, Kra and others; see [24] and references therein. Since the focus of this short survey is on the classical cusp forms on $\mathcal{H}$ defined by Poincaré series, we will not discuss in more detail the abovementioned and miscellaneous other types of Poincaré series, such as Poincaré series on $\mathrm{SU}(n,1)/\mathrm{U}(n)$ [14] or Poincaré series on adelic groups [27], whose construction and non-vanishing are also intriguing and only partially investigated research problems.

7 Linear Relations Among Poincaré Series

A natural generalization of the question of non-vanishing of Poincaré series is the problem of determining all linear relations between the Poincaré series of a fixed type. This is the first of the three open questions about the classical Poincaré series stated by Iwaniec in [17, p. 54]. In the literature, there exist several approaches and results related to this problem.

Aside from the aforementioned asymptotic orthogonality of Fourier coefficients of the classical Poincaré series (see Sect. 3.4), which generalizes to the Siegel Poincaré series [20] and Hilbert Poincaré series [21] of exponential type, another interesting approach to the problem in question is given by the Petersson principal parts condition, which relates linear relations among the classical Poincaré series with the principal parts of certain weakly holomorphic modular forms (see [5, 37, 42] and [6]) and extends into the following result on harmonic weak Maass forms.

Let $N \in \mathbb{Z}_{>0}$. A harmonic weak Maass form of weight $k \in \mathbb{Z}$ for $\Gamma_0(N)$ is a smooth function $f : \mathcal{H} \to \mathbb{C}$ with the following properties:

(H1) $f|_k\gamma = f$ for all $\gamma \in \Gamma_0(N)$.

(H2) $\left(-y^2\left(\partial_x^2 + \partial_y^2\right) + iky\left(\partial_x + i\partial_y\right)\right) f = 0$.

(H3) There exist a polynomial $P_{f,\infty}\left(q^{-1}\right) \in \mathbb{C}\left[q^{-1}\right]$ (known as the extended principal part of f at ∞) and a real number $\varepsilon > 0$ such that

$$f(z) - P_{f,\infty}\left(e^{-2\pi i z}\right) = O\left(e^{-\varepsilon y}\right) \quad \text{when } y \to \infty \text{ (uniformly in } x).$$

Analogous conditions are required at all cusps.

If f is moreover holomorphic on $\mathcal{H}$, we say that f is a weakly holomorphic modular form. Let $H_k^\infty(\Gamma_0(N))$ denote the space of harmonic weak Maass forms of weight k for $\Gamma_0(N)$ with constant extended principal part at all cusps that are non-equivalent to ∞.

Theorem 13 ([42, Thm. 2.2(1)–(2)]) *Let $k \in \mathbb{Z}_{\geq 2}$ and $N \in \mathbb{Z}_{>0}$. For every polynomial $P\left(q^{-1}\right) = \sum_{n=1}^{d} \frac{\alpha_n}{n^{k-1}} q^{-n} \in \mathbb{C}\left[q^{-1}\right]$, there exists a unique harmonic weak Maass form $f_P \in H^{\infty}_{2-k}(\Gamma_0(N))$ whose principal part at ∞ is $P\left(q^{-1}\right)$. The function f_P is a weakly holomorphic modular form if and only if*

$$\sum_{n=1}^{d} \overline{\alpha_n} \Psi_{n,k,N} = 0.$$

In recent years, several more new perspectives and results on linear relations among the classical Poincaré series have been discovered. For instance, in [9], algebraic properties of the coefficients in linear relations among the classical Poincaré series are investigated, while [2] contain results about linear independence of the Siegel Poincaré series of exponential type.

Acknowledgments I would like to thank the anonymous referees for their valuable suggestions that helped to improve the content of the paper.

References

1. S. Baba, M.R. Murty, Irreducibility of Hecke polynomials. Math. Res. Lett. **10**(5–6), 709–715 (2003)
2. S. Böcherer, S. Das, Linear independence of Poincaré series of exponential type via non-analytic methods. Trans. Am. Math. Soc. **367**(2), 1329–1345 (2015)
3. J.H. Bruinier, K. Ono, R. Rhoades, Differential operators and harmonic weak Maass forms. Math. Ann. **342**, 673–693 (2008)
4. S. Choi, B.-H. Im, A note on the non-vanishing of Poincaré series for the Fricke group $\Gamma_0^+(p)$. Proc. Jpn. Acad., Ser. A **95**(3), 24–29 (2019)
5. S. Choi, C.H. Kim, Valence formulas for certain arithmetic groups and their applications. J. Math. Anal. Appl. **420**(1), 447–463 (2014)
6. Y. Choie, W. Kohnen, K. Ono, Linear relations between modular form coefficients and non-ordinary primes. Bull. Lond. Math. Soc. **37**(3), 335–341 (2005)
7. U. Christian, Über Hilbert-Siegelsche Modulformen und Poincarésche Reihen. Math. Ann. **148**, 257–307 (1962)
8. S. Das, Nonvanishing of Jacobi Poincaré series. J. Aust. Math. Soc. **89**(2), 165–179 (2010)
9. S. Das, S. Ganguly, Linear relations among Poincaré series. Bull. Lond. Math. Soc. **44**(5), 988–1000 (2012)
10. S. Das, S. Ganguly, Nonvanishing of Poincaré series on average. Int. J. Number Theory **9**(1), 1–8 (2013)
11. S. Das, W. Kohnen, Nonvanishing of Koecher-Maass series attached to Siegel cusp forms. Adv. Math. **281**, 624–669 (2015)
12. S. Das, J. Sengupta, Nonvanishing of Siegel Poincaré series. Math. Z. **272**(3–4), 869–883 (2012)
13. M. Derickx, M. Van Hoeij, J. Zeng, Computing Galois representations and equations for modular curves $X_H(\ell)$. arXiv:1312.6819v2
14. T. Foth, Bohr-Sommerfeld tori and relative Poincaré series on a complex hyperbolic space. Commun. Anal. Geom. **10**(1), 151–175 (2002)
15. E. Gaigalas, Poincaré series. Litovsk. Mat. Sb. **24**(3), 70–73 (1984)

16. A. Hamieh, W. Raji, Non-vanishing of L-functions of Hilbert modular forms inside the critical strip. Acta Arith. **185**(4), 333–346 (2018)
17. H. Iwaniec, *Topics in Classical Automorphic Forms*. Graduate Studies in Mathematics, vol. 17 (American Mathematical Society (AMS), Providence, RI, 1997)
18. W. Kohnen, Nonvanishing of Hecke L-functions associated to cusp forms inside the critical strip. J. Number Theory **67**(2), 182–189 (1997)
19. E. Kowalski, O. Robert, J. Wu, Small gaps in coefficients of L-functions and $\mathfrak{B}$-free numbers in short intervals. Rev. Mat. Iberoam. **23**(1), 281–326 (2007)
20. E. Kowalski, A. Saha, J. Tsimerman, A note on Fourier coefficients of Poincaré series. Mathematika **57**(1), 31–40 (2011)
21. M. Kumari, Non-vanishing of Hilbert Poincaré series. J. Math. Anal. Appl. **466**(2), 1476–1485 (2018)
22. D.H. Lehmer, The vanishing of Ramanujan's function $\tau(n)$. Duke Math. J. **14**, 429–433 (1947)
23. J. Lehner, On the non-vanishing of Poincaré series. Proc. Edinb. Math. Soc., II. Ser. **23**, 225–228 (1980)
24. M. Masumoto, A characterization of the kernel of the Poincaré series operator. Trans. Am. Math. Soc. **300**, 695–704 (1987)
25. T. Miyake, *Modular Forms*. Translated from the 1976 Japanese original by Yoshitaka Maeda. Reprint of the first 1989 English edition. Springer Monographs in Mathematics (Springer, Berlin, 2006)
26. C.J. Mozzochi, On the non-vanishing of Poincaré series. Proc. Edinb. Math. Soc., II. Ser. **32**(1), 131–137 (1989)
27. G. Muić, On a construction of certain classes of cuspidal automorphic forms via Poincaré series. Math. Ann. **343**(1), 207–227 (2009)
28. G. Muić, On the cuspidal modular forms for the Fuchsian groups of the first kind. J. Number Theory **130**(7), 1488–1511 (2010)
29. G. Muić, On the non-vanishing of certain modular forms. Int. J. Number Theory **7**(2), 351–370 (2011)
30. G. Muić, On the analytic continuation and non-vanishing of L-functions. Int. J. Number Theory **8**(8), 1831–1854 (2012)
31. K. Ono, Algebraicity of harmonic Maass forms. Ramanujan J. **20**, 297–309 (2009)
32. C. O'Sullivan, M.S. Risager, Non-vanishing of Taylor coefficients and Poincaré series. (English) Ramanujan J. **30**(1), 67–100 (2013)
33. S. Pandey, B. Sahu, Nonvanishing of kernel functions and Poincaré series for Jacobi forms. J. Math. Anal. Appl. **515**(2), 12 p. (2022)
34. H. Petersson, Über die Entwicklungskoeffizienten der automorphen Formen. Acta Math. **58**(1), 169–215 (1932)
35. H. Petersson, Über eine Metrisierung der automorphen Formen und die Theorie der Poincaréschen Reihen. Math. Ann. **117**, 453–537 (1940)
36. H. Petersson, Einheitliche Begründung der Vollständigkeitssätze für die Poincaréschen Reihen von reeller Dimension bei beliebigen Grenzkreisgruppen von erster Art. Abh. Math. Semin. Hansische Univ. **14**, 22–60 (1941)
37. H. Petersson, Über automorphe Formen mit Singularitäten im Diskontinuitätsgebiet. Math. Ann. **129**, 370–390 (1955)
38. H. Poincaré, Mémoire sur les fonctions fuchsiennes. Act. Math. **1**, 193–294 (1883)
39. A. Raghuram, *Nonvanishing of L-functions of cusp forms inside the critical strip*, in *Number Theory*, ed. by S.D. Adhikari, et al. Proceedings of the International Conference on Analytic Number Theory with Special Emphasis on L-Functions held at the Institute of Mathematical Sciences, Chennai, India, January 2002. Ramanujan Mathematical Society Lecture Notes Series, vol. 1, pp. 97–105 (2005)
40. B. Ramakrishnan, K.D. Shankhadhar, *Nonvanishing of L-functions associated to cusp forms of half-integral weight*, in *Automorphic Forms*, ed. by B. Heim, et al., Research in number theory from Oman. Proceedings of the International Conference on Automorphic Forms and Number Theory, Muscat, Oman, February 18–22, 2012. Springer Proceedings in Mathematics & Statistics, vol. 115, pp. 223–231 (2014)

41. R.A. Rankin, The vanishing of Poincaré series. Proc. Edinburgh Math. Soc. **23**(2), 151–161 (1980)
42. R.C. Rhoades, Linear relations among Poincaré series via harmonic weak Maass forms. Ramanujan J. **29**(1–3), 311–320 (2012)
43. K.D. Shankhadhar, On the non-vanishing of Jacobi Poincaré series. Ramanujan J. **43**(1), 1–14 (2017)
44. M. Schwagenscheidt, Nonvanishing and central critical values of twisted L-functions of cusp forms on average. Funct. Approx. Comment. Math. **55**(1), 45–58 (2016)
45. J.R. Smart, A basis theorem for cusp forms on groups of genus zero. Michigan Math. J. **10**, 375–380 (1963)
46. S. Žunar, On the non-vanishing of L-functions associated to cusp forms of half-integral weight. Ramanujan J. **51**(3), 455–477 (2020)
47. S. Žunar, Non-vanishing of vector-valued Poincaré series. J. Number Theory **224**, 217–242 (2021)
48. S. Žunar, On a family of Siegel Poincaré series. Int. J. Number Theory (in press); arXiv:2210.07192

The manufacturer's authorised representative in the EU is Springer Nature Customer Service Centre GmbH, Europaplatz 3, 69115 Heidelberg, Germany. If you have any concerns regarding our products, please contact ProductSafety@springernature.com

Printed and bound by CPI Group (UK) Ltd, Croydon, CR0 4YY
01/10/2026
02225904-0014